MW00965333

Soil Biology

Volume 32

Series Editor
Ajit Varma, Amity Institute of Microbial Technology,
Amity University Uttar Pradesh, Noida, UP, India

For further volumes:
http://www.springer.com/series/5138

Ebrahim Mohammadi Goltapeh •
Younes Rezaee Danesh • Ajit Varma
Editors

Fungi as Bioremediators

 Springer

Editors

Ebrahim Mohammadi Goltapeh
College of Agriculture, Dept. of Plant
    Pathology
Tarbiat Modares University
Tehran, Iran

Younes Rezaee Danesh
Faculty of Agriculture, Dep Plant Protection
Urmia University
Urmia, Iran

Ajit Varma
Amity Institute of Microbial Technology
Amity University Uttar Pradesh
Noida, Uttar Pradesh
India

ISSN 1613-3382
ISBN 978-3-642-33810-6   ISBN 978-3-642-33811-3 (eBook)
DOI 10.1007/978-3-642-33811-3
Springer Heidelberg New York Dordrecht London

Library of Congress Control Number: 2013930138

Printed on acid-free paper

Springer is part of Springer Science+Business Media (www.springer.com)

Preface

Since earliest recorded times, humans have manipulated their surroundings in order to extract and exploit resources. In the modern era, resource extraction, processing, and utilization have increased tremendously. Several kinds of waste materials are produced, in which most of them are now classified as hazardous. While the side effects of industrial development increased, nations are also experiencing soil, groundwater, as well as surface water contamination. Many of the contaminants have an immediate threat to public health and environment. Unfortunately, there is no standard instruction that could be applied in remediation of a contaminated site, because each site has its own distinct feature, which will vary in terms of toxicity, volatility, mobility, and so on.

Regarding to these considerations, contaminants' behavior should be determined in order to removing them efficiently with minimum damage to the site. The purpose of soil remediation is not only to enhance the degradation, transformation, or detoxification of pollutants but also to protect the quality and capacity of the soil to function within ecosystem boundaries, to maintain environmental quality, and to sustain biological productivity. It is difficult to evaluate this market with any specificity, but the international market for remediation is estimated to be around US$ 25–30 billion. It is challenging to establish such estimates, as many countries have not undertaken comprehensive identification of contaminated sites. The USA, Canada, Western Europe, Japan, and Australia are considered to be the dominant international markets for remediation, with an established presence of a large number of environmental companies, products, and services. Emerging economies of some more developed Asian, Eastern European, and Latin American countries will represent significant medium-term remedial market opportunities.

There are several different remediation strategies used around the world to treat soil contaminated with toxic metals and/or organic chemicals. Three widely used strategies are (1) immobilization or retention of toxicants within a confined area, (2) removal of contaminants from the soil, (3) destruction of organic pollutants by chemical, physical, or biological means. These strategies either individually or in combination with each other have been routinely implemented by the remediation industry to successfully treat contaminated soil. Biological remediation technologies require knowledge of interdisciplinary sciences, involving microbiology, chemistry, hydrogeology, engineering, soil and plant sciences, geology, and ecology. Biological processes are typically implemented at a relatively low cost, and biological remediation methods have been successfully used to treat polluted soils, oily sludges, and groundwater contaminated by petroleum hydrocarbons, solvents, pesticides, and other chemicals.

Bioremediation started over 50 years ago with research examining the fate of pesticides in agricultural soils. Scientists began to use fungi and bacteria for the degradation of xenobiotic organic compounds toward the middle of the twentieth century. The use of bacteria showed fast and promising results, but research on evaluating fungi has lagged behind. This does not mean that fungi are not suitable organisms or that they function less satisfactorily than bacteria in degrading such compounds. The participation of fungi in bioremediation is now well established in all ecosystems. Mycoremediation is one of the most complex areas in applied remediation engineering. During the past two decades, many fungal scientists and engineers have wanted to try using fungi in the degradation of organic compounds, and for those who did try using them, good results were obtained. The discovery of the value of white-rot fungi in bioremediation has brought greater success and has thus stimulated research throughout the world. A new era in the use of fungal technologies for the degradation of organic compounds has begun. Thus, the need has arisen for a book that discusses the unique role of fungi in bioremediation.

The prime objective of this book is to highlight the potential of filamentous fungi in bioremediation and to discuss the physiology, chemistry, and biochemistry of organic and inorganic pollutant transformations. The chapters are written by leading international authorities in their fields and represent the latest and most complete synthesis of this subject area. The state of the science described here represents pioneer work that focuses on the new and exciting field of mycoremediation. The book contains elements from all scientific and engineering disciplines known globally and lays a strong foundation in the subject that will serve to connect knowledge developed in both the twentieth and twenty-first centuries. The book is encyclopedic in scope and presents various types of fungi and the associated fungal processes used to clean up wastes and wastewaters in contaminated environments. The book covers aspects related to degradative fungi, biochemistry, enzymology, reactor engineering, genetic engineering, ecology of biodegradation, and practical applications. The knowledge flows broadly from fundamental to practical aspects, making it useful to learn and apply bioremediation holistically. The book not only contains an interwoven synthesis and historical perspective of the technology but also provides "slow-release nutrition" for inventions and new frontiers for future research. The latest advances in genetic engineering and molecular biotechnologies that will be useful for the creation of suitable fungi capable of faster detoxification of these compounds are also described.

The book is intended to reach a wide audience, including managers and leaders in research and the practice of mycoremediation, and should be very useful as a reference tool for practicing engineers, scientists, waste site managers, and regulatory experts. It will also provide useful information for experts in allied fields, such as botany, mycology, geology, ecology, fungal biochemistry, genetics, enzymology, metabolic engineering, environmental microbiology, and biotechnology. This should be a leading source for graduate and undergraduate students interested in understanding the capacities and processes of fungal biodegradation. Graduate students can conduct experiments or research in the laboratory or apply fungi in bioremediation at contaminated sites without seeking special guidance. The work will also serve as a handbook for the creation of new designs and components for mycoremediation processes.

The book will stimulate thought and greater research in the wider context of mycoremediation processes in the coming decades. Hazardous wastes and wastewaters constitute a problem of modern civilization that will not go away for centuries. New wastes and wastewaters are being generated every year with our growing industrialization. A day will come when fungi will play a greater role in the transformation and detoxification of hazardous wastes and wastewaters than at present.

Experts in the area of environmental microbiology, biotechnology, and bioremediation, from diverse institutions worldwide, have contributed to this book. We would like to express our sincere appreciation to each contributor for his/her work and for their patience and attention to detail during the entire production process. We sincerely hope these eminent contributors will encourage us in the future as well, in the greatest interest of academia.

The editors would like to thank Dr. Ashok K Chauhan (Founder President, Ritnand Balved Education Foundation), Sri Atul Chauhan (Chancellor, Amity University Uttar Pradesh), and Sri Aseem Chauhan (Chancellor, Amity University Rajasthan) for their encouragement and inspiration. We are extremely grateful to the staff members of Springer Heidelberg, especially Hanna Hensler-Fritton and Jutta Lindenborn, for their continued interest, critical evaluation, constructive criticism, and support. We wish to acknowledge the help and support given to us by our faculty colleagues, family members, and friends for their constant encouragement. Special thanks are due to our Ph.D. student Ms. Ruchika Bajaj for day to day monitoring the progress of the book and willing helping me.

Tehran, Iran   Ebrahim M. Goltapeh
Urmia, Iran   Younes R. Danesh
Noida, Uttar Pradesh, India   Ajit Varma

Contents

Part I
Ecophysiology of Fungal Bioremediation

Chapter 1
An Introduction to Bioremediation

Babak Pakdaman Sardrood, Ebrahim Mohammadi Goltapeh, and Ajit Varma

1.1 Introduction

The explosive rise of global population has led to the increased exploitation of natural resources and sources to respond to the high demands of the population for food, energy, and all other requirements. Industrial revolution was a response to these requirements; however, it has resulted in the production of huge number of various organic and inorganic chemicals that have directly and indirectly led to the prolonged pollution of the habitats. The duration of the contamination is regarded to be because of their difficult biodegradability. The trend of environmental pollution is so fast and vast that the detectable rates of contamination are even encountered in the farthest ocean waters. Based on the estimations made by the environmental protection agency (EPA) only around 10 % of all wastes were safely disposed off (Chaudhry 1994; Reddy and Mathew 2001).

In addition to the pollutants and toxicants released from industries that continuously affect the environment, an abstracted review of news reveals the occasional but serious occurrence of environmental disasters such as the Exxon Valdez oil spill, the Union-Carbide (Dow) Bhopal disaster, large-scale contamination of the

B.P. Sardrood
Department of Plant Pathology, Ramin Agricultural and Natural Resources University, Mollasani, Ahwaz, Iran
e-mail: bpakdaman@yahoo.com

E.M. Goltapeh (✉)
Department of Plant Pathology, Agricultural Faculty, Tarbiat Modares University,
PO Box 14115-336, Tehran, Iran
e-mail: emgoltapeh@yahoo.com; emgoltapeh@modares.ac.ir

A. Varma
Amity Institute of Microbial Technology (AIMT), Amity Science, Technology & Innovation Foundation (ASTIF), Amity University Uttar Pradesh, Noida E-3 Block, Fourth Floor, Sector 125, Noida 201303, Uttar Pradesh, India
e-mail: ajitvarma@amity.edu

E.M. Goltapeh et al. (eds.), *Fungi as Bioremediators*, Soil Biology 32,
DOI 10.1007/978-3-642-33811-3_1, © Springer-Verlag Berlin Heidelberg 2013

Rhine River, the progressive deterioration of the aquatic habitats and conifer forests in the Northeastern US, Canada, and some parts of Europe, or the release of radioactive material in the Chernobyl accident, and most recently the crises resulted from crude oil pollution of Mexico gulf waters and the leakage of the radioactive materials from Fukushima reactor in Japan. The contaminants known to be biologically degraded by microorganisms so far known and applied in bioremediation (bioremediants) have been categorized into five groups (Hickey and Smith 1996). These include:

(a) Halogenated aromatic hydrocarbons
(b) Munitions wastes
(c) Organic solvents
(d) Pesticides
(e) Polyaromatic hydrocarbon (PAH) (creosote oily wastes)

PAH, pentachlorophenols (PCP), polychlorinated biphenyls (PCB), 1,1,1-trichloro-2,2-bis (4-chlorophenyl) ethane (DDT), 2-benzene, toluene, ethylbenzene, and xylene (BTEX), and trinitrotoluene (TNT) are persistent pollutants in the environment known to exert carcinogenic and/or mutagenic impacts and have been classified as priority pollutants by EPA. It has cost around one trillion USD to decontaminate toxic waste sites in the USA using traditional waste disposal methods such as incineration and landfilling (Reddy and Mathew 2001).

Bioremediation is regarded to be an effective and in the mean time an economic method for the decontamination of environment.

1.2   Role of Environmental Biotechnology in Pollution Management

Biotechnology can be used to assess the well being of ecosystems, transform pollutants into benign substances, generate biodegradable materials from renewable sources, and develop environmentally safe manufacturing and disposal processes.

Environmental biotechnology takes advantage of appropriately qualified living organisms and employs genetic engineering to improve the efficiency and cost, which are key factors in the future widespread exploitation of organisms to reduce the environmental burden of toxic substances.

In view of the urgent need of an efficient environmental biotechnological process, researchers have devised a technique called bioremediation, which is an emerging approach to rehabilitating areas fouled by pollutants or otherwise damaged through ecosystem mismanagement.

1.3   Bioremediation

The term of bioremediation has been made of two parts: "bios" means life and refers to living organisms and "to remediate" that means to solve a problem. "Bioremediate" means to use biological organisms to solve an environmental problem such as contaminated soil or groundwater. Bioremediation is the use of living microorganisms to degrade environmental pollutants or to prevent pollution. In other words, it is a technology for removing pollutants from the environment thus restoring the original natural surroundings and preventing further pollution (Sasikumar and Papinazath 2003).

Bioremediation could simply be defined as a biological process of the decontamination of contaminated environment. The environment may be either terrestrial, aqueous, or both. However, a more comprehensive definition is presented below:

Bioremediation is a means of cleaning up contaminated environments by exploiting the diverse metabolic abilities of microorganisms to convert contaminants to harmless products by mineralization, generation of carbon (IV) oxide and water, or by conversion into microbial biomass (Baggott 1993; Mentzer and Ebere 1996).

A point to emphasize here is that bioremediation and biodegradation should not be confused with each other. Bioremediation as a technique may include biodegradation as only one of the mechanisms involved or applied in the process of bioremediation. Only some of contaminants are biodegradable, and only some of microorganisms can degrade a fraction of contaminants (Walsh 1999). Therefore, it would be of worth to study the biodegrading potential of microorganisms.

Despite the fact that microorganisms have been used for the treatment and transformation of waste products for at least a century, bioremediation is considered as a new technology for eco-friendly decontamination of polluted environment. As a popular case of the application of this technology, municipal waste water is microbiologically decontaminated under controlled conditions so that dependent upon the metabolic activities of microorganisms, different systems of activated sludge and fixed films are applied in waste water treatment facilities (King et al. 1992). Wastes and pollutions can be permanently eliminated. Also, lasting liabilities are eliminated taking advantage of less expensive but in the mean time more long-standing biological systems. Furthermore, bioremediation methods could be applied in an integrated manner together and coupled with other treatment approaches. Bioremediation is a natural process and is therefore perceived by the public as an acceptable waste treatment process for contaminated material such as soil. Microbes able to degrade the contaminant increase in numbers when the contaminant is present; when the contaminant is degraded, the biodegradative population declines. The residues for the treatment are usually harmless products and include carbon dioxide, water, and cell biomass.

Theoretically, there are enough bioremediants in nature that can be applied against a broad range of pollutants and bioremediation can be considered as a useful technique for the complete destruction of a wide variety of contaminants. Many compounds that are legally regarded as detrimental and dangerous can be biotransformed to harmless products. This eliminates the chance of future liability associated with treatment and disposal of contaminated material. Instead of transferring contaminants from one environmental medium to another, for example, from land to water or air, the complete destruction of target pollutants is possible. Bioremediation can save life web and prohibit the passage of dangerous and risky contaminants from an ecosystem to another. Bioremediation can often be carried out on site, often without causing a major disruption of normal activities. This also eliminates the need to transport quantities of waste off site and the potential threats to human health and the environment that can arise during transportation. Bioremediation can prove less expensive than other technologies used for clean-up of hazardous waste (Vidali 2001).

However, bioremediation technology suffers from two drawbacks. One is that only a few bacteria and fungi act on a broad range of organic compounds. So far no organism has been known enough omnivorous to destroy a large percentage of the natural chemicals exist. This drawback may be bypassed through screenings for the discovery of new microbial species and detection of the potential of microorganisms in one hand, and the synchronized or successive application of the microorganisms that complete the bioactivity of each other.

Another drawback to bioremediation is that it takes a long period of time to act and impose its effect. There are some solutions to get rid of such a limitation. Genetic manipulation techniques have led to an invaluable opportunity to obtain new strains of enhanced or new bioremediation activity. Another solution may be the addition of some enhancers in an environment or in a formulation to biochemically fortify certain bioremediation pathway(s). The third solution is the synchronous use of one or more microorganisms that directly and/or indirectly increase the bioactivity of a bioremediant. Increase of the bioremediant population is the fourth and the simplest solution.

Bioremediation is limited to those compounds that are biodegradable. Not all compounds are susceptible to rapid and complete degradation, and there are some concerns that the products of biodegradation may be more persistent or toxic than the parent compounds. Biological processes are often highly specific. Important site factors required for successful bioremediation include the presence of metabolically capable microbial populations, suitable environmental growth conditions, and appropriate levels of nutrients and contaminants. It is difficult to extrapolate from bench and pilot-scale studies to full-scale field operations. Research is needed to develop and engineer bioremediation technologies that are appropriate for sites with complex mixtures of contaminants that are not evenly dispersed in the environment. Contaminants may be present as solids, liquids, and gases. Bioremediation often takes longer than other treatment options, such as excavation and removal of soil or incineration. Regulatory uncertainty remains regarding acceptable performance criteria for bioremediation.

There is no accepted definition of "clean," evaluating performance of bioremediation is difficult, and there are no acceptable endpoints for bioremediation treatments (Vidali 2001).

Bioremediation although considered as a reliable technique in the middle of present environmental problems, however, it can also be considered problematic because, while additives applied to promote the activity of the particular micro-organism(s) may disrupt other organisms inhabiting same environment when done in situ. Even if genetically modified microorganisms are released into the environment after a certain point of time it becomes difficult to remove them. Bioremediation is generally very costly, is labor intensive, and can take several months for the remediation to achieve acceptable levels. Another problem regarding the use of in situ and ex situ processes is that it is capable of causing far more damage than the actual pollution itself.

Nutrient imbalance can hinder biodegradation. Inadequate provision of nitrogen, phosphorus, potassium, and sulfur (which is probably the most important and the most easily modified of all the factors) could limit the rate of hydrocarbon degradation in the terrestrial environment (McGill and Nyborg 1975).

In the agreement to the United States Office of Technology Assessment (1990), the amendment of limiting nutrients to the spill site is necessary. There are sufficient hydrocarbon-utilizing microorganisms in soil to start bioremediation as soon as nutrient limitation is alleviated (Stone et al. 1942). The natural phospholipids, soybean lecithin and ethyl allophanate, are respectively the best available phosphorus and nitrogen sources, for the microbial bioremediants of oil contaminations (Olivieri et al. 1978).

From the view point of future prospects of bioremediation, it seems that the development of our knowledge of microbial populations, their interactions to the natural environment and contaminants, the increase of their genetic capabilities to degrade contaminants, the long-term field studies of new economical bioremediation techniques can increase the potential for significant advances. There is no doubt that bioremediation is the need of present world and can lead to protection and preservation of natural resources we have browsed from the next generations.

1.4   Types of Bioremediation

On the basis of place where wastes are removed, there are principally two ways of bioremediation:

1.4.1   In Situ Bioremediation

Most often, in situ bioremediation is applied to eliminate the pollutants in contaminated soils and groundwater. It is a superior method for the cleaning of contaminated environments because it saves transportation costs and uses harmless microorganisms to eliminate the chemical contaminations. These microorganisms are better to be of positive chemotactic affinity toward contaminants. This feature increases the probability of the bioremediation in close points where bioremediants have not distributed. Also, the method is preferred as it causes the least disruption of the contaminated area. This would be of much relevance either where the least investment and pollution are favored (for example in factories) or in areas contaminated with dangerous contaminants (for example in areas contaminated with chemical or radioactive materials).

Another advantage of in situ bioremediation is the feasibility of synchronous treatment of soil and groundwater. However, in situ bioremediation posses some disadvantages: the method is more time-consuming compared to other remedial methods, and it leads to a changed seasonal variation in the microbial activity because of the direct exposure to the variations in uncontrollable environmental factors, and the use of additives may lead to additional problems. The yield of bioremediation is determined by the kind of waste materials, namely if wastes could provide the required nutrients and energy, then microorganisms would be able to bioremediate. However, in the absence of favorable wastes, the loss of bioactivity may be compensated through stimulation of native microorganisms. Another choice of less preference is to apply genetically engineered microorganisms.

Two types of in situ bioremediation are distinguished based on the origin of the microorganisms applied as bioremediants:

(i) *Intrinsic bioremediation*—This type of in situ bioremediation is carried out without direct microbial amendment and through intermediation in ecological conditions of the contaminated region and the fortification of the natural populations and the metabolic activities of indigenous or naturally existing microfauna by improving nutritional and ventilation conditions.

(ii) *Engineered in situ bioremediation*—This type of bioremediation is performed through the introduction of certain microorgansims to a contamination site. As the conditions of contamination sites are most often unfavorable for the establishment and bioactivity of the exogenously amended microorganisms, therefore here like intrinsic bioremediation, the environment is modified in a way so that improved physico-chemical conditions are provided. Oxygen, electron acceptors, and nutrients (for example nitrogen and phosphorus) are required to enhance microbial growth.

1.4.2   Ex Situ Bioremediation

The process of bioremediation here takes place somewhere out from contamination site, and therefore requires transportation of contaminated soil or pumping of groundwater to the site of bioremediation. This technique has more disadvantages than advantages.

Depending on the state of the contaminant in the step of bioremediation, ex situ bioremediation is classified as:

(i) Solid phase system (including land treatment and soil piles)—The system is used in order to bioremediate organic wastes and problematic domestic and industrial wastes, sewage sludge, and municipal solid wastes. Solid-phase soil bioremediation includes three processes including land-farming, soil biopiling, and composting.

(ii) Slurry phase systems (including solid–liquid suspensions in bioreactors)— Slurry phase bioremediation is a relatively more rapid process compared to the other treatment processes.

Contaminated soil is mixed with water and other additives in a large tank called a bioreactor and intermingled to bring the indigenous microorganisms in close contact with soil contaminants. Nutrients and oxygen are amended, and the conditions in the bioreactor are so adjusted that an optimal environment for microbial bioremediation is provided. After completion of the process, the water is removed, and the solid wastes are disposed off or processed more to decontaminate remaining pollutants.

1.5   Bioremediation Techniques

There are several bioremediation techniques, some of them have been listed as follows (Baker and Herson 1994):

1.5.1   Bioaugmentation

The addition of bacterial cultures to a contaminated medium used frequently in in situ processes. Two factors limit the use of added microbial cultures in a land treatment unit: (a) nonindigenous cultures rarely compete well enough with an indigenous population to develop and sustain useful population levels and (b) most soils with long-term exposure to biodegradable waste have indigenous microorganisms that are effective degraders if the land treatment unit is well managed (Vidali 2001).

1.5.2   Biofilters

The use of microbial stripping columns used to treat air emissions.

1.5.3   Bioreactors

The use of biological processes in a contained area or reactor for biological treatment of relatively small amounts of waste. This method is used to treat slurries or liquids. Slurry reactors or aqueous reactors are used for ex situ treatment of contaminated soil and water pumped up from a contaminated plume.

Bioremediation in reactors involves the processing of contaminated solid material (soil, sediment, sludge) or water through an engineered containment system. A slurry bioreactor may be defined as a containment vessel and apparatus used to create a three-phase (solid, liquid, and gas) mixing condition to increase the bioremediation rate of soil-bound and water-soluble pollutants as a water slurry of the contaminated soil and biomass (usually indigenous microorganisms) capable of degrading target contaminants. In general, the rate and extent of biodegradation are greater in a bioreactor system than in situ or in solid-phase systems because the contained environment is more manageable and hence more controllable and predictable. Despite the advantages of reactor systems, there are some disadvantages. The contaminated soil requires pretreatment (e.g., excavation) or alternatively the contaminant can be stripped from the soil via soil washing or physical extraction (e.g., vacuum extraction) before being placed in a bioreactor (Vidali 2001). Bioreactors have been used to treat soil and other materials contaminated with petroleum residues (McFarland et al. 1992; Déziel et al. 1999).

1.5.4   Biostimulation

The stimulation of the indigenous microbial populations in soils and/or ground water. This process may be done either in situ or ex situ.

1.5.5   Bioventing

The process of drawing oxygen through the contaminated medium to stimulate microbial growth and activity.

Bioventing is the most common in situ treatment and involves supplying air and nutrients through wells to contaminated soil to stimulate the indigenous bacteria. Bioventing employs low air flow rates and provides only the amount of oxygen necessary for the biodegradation while minimizing volatilization and release of contaminants to the atmosphere. It works for simple hydrocarbons and can be used where the contamination is deep under the surface (Vidali 2001). In many soils effective oxygen diffusion for desirable rates of bioremediation extend to a range of only a few centimeters to about 30 cm into the soil, although depths of 60 cm and greater have been effectively treated in some cases (Vidali 2001).

1.5.6   Composting

An aerobic and thermophillic process that mixes contaminated soil with a bulking agent. Composting may be performed using static piles, aerated piles, or continuously fed reactors. Composting is a technique that involves combining contaminated soil with nonhazardous organic amendments such as manure or agricultural wastes. The presence of these organic materials supports the development of a rich microbial population and elevated temperature characteristic of composting (Vidali 2001). Composting is a process by which organic wastes are degraded by microorganisms, typically at elevated temperatures. Typical compost temperatures are in the range of 55–65 °C. The increased temperatures result from heat produced by microorganisms during the degradation of the organic material in the waste. Windrow composting has been demonstrated using the following basic steps. First, contaminated soils are excavated and screened to remove large rocks and debris (Antizar-Ladislao et al. 2007, 2008).

The soil is transported to a composting pad with a temporary structure to provide containment and protection from weather extremes. Amendments (straw, alfalfa, manure, agricultural wastes, and wood chips) are used for bulking agents and as a supplemental carbon source. Soil and amendments are layered into long piles, known as windrows. The windrow is thoroughly mixed by turning with a commercially available windrow turning machine. Moisture, pH, temperature, and explosives concentration are monitored. At the completion of the composting period, the windrows would be disassembled and the compost is taken to the final disposal area.

1.5.7   Landfarming/Land Treatment/Prepared Bed Bioreactors

Solid phase treatment system for contaminated soil that may be applied as an in situ process or ex situ in a soil treatment cell. Landfarming is a simple bioremediation technique in which contaminated soil is excavated and spread over a prepared bed and periodically tilled until pollutants are degraded. The goal is to stimulate indigenous biodegradative microorganisms and facilitate their aerobic degradation of contaminants. In general, the practice is limited to the treatment of superficial 10–35 cm of soil (Vidali 2001).

Since landfarming has the potential to reduce monitoring and maintenance costs, as well as clean-up liabilities, it has received much attention as a disposal alternative (Vidali 2001). Spilled oil and wood-preserving wastes have been bioremediated by landfarming treatments (Haught et al. 1995; Margesin and Schinner 1999).

1.5.8   Biopiling

Biopiles are a hybrid of landfarming and composting. Essentially, engineered cells are constructed as aerated composted piles. Adding compost to contaminated soil enhances bioremediation because of the structure of the organic compost matrix (Kästner and Mahro 1996). Compost enhances the oxidation of aromatic contaminants in soil to ketones and quinones, which eventually disappear (Wischmann and Steinhart 1997).

Typically used for treatment of surface contamination with petroleum hydrocarbons they are a refined version of landfarming that tend to control physical losses of the contaminants by leaching and volatilization. Biopiles provide a favorable environment for indigenous aerobic and anaerobic microorganisms (Vidali 2001).

Biopile treatment is a full-scale technology in which excavated soils are mixed with soil amendments, placed on a treatment area, and bioremediated using forced aeration. The contaminants are reduced to carbon dioxide and water. The basic biopile system includes a treatment bed, an aeration system, an irrigation/nutrient system, and a leach ate collection system. Moisture, heat, nutrients, oxygen, and pH are controlled to enhance biodegradation. The irrigation/nutrient system is buried under the soil to pass air and nutrients either by vacuum or positive pressure. Soil piles can be up to 20 ft high and may be covered with plastic to control runoff, evaporation, and volatilization, and to promote solar heating. If volatile organic compounds (VOCs) in the soil volatilize into the air stream, the air leaving the soil may be treated to remove or destroy the VOCs before they are discharged into the atmosphere. Treatment time is typically 3–6 months (Prasad Shukla et al. 2010; Wu and Crapper 2009).

1.6 Organisms Involved in Bioremediation Process

Organisms that are due to be applied in bioremediation shall fulfill the following requirements (Alexander 1994) (a) The organisms will have the effective enzymes important in bio-remediation; (b) The organism shall be able to live and demonstrate its bioactivity under conditions of pollution; (c) The organism must be able to get access to the contaminant that may be not soluble in aqueous environments or severely adsorbed to solid surfaces; (d) the substrate site of the contaminant must be accessible for the active site of the enzyme of role in bioremediation; (e) contaminant and the enzymatic system must come in close contact somewhere in or out of the cell; and finally (f) appropriately favorable environmental conditions must exist or be provided to arise the population of the potential bioremediant.

The successful occurrence of bioremediation would be dependent on the provision of the conditions mentioned above; however, various types of uni-/multicellular organisms have the required potentials to be applied in bioremediation processes. Indeed species of plants, bacteria, and fungi may be used to eliminate pollutants. However, microorganisms are of the highest bioremediation potentials as they are natural decomposers in different ecosystems and can easily proliferate. Microorganisms as fungi and bacteria degrade and break down the molecules of natural and or synthetic origins. Their high rate of generation, chemotactism, complex enzymatic, and secretion systems make them valuable replacements for other chemical and or physical remediation agents. A continuous study to identify and select new species and strains for bioremediation processes is highly required. The recent discovery of a bacterial species *Geobacter metallireducens* is a good proof for the necessity of such studies. The bacterium reduces and removes radioactive uranium from drainage waters in mining operations and from contaminated ground waters. Even dead microbial cells can be useful in bioremediation technologies. The bacterium reduces and removes radioactive uranium from drainage waters in mining operations and from contaminated ground waters. Even dead microbial cells can be useful in bioremediation technologies. Highly toxic heavy metals are reduced and fixed by secretions of some bacteria and algae and are omitted from the flow of food materials in the ecosystem. Similarly, plants like locoweed can absorb and bioaccumulate high amounts of selenium in their tissues in a form that is not hazardous till consumption of plant tissues.

Various microorganisms like various corynebacteria, mycobacteria, pseudomonads, and some of yeasts have been known to act as nontoxic bioemulsifiers and to eliminate oil slicks and petroleum pollutions through biodegradation of oil hydrocarbons metabolized as sources of energy and carbon. Similarly, some microorganisms have the potential to degrade and or metabolize synthetic compounds (as remnants of pesticides in agroecosystems) collectively called as xenobiotics. Fungi as well as some anaerobic bacteria can degrade dyes. Bioremediation is a process performed through different mechanisms such as biosorption, biodegradation, bioaccumulation, and metabolism (biotransformation, detoxification, and ...) of the contaminant molecules. Dependent on the type of the organism, some terms are used to specify the bioremediation. Phytoremediation is referred to the type of bioremediation that is relied on plants and algae as bioremediants. Mycoremediation is a type of bioremediation where fungi serve as bioremediants. In this book and in the following of this chapter, mycoremediation will be the only subject of consideration.

1.7 Mycoremediation

The history of fungal bioremediation or mycoremediation goes back to only a couple of decades ago. Despite of youth of the science of mycoremediation, this new branch of environmental biotechnology attracts a daily increasing attention of scientists that is because of the exceptional characteristics of mycoremediants, themselves. Fungi are equipped with a well-developed enzymatic system that awards them the ability to grow well on a broad range of natural as well as synthetic substrates. Fungi produce and secrete higher rates of different extracellular enzymes into their peripheral environment and degrade various substrates to small molecules that can be absorbed by and metabolized in their cells.

The fact that fungi are exogastric creates an opportunity for metabolism of new substrates such as many nonpolar, nonsoluble toxic compounds that are not amenable to intracellular processes such as cytochrome P450 (Reddy and Mathew 2001; Levin et al. 2003).

The evolutionary notable characteristic of fungal morphology, production, and ramification of cylindrical strands (hyphae, in singular form: hypha) of tip growth enables them to search for new yet noncolonized sources of material and energy and penetrate the useable substrates. Beside its physically positive impact in the enforced occupation of the inner parts of the penetrated substrates, the hyphal morphology of fungi with strong flow of cytoplasmic stores of material and energy toward hyphal tips enables fungi to get access to the inner layers of the substrates that are not in direct contact with an aqueous environment. The diversity of the produced and secreted enzymes besides hyphal growth of fungi gives them the ability to overcome the problem that is often encountered with the substrate sites covered with other molecules that inhibit fungal enzyme active site from reaching the substrate site. Another advantage of mycoremediants is that the enzymes of importance in bioremediation are stimulated under nutrient deficiency conditions (Mansur et al. 2003; Aust et al. 2003). Fungal growth rate is also reasonably enough fast for applications in bioremediation processes. In addition to the ability to penetrate contaminating substrates, fungi are regarded superior to bacteria in that they can grow under environmentally stressed conditions (low pH and poor nutrient status), where bacteria are expected to be of limited bioactivity (Davis and Westlake 1979). Fungi are able to survive, grow, and develop under toxic conditions intolerable for most bacteria (Aust et al. 2003).

Most fungi are of short life cycles and higher rates of sporulation and therefore can raise their populations in a rather short time span. For instance, yeast populations in a fresh water stream increased by several orders of magnitude in the 5 days after an oil spill (Jones 1976). Fungal degradation may proceed more rapidly than bacterial degradation, with complexation suggested as the main mechanism of calcium mobilization (Gadd 2010). Moreover, the fungi can be easily transported, genetically engineered, and scaled up.

All these features encourage us to consider fungi as the organisms preferred for applications in bioremediation. Fungal bioremediation is subject to the prevailing temperature, moisture, and soil conditions (Kearney and Wauchope 1998). The soil pH, water availability, nutritional status, and oxygen levels vary and may not always are optimal for the growth of white rot fungi (Philippoussis et al. 2002; Zervakis et al. 2001) or extracellular enzyme production for pollutant transformation (Gadd 2001). Hence the kinetics of pesticides degradation in the soil is commonly biphasic with a very rapid degradation rate in the beginning followed by a very slow prolonged dissipation. The remaining residues are often quite resistant to degradation (Alexander 1994). Among environmental parameters, the availability of water in soil may be a very important factor affecting the success of bioremediation, as water availability affects oxygen supply and thus fungal growth and enzyme production (Philippoussis et al. 2001; Marin et al. 1998). In addition to affecting microbial behavior, water availability affects contaminant binding and distribution in the soil. The behavior of organic compounds in water plays a very important role in their accessibility for microbial utilization in the environment (Atagana et al. 2003). Other factors effective on biodegradation in soil include chemical nature, concentration of the contaminant, soil type, amount of soil organic matter, and microbial community structure and activity (Schoen and Winterlin 1987). Reportedly, degradation of a diverse group of organic contaminants is dependent on the nonspecific and non-stereoselective ligninase not specifically induced by the pollutants and produced under substrate limiting growth conditions (Singh and Kuhad 2000).

Many fungal genera have been identified that include species able to metabolize hydrocarbons and some of them may be applied in bioremediation of oil-polluted regions. These fungal genera include: *Acremonium* (Llanos and Kjøller 1976), *Aspergillus* (Bartha and Atlas 1977; Obire et al. 2008), *Aureobasidium* (Bartha and Atlas 1977), *Candida* (Bartha and Atlas 1977; Obire et al. 2008), *Cephalosporium* (Bartha and Atlas 1977; Obire et al. 2008), *Cladosporium* (Walker et al. 1973; Bartha and Atlas 1977; Obire et al. 2008), *Cunninghamella* (Bartha and Atlas 1977), *Fusarium* (Llanos and Kjøller 1976; Obire et al. 2008), *Geotrichum* (Obire et al. 2008), *Gliocladium* (Llanos and Kjøller 1976), *Graphium* (Llanos and Kjøller 1976), *Hansenula* (Bartha and Atlas 1977), *Mortierella* (Llanos and Kjøller 1976), *Mucor* (Obire et al. 2008), *Paecilomyces* (Llanos and Kjøller 1976), *Penicillium* (Llanos and Kjøller 1976; Bartha and Atlas 1977; Obire et al. 2008), *Rhodosporidium* (Ahearn and Meyers 1976; Bartha and Atlas 1977), *Rhodotorula* (Bartha and Atlas 1977; Obire et al. 2008), *Saccharomyces* (Bartha and Atlas 1977), Sphaeropsidales (Llanos and Kjøller 1976), *Sporobolomyces* (Bartha and Atlas 1977), *Torulopsis* (Bartha and Atlas 1977), *Trichoderma* (Hadibarata and Tachibana 2009; Llanos and Kjøller 1976; Obire et al. 2008), *Trichosporon* (Ahearn and Meyers 1976; Bartha and Atlas 1977).

Obire (1988) found several oil-degrading aquatic yeast species belonged to the genera *Candida*, *Rhodotorula*, *Saccharomyces*, and *Sporobolomyces* (yeasts), and among filamentous fungi, *Aspergillus niger*, *Aspergillus terreus*, *Blastomyces* sp., *Botryodiplodia theobromae*, *Fusarium* sp., *Nigrospora* sp., *Penicillium chrysogenum*, *Penicillium glabrum*, *Pleurofragmium* sp., and *Trichoderma harzianum*. Annual application of various fungicides, herbicides, and insecticides at practical doses has a negative impact on the dynamics of nitrate nitrogen, mobile phosphorus, and exchangeable potassium in different soil layers (Ivanov 1974). The ecological targets include microorganisms involved in nitrogen and carbon transformations: low application doses of some agrochemicals such as simazine, atrazine, and zeazine decrease activity of urease and other soil enzymes and promote the accumulation of phytotoxic bacteria and fungi (Grodnitskaya and Sorokin 2006). *Trichoderma viride* degrades insecticides fenitrothion and fenitrooxon (Baarschers and Heitland 1986), and some of its strains are able to degrade malathion through carboxylesterase(s) (Matsumura and Boush 1966). Mukherjee and Gopal (1996) compared the potential of two soil fungi, *Aspergillus niger* and *Trichoderma viride* for the degradation of chlorpyrifos and found that *T. viride* was more active, 95.7 of chlorpyrifos degraded in the presence of *T. viride* as compared to 72.3 % in the presence of *A. niger* by day 14. The toxic metabolite of chlorpyrifos, 3,5,6-trichloropyridinol was not detected during the 14-day incubation period. *Trichoderma harzianum* has been found to degrade DDT, dieldrin, endosulfan, pentachloronitrobenzene, and pentachlorophenol but not hexachlorocyclohexane. The fungus degrades endosulfan under various nutritional conditions throughout its growth stages. Endosulfan sulfate (the first fungal meta-bolite) and endosulfan diol have been detected as the major fungal metabolites of endosulfan mainly resulted from oxidative system (Katayama and Matsumura 1993). Askar et al. (2007) have indicated the possibility of the herbicide bromoxynil biodegradation through treatments with *T. harzianum* and *T. viride* aimed to prohibit underground and surface water sources. The ability of *Trichoderma* species to survive and grow in the agar media amended with the recommended doses of prevalent herbicides (Pakdaman et al. 2002) seems to be very promising in the cleanup of agroecosystems and omission of agrochemical xenobiotics that other-wise could hurt future agricultural crops. The application of *Trichoderma* species would not only clean the agricultural ecosystems but also would lead to increased yields of future crops as the result of the biological control of plant pathogens (Pakdaman and Goltapeh 2006) and induction of plant defensive system. Further-more, *Trichoderma* can be applied in the bioremediation of soils contaminated with toxic metabolites from toxigenic microorganisms (Grodnitskaya and Sorokin 2006). *Trichoderma* is regarded as a well-known biological control agent that can be applied instead of a range of agrochemicals from fungicides, nematicides, acaricides, and insecticides, while the fungus is also a mycoremediant. The fungus while it enhances plant growth and development, behaves friendly to the free-living plant growth-promoting fungi *Piriformospora indica* and *Sebacina vermifera* (Ghahfarokhy et al. 2011) indicating the possibility of the use of all three fungi in integrated programs for organic agricultural systems.

Lentinus edodes, the gourmet mushroom has been shown to possess the capacity for removing more than 60 % of pentachlorophenol from soil (Pletsch et al. 1999) and appears to remain active at lower temperatures typically encountered with temperate soils of central and Northern Europe (Okeke et al. 1996).

A number of fungal strains (in the parenthesis) have been isolated from oil refinery soils in Poland (IETU 1999; Ulfig et al. 1997, 1998): *Aphanoascus reticulisporus* (3), *Aphanoascus keratinophilum* (6), *Candida famata* (14), *Exophiala* sp. (6), *Fusarium* sp. (11), *Geomyces pannorum* (3), *Geotrichum candidum* (3), *Microsporum gypseum* (5), *Paecilomyces lilacinus* Fungi (3), *Penicillium* sp. (4), *Phialophora* sp. (2), *Phoma* sp. (3), *Pseudallescheria boydii* (4), *Scopulariopsis brevicaulis* (1), *Trichophyton ajelloi* (15), and *Trichophyton terrestre* (4).

Phanerochaete chrysosporium and other white rot fungi are able to degrade a broad range of structurally diverse xenobiotics ranging from the insecticides DDT and lindane to wood-preserving chemicals (Kirk et al. 1992), including PCP and the creosote components anthracene and phenanthrene, to polychlorinated biphenyls and dioxins. Other compounds degraded by white rot fungi include 2,3,7,8-TCDD, 3,4,3′,4-TCB, benzo(α)pyrene, Aroclor 1254, 4-chloroaniline, 3,4-dichloroaniline, chloroaniline-lignin conjugates, benzo(α)pyrene, pentachlorophenol, triphenyl-methane dyes, crystal violet, pararosaniline, cresol red, bromophenol blue, ethyl violet, malachite green, brilliant green, 2,4,5-trichlorophenoxyacetic acid, phenanthrene, polycyclic aromatics, anthracene, fluoranthene, benzo(β) fluoranthene, benzo(k)fluoranthene, benzo(α)pyrene, indeno(ghi)pyrene, benzoperylene, azo and heterocyclic dyes, Orange II, Tropaeolin O, Congo red, Azure B, and trinitrotoluene (Bumpus et al. 1985; Eaton 1985; Arjmand and Sandermann 1985, 1986; Haemmerli et al. 1986; Bumpus and Aust 1987; Kohler et al. 1988; Mileski et al. 1988; Lamar et al. 1990; Lin et al. 1990; Bumpus and Brock 1988; Ryan and Bumpus 1989; Bumpus 1989; Huttermann et al. 1989; Cripps et al. 1990; Fernando et al. 1990). In a series of experiments, from laboratory bench-scale to full-scale field demonstrations, Haught et al. (1995) demonstrated the potential of *Phanerochaete chrysosporium* and *P. sordida* in PAH biodegradation. However, white rot fungi are of no considerable ability in the removal of high-molecular weight PAHs (five rings and above). A pilot-scale reactor system was developed that combined extraction of PAH-contaminated soil with a physically separate fungal bioreactor containing *P. chrysosporium* (May et al. 1997). The extraction of high-molecular weight PAHs from the soil led to their further bioavailability for the fungus and to high degradation rates. In another study, *P. sordida* transformed PAHs with three and four rings in creosote-contaminated soil, but five- and six-ring PAHs were not degraded (Davis et al. 1993).

The edible oyster mushroom *Pleurotus ostreatus* is able to degrade 80 % of the total PAHs in soil within 35 days (Bogan and Lamar 1999). The comparison of the abilities of *P. ostreatus*, *P. chrysosporium*, and *Trametes versicolor* in the biodegradation of PAHs, and in the in solidum production of ligninolytic enzymes revealed the superiority of *P. ostreatus* in the colonization of sterilized soil from straw-grown inocula and in the degradation of anthracene, phenanthrene, and pyrene. *P. ostreatus* and *T. versicolor* produced similar rates of manganese peroxidase and laccase in soil but *P. chrysosporium* produced only extremely very low rates of these enzymes (Novotny et al. 1999). In aged soil contaminated with creosote, *P. ostreatus* degraded approximately 40 % of the benzo[α]pyrene present after 12 weeks of incubation (Eggen and Majcherczyk 1998; Eggen and Sveum 1999). However, degradation severely came down to around 1 % when spent mushroom compost containing *P. ostreatus* was supplemented with fish oil and used for a soil contaminated with creosote. After 7 weeks, approximately 89 % of the three-ring PAHs, 87 % of the four-ring PAHs, and 48 % of the five-ring PAHs had been degraded (Eggen 1999). Removal of 86 % of the priority PAHs was reported. However, the use of ligninolytic fungi for remediation of PAH-contaminated soil has not always given promising results.

In a bench-scale test, *P. ostreatus* effectively decreased the decreasing concentrations of the pesticide lindane from 345 to 30 mg l^{-1}, within 45 days (http://www.earthfax.com). Subsequent pilot-scale tests utilizing macroscale plots with capacities of about 2 cubic yards, lindane concentrations decreased from 558 to 37 mg l^{-1} in 274 days. Following performance of the pilot-scale tests, approximately 750 tons of contaminated soil was excavated. The contaminated soil was mixed with 16 % (w/w) fungal inoculum (i.e., sawdust and cottonseed hulls thoroughly colonized with *P. ostreatus*). Initial lindane concentrations ranged from 7.1 to 37 mg l^{-1} averaging 21 mg l^{-1}. After 24 months of treatment, lindane concentrations decreased by 97 % to 0.57 mg l^{-1}, achieving the industrial treatment goal of 4.4 mg l^{-1} and almost also reaching the residential risk-based concentration of 0.49 mg l^{-1} (http://www.earthfax.com). The fungal mycelium effectively colonizes natural soil (Lang et al. 2000) and its temperature requirements are considerably lower than that of *P. chrysosporium* (Hestbjerg et al. 2003), as it is active at 8 °C (Heggen and Sveum 1999). Novotny et al. (1999) has described *P. ostreatus* as a suitable candidate to apply for the clean-up of soils contaminated with recalcitrant pollutants because of its capability of robust growth and efficient extracellular enzyme production in soil even in the presence of pollutants such as PAHs. They suggested that mycelial growth through contaminated soil and efficient enzyme expression were the key to removal of the pollutant molecules from the bulk soil. The production and activity of these enzymes in contaminated soil under field conditions are two prerequisites for successful application of white rot fungi in soil bioremediation (Lang et al. 1998). Beyond introduction of white rot fungi in natural soil, enhanced degradation of pesticide molecules requires effective growth and competition with indigenous microorganisms (Canet et al. 2001). Therefore, *P. ostreatus* as a mycoremediant applicable in the chemical preparation of pesticide-contaminated agricultural lands, and in the meantime as a reliable fungal biological control agent for anti-nematode disinfestations (Palizi et al. 2006, 2007, 2009) can be regarded as an effective control tool that can be practically applied instead of abrogated nematicides.

Fragoeiro (2005) studied the bioremediative potential of eight isolates (*Phanerochaete chrysosporium* R170, *Pleurotus ostreatus* R14, *Trametes versicolor* R26 and R101, *Polystictus sanguineus* R29, *Pleurotus cystidiosus* R46, *Pleurotus sajor-caju* R139, *Trametes socotrana* R100) of white rot fungi on soil extract agar amended individually and as a mixture with 0, 5, 10, and 20 mg l^{-1} simazine, trifluralin and dieldrin under two different water regimes (-0.7 and -2.8 MPa water potential). She found that the best isolates were *T. versicolor* (R26 and R101) and *P. ostreatus*, exhibiting good tolerance to the pesticides and water stress and the ability to degrade lignin and produce laccase in the presence of these pesticides. As a result, the activity of those three isolates plus *P. chrysosporium* (well described for its bioremediation potential) was examined in soil extract broth in relation to differential degradation of the pesticide mixture at different concentrations (0–30 mg l^{-1}) under different osmotic stress levels (-0.7 and -2.8 MPa). Enzyme production, relevant to P and N release (phosphomonoesterase, protease), carbon cycling (β-glucosidase, cellulase) and laccase, involved in lignin degradation was quantified. The results suggested that the test isolates have the ability to degrade different groups of pesticides, supported by the capacity for expression of a range of extracellular enzymes at both -0.7 and -2.8 MPa water potential. *P. chrysosporium* and *T. versicolor* R101, were able to degrade this mixture of pesticides independently of laccase activity, whereas *P. ostreatus* and *T. versicolor* R26 showed higher production of this enzyme. Complete degradation of dieldrin and trifluralin was observed, while about 80 % of the simazine was degraded regardless of osmotic stress treatment in a nutritionally poor soil extract broth. The results with toxicity test (Toxalert®10), suggested the pesticides were metabolized. Therefore the capacity for the degradation of high concentrations of mixtures of pesticides and the production of a range of enzymes, even under osmotic stress conditions suggested potential applications in soil. Subsequently, microcosm studies of soil artificially contaminated with a mixture of pesticides (simazine, trifluralin, and dieldrin, 5 and 10 mg kg^{-1} soil) inoculated with *P. ostreatus*, *T. versicolor* R26 and *P. chrysosporium* and grown on wood chips and spent mushroom compost (SMC) were examined for biodegradation capacity at 15 °C. The three test isolates successfully grew and produced extracellular enzymes in soil. Respiratory activity was enhanced in soil inoculated with the test isolates and was generally higher in the presence of the pesticide mixture, which suggested increased mineralization. Cellulase and dehydrogenase were also higher in inoculated soil than in the control especially after 12 weeks incubation. Laccase was produced at very high levels, only when *T. versicolor* R26 and *P. ostreatus* were present. The greatest degradation for the three pesticides was achieved by *T. versicolor* R26, after 6 weeks with degradation rates for simazine, trifluralin, and dieldrin 46, 57, and 51 % higher than in natural soil and by *P. chrysosporium*, after 12 weeks, with degradation rates 58, 74, and 70 % higher than the control. The amendment of soil with SMC also improved pesticide degradation (17, 49, and 76 % increase in degradation of simazine, trifluralin, and dieldrin compared with the control).

The bracket-like polypore fungus, *Ganoderma australe* can also degrade lindane in liquid agitated cultures (Dritsa et al. 2005). The enzymes produced by these fungi are lignin peroxidase, manganese peroxidase, and laccases, which are frequently referred to as lignin-modifying enzymes (LMEs), and are highly induced in the presence of wheat bran. Rigas et al. (2007) studied the bioremediation of a soil contaminated by lindane utilizing the fungus *Ganoderma australe via* response surface methodology and identified and evaluated five parameters of determinative impacts on the bioremediation process effectiveness of the solid-state system and concluded that the most important response for bioremediation purposes was biodegradation/biomass maximized at the factors levels: temperature 17.3 °C, moisture 58 %, straw content 45 %, lindane content 13 ppm, and nitrogen content 8.2 ppm.

The degradation of simazine by *Penicillium steckii* in soil samples from areas of the herbicide application has been reported (Kodama et al. 2001).

Species from the genera *Aspergillus*, *Alternaria*, and *Cladosporium* are able to colonize samples of concrete applied as radioactive waste barrier in the Chernobyl reactor, can leach iron, aluminum, silicon, and calcium, and re-precipitate silicon and calcium oxalate in their microenvironments (Fomina et al. 2007). A study concerning the metabolism of polyphosphate in *Trichoderma harzianum*, a biocontrol agent with innate resistance against most chemicals used in agriculture, including metals, when grown in the presence of different concentrations of cadmium, has indicated the biomass production is affected by the concentration of metal used. Control cultures were able to accumulate polyphosphate under the conditions used. Moreover, the presence of cadmium induced a reduction in polyphosphate content related to the concentration used. The morphological/ultrastructural aspects have been characterized by using optical and scanning electron microscopy and were affected by the heavy metal presence and concentration. The efficiency of cadmium removal has revealed the potential of *Trichoderma harzianum* for use in remediation. The data indicate the potential for polyphosphate accumulation by the fungus, as well as its degradation related to tolerance/survival in the presence of cadmium ions (De Freitas et al. 2011).

Fungi not only lonely but also in association with other organisms (algae, bacteria, plants) can exert their positive bioremediative impacts on the environment. As an example, mycorrhizal associations may be applied in metal cleanup in the general area of phytoremediation (Van der Lelie et al. 2001; Rosen et al. 2005; Gohre and Paszkowski 2006). Mycorrhizas can increase phytoextraction directly or indirectly by increasing plant root growth and development. Plants inoculated with mycorrhizas isolated from metal-contaminated environments indicate increased phytoaccumulation of metals. However, many complicating determinant factors such as metal tolerance of fungal strains, their mycorrhizal status, and the nutritional status of polluted soils affect the final output of mycorrhization (Meharg 2003). Mycorrhizas decrease the absorption of metals by plants (Tullio et al. 2003). Arbuscular mycorrhizas (AM) reduce the transfer of heavy metals like zinc to shoots of their host plants in moderately zinc-contaminated soils, where metals are bound and trapped in mycorrhizal structures and immobilized in the mycorrhizosphere (Christie et al. 2004). However, such mutual relationships between mycobiont and phytobiont in metal-contaminated regions are not always of beneficial consequences and dependant on the local conditions may lead to neutral and or even detrimental results (Meharg and Cairney 2000). A protective metal-trapping effect of ectomycorrhizal fungi has been postulated (Leyval et al. 1997). A copper-adapted *Suillus luteus* isolate has been shown to protect pine seedlings against elevated toxic concentrations of copper ions. Such a metal-adapted *Suillus–Pinus* combination would expectedly be useful in large-scale reclamation under phytotoxic conditions enfaced in metal-contaminated and industrial sites (Adriaensen et al. 2005). Persistent fixation of Cd (II) and Pb (II) through the formation of an efficient ectomycorrhizal barrier has been indicated to reduce the translocation of the heavy metals into tissues of birch trees (Krupa and Kozdroj 2004). Such findings may be of practical applications in soil mycoremediation and re-vegetation (Gadd 2010). Natural soil organic compounds such as the insoluble glycoprotein glomalin abundantly produced on the hyphae of arbuscular mycorrhizal fungi can sequester and stabilize potentially toxic metals such as copper, cadmium, lead, and manganese and therefore, may be regarded as a useful biostabilizer in mycoremediation of heavy metal contaminated regions (Gonzalez-Chavez et al. 2004). Increased uranium concentration and content in roots and decreased concentrations of uranium in shoots have been reported as the effect of *Glomus intraradices*. AM fungi and root hairs improve not only phosphorus acquisition but also root absorption of uranium, and the mycorrhiza generally decreases root to shoot transfer of uranium (Rufyikiri et al. 2004; Chen et al. 2005a, b). With ericaceous mycorrhizas in *Calluna*, *Erica*, and *Vaccinium* spp. grown on Cu- and Zn-contaminated and/or naturally metalliferous soils, the fungi clearly prevent upward transfer of metals to plant shoots (Bradley et al. 1981, 1982). As ericaceous plants commonly grow on nutrient-deficient soils, the mycorrhiza may additionally benefit these plants through enhanced absorption of soil nutrients (Smith and Read 1997). Thus, development of stress-tolerant plant–mycorrhizal associations may be a promising strategy for phytoremediation and soil amelioration (Schutzendubel and Polle 2002). The efficiency of such symbiotic associations in ericaceous plants is so high that these plants can naturally colonize harshly polluted areas (Cairney and Meharg 2003). The ectomycorrhizal fungi *Suillus granulatus* and *Pisolithus tinctorius* can promote the release of cadmium and phosphorus from rock phosphate (Leyval and Joner 2001) while the ericoid mycorrhizal fungus *Oidiodendron maius* can solubilize zinc oxide and phosphate (Martino et al. 2003). Experimental studies on ericoid mycorrhizal and ectomycorrhizal fungi have showed that many species are able to solubilize zinc, cadmium, copper phosphates, and lead chlorophosphate (pyromorphite) and release phosphate and metals (Fomina et al. 2004). However, it has been demonstrated that mycorrhization is sometimes of neutral effects, for example non-mycorrhizal *Pinus sylvestris* and pines infected with the ectomycorrhizal *Paxillus involutus* are able to enhance zinc phosphate dissolution, resist to metal toxicity, and acquire the mobilized phosphorus, increase the amount of phosphorus in shoots while zinc phosphate is present in the growth matrix (Fomina et al. 2006).

Synergistic effects have been recorded between fungi and bacteria during bioremediation processes, for example, besides direct degradation of hydrocarbons, fungal mycelia can penetrate oil and increase the surface area available for biodegradation and bacterial attack. Also, it has been reported that despite of initial degradation of a synthetic petroleum mixture by bacteria, the rate of biodegradation increased up to twice as the result of the synchronous activity of both bacteria and fungi. Martens and Zadrazil (1998) screened a variety of wood-rotting fungi for their ability to degrade PAHs in a bioreactor containing straw and soil. A higher degradation rate (40–58 % of the applied $[^{14}C]$-PAH as $^{14}CO_2$) was observed in microcosms containing fungal strains that did not colonize the soil than in those inoculated with the soil-colonizing fungi. An explanation for the difference was that the indigenous soil bacteria were stimulated by compounds produced during the lysis of straw by non-colonizing fungi, which provided carbon sources to enhance bacterial growth and PAH degradation.

1.8   Conclusion and Perspective

With the increasing population of the world, the science and technology of bioremediation is going to become a necessity of today modern life. Fortunately considering the youth of this interdisciplinary science, there have been significant progresses in the field. The diversity of bioremediants, the multiplicity and diversity of available techniques, the variation of the substrates used by bioremediants in different types of aqueous and terrestrial habitats all seem as good signs of this well-promising science and technology. Therefore, bioremediation will expectedly be of rising number of applications in different environments from battle fields to rural, urban, and industrial areas. Fungi as a huge group of unicellular–filamentous microorganisms of high rate of biodiversity being isolated from different environments are rightfully considered as a potent group of bioremediants and with an eye to the recent advances in genetic and metabolic engineering, it seems that fungi will have much more share in the bioremediation of pollutants and wastes.

Based on what mentioned above, it is well expectable that bioremediation and, in a narrower sense, mycoremediation would expand to more specific scientific branches in near future in order to be able to quickly respond to the challenges of current and future world.

References

Adriaensen K, Vralstad T, Noben JP, Vangronsveld J, Colpaert JV (2005) Copper-adapted *Suillus luteus*, a symbiotic solution for pines colonizing Cu mine spoils. Appl Environ Microbiol 71:7279–7284

Ahearn DG, Meyers SP (1976) Fungal degradation of oil in the marine environment. In: Gareth Jones EB (ed) Recent advances in aquatic mycology. Elek, London, pp 127–130

Alexander M (1994) Biodegradation and bioremediation. Academic, Boston, MA

Antizar-Ladislao B, Beck AJ, Spanova K, Lopez-Real J, Russell NJ (2007) The influence of different temperature programmes on the bioremediation of polycyclic aromatic hydrocarbons (PAHs) in a coal-tar contaminated soil by in-vessel composting. J Hazard Mater 14:340–347

Antizar-Ladislao B, Spanova K, Beck AJ, Russell NJ (2008) Microbial community structure changes during bioremediation of PAHs in an aged coal-tar contaminated soil by in-vessel composting. Int Biodeter Biodegr 61:357–364

Arjmand M, Sandermann H (1985) Mineralization of chloroaniline/lignin conjugates and of free chloroanilines by the white rot fungus *Phanerochaete chrysosporium*. J Agric Food Chem 33:1055–1060

Arjmand M, Sandermann H (1986) Plant biochemistry of xenobiotics. Mineralization of chloroaniline/lignin metabolites from wheat by the white rot fungus, Phanerochaete chrysosporium. Zeitschrift der Naturforschung 41c:206–214

Askar AI, Ibrahim GH, Osman KA (2007) Biodegradation kinetics of bromoxynil as a pollution control technology. Egypt J Aquat Res 33:111–121

Atagana H, Haynes R, Wallis F (2003) The use of surfactants as possible enhancers in bioremediation of creosote contaminated soil. Water Air Soil Pollut 142:137–149

Aust SD, Swaner PR, Stahl JD (2003) Detoxification and metabolism of chemicals by white-rot fungi. In: Zhu JJPC, Aust SD, Lemley Gan AT (eds) Pesticide decontamination and detoxification. Oxford University Press, Washington, D.C, pp 3–14

Baarschers WH, Heitland HS (1986) Biodegradation of fenitrothion and fenitrooxon by the fungus *Trichoderma viride*. J Agric Food Chem 34:707–709

Baggott J (1993) Biodegradable lubricants. A paper presented at the Institute of Petroleum Symposium: "Life cycle analysis and eco-assessment in the oil industry," Nov. 1992. Shell, England

Baker K, Herson D (1994) Bioremediation. McGraw-Hill, New York

Bartha R, Atlas RM (1977) The microbiology of aquatic oil spills. Adv Appl Microbiol 22:225–266

Bogan BW, Lamar RT (1999) Surfactant enhancement of white-rot fungal PAH soil remediation. In: Leason A, Allman BC (eds) Phytoremediation and innovative strategies for specialized remedial applications. The fifth international *in situ* and on-site bioremediation symposium, San Diego, CA, 19–22 Apr 1999. Battelle, Columbus, OH, pp 81–86

Bradley R, Burt AJ, Read DJ (1981) Mycorrhizal infection and resistance to heavy metals. Nature 292:335–337

Bradley B, Burt AJ, Read DJ (1982) The biology of mycorrhiza in the Ericaceae. VIII. The role of mycorrhizal infection in heavy metal resistance. New Phytol 91:197–209

Bumpus JA (1989) Biodegradation of polycyclic aromatic hydrocarbons by *Phanerochaete chrysosporium*. Appl Environ Microbiol 55:154–158

Bumpus JA, Aust SD (1987) Biodegradation of environmental pollutants by the white rot fungus *Phanerochaete chrysosporium*. Bioessays 6:166–170

Bumpus JA, Brock BJ (1988) Biodegradation of crystal violet by the white rot fungus, *Phanerochaete chrysosporium*. Appl Environ Microbiol 54:1143–1150

Bumpus JA, Tien M, Wright D, Aust SD (1985) Oxidation of persistent environmental pollutants by a white rot fungus. Science 228:1434–1436

Cairney JWG, Meharg AA (2003) Ericoid mycorrhiza: a partnership that exploits harsh edaphic conditions. Eur J Soil Sci 54:735–740

Canet R, Birnstingl J, Malcolm D, Lopez-Real J, Beck A (2001) Biodegradation of polycyclic aromatic hydrocarbons (PAHs) by native microflora and combinations of white rot fungi in a coal-tar contaminated soil. Bioresour Technol 76:113–117

Chaudhry GR (1994) Biological degradation and bioremediation of toxic chemicals. Dioscorides, Portland, OR, 515 p

Chen BD, Jakobsen I, Roos P, Zhu YG (2005a) Effects of the mycorrhizal fungus *Glomus intraradices* on uranium uptake and accumulation by *Medicago truncatula* L. from uranium-contaminated soil. Plant Soil 275:349–359

Chen BD, Zhu YG, Zhang XH, Jakobsen I (2005b) The influence of mycorrhiza on uranium and phosphorus uptake by barley plants from a field-contaminated soil. Environ Sci Pollut Res Int 12:325–331

Christie P, Li XL, Chen BD (2004) Arbuscular mycorrhiza can depress translocation of zinc to shoots of host plants in soils moderately polluted with zinc. Plant Soil 261:209–217

Cripps C, Bumpus JA, Aust SD (1990) Biodegradation of azo and heterocyclic dyes by *Phanerochaete chrysosporium*. Appl Environ Microbiol 56:1114–1116

Davis JB, Westlake DWS (1979) Crude oil utilization by fungi. Can J Microbiol 25:146–156

Davis MW, Glaser JA, Evans JW, Lamar RT (1993) Field evaluation of the lignin-degrading fungus *Phanerochaete sordida* to treat creosote-contaminated soil. Environ Sci Technol 27:2572–2576

De Freitas AL, de Moura GF, de Lima MAB, de Souza PM, da Silva CAA, de Campos GMT, do Nascimento AE (2011) Role of the morphology and polyphosphate in *Trichoderma harzianum* related to cadmium removal. Molecules 16:2486–2500

Déziel E, Comeau Y, Villemur R (1999) Two-liquid-phase bioreactors for enhanced degradation of hydrophobic/toxic compounds. Biodegradation 10:219–233

Dritsa V, Rigas F, Avramides EJ, Hatzianestis I (2005) Biodegradation of lindane in liquid cultures by the polypore fungus *Ganoderma australe*. In: 3rd European bioremediation conference, Chania

Eaton DC (1985) Mineralization of polychlorinated biphenyls by *Phanerochaete chrysosporium*, a ligninolytic fungus. Enzyme Microb Technol 7:194–196

Eggen T (1999) Application of fungal substrate from commercial mushroom production – *Pleurotus ostreatus* – for bioremediation of creosote contaminated soil. Int Biodeter Biodegr 44:117–126

Eggen T, Majcherczyk A (1998) Removal of polycyclic aromatic hydrocarbons (PAH) in contaminated soil by white rot fungus *Pleurotus ostreatus*. Int Biodeter Biodegr 41:111–117

Eggen T, Sveum P (1999) Decontamination of aged creosote polluted soil: the influence of temperature, white rot fungus *Pleurotus ostreatus*, and pretreatment. Int Biodeter Biodegr 43:125–133

Fernando J, Bampus JA, Aust SD (1990) Biodegradation of TNT (2, 4, 6-trinitrotoluene) by *Phanerochaete chrysosporium*. Appl Environ Microbiol 56:1666–1671

Fomina MA, Alexander IJ, Hillier S, Gadd GM (2004) Zinc phosphate and pyromorphite solubilization by soil plant-symbiotic fungi. Geomicrobiol J 21:351–366

Fomina M, Charnock JM, Hillier S, Alexander IJ, Gadd GM (2006) Zinc phosphate transformations by the *Paxillus involutus*/pine ectomycorrhizal association. Microb Ecol 52:322–333

Fomina M, Podgorsky VS, Olishevska SV, Kadoshnikov VM, Pisanska IR, Hillier S, Gadd GM (2007) Fungal deterioration of barrier concrete used in nuclear waste disposal. Geomicrobiol J 24:643–653

Fragoeiro SI de S (2005) The use of fungi in bioremediation of pesticides. PhD thesis, Applied Mycology Group Institute of Bioscience and Technology, Cranfield University, Cranfield

Gadd GM (ed) (2001) Fungi in bioremediation. Cambridge University Press, Cambridge

Gadd GM (2010) Metals, minerals and microbes: geomicrobiology and bioremediation. Microbiology 156:609–643

Ghahfarokhy MR, Goltapeh EM, Purjam E, Pakdaman BS, Modarres Sanavy SAM, Varma A (2011) Potential of mycorrhiza-like fungi and *Trichoderma* species in biocontrol of take-all disease of wheat under greenhouse condition. J Agric Technol 7:185–195

Gohre V, Paszkowski U (2006) Contribution of the arbuscular mycorrhizal symbiosis to heavy metal phytoremediation. Planta 223:1115–1122

Gonzalez-Chavez MC, Carrillo-Gonzalez R, Wright SF, Nichols KA (2004) The role of glomalin, a protein produced by arbuscular mycorrhizal fungi, in sequestering potentially toxic elements. Environ Pollut 130:317–323

Grodnitskaya ID, Sorokin ND (2006) Use of micromycetes *Trichoderma* for soil bioremediation in tree nurseries. Biol Bull 33:400–403

Hadibarata T, Tachibana S (2009) Microbial degradation of n-eicosane by filamentous fungi. In: Obayashi Y, Isobe T, Subramanian A, Suzuki S, Tanabe S (eds) Interdisciplinary studies on environmental chemistry – environmental research in Asia. Terrapub, Tokyo, pp 323–329

Haemmerli SD, Liesola MSA, Sanglard D, Feichter A (1986) Oxidation of benzo(a)pyrene by extracellular ligninases from *Phanerochaete chrysosporium*. J Biol Chem 261:6900–6903

Haught RC, Neogy R, Vonderhaar SS, Krishnan ER, Safferman SI, Ryan J (1995) Land treatment alternatives for bioremediating wood preserving wastes. Hazard Waste Hazard Mater 12:329–344

Heggen T, Sveum P (1999) Decontamination of age creosote polluted soil: the influence of temperature, white rot fungus *Pleurotus ostreatus*, and pre-treatment. Int Biodeter Biodegr 43:125–133

Hestbjerg H, Willumsen P, Christensen M, Andersen O, Jacobsen C (2003) Bioaugmentation of tar-contaminated soils under field conditions using *Pleurotus ostreatus* refuse from commercial mushroom production. Environ Toxicol Chem 22:692–698

Hickey RF, Smith G (1996) Biotechnology in industrial waste treatment and bioremediation. Lewis, New York

Huttermann A, Trojanowski J, Loske D (1989) Process for the decomposition of complex aromatic substances in contaminated soils/refuse matter with microorganisms. German Patent DE3,731,816

IETU (1999) Comprehensive report of remediation applications at an oil refinery in southern Poland. Report prepared for U.S. DOE, FETC

Ivanov AI (1974) Effect of simazine on soil feeding conditions. Agrokhimiya 3:113–115

Jones EBG (1976) Recent advances in aquatic mycology. Elek, London, 749p

Kästner M, Mahro B (1996) Microbial degradation of polycyclic aromatic hydrocarbons in soils affected by the organic matrix of compost. Appl Microbiol Biotechnol 44:668–675

Katayama A, Matsumura F (1993) Degradation of organochlorine pesticides, particularly endosulfan, by *Trichoderma harzianum*. Environ Toxicol Chem 12:1059–1065

Kearney P, Wauchope R (1998) Disposal options based on properties of pesticides in soil and water. In: Kearney P, Roberts T (eds) Pesticide remediation in soils and water, Wiley series in agrochemicals and plant protection. Wiley, New York

King RB, Gilbert ML, Sheldon JK (1992) Practical environmental bioremediation. Lewis, New York

Kirk TK, Lamar RT, Glaser JA (1992) The potential of white-rot fungi in bioremediation. In: Mongkolsuk S, Lovett PS, Trempy JE (eds) Biotechnology and environmental science – molecular approaches. Proceedings of an international conference on biotechnology and environmental science: molecular approaches, Bangkok, 21–24 Aug 1990. Plenum, New York, pp 131–138

Kodama T, Ding L, Yoshida M, Yajima M (2001) Biodegradation of a s-triazine herbicide, simazine. J Mol Catal B Enzym 11:1073–1078

Kohler A, Jager A, Willerhausen H, Graf H (1988) Extracellular ligninase of *Phanerochaete chrysosporium* Burdsall has no role in the degradation of DDT. Appl Microbiol Biotechnol 29:616–620

Krupa P, Kozdroj J (2004) Accumulation of heavy metals by ectomycorrhizal fungi colonizing birch trees growing in an industrial desert soil. World J Microbiol Biotechnol 20:427–430

Lamar RT, Glaser JA, Kirk TK (1990) Fate of pentachlorophenol (PCP) in sterile soils inoculated with the white-rot basidiomycete *Phanerochaete chrysosporium*: mineralization, volatilization and depletion of PCP. Soil Biol Biochem 22:433–440

Lang E, Nerud F, Zadrazil F (1998) Production of ligninolytic enzymes by *Pleurotus* sp. and *Dichomitus squalens* in soil and lignocellulose substrate as influenced by soil microorganisms. FEMS Microbiol Lett 167:239–244

Lang E, Gonser A, Zadrazil F (2000) Influence of incubation temperature on activity of ligninolytic enzymes in sterile soil by *Pleurotus* sp. and *Dichomitus sqalens*. J Basic Microbiol 40:33–39

Levin L, Viale A, Forchiassin A (2003) Degradation of organic pollutants by the white rot basidiomycete *Trametes trogii*. Int Biodeter Biodegr 52:1–5

Leyval C, Joner EJ (2001) Bioavailability of heavy metals in the mycorrhizosphere. In: Gobran GR, Wenzel WW, Lombi E (eds) Trace elements in the rhizosphere. CRC, Boca Raton, FL, pp 165–185

Leyval C, Turnau K, Haselwandter K (1997) Effect of heavy metal pollution on mycorrhizal colonization and function: physiological, ecological and applied aspects. Mycorrhiza 7:139–153

Lin J-E, Wang HY, Hickey RF (1990) Kinetics of pentachlorophenol by *Phanerochaete chrysosporium*. Biotechnol Bioeng 35:1125–1134

Llanos C, Kjøller A (1976) Changes in the flora of soil fungi following oil waste application. Oikos 27:377–382

Mansur MME, Arias JL, Copa-Patino M, Flärdh M, González AE (2003) The white-rot fungus *Pleurotus ostreatus* secretes laccase isozymes with different substrate specificities. Mycologia 95:1013–1020

Margesin R, Schinner F (1999) Biological decontamination of oil spills in cold environments. J Chem Technol Biotechnol 74:381–389

Marin S, Sanchis V, Ramos A, Magan N (1998) Effect of water activity on hydrolytic enzyme production by *Fusarium moniliforme* and *Fusarium proliferatum* during colonisation of maise. Int J Food Microbiol 42:185–194

Martens R, Zadrazil F (1998) Screening of white-rot fungi for their ability to mineralize polycyclic aromatic hydrocarbons in soil. Folia Microbiol 43:97–103

Martino E, Perotto S, Parsons R, Gadd GM (2003) Solubilization of insoluble inorganic zinc compounds by ericoid mycorrhizal fungi derived from heavy metal polluted sites. Soil Biol Biochem 35:133–141

Matsumura F, Boush GM (1966) Malathion degradation by *Trichoderma viride* and a *Pseudomonas* species. Science 153:1278–1280

May R, Schröder P, Sandermann H Jr (1997) *Ex-situ* process for treating PAH-contaminated soil with *Phanerochaete chrysosporium*. Environ Sci Technol 31:2626–2633

McFarland MJ, Qiu XJ, Sims JL, Randolph ME, Sims RC (1992) Remediation of petroleum impacted soils in fungal compost bioreactors. Water Sci Technol 25:197–206

McGill WB, Nyborg M (1975) Reclamation of wet forest soils subjected to oil spills. Alberta Institute of Pedology, Edmonton, AB, Publ. no. G–75–1

Meharg AA (2003) The mechanistic basis of interactions between mycorrhizal associations and toxic metal cations. Mycol Res 107:1253–1265

Meharg AA, Cairney JWG (2000) Co-evolution of mycorrhizal symbionts and their hosts to metal-contaminated environments. Adv Ecol Res 30:69–112

Mentzer E, Ebere D (1996) Remediation of hydrocarbon contaminated sites. A paper presented at 8th biennial international seminar on the petroleum industry and the Nigerian environment, November, Port Harcourt

Mileski GJ, Bumpus JA, Jurek MA, Aust SD (1988) Biodegradation of pentachlorophenol by the white rot fungus *Phanerochaete chrysosporium*. Appl Environ Microbiol 55:2885–2889

Mukherjee I, Gopal M (1996) Degradation of chlorpyrifos by two soil fungi *Aspergillus niger* and *Trichoderma viride*. Toxicol Environ Chem 57:145–151

Novotny C, Erbanova P, Sasek V, Kubatova A, Cajthaml T, Lang E, Krahl J, Zadrazil F (1999) Extracellular oxidative enzyme production and PAH removal in soil by exploratory mycelium of white rot fungi. Biodegradation 10:159–168

Obire O (1988) Studies on the biodegradation potentials of some microorganisms isolated from water systems of two petroleum producing areas in Nigeria. Niger J Bot 1:81–90

Obire O, Anyanwu EC, Okigbo RN (2008) Saprophytic and crude oil-degrading fungi from cow dung and poultry droppings as bioremediating agents. Int J Agric Technol 4:81–89

Office of Technology Assessment (OTA) (1990) Bioremediation of marine oil spill. Workshop. Government Printing Press, Washington, DC, vol 4, pp 1–30

Okeke B, Smith J, Paterson A, Watson-Craik I (1996) Influence of environmental parameters on pentachlorophenol biotransformation in soil by *Lentinula edodes* and *Phanerochaete chrysosporium*. Appl Microbiol Biotechnol 45:263–266

Olivieri R, Robertiello A, Degen L (1978) Enhancement of microbial degradation of oil pollutants using lipophilic fertilizers. Mar Pollut Bull 9:217–220

Pakdaman BS, Goltapeh EM (2006) An *in vitro* study on the possibility of rapeseed white stem rot disease control through the application of prevalent herbicides and *Trichoderma* species. Pak J Biol Sci 10:7–12

Pakdaman BS, Khabbaz H, Goltapeh EM, Afshari HA (2002) *In vitro* studies on the effects of sugar beet field prevalent herbicides on the beneficial and deleterious fungal species. Pak J Plant Pathol 1:23–24

Palizi P, Goltapeh EM, Pourjam E (2006) Nematicidal activity of culture filtrate of seven *Pleurotus* species on *Pratylenchus vulnus*. In: 17th Iranian plant protection congress, vol 2, p 457

Palizi P, Goltapeh EM, Pourjam E, Safaie N (2007) The relationship of oyster mushrooms fatty acid profile and their nematicidal activity. In: 5th national biotechnology congress of Iran, p 762

Palizi P, Goltapeh EM, Pourjam E, Safaie N (2009) Potential of oyster mushrooms for the biocontrol of sugar beet nematode (*Heterodera schachtii*). J Plant Prot Res 49:27–33

Philippoussis A, Diamantopoulou P, Euthimiadou H, Zervakis G (2001) The composition and porosity of lignocellulosic substrates influence mycelium growth and respiration rates of *Lentinus edodes*. Int J Med Mushrooms 3:198

Philippoussis A, Diamantopoulou P, Zervakis G (2002) Monitoring of mycelium growth and fructification of Lentinula edodes on several agricultural residues. In: Sanchez JE, Huerta G, Montiel E (eds) Mushroom biology and mushroom products. UAEM, Cuernavaca, pp 279–287

Pletsch M, de Araujo B, Charlwood B (1999) Novel biotechnological approaches in environmental remediation research. Biotechnol Adv 17:679–687

Prasad Shukla K, Kumar NS, Sharma S (2010) Bioremediation: developments, current practices and perspectives. Genet Eng Biotechnol J GEBJ-3

Reddy CA, Mathew Z (2001) Bioremediation potential of white rot fungi. In: Gadd GM (ed) Fungi in bioremediation. Cambridge University Press, Cambridge

Rigas F, Papadopoulou K, Dritsa V, Doulia D (2007) Bioremediation of a soil contaminated by lindane utilizing the fungus *Ganoderma australe* via response surface methodology. J Hazard Mater 140:325–332

Rosen K, Zhong WL, Martensson A (2005) Arbuscular mycorrhizal fungi mediated uptake of Cs-137 in leek and ryegrass. Sci Total Environ 338:283–290

Rufyikiri G, Huysmans L, Wannijn J, Van Hees M, Leyval C, Jakobsen I (2004) Arbuscular mycorrhizal fungi can decrease the uptake of uranium by subterranean clover grown at high levels of uranium in soil. Environ Pollut 130:427–436

Ryan TP, Bumpus JA (1989) Biodegradation of 2, 4, 5-trichlorophenoxyacetic acid in liquid culture and in soil by the white rot fungus *Phanerochaete chrysosporium*. Appl Microbiol Biotechnol 31:302–307

Sasikumar CS, Papinazath T (2003) Environmental management: bioremediation of polluted environment. In: Bunch MJ, Suresh VM, Kumaran TV (eds) Proceedings of the third international conference on environment and health, Chennai, India, 15–17 Dec 2003. Department of Geography, University of Madras and Faculty of Environmental Studies, York University, Chennai, pp 465–469

Schoen S, Winterlin W (1987) The effects of various soil factors and amendments on the degradation of pesticide mixture. J Environ Sci Health 22:347–377

Schutzendubel A, Polle A (2002) Plant responses to abiotic stresses: heavy metal-induced oxidative stress and protection by mycorrhization. J Exp Bot 53:1351–1365

Singh BK, Kuhad RC (2000) Degradation of insecticide lindane (γ-HCH) by white rot fungus *Cyathus bulleri* and *Phanerochaete sordida*. Pest Manag Sci 56:142–146

Smith SE, Read DJ (1997) Mycorrhizal symbiosis, 2nd edn. Academic, San Diego, CA

Stone RW, Fenske MR, White AGC (1942) Bacteria attacking petroleum and oil fractions. J Bacteriol 44:169–178

Tullio M, Pierandrei F, Salerno A, Rea E (2003) Tolerance to cadmium of vesicular arbuscular mycorrhizae spores isolated from a cadmium-polluted and unpolluted soil. Biol Fertil Soils 37:211–214

Ulfig K, Płaza G, Hazen TC, Fliermans CB, Franck MM, Lombard KH (1997). Bioremediation treatability and feasibility studies at a Polish petroleum refinery. In: Proceedings Warsaw '96, Florida State University Press

Ulfig K, Płaza G, Lukasik K, Krajewska J, Mańko T, Wypych J, Dziewięcka B, Worsztynowicz A (1998) Selected filamentous fungi as bioindicators of leachate toxicity and bioremediation progress. In: National scientific-technical symposium "soil bioremediation", Wisła-Bukowa

Van der Lelie D, Schwitzguebel JP, Glass DJ, Vangronsveld J, Baker A (2001) Assessing phytoremediation's progress in the United States and Europe. Environ Sci Technol 35:446A–452A

Vidali M (2001) Bioremediation. An overview. Pure Appl Chem 73:1163–1172

Walker JD, Cofone L, Cooney JI (1973) Microbial petroleum degradation: the role of *Cladosporium resinae* on prevention and control of oil spills. In: API/EPA/USLG conference, Washington, DC

Walsh JB (1999) A feasibility study of bioremediation in a highly organic soil. A thesis submitted to the faculty of the Worcester Polytechnic Institute in partial fulfillment of the requirements for the Degree of Master of Science in Environmental Engineering, May 1999

Wischmann H, Steinhart H (1997) The formation of PAH oxidation products in soils and soil/compost mixtures. Chemosphere 35:1681–1698

Wu T, Crapper M (2009) Simulation of biopile processes using a hydraulics approach. J Hazard Mater 17:1103–1111

Zervakis G, Philippoussis A, Ioannidou S, Diamantopoulou P (2001) Mycelium growth kinetics and optimal temperature conditions for the cultivation of edible mushroom species on lignocellulosic substrates. Folia Microbiol 46:231–234

Chapter 2
The Bioremediation Potential of Different Ecophysiological Groups of Fungi

Antonella Anastasi, Valeria Tigini, and Giovanna Cristina Varese

2.1 Introduction on the Different Ecophysiological Groups of Fungi and on Their Potential in the Bioremediation of Soil

Fungi are unique organisms due to their morphological, physiological, and genetic features; they are ubiquitous, able to colonize all matrices (soil, water, air) in natural environments, in which they play key roles in maintaining the ecosystems equilibrium. If we consider the different matrices, the air is an important vehicle for the dissemination of fungal propagules (conidia, spores, hyphae), which represent the main component of the bioaerosol. The aquatic environment, both marine and freshwater, is steadily colonized by fungi, too; however, the main habitat of these organisms is soil. Within this complex and heterogeneous matrix, fungi are found in a variety of ecological niches, from the arctic tundra to the desert sand dunes, where they play key roles for the maintenance of ecosystems.

The tremendous evolutionary success of this heterogeneous group of organisms is evidenced by the high number of species, the diversity of niches, and habitats occupied, the ability to establish symbiosis (both mutualistic and pathosistic) with other organisms, mainly plants and animals, and to survive under restrictive conditions for most other organisms. In terms of biodiversity, fungi are probably the second most common group of organisms on our planet, with an estimated 1.5 million species, second only to arthropods. However, this enormous biodiversity is still largely unexplored, as currently described species are about 100,000.

All fungi are heterotrophic and obtain the organic substance necessary for their growth through three different ways: saprophytism, parasitism (pathosistic symbiosis), and mutualistic symbiosis. Saprotrophic fungi play a key role in the decomposition of organic matter and, therefore, in the circulation of the elements,

A. Anastasi • V. Tigini • G.C. Varese (✉)

Department of Life science and Systems Biology, University of Turin, Viale Mattioli, 25, 10125 Turin, Italy

e-mail: antonella.anastasi@unito.it; valeria.tigini@unito.it; cristina.varese@unito.it

E.M. Goltapeh et al. (eds.), *Fungi as Bioremediators*, Soil Biology 32,
DOI 10.1007/978-3-642-33811-3_2, © Springer-Verlag Berlin Heidelberg 2013

both in natural and anthropic environments. However, many other fungi are biotrophs, and in this role a number of successful groups form symbiotic associations with plants (mycorrhizae), algae, animals (especially arthropods), and prokaryotes. Besides, fungi are among the most important plant pathogens, including rusts and smuts, and can be parasites of animals, humans included.

Virtually all natural organic compounds can be degraded by one or more fungal species thanks to the production of a variety of enzymes such as amylases, lipases, and proteases that allow them to use substrates as starches, fats, and proteins. A more limited number of species can use pectines, cellulose, and hemicellulose as carbon sources. Finally, some fungi are the main degraders of natural polymers particularly complex and resistant to microbial attack, such as keratin, chitin, and lignin. Due to the high nonspecificity of the enzymes involved in the degradation of lignin, wood fungi can attack numerous aromatic and aliphatic xenobiotic compounds, including environmental pollutants such as polycyclic aromatic hydrocarbons (PAHs), polychlorinated biphenyls (PCBs), pesticides, and herbicides. These capabilities make them organisms of great interest for potential use in environmental bioremediation.

Also symbiotic fungi can be candidates for use in remediation of soils. Mycorrhizal fungi occupy the structural and functional interface between decomposition and primary production and play a key role in many soil ecosystems. They mobilize nutrients from organic sources and appear to possess well-developed saprotrophic capabilities (i.e., oxidative and hydrolytic enzymes).

Along with the ability to degrade a wide range of pollutants, microorganisms useful in soil bioremediation should possess several other features, including the capability to extensively colonize the soil matrix, resist at high concentrations of toxic compounds, survive over a long period in restrictive conditions, and compete with the other components of the soil microbiota. In the following parts of the chapter, we will explore the specific characteristics of different groups of fungi (Fig. 2.1), with regard to the main aspects that make an organism a useful bioremediator. In particular, we will focus on saprotrophs as wood fungi that are known for over 50 years for their ability to degrade a wide range of pollutants and for their potential use as bioremediators. Moreover, thanks to their better adaptability to soil, the role of litter and soil fungi will also be highlighted. Besides these ecophysiological groups of saprotrophs, we will also consider the potential use as bioremediators of mycorrhizal fungi that have the advantage of being distributed throughout the soil by roots and provided with a long-term supply of photosynthetic carbon from their hosts.

2.2 Wood Degrading Fungi, Their Main Ecophysiological Features and Their Potential in Soil Bioremediation

Wood is composed up to 30 % of lignin, a complex phenyl propane polymer that coats cell wall polysaccharides and chemically combines with them conferring resistance against microbial degradation. Wood degrading fungi are the only

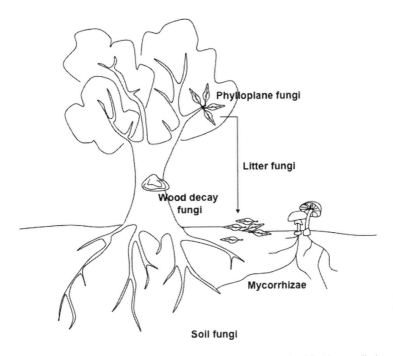

Fig. 2.1 Different ecophysiological groups of fungi potentially involved in bioremediation

organisms able to completely mineralize lignin and this degradation is probably the most important process for carbon recycling in nature. According to the appearance of the rotten wood, ligninolytic fungi are distinguished as white (WRF), brown (BRF), and soft-rot (SRF) fungi. In the following paragraphs we will focus on the first two groups only.

2.2.1   White-Rot Fungi

Fungi causing white rot are represented in all the main groups of the Basidiomycota (Agaricomycotina), and in some of the Ascomycota, namely the Xylariaceae. In common usage, the term "white rot" has been traditionally used to describe forms of wood decay in which the wood assumes a bleached appearance, with a spongy, stringy, or laminated structure, and where lignin as well as cellulose and hemicellulose is broken down (Anastasi et al. 2009a). WRF, thanks to their extracellular, oxidative enzymes, are able to completely mineralize lignin and carbohydrate components of wood to CO_2 and H_2O. Interestingly, these fungi do not use lignin as a carbon source for growth; instead, they degrade lignin to obtain the cellulose that is toward the interior of the wood fiber.

WRF can be used to detoxify or remove various aromatic pollutants and xenobiotics found in contaminated soil (Schauer and Borriss 2004). Actually, their ligninolytic enzymes are nonspecific, non-stereoselective, and effective against a broad spectrum of aromatic compounds. The first observation of aromatic pollutants breakdown by WRF dates back 50 years (Lyr 1963), and the connection of this process with ligninolytic metabolism was observed since 1980s in *Phanerochaete chrysosporium*, the model fungus for ligninolysis (Bumpus et al. 1985). From that time, it was shown that an impressive array of organopollutants can be degraded by ligninolytic fungi, including PAHs, PCBs, nitroaromatics, pesticides, herbicides, dyes, etc. (Rabinovich et al. 2004). The potential of WRF to degrade persistent organic pollutants (POPs) in both artificially spiked and industrially contaminated soils has been widely demonstrated, and dense mycelial growth coupled with significant enzyme production are considered key aspects for effective bioremediation (Borràs et al. 2010).

2.2.2 Enzymatic Mechanisms of Degradation of Organopollutants in White-Rot Fungi

WRF produce a battery of lignin degrading enzymes that catalyze oxidation of xenobiotics in addition to their ability to degrade lignin. They consist of peroxidases, laccases, and other enzymes involved in the formation of radicals, ROS and H_2O_2, that cleave the carbon–carbon and carbon–oxygen bonds of the lignin/xenobiotic by means of a free radical mechanism. This free radical mechanism provides the basis for the nonspecific nature of degradation of a variety of structurally diverse pollutants, obviating the need for these organisms to be adapted to the chemical being degraded (Reddy and Mathew 2001).

2.2.2.1 Peroxidases

Fungal heme-containing peroxidases, i.e., lignin (LiP), manganese (MnP), and versatile (VP) peroxidases, use hydrogen peroxide to promote the one-electron oxidation of chemicals to free radicals; they are involved in the biodegradation of lignocelluloses and participate to the bioconversion of diverse recalcitrant compounds. The main features of fungal heme-containing peroxidases are shown in Table 2.1. The catalytic cycle is similar in all peroxidases. In the resting state, the heme iron is in the ferric state; hydrogen peroxide oxidizes the ferric enzyme by two electrons forming a ferryl (Fe IV) π-porphyrin cation radical, named compound I. A chemical can then be oxidized by one electron to a radical, and compound I can be reduced by one electron to compound II; a subsequent oxidation of another molecule by compound II returns the enzyme to its ferric resting state (Vidossich et al. 2010).

Table 2.1 Properties of fungal heme peroxidases

	E.C.	Molecular weight (kDa)	pI	Substrates	Redox potential (V)
LiP	1.11.1.14	35–48	3.1–4.7	Non-phenolic (veratryl alcohol) and phenolic aromatic compounds, organopollutants (PAHs, chlorophenols, nitroaromatics, dyes, and explosives)	1.4–1.5
MnP	1.11.1.13	38–50	2.9–7.1	Manganese, phenols, and non-phenolic lignin moieties via lipid radicals	1.0–1.2
VP	1.11.1.16	43–45	3.4–3.9	Non-phenolic, phenolic, and dye substrates, manganese	1.4–1.5

Mainly two aspects in the molecular structure of lignin peroxidases provide them their unique catalytic properties: a heme environment, conferring high redox potential to the oxo-ferryl complex, and the existence of specific binding sites for oxidation of their characteristic substrates, including non-phenolic aromatics in the cases of LiP, Mn^{2+} in the case of MnP, and both types of compounds in the case of VP (Martínez et al. 2005). Concerning the heme environment, an interesting aspect of the molecular structure of these enzymes is the position of the so-called proximal histidine, the Nε of the side-chain of a histidine residue. Peroxidases differ mainly in the position of this iron ligand; in ligninolytic peroxidases it is displaced away from the heme iron, increasing its electron deficiency and increasing the redox potential of the oxo-ferryl complex (Martínez et al. 2005).

The aromatic substrates binding site and the Mn binding site were first identified in LiP and MnP and then confirmed in VP, whose hybrid molecular architecture combines the properties of LiP and MnP. Mn binding site (2 Glu, 1 Asp) is situated near the cofactor and this enables direct electron transfer to the heme. By contrast, veratryl alcohol and other LiP/VP aromatic substrates, that cannot penetrate inside the protein to transfer electrons directly to the cofactor, are oxidized at the enzyme surface by a catalytically active tryptophan (Trp-164) and electrons are transferred to the heme by a protein pathway (long-range electron transfer, LRET) (Martínez et al. 2005).

Besides these "classic" fungal peroxidases, most recently several isoforms of a novel heme-thiolate peroxidase have been characterized from the wood-colonizing basidiomycete *Agrocybe aegerita*, which later turned out to be a true peroxygenase, efficiently transferring oxygen from peroxide to various organic substrates including aromatic, heterocyclic, and aliphatic compounds (Kinne et al. 2010). Due to their unique ability to epoxidize and hydroxylate aromatic rings by means of hydrogen peroxide, these enzymes are nowadays mostly referred to as aromatic peroxygenases (Hofrichter et al. 2010). The broad substrate range and ubiquitous distribution of these enzymes may indicate an important environmental role for them in the oxidation of both natural and anthropogenic aromatics, as recently demonstrated for toluene (Kinne et al. 2010).

Dye peroxidase (DyPs) are a new superfamily of heme peroxidases, found in fungi and bacteria, that show a little sequence similarity to classical fungal peroxidases and lack the typical heme-binding region (Hofrichter et al. 2010). These enzymes oxidize various dyes, in particular anthraquinone derivatives, but no report is found on their potential in bioremediation of soil contaminants.

2.2.2.2 Laccases

Laccase activity has been demonstrated in many fungal species belonging to Ascomycota and Basidiomycota and in particular in wood rotting fungi; almost all species of WRF were reported to produce laccase to various degrees (Baldrian 2006). Laccases (E.C. 1.10.3.2) are blue copper oxidoreductases, typically with molecular masses of approximately 60–70 kDa and pI of 4, able to catalyze the one-electron oxidation of polyphenols, methoxy-substituted phenols, aromatic diamines, and other compounds using O_2 as oxidant (Canas and Camarero 2010). Recently laccases have attracted attention for their ability to degrade recalcitrant pollutants, as pyrene (Anastasi et al. 2009b), benzo[a]pyrene (Li et al. 2010), and chrysene (Nikiforova et al. 2010).

Laccases contain four copper atoms in their active site and catalyze a one-electron oxidation concomitantly with the four-electron reduction of molecular oxygen to water; after four electrons have been received by a laccase molecule, the enzyme reduces O_2 to H_2O, returning to the native state (Lundell et al. 2010). The fact that laccases can use O_2 as the final electron acceptor represents a considerable advantage for industrial and environmental applications compared with peroxidases, which require a continuous supply of H_2O_2. The possibility of increasing the production of laccase by the addition of inducers to fungal cultures and a relatively simple purification process are other advantages (Baldrian 2006). Taking into account that the advantage of peroxidases is their higher redox potential, to engineer the active site of laccases to obtain high redox potential variants would be of considerable biotechnological interest (Martínez et al. 2005).

Reactions catalyzed by laccases in technological or synthetic applications can be of two types. In the simplest case, the substrate molecules are oxidized to the corresponding radicals by direct interaction with the catalytic core of the enzyme. Frequently, however, the substrates cannot be oxidized directly by the enzyme, either because they are too large to penetrate into the enzyme active site or because they have a particular high redox potential (Riva 2006). By mimicking nature, it is possible to overcome this limitation with the addition of so-called "chemical mediators," which are suitable compounds that act as intermediate substrates for the laccase, whose oxidized radical forms are able to interact with the bulky or high redox-potential substrates targets (Riva 2006). Typical mediators are ABTS, 1-hydroxybenzotriazole, and violuric acid, though several other compounds may also be used. The use of eco-friendly mediators easily available from lignocellulose, could contribute to the environmental application of laccases (Canas and Camarero 2010).

2.2.3 Brown-Rot Fungi

Brown rot is a kind of wood decay caused exclusively by Basidiomycota, namely Agaricomycetes. This class encompasses many orders and families, though the overwhelming majority of the BRF belongs to the Agaricales, Hymenochaetales, Gloeophyllales, and Polyporales. Interestingly, only 6 % of all the known wood decay fungi are now known to cause a brown rot and are almost exclusively associated with conifers (Hibbett and Donoghue 2001). BRF degrade cellulose and hemicellulose present in wood after only a partial modification of lignin (demethylation, partial oxidation, and depolymerization) by a nonenzymatic Fenton-type catalytic system. Because of the preferential degradation of polymers, the decayed wood loses its inherent strength and acquires a brittle consistency, breaks up like cubes, and finally crumbles into powder. The modified lignin remaining gives the decayed wood its characteristic color (dark brown) and consistency.

2.2.4 Mechanisms of Degradation of Pollutants in Brown-Rot Fungi

In 1974, Koenigs hypothesized that BRF use, unlike WRF, a Fenton-type catalytic system, based on free-radical reactions initiated by hydroxyl radicals. These hydroxyl radicals are formed according the following formula:

$$H_2O_2 + Fe^{2+} \rightarrow OH^- + OH^\bullet + Fe^{3+}$$

The pathways leading to H_2O_2 and Fe^{2+} generation remain not fully understood. Kremer and Wood (1992) supposed the production of Fenton reagents by cellobiose oxidase: this enzyme reduces efficiently many Fe^{3+} complexes. Hyde and Wood (1997) proposed the production of Fenton reagents by the BRF C. puteana, involving the cellobiose dehydrogenase. The enzyme purified from this fungus has been shown to couple oxidation of cellodextrins to conversion of Fe^{3+} to Fe^{2+}.

BRF are today studied for the metal removal from wood wastes treated with chromated copper arsenate due to their copper tolerance and production of high levels of oxalic acid (Kim et al. 2009). However, they are also known for the degradation activity against organic contaminants. Martens et al. (1996) reported the degradation by *Gloeophyllum striatum* of a fluoroquinolone antibacterial drug (enrofloxacin); this and other similar antibiotics may find a way to contaminate soils, particularly in areas of large animal operations. Later, Schlosser et al. (2000) investigated the degradation pathway of 2,4-dichlorophenol by *G. striatum*; the authors found that identical metabolites to those formed in vitro with Fenton's reagent were produced by the fungal degradation. Andersson et al. (2003) made a comparison in the treatment of a PAHs contaminated soil between the WRF *Pleurotus ostreatus* and the BRF *Antrodia vaillantii*: the BRF resulted in a better degradation than *P. ostreatus*, without metabolite accumulation and without a negative effect on the indigenous soil microorganisms. More recently, 12 species of BRF have been investigated for their ability to degrade 1,1,1-trichloro-2,2-bis (4-chlorophenyl) ethane (DDT); *Gloeophyllum trabeum*, *Fomitopsis pinicola*, and *Daedalea dickinsii* showed a high ability to degrade DDT via a chemical Fenton reaction (Purnomo et al. 2008).

Apart from the degradation of the pollutants, one important aspect for bioremediation efficacy is the ability of the inoculated fungi to colonize the soil matrix. In 2001, Andersson and collaborators showed that the BRF *A. vaillantii* was able to efficiently colonize contaminated soil. These findings about BRF may be important for fungal soil bioremediation technology and should be emphasized in further studies.

2.2.5 *Applications of Wood Degrading Fungi in Bioremediation*

There has been a plethora of review articles with respect to mechanisms of degradation of POPs by wood fungi, mainly focusing on the species *Phanerochaete chrysosporium*, *Pleurotus ostreatus*, *Trametes versicolor*, *Bjerkandera adusta*, *Lentinula edodes*, and *Irpex lacteus* (Singh 2006). Along with their ability to degrade POPs, these fungi possess a number of advantages not associated with other bioremediation systems. Because key components of the lignin-degrading system are extracellular, the fungi can degrade very insoluble chemicals such as lignin or many of the hazardous environmental pollutants. Moreover, the extracellular system enables fungi to tolerate considerably high concentrations of toxic pollutants (Reddy and Mathew 2001).

However, up to now most of the research on fungal bioremediation has been conducted on laboratory scale, so further work is required to study these capacities taking into account the natural variables and their applicability in large-scale contaminated fields (Pinedo-Rivilla et al. 2009).

Since soil is not their natural environment, wood fungi application in soil bioremediation is necessarily by means of a bioaugmentation. Bioaugmentation has been proven successful in cleaning up of sites contaminated with several organic compounds but still faces many environmental problems. One of the most difficult issues is survival of strains introduced in the non-native soil environment.

Many studies have shown that both abiotic and biotic factors influence the growth of wood fungi in soil, and hence the effectiveness of bioaugmentation (Andersson et al. 2003; Baldrian 2008). Soils generally contain less nutrients than wood; hence the addition of C and N sources, preferably in the form of lignocellulose, are necessary (Baldrian 2008). Corn cobs, wheat, straw, wood chips, and bark have been frequently used for introduction of pre-grown wood fungi into soil, acting both as carriers and as nutrient sources (Covino et al. 2010). Generally, the larger the inoculum biomass, the faster and more successful is the establishment of the fungus in the soil.

The filamentous growth of basidiomycetes forming mycelial cords is a significant advantage for soil colonization, in which nutrient resources distribution is not homogeneous (Baldrian 2008). Actually, mycelial cords act as roads for nutrient transport, allowing the proliferation through patches low in nutrients and creating a network interconnecting the resource units (Baldrian 2008). Moreover, the hyphal growth mode of fungi makes them able to extensively penetrate into soil and serve, at the same time, as dispersion vectors of indigenous pollutant-degrading bacteria (Kohlmeier et al. 2005).

Temperature has a considerable effect on the ability of wood fungi to grow and degrade organopollutants, and, in general, most contaminated sites will not be at the optimum temperature for bioremediation during every season of the year. High temperatures generally increase the growth rate, and consequently the metabolic activity, and very slow or no growth is detected at temperatures below 10 °C (Baldrian 2008).

Fungal colonization is also affected by soil texture, pH, and presence of inhibitory compounds. Overall it is clear that, however efficient a fungal soil inoculant may be in transforming POPs in the laboratory, the chemical and physical restrictions encountered in the heterogenous soil environment will prevent complete pollutant transformation. Long periods of contact of POPs with soil constituents allow more time for sorption reactions to occur, consequently reducing pollutants bioavailability (Singleton 2001). Recently, Covino et al. (2010) demonstrated that degradation of PAHs by *Lentinus tigrinus* and *Irpex lacteus* was negatively correlated with their organic sorption coefficients and hydrophobicity. By physical and chemical (i.e., surfactant addition) means it is possible to increase POPs bioavailability; however, the extra cost involved must be balanced with the level of clean-up required (Singleton 2001).

Along with physical and chemical factors, another important aspect affecting the colonization of soil by ligninolytic fungi is the presence of indigenous soil organisms. Actually, when fungi enter nonsterile soils, they must compete with the indigenous soil microbiota for nutrients and protect the mycelium against microbial attack (McErlen et al. 2006). Wood fungal species differ significantly in their ability to colonize nonsterile soil, and they can be classified into strong competitors (i.e., *Pleurotus* spp., *Phanerochaete* spp., *Trametes versicolor*) and weak competitors (*Ganoderma applanatum*, *Dichomitus squalens*) (Baldrian 2008); however, it should be considered that colonization ability may vary even within a single species and is partly dependent on soil type (Baldrian 2008).

Interactions between wood fungi and soil bacteria are mostly combative: the fungal growth inhibition by bacteria is mediated by the production of phenazine derivatives, antifungal antibiotics, or by mycophagy; on the other hand the suppression of bacteria by fungi could be due to oxidoreductase activity, production of hydroxyl radicals, and of antibiotic compounds (Baldrian 2008).

Successful colonization of nonsterile soils by WRF demonstrates the ability to overcome the adverse effect of soil microflora: *T. versicolor* was shown to attack soil bacteria, i.e., *Pseudomonas* (Thorn and Tsuneda 1992), and *P. ostreatus* was able, during degradation of PAHs in nonsterile, artificially contaminated soils, to inhibit the growth of indigenous bacteria and change the composition of the bacterial community (Andersson et al. 2000).

More recently, the effect of bacterial stress on *I. lacteus* and *T. versicolor* was investigated (Borràs et al. 2010). The bacterial stress was represented either by the innate soil microflora or a defined mixture of the soil bacteria *Pseudomonas aeruginosa* and *Rhodococus erythropolis* inoculated into the sterile soil. The effect was measured by the efficiency of removal of 16 PAHs spiked into soil, by the ability of the fungi to colonize the soil and produce MnP and laccases. This comparative study demonstrated a significant decrease of degradation of total PAHs in the presence of live bacteria. On the other hand, the bacteria tested did not affect the capability of the two fungal organisms to colonize soil and did not influence the fungal growth yields and the extracellular enzyme levels.

However, the interaction with the indigenous microflora may also be beneficial for the process of bioremediation. Metabolites produced by oxidation of high-molecular weight POPs by WRF can be mineralized by indigenous microflora, with an overall increase of the biodegradation rate. Recently, Cea et al. (2010) demonstrated that members of the phylum Proteobacteria (*Xanthomonadaceae*, *Burkholderiaceae*, *Enterobacteriaceae*) actively participate in the remediation of chlorophenols in soils together with the WRF *Anthracophyllum discolor*.

2.3   Litter Fungi

Litter decomposing fungi (LDF) are another ecophysiological group of basidiomycetes, which play a pivotal role in the ecology of forests since they are deeply involved in wood and litter decomposition, humification, and mineralization of soil organic matter. A major part of the total litter input to forest soils is non-woody plant residues such as leaves, fruit, and reproductive structures. Plant litter consists of several groups of compounds and the relative amounts of the different constituents vary depending on the plant part (leaves, stem, etc.) and species. Broadly, the dominant C-rich components are soluble organic compounds (sugars, low molecular weight phenolics, hydrocarbons, and glycerides), hemicellulose, cellulose, and lignin (Berg and McClaugherty 2003).

Basidiomycetous litter fungi are considered especially important due to their production of ligninolytic enzymes essential for degradation of plant materials (Osono and Takeda 2002). Ascomycetous fungi constitute a large part of the fungal community in forest litter (Lindahl et al. 2007), although the lignin degrading capacity of most ascomycetes appears to be limited (Osono and Takeda 2002). The community of saprotrophic litter decomposers has been observed to be restricted to the upper litter layer, whereas the fungal community in the deeper layers is typically dominated by mycorrhizal fungi (Lindahl et al. 2007).

When the newly shed litter reaches the forest floor, it is already colonized by phylloplane and endophytic fungi. The ecological role of endophytes is not clear, but several of them remain in the dead litter and some have saprotrophic capabilities

(Osono 2006). In this early stage of decomposition, mainly soluble sugars and other low molecular weight compounds are lost from the litter and enzymes linked to soluble saccharides are also prominent but then rapidly decline. Some endophytic fungi in the early community also have cellulolytic capacities and have been observed to cause significant mass loss in laboratory experiments (Korkama-Rajala et al. 2008). Knowledge of the functional capacities of these fungi is, however, very limited.

The first fungal community in the recently shed litter is soon accompanied by early basidiomycetous fungi. Species within the genera *Athelia*, *Marasmius*, and *Sistotrema* are frequently found (Lindahl et al. 2007; Korkama-Rajala et al. 2008). During the second phase of litter decomposition, cellulolytic enzymes are active and the main degradation of the polymer occurs. Laccase activity can also be observed relatively early in decomposition of litter with high contents of phenolic compounds. Typical litter basidiomycetes such as species of the genera *Mycena*, *Clitocybe*, and *Collybia* are prominent during this stage of decomposition (Osono 2007).

In later stages of decomposition, the typical LDF appear to be absent and, instead, mycorrhizal fungi have been found to dominate the fungal community in the humus layer of both deciduous and coniferous forests (Lindahl et al. 2007). Due to problems with culturability of most mycorrhizal fungi, this functional group escapes detection by traditional, culture-based methods.

2.3.1 Litter Fungi in Soil Bioremediation

Several papers report that LDF, almost exclusively basidiomycetes, produce a wide variety of oxidoreductases, most frequently laccases and MnP, that, similarly to wood degrading fungi, could be involved in the degradation of different POPs (Casieri et al. 2010).

From a historical point of view, the first report of the ability to degrade xenobiotics by this group of fungi date back to 1999 when Wunch and collaborators reported the degradation of benzo[*a*]pirene by *Marasmiellus troyanus* in liquid culture. The same authors confirmed the degradative capability of the fungus in a bioaugmentation experiment carried out in nonsterile soil microcosms; however, no mechanism of action was suggested (Wunch et al. 1999).

The potential application of LDF in soil bioremediation for degradation of recalcitrant organopollutants was demonstrated by Steffen et al. who showed the bioconversion of high molecular mass PAHs by selected LDF in both liquid cultures (2002) and contaminated nonsterile soils (2007). The authors pointed out that several LDF were able to convert different POPs to some extent and proposed that MnP was the key enzyme in the degradation process. They also observed considerable differences in the degradation activity by different species and showed that *Stropharia rugosoannulata* and *S. coronilla* were the most efficient degraders. Very recently the involvement of laccases from *Marasmium quercophilus*

(Farnet et al. 2009) and peroxinagenes from *Coprinellus radians* (Aranda et al. 2010) in the degradation of PAHs, methylnaphthalenes, and dibenzofurans was also reported.

Hence, the literature demonstrates that selected litter-decomposing basidomycetes have a well-founded potential to be used in bioremediation towards different POPs especially if suitable carriers for inoculation and production of the oxidative enzymes are selected (Valentín et al. 2009). The abilities of LDF to colonize the soil, to survive there over long periods, and to compete with other microorganisms should be considered as ecological features that can make them even more suitable for bioremediation applications compared with WDF, which usually prefer to colonize compact woods (logs, trunks, etc.) and have poor capability to grow in different niches such as soil. Moreover, the efficiency of LDF varies among the different species, and the degradation potential of LDF may be even considerably higher since only a limited number of species has been tested so far (Steffen et al. 2007; Casieri et al. 2010).

2.4 Soil Fungi

The spatial distribution and metabolic activity of soil microorganisms is heterogeneous and closely correlated to organic matter availability, which is often concentrated in hot spots. In this contest, fungi with high sporulation capability, such as mitosporic fungi (Deuteromycetes) and Zygomycetes, can easily straggle to reach the suitable environmental condition and nutrients for their growth. When the environment is unsuitable for growth (low nutrients, presence of metabolites, etc.), their asexual spores, called conidia, show an exogenously imposed dormancy induced by the phenomenon termed fungistasis, allowing the fungus to survive (Deacon 2006).

Some few investigations have focused on the spatial heterogeneity of pollutants biodegradation, finding that it mainly occurs between the topsoil (an organic rich layer which corresponds to the root zone), the rhizosphere, and the rhizoplane (Gonod et al. 2003). Since bioremediation activity in this microhabitat seems to be mainly ascribable to the action of its fungal component (Bengtsson et al. 2010), it is of crucial importance to deepen the ecological and physiological characteristics of fungi that live in this part of the soil.

The loss of organic and inorganic carbon from roots into soil underpins nearly all the major changes that occur in the rhizosphere. Thus, the rhizosphere is characterized by high microbial activity due to the rich supply of organic carbon compounds derived from the root. Microbial populations in the rhizosphere may be 10–100 times higher than in the bulk mineral soil. The difference between the soil and rhizosphere–rhizoplane mycobiota is substantial: diminished species variability, changes in the representation of many species, increased mycelial biomass, and increased mean radial growth rate of typical species of micromycetes (Kurakov et al. 1994).

Fungi play several ecological roles in rhizosphere, such as symbiontic, pathosistic, and saprotrophic ones. In particular the last one is strongly related to the bioremediation potentiality of rhizosphere fungi. In all the phases of the sequence of the organic matter decomposition, fungi play an important role with an overlapping sequence of activities with different patterns of behavior. On account of the fungal abilities to degrade particular types of substrates, they can be divided in pioneer saprotrophic fungi, polymer-degrading fungi, and fungi that degrade recalcitrant polymers (Deacon 2006).

Among pioneer saprotrophic fungi there are several species of *Mucor*, *Cunninghamella*, *Rhizopus*, and other Zygomycota, which utilize sugars and other simple soluble nutrients. These pioneer fungi grow rapidly, have a short exploitative phase, and a high competitive ability; they are generally characterized by high tolerance to environmental stresses such as the presence of pollutants, in fact, they are common in polluted soils (Tigini et al. 2009). Independently by their degradation capabilities, these fungi are capable to decrease the concentration of organic pollutants such as PAH, by accumulating them in intracellular lipid vesicles. Moreover, these vesicles could have a role in biodegradation too (Verdin et al. 2004).

This accumulation activity can also be common to polymer-degrading fungi that have an extended phase of growth on the major structural polymers such as cellulose, hemicelluloses, or chitin (Verdin et al. 2004). These fungi tend to defend the resource against potential invaders, either by sequestering critically limiting nutrients or by producing inhibitory metabolites (Deacon 2006). Species belonging to the genera *Trichoderma*, *Fusarium*, *Penicillium*, *Stachybotrys*, *Aspergillus*, *Cladosporium*, *Mortierella*, *Beauveria*, *Engyodontium* are some examples of this kind of fungi that have been recently described as tolerant to pollutants such as PCBs, chlorobenzoic acids (CBA), and endosulfan and that are indicated as potential bioremediation agents in soil (Garon et al. 2000; Tigini et al. 2009; Pinedo-Rivilla et al. 2009).

Fungi that degrade recalcitrant polymers often predominate in the later stages of decomposition. The ecological success of fungi that develop later in the decomposition sequence is related to their specialized ability to degrade polymers such as lignin and keratin that most other fungi cannot utilize. Among them, *Fusarium*, *Penicillium*, *Aspergillus*, *Paecilomyces*, *Microsporum*, *Acremonium*, and *Geomyces* are often reported as xenobiotics degrading (Pinedo-Rivilla et al. 2009; Tigini et al. 2009).

When we approach to a polluted area, however, this theoretical sequence can be substantially modified. Actually, one of the main effects of the industrial pollution on soil is the significant decline of biodiversity of the native microbial population at both the genetic and the functional point of view. Environmental pollution adversely affects many levels of the ecosystem organization. It might affect the efficiency of using the available resources, making the system more sensitive to subsequent stresses, or might lead to the development of tolerance making the system more resistant to additional stresses.

In contaminated soils, the fungal community is generally remarkably reduced to few fungal species, which are tolerant to pollutants (Tigini et al. 2009). In such environments microorganisms capable of pollutants removal can constitute up to 100 % of the viable ones (Venosa and Zhu 2003). In soils with relatively low levels of pollution, particular groups of fungi can become the dominant microbial population: opportunistic and phytotoxic fungi can increase as a consequence of soil contamination and decrease after soil remediation due to the adaptation of these fungi to the surrounding environment and the use of pollutants as a carbon source (Kireeva et al. 2008).

In any case, the loss of biodiversity in soil due to pollution has quite always a negative effect on the removal of xenobiotics; actually, given the complementary action that often exercise microorganisms consortia, the presence of different species ensures a greater effectiveness in soil bioremediation.

2.4.1 Mechanisms of Pollutants Degradation by Soil Fungi

Biodegradation strategies are generally divided in two categories (1) the target compound is directly used as a carbon source and (2) the target compound is enzymatically attacked but is not used as a carbon source (cometabolism). Although mitosporic fungi participate both strategies, they are often reported as more proficient at cometabolism. In the rhizosphere, in particular, the plant root often provides C sources essential for fungal degradation of more complex molecules. In last years, however, there is an increasing of demonstrations of direct metabolism of pollutants by mitosporic fungi and Zygomycetes.

Saprotrophic fungi produce a group of extra- or intracellular enzymes that include proteases, carbohydratases (e.g., cellulases, amylases, xylanases, etc.), esterases, phosphatases and phytases, which are physiologically necessary to living organisms. Due to their intrinsic low substrate specificity, hydrolases may play a pivotal role in the bioremediation of several pollutants including insoluble wastes.

The nonspecific ligninolytic systems (e.g., laccases and peroxidases), quite always associated with WRF, are often reported as putative enzymes involved in xenobiotics degradation by mitosporic fungi, too. However, while most of the WRF require a high level of consumption of easily metabolized cosubstrates to carry out lignin transformations (up to 20-fold the degraded lignin weight), lignin transformation with a low cosubstrate requirement was observed in *Penicillium chrysogenum* (Rodriguez et al. 1994).

LiP, MnP, laccases, cellulases, and hemicellulases can work synergistically in the decomposition of some pollutants. In some cases, the involvement of extracellular oxidative enzymes in the degradation of pollutants is not so clear because of the lack of correlation between pollutants removal and enzymes production (Tigini et al. 2009). Nevertheless, the enzyme detection in the degradation of pollutants is not sufficient to demonstrate their involvement in the depletion of toxic molecules, and the use of a inhibitor can be useful to verify the decrease of degradation activity.

Other than extracellular enzymes in lignin catabolism, a mechanism involved in PAH biodegradation could be the cytP450 system, which was extensively studied in *Cunninghamella* species (Cerniglia and Sutherland 2001). Non-ligninolytic fungi often metabolize PAHs with a mechanism that suggests the hydroxylation by a cytP450 monooxygenase followed by conjugation with sulfate ion. This was true for the degradation of pyrene and BaP by *Aspergillus terreus*, which metabolized these pollutants mainly in pyrenylsulfate and benzo(*a*)pyrenylsulfate, respectively (Capotorti et al. 2004).

Monooxygenases resulted involved in the degradation of fluorene, another very toxic PAH, by most of the micromycetes isolated from soil (Garon et al. 2000). Among them, the best degraders were three strains belonging to *Cunninghamella* genus and several species reported for the first time as fluorene degraders: *A. terreus*, *Colletotrichum dematium*, *Cryphonectria parasitica*, *Cunninghamella blakesleeana*, *C. echinulata*, *Drechslera spicifera*, *Embellisia annulata*, *Rhizoctonia solani*, and *Sporormiella australis*.

2.5 Mycorrhizal Fungi

Mycorrhizae are symbioses between plant roots and an array of soil-inhabiting filamentous fungi. These associations are virtually ubiquitous and generally considered mutualistics as they are based on a bidirectional exchange of nutrients that is essential to the growth and survival of both partners (Robertson et al. 2007). The fungal partner captures nitrogen (N), phosphorus (P), and other nutrients from the soil environment and exchanges them with the plant partner for photosynthetically derived carbon (C) compounds that feeds fungal metabolism.

Several types of mycorrhizal associations have been classified according to the fungus involved and the resulting structures produced by the root–fungus association: ectomycorrhizas (ECM), ericoid mycorrhizas (ERM), ectendomycorrhizas, arbuscular mycorrhizas (AM), arbutoid mycorrhizas (ARM), monotropoid mycorrhizas, and orchid mycorrhizas. Most trees in boreal forests form ECM, whereas the major constituents of the understorey vegetation usually form AM, ERM or ARM. Their role in nutrient transport in ecosystem and protection of plants against environmental and cultural stress has long been known.

2.5.1 Mycorrhizal Fungi in Soil Bioremediation

Phytoremediation is widely applied as a catch-all term for the use of plants to soil bioremediation. The term is certainly suited to hyperaccumulation of metals by plants, since the plant tissues are the repository of the pollutants. Where plants are used to remediate POPs, however, it would be better to use the term "rhizosphere/ mycorrhizosphere remediation," because POP degrading activity will, in most scenarios, occur in the rhizosphere/mycorrhizosphere, rather than in the plant per se. Several papers report that the mycorrhizosphere microbial community (including nonsymbiotic fungi and bacteria) may act in concert with mycorrhizal fungi to degrade POPs. Enhanced degradation or mineralization in the rhizosphere has been demonstrated for a range of pesticides, PAHs, oil, surfactants, PCBs, and chlorinated alkanes in both woody and herbaceous plants (Gao et al. 2011; Teng et al. 2010). This enhanced rhizosphere degradation is generally related to the plant-stimulated microbial activity since rhizosphere microorganisms usually do not degrade POPs to yield energy, rather they may cometabolize them as a consequence of utilizing plant-derived cyclic compounds. Moreover, several recent papers underlined the importance of synergistic interactions among different species of AM fungi (Gao et al. 2011) or between AM fungi and rhizobia (Teng et al. 2010) to enhance degradation of POPs in soils.

It is unknown how the carbon contributions of the phytobiont influenced fungal responses or how synergistic or antagonistic interactions between mycorrhizas and other microorganisms altered their ability to mineralize or degrade organic pollutants. Results from recent field studies show that, with a high host carbon demand or with a decreased host photosynthetic potential, ECM fungi exhibit a saprotrophic behavior (Cullings et al. 2010). Mycorrhizal fungi may benefit plants that grow in contaminated soils providing greater access to water and nutrients and possibly protecting them from direct contact with toxic contaminants (Robertson et al. 2007). However, in addition to other biological (e.g., bacterial plasmid transfer) and physical (e.g., pollutants drawn into the rhizosphere by the transpiration stream, alteration of soil structure, translocation of hydrophobic organic pollutants by mycorrhizal hyphae from soil to plant root) factors, also the direct degradation capability of mycorrhizal fungi may also play a role metabolizing, removing or immobilizing POPs, or transforming, and altering their mobility and toxicity.

According to several authors, in fact, many ECM and ERM fungi have retained some ability to degrade organic pollutants (Casieri et al. 2010) thanks to the production of extracellular enzymes able to directly or indirectly oxidize aromatic rings. The capability to degrade POPs of mycorrhizal fungi, and especially of ECM and ERM fungi, seems mainly related to the production of polyphenol oxidases (e.g., laccase, catechol oxidase, and tyrosinase) and peroxidases. Many papers report the production of these enzymes in axenic cultures (Burke and Cairney 2002; Casieri et al. 2010) and the widespread occurrence of laccases and peroxidases genes in many ECM fungi has been recently demonstrated (Lundell et al. 2010). However, the real degradative potential of mycorrhizal fungi (when in symbiosis with the plant) and the impact of mycorrhizal formation on the secretion of exoenzymes by the host plant and the mycobiont is almost unknown Courty et al. (2011) showed that the colonization of poplar roots by *Laccaria bicolor* dramatically modified their ability to secrete enzymes involved in organic matter breakdown or organic phosphorus mobilization, and they also demonstrated that the level of enzymes secreted by the ectomycorrhizal root tips is under the genetic control of the host (different genotypes).

Few studies have considered mycorrhizal fungi in symbiosis with plants for the degradation of organic pollutants and the results are often contradictory. Günter et al. (1998), working in axenic conditions, demonstrated that, in symbiosis with Scots pine, *S. granulatus* and *P. involutus* increased the level of peroxidases in the fungus/root homogenate and in the nutrient solution of the mycorrhizal plants and Meharg et al. (1997) reported that the degradation of 2,4-dichlorophenol by mycorrhizal pine roots under aseptic conditions was higher than in pure cultures of the same mycorrhizal fungi growing on expanded clay.

For more recalcitrant pollutants like PAHs, Genney et al. (2004) reported that the ECM had no impact on mineralization or volatilization of naphthalene and retarded the degradation of fluorene spiked into forest soil and Koivula et al. (2004) showed that the presence of pine and its mycorrhizal fungus had no significant effect on the pirene mineralization yields. More recently, Joner et al. (2006) reported a consistent negative effect of mycorrhizal inoculation on PAHs degradation; whereas Gunderson et al. (2007) showed how ECM colonization of hybrid poplar in diesel contaminated soils increased fine root production and whole-plant biomass but inhibited removal of total petroleum hydrocarbon from a diesel contaminated soil.

In conclusion, the majority of mycorrhizal fungi appear to be able to degrade a range of contaminants. To date only a limited number of mycorrhizal fungal taxa (and only single isolates of each) have been investigated with respect to their abilities to degrade organic pollutants. Given that well in excess of thousands of fungal species are likely to exist worldwide and that considerable physiological variation exists even between individual isolates of a single species (Casieri et al. 2010), only a fraction of the potential of mycorrhizal fungi to degrade pollutants has so far been determined. Further studies are necessary to understand the impact of POPs contamination on soil organisms and the physiological mechanisms of nutrient exchange among the different organisms, mycorrhizal, and plant partners included. A more thorough knowledge of which organisms are likely to survive and compete in various ecosystems is required, as well as a better understanding of whether certain types of fungal associations with different plant hosts gain in ecological importance following disturbance events.

2.6 Conclusions

The importance of developing multidisciplinary approaches to solve problems related to anthropogenic pollution is now clearly appreciated by the scientific community. Several gaps in knowledge still exist and scientists, industrialists, and government officials should collaborate to guide future research in both ecological and sustainable management contexts so that we will be better prepared to manage systematic and occasional contaminations.

The taxonomic, genetic, and functional diversity of fungi potentially useful in soil bioremediation is immense and continues to expand. It is also clear that the different ecophysiological groups of fungi are strictly interconnected creating communities that underpin survival and productivity of the different ecosystems. In addition, it appears that the redundancy of the biodegradation capacity of the different groups of fungi is essential to compensate for the depletion of microbial communities due to soil contamination and, hence, it is a key aspect for the ecosystem recovery.

The recent literature shows that each ecophysiological group of fungi may play a direct role in biodegradation of complex organic substrates via enzymatic catabolism. However, most fungi have been examined in isolation from an ecosystem context, thereby excluding interactions among the different organisms and the soil environment. Thus, further efforts are needed to better understand how the soil ecosystem works as a whole; a better knowledge of the complexity of this heterogeneous environment, and of the interactions between the different organisms present, will make it possible to formulate more effective bioremediation strategies.

References

Anastasi A, Vizzini A, Prigione V, Varese GC (2009a) Wood degrading fungi: morphology, metabolism and environmental applications. In: Varma A, Chauhan AK (eds) A textbook of molecular biotechnology. I.K. International, New Delhi, pp 957–993

Anastasi A, Coppola T, Prigione V, Varese GC (2009b) Pyrene degradation and detoxification in soil by a consortium of basidiomycetes isolated from compost: role of laccases and peroxidases. J Hazard Mater 165:1229–1233

Andersson BE, Welindera L, Olsson PA, Olsson S, Henrysson S (2000) Growth of inoculated white-rot fungi and their interactions with the bacterial community in soil contaminated with polycyclic aromatic hydrocarbons, as measured by phospholipid fatty acids. Bioresour Technol 73:29–36

Andersson BE, Tornberg K, Henrysson T, Olsson S (2001) Three-dimensional outgrowth of a wood-rotting fungus added to a contaminated soil from a former gasworks site. Bioresour Technol 78:37–45

Andersson BE, Lundstedt S, Tornberg K, Schnurer Y, Lars G, Berg O, Mattiasson B (2003) Incomplete degradation of polycyclic aromatic hydrocarbons in soil inoculated with wood-rotting fungi and their effect on the indigenous soil bacteria. Environ Toxicol Chem 22:1238–1243

Aranda E, Ullrich R, Hofrichter M (2010) Conversion of polycyclic aromatic hydrocarbons, methyl naphthalenes and dibenzofuran by two fungal peroxygenases. Biodegradation 21:267–281

Baldrian P (2006) Fungal laccases – occurrence and properties. FEMS Microbiol Rev 30:215–242

Baldrian P (2008) Wood-inhabiting ligninolytic basidiomycetes in soils: ecology and constraints for applicability in bioremediation. Fungal Ecol 1:4–12

Bengtsson G, Törneman N, Yang X (2010) Spatial uncoupling of biodegradation, soil respiration, and PAH concentration in a creosote contaminated soil. Environ Pollut 158:2865–2871

Berg B, McClaugherty C (2003) Plant litter – decomposition, humus formation, carbon sequestration. Springer, Berlin, 352 pp

Borràs E, Caminal G, Sarrà M, Novotny C (2010) Effect of soil bacteria on the ability of polycyclic aromatic hydrocarbons (PAHs) removal by Trametes versicolor and Irpex lacteus from contaminated soil. Soil Biol Biochem 42:2087–2093

Bumpus JA, Tien M, Wright D, Aust SD (1985) Oxidation of persistent environmental pollutants by a white rot fungus. Science 228:1434–1436

Burke RM, Cairney JWG (2002) Laccases and other polyphenol oxidases in ecto- and ericoid mycorrhizal fungi. Mycorrhiza 12:105–116

Canas AI, Camarero S (2010) Laccases and their natural mediators: biotechnological tools for sustainable eco-friendly processes. Biotechnol Adv 28:694–705

Capotorti G, Digianvincenzo P, Cesti P, Bernardi A, Guglielmetti G (2004) Pyrene and benzo(a) pyrene metabolism by an Aspergillus terreus strain isolated from a polycylic aromatic hydrocarbons polluted soil. Biodegradation 15:79–85

Casieri L, Anastasi A, Prigione V, Varese GC (2010) Survey of ectomycorrhizal, litter-degrading, and wood-degrading Basidiomycetes for dye decolorization and ligninolytic enzyme activity. Anton Leeuw Int J G 98:483–504

Cea M, Jorquera M, Rubilar O, Langer H, Tortella G, Diez MC (2010) Bioremediation of soil contaminated with pentachlorophenol by *Anthracophyllum discolor* and its effect on soil microbial community. J Hazard Mater 181:315–323

Cerniglia CE, Sutherland JB (2001) Bioremediation of polycyclic aromatic hydrocarbons by ligninolytic and nonligninolytic fungi. In: Gadd GM (ed) Fungi in bioremediation. Cambridge University Press, Cambridge, pp 136–187

Courty PE, Labbé J, Kohler A, MarcxaisB BC, Churin JL, Garaye J, Le Tacon F (2011) Effect of poplar genotypes on mycorrhizal infection and secreted enzyme activities in mycorrhizal and non-mycorrhizal roots. J Exp Bot 62:249–260

Covino S, Cvancarova M, Muzikar M, Svobodova K, D'annibale A, Petruccioli M, Federici F, Kresinova Z, Cajthaml T (2010) An efficient PAH-degrading *Lentinus (Panus) tigrinus* strain: effect of inoculum formulation and pollutant bioavailability in solid matrices. J Hazard Mater 183:669–676

Cullings K, Ishkhanova G, Ishkhanov G, Henson J (2010) Induction of saprophytic behavior in the ectomycorrhizal fungus *Suillus granulatus* by litter addition in a *Pinus contorta* (Lodgepole pine) stand in Yellowstone. Soil Biol Biochem 42:1176–1178

Deacon J (2006) Fungal biology. Blackwell, Malden, MA, pp 1–371

Farnet AM, Gil G, Ruaudel F, Chevremont AC, Ferre E (2009) Polycyclic aromatic hydrocarbon transformation with laccases of a white-rot fungus isolated from a Mediterranean schlerophyllous litter. Geoderma 149:267–271

Gao Y, Cheng Z, Ling W, Huang J (2011) Arbuscular mycorrhizal fungal hyphae contribute to the uptake of polycyclic aromatic hydrocarbons by plant roots. Bioresour Technol 101:6895–6901

Garon D, Krivobok S, Seigle-Murandi F (2000) Fungal degradation of fluorene. Chemosphere 40:91–97

Genney DR, Alexander IJ, Killham K, Meharg AA (2004) Degradation of the polycyclic aromatic hydrocarbon (PAH) fluorene is retarded in a Scots pine ectomycorrhizosphere. New Phytol 163:641–649

Gonod LV, Chenu C, Soulas G (2003) Spatial variability of 2,4-dichlorophenoxyacetic acid (2,4-D) mineralisation potential at the millimetre scale in soil. Soil Biol Biochem 35:373–382

Gunderson JJ, Knight JD, Van Rees KCJ (2007) Impact of ectomycorrhizal colonization of hybrid poplar on the remediation of diesel-contaminated soil. J Environ Qual 36:927–934

Günter T, Perner B, Gramss G (1998) Activities of phenol oxidizing enzymes of ectomycorrhizal fungi in axenic culture and in symbiosis with Scots pine (*Pinus sylvestris* L.). J Basic Microbiol 38:197–206

Hibbett DS, Donoghue MJ (2001) Analysis of character correlations among wood decay mechanisms, mating systems, and substrate ranges in homobasidiomycetes. Syst Biol 50:215–242

Hofrichter M, Ullrich R, Pecyna MJ, Liers C, Lundell T (2010) New and classic families of secreted fungal heme peroxidases. Appl Microbiol Biotechnol 87:871–897

Hyde SM, Wood PM (1997) A mechanism for production of hydroxyl radicals by the brownrot fungus *Coniophora puteana*: Fe(III) reduction by cellobiose dehydrogenase and Fe(II) oxidation at a distance from the hyphae. Microbiology 143:259–266

Joner EJ, Leyval C, Colpaert JV (2006) Ectomycorrhizas impede phytoremediation of polycyclic aromatic hydrocarbons (PAHs) both within and beyond the rhizosphere. Environ Pollut 142:34–38

Kim G, Choia Y, Kim J (2009) Improving the efficiency of metal removal from CCA-treated wood using brown rot fungi. Environ Technol 30:673–679

Kinne M, Zeisig C, Ullrich R, Kayser G, Hammel KE, Hofrichter M (2010) Stepwise oxygenations of toluene and 4-nitrotoluene by a fungal peroxygenase. Biochem Biophys Res Commun 397:18–21

Kireeva NA, Bakaeva MD, Galimzianova NF (2008) Evaluation of the effect of various methods of oil-polluted soil bioremediation on micromycete complexes. Prikl Biokhim Mikrobiol 44:63–68

Koenigs JW (1974) Hydrogen peroxide and iron: a proposed system for decomposition of wood by brown rot basidiomycetes. Wood Fiber 6:66–79

Kohlmeier S, Smits TMH, Ford RM, Keel C, Harms H, Lukas YW (2005) Taking the fungal highway: mobilization of pollutant-degrading bacteria by fungi. Environ Sci Technol 39:4640–4646

Koivula TT, Salkinoja-Salonen M, Peltola R, Romantschuk M (2004) Pyrene degradation in forest humus microcosms with or without pine and its mycorrhizal fungus. J Environ Qual 33:45–53

Korkama-Rajala T, Mueller MM, Pennanen T (2008) Decomposition and fungi of needle litter from slow- and fast-growing norway spruce (*Picea abies*) clones. Microb Ecol 56:76–89

Kremer SM, Wood PM (1992) Evidence that cellobiose oxidase from *Phanerochaete chrysosporium* is primarily an Fe(III) reductase. Kinetic comparison with neutrophil NADPH oxidase and yeast flavocytochrome-B2. Eur J Biochem 205:133–138

Kurakov AV, Than HTH, Belyuchenko IS (1994) Microscopic fungi of soil, rhizosphere and rhizoplane of cotton plant and tropical grasses introduced in the south of Tajikistan. Microbiology 63:624–629

Li X, Lin X, Yin R, Wu Y, Chu H, Zeng J, Yang T (2010) Optimization of laccase-mediated benzo [a]pyrene oxidation and the bioremedial application in aged polycyclic aromatic hydrocarbons-contaminated soil. J Health Sci 56:534–540

Lindahl BD, Ihrmark K, Boberg J, Trumbore SE, Högberg P, Stenlid J, Finlay RD (2007) Spatial separation of litter decomposition and mycorrhizal nitrogen uptake in a boreal forest. New Phytol 173:611–620

Lundell TK, Makela MR, Hilden K (2010) Lignin-modifying enzymes in filamentous basidiomycetes – ecological, functional and phylogenetic review. J Basic Microbiol 50:5–20

Lyr H (1963) Enzymatische detoxification chlorieter phenole. Phytopathol Z 47:73–83

Martens R, Wetzstein HG, Zadrazil F, Capelari M, Hoffmann P, Schmeer N (1996) Degradation of fluoroquinolone Enrofloxacin by wood-rotting fungi. Appl Environ Microbiol 62:4206–4209

Martínez AT, Speranza M, Ruiz-Dueñas FJ, Ferreira P, Camarero S, Guillén F, Gutiérrez A, Martínez MJ, del Río JC (2005) Biodegradation of lignocellulosics: microbial, chemical, and enzymatic aspects of the fungal attack of lignin. Int Microbiol 8:195–204

McErlen C, Marchant R, Banat IM (2006) An evaluation of soil colonisation potential of selected fungi and their production of ligninolytic enzymes for use in soil bioremediation applications. Anton Leeuw Int J G 90:147–158

Meharg AA, Cairney JWG, Maguire N (1997) Mineralization of 2,4-dichlorophenol by ectomycorrhizal fungi in axenic culture and in symbiosis with pine. Chemosphere 34:2495–2504

Nikiforova SV, Pozdnyakova NN, Makarov OE, Chernyshova MP, Turkovskaya OV (2010) Chrysene bioconversion by the white rot fungus *Pleurotus ostreatus* D1. Microbiology 79:456–460

Osono T (2006) Role of phyllosphere fungi of forest trees in the development of decomposer fungal communities and decomposition processes of leaf litter. Can J Microbiol 52:701–716

Osono T (2007) Ecology of ligninolytic fungi associated with leaf litter decomposition. Ecol Res 22:955–974

Osono T, Takeda H (2002) Comparison of litter decomposing ability among diverse fungi in a cool temperate deciduous forest in Japan. Mycologia 94:421–427

Pinedo-Rivilla C, Aleu J, Collado IG (2009) Pollutants biodegradation by fungi. Curr Org Chem 13:1194–1214

Purnomo AF, Kamei I, Kondo R (2008) Degradation of 1,1,1-trichloro-2,2-bis (4-chlorophenyl) ethane (DDT) by brown-rot fungi. J Biosci Bioeng 105:614–621

Rabinovich ML, Bolobova AV, Vasilchenko LG (2004) Fungal decomposition of natural aromatic structures and xenobiotics: a review. Appl Biochem Microbiol 40:1–17

Reddy CA, Mathew Z (2001) Bioremediation potential of white rot fungi. In: Gadd GM (ed) Fungi in bioremediation. Cambridge University Press, Cambridge, pp 52–78

Riva S (2006) Laccases: blue enzymes for green chemistry. Trends Biotechnol 24:219–226

Robertson SJ, McGill WB, Massicotte HB, Rutherford M (2007) Petroleum hydrocarbon contamination in boreal forest soils: a mycorrhizal ecosystems perspective. Biol Rev 82:213–240

Rodriguez A, Carnicero A, Perestelo F, de la Fluente G, Milstein O, Falcón MA (1994) Effect of *Penicillium chrysogenum* on lignin transformation. Appl Environ Microbiol 60:2971–2976

Schauer F, Borriss R (2004) Biocatalysis and biotransformation. In: Tkacz JS, Lane L (eds) Advances in fungal biotechnology for industry, agriculture, and medicine. Kluwer/Plenum, New York

Schlosser D, Fahr K, Karl W, Wetzstein HG (2000) Hydroxylated metabolites of 2,4-dichlorophenol imply a Fenton-type reaction in Gloeophyllum striatum. Appl Environ Microbiol 66:2479–2483

Singh H (2006) Fungal metabolism of polycyclic aromatic hydrocarbons. In: Singh H (ed) Mycoremediation, fungal bioremediation. Wiley, Hoboken, NJ, pp 283–356

Singleton I (2001) Fungal remediation of soils contaminated with persistent organic pollutants. In: Gadd GM (ed) Fungi in bioremediation. Cambridge University Press, Cambridge, pp 79–96

Steffen KT, Hatakka A, Hofrichter M (2002) Removal and mineralization of polycyclic aromatic hydrocarbons by litter-decomposing basidiomycetous fungi. Appl Microbiol Biotechnol 60:212–217

Steffen KT, Schubert S, Tuomela M, Hatakka A, Hofrichter M (2007) Enhancement of bioconversion of high-molecular mass polycyclic aromatic hydrocarbons in contaminated non-sterile soil by litter-decomposing fungi. Biodegradation 18:359–369

Teng Y, Luo Y, Sun X, Tu C, Xu L, Liu W, Li Z, Christie P (2010) Influence of arbuscular mycorrhiza and rhizobium on phytoremediation by alfalfa of an agricultural soil contaminated with weathered PCBs: a field study. Int J Phytoremediation 12:516–533

Thorn RG, Tsuneda A (1992) Interactions between various wood-decay fungi and bacteria: antibiosis, attack, lysis or inhibition. Rep Tottori Mycol Inst 30:13–20

Tigini V, Prigione V, Di Toro S, Fava F, Varese GC (2009) Isolation and characterisation of polychlorinated biphenyl (PCB) degrading fungi from a historically contaminated soil. Microb Cell Fact 8:5

Valentín L, Kluczek-Turpeinen B, Oivanen P, Hatakka A, Steffen K, Tuomela M (2009) Evaluation of basidiomycetous fungi for pretreatment of contaminated soil. J Chem Technol Biotechnol 84:851–858

Venosa AD, Zhu X (2003) Biodegradation of crude oil contaminating marine shorelines and freshwater wetlands. Spill Sci Technol Bull 8:163–178

Verdin A, Sahraoui AL-H, Durand R (2004) Degradation of benzo[a]pyrene by mitosporic fungi and extracellular oxidative enzymes. Int Biodeter Biodegr 53:65–70

Vidossich P, Alfonso-Prieto M, Carpena X, Fita I, Loewen P, Rovira C (2010) The dynamic role of distal side residues in heme hydroperoxidase catalysis. Interplay between X-ray crystallography and ab initio MD simulations. Arch Biochem Biophys 500:37–44

Wunch KG, Alworth WL, Bennett JW (1999) Mineralization of benzo[a]pyrene by *Marasmiellus troyanus*, a mushroom isolated from a toxic waste site. Microbiol Res 154:75–79

Chapter 3
Fungal Wood Decay Processes as a Basis for Bioremediation

Barbara Piškur, Miha Humar, Ajda Ulčnik, Dušan Jurc, and Franc Pohleven

3.1 Introduction

Soils are complex and heterogeneous ecosystems, composed of inorganic particles, organic components in different decomposition stages, water, gases and organisms that have an important function in nutrient cycling (Killham 1995; Atlas and Bartha 1998). Unfavourable physical and chemical properties of soils represent an environmental and also an economical problem. Numerous erosion processes can result in water and air contamination and in formation of large areas of unproductive soils, with low amounts of organic substances. Natural succession of different organisms in such soils is slow and also renders the establishment of vegetation overgrowth.

Fungi and their interactions with other organisms are necessary for stable functioning of ecosystems and biosphere. Fungal activities are unique and indispensable in biogeochemical nutrient cycles and represent a genetic reservoir with enormous potential to restore and conserve the environmental equilibrium (Hawksworth and Colwell 1992). Wood decay fungi, especially white-rot fungi, are the only known organisms, which can degrade all components of wood cell walls, including lignin (Zabel and Morrell 1992; de Boer et al. 2005). Wood degradation products have an important function in formation of organic soil components and can positively influence the natural succession pattern (Ponge 2005). Organic matter significantly influences on the physical, chemical and biological properties of the soil. Humus effects the uptake of micronutrients by plants and influences on the numerous soil characteristics (e.g. colour, water

B. Piškur (✉) • D. Jurc
Department of Forest Protection, Slovenian Forestry Institute, Večna pot 2, 1000 Ljubljana, Slovenia
e-mail: barbara.piskur@gozdis.si; dusan.jurc@gozdis.si

M. Humar • A. Ulčnik • F. Pohleven
Department of Wood Science and Technology, Biotechnical Faculty, University of Ljubljana, Jamnikarjeva 101, 1000 Ljubljana, Slovenia
e-mail: miha.humar@bf.uni-lj.si; ajda.ulcnik@bf.uni-lj.si; franc.pohleven@bf.uni-lj.si

E.M. Goltapeh et al. (eds.), *Fungi as Bioremediators*, Soil Biology 32,
DOI 10.1007/978-3-642-33811-3_3, © Springer-Verlag Berlin Heidelberg 2013

retention, soil structure stabilisation, increased permeability, buffer effect, cation exchange capacity, combines with organic molecules) (Stevenson 1982).

The term mycoremediation usually refers to the exploitation of a unique fungal capacity to break down various organopollutants or to remove heavy metals from contaminated substrates but was also expanded on application of fungi to revitalise degraded and organically depleted areas (Pointing 2001; Bennet et al. 2002; Humar and Pohleven 2003a; Šašek et al. 2003; Singh 2006; Vidic 2008; Piškur et al. 2009). However, bioleaching is nowadays a more established term for processes of heavy metals removal from contaminated substrates (Amartey et al. 2007). Mycoremediation principles are becoming more and more popular and are extensively studied. Mycelial growth enables fungi to be successful biodegraders compared to other organisms, for example, bacteria—fungi can colonise insoluble substrates and their mycelium represents a larger contiguous area with the substratum. Fungi can tolerate higher concentrations of toxic substances due to secreted extracellular enzymes (Bennet et al. 2002).

Ligninolytic enzymatic system of white-rot fungi can, beside wood components, also transform and/or mineralise a large spectrum of organic pollutants (Pointing 2001). The white-rot fungus *Phanerochaete chrysosporium* was noticed in the 1980s for its ability to degrade various substances: polyaromatic carbohydrates, chlorinated phenols, nitroaromatic substances, dyes and other environmental pollutants (Bumpus and Aust 1987; Aitken et al. 1989; Aust 1990; Tucker et al. 1995). More than 1,500 species of white-rot fungi and also numerous other fungi (like brown rots) represent a great potential for mycoremediation, which has not yet been fully evaluated (Bennet et al. 2002). Some of wood brown-rot fungi were successfully tested for bioleaching, detoxification and recycling of waste wood, impregnated with formulations based on copper, chrome, arsenic and boron (Humar et al. 2002a, b; Humar and Pohleven 2003b).

It is essential to understand basic processes involved in the wood degradation and succession of organisms to get a correct insight into mycoremediation principles. The main purpose of this chapter is to introduce the basis of wood degradation and to emphasise the influence of exogenously added fungi on the development and stability of indigenous microbial communities. The exploitation of fungi for bioremediation is extensively studied, but unfortunately only a minor part of published papers could be considered here and it is on the reader to dig deeper into the mycoremediation "know why".

3.2   Wood Substrate as an Organic Amendment

Wood, grasses and the majority of plant litter represents the major part of earth's biomass and are with one name called lignocellulose. A greater portion of lignocelluloses is constituted from cellulose, hemicelluloses and lignin. Some authors (Fengel and Wegener 1989) estimate that there is around $2.5–4 \times 10^{11}$ tons of cellulose and approximately the same amount of lignin, which altogether represents 70 % of all organic carbon on the earth. Photosynthesis and degradation of lignocelluloses are two important processes in the global carbon cycle of nutrients (Tuomela et al. 2000). Biotic wood degradation has been anthropogenically characterised as a harmful process, but saprophytic fungi (and their interactions with other organisms) are indispensable for normal functioning of all earth's ecosystems (Rayner and Boddy 1988).

Coniferous and non-coniferous woods differ in structure and quantity of hemicelluloses and lignin (Torelli 1986; Eaton and Hale 1993). Average chemical composition of different wood sources is 50 % of carbon, 43 % of oxygen and 6 % of hydrogen. The remaining percentage is represented by nitrogen and ash (Fengel and Wegener 1989). According to prCENT/TS 14961 (2004), carbon content in European non-coniferous woods is expected to be 49 % and in coniferous woods 51 % on a dry, ash free mass. The nitrogen content is around 0.1 %. Macromolecules (cellulose, hemicelluloses and lignin) are present at both wood groups, while the content and types of low molecular substances are variable. In general, they can be classified into organic (extractives) and inorganic substances (ash) (Fengel and Wegener 1989). Extractives include different waxes, fats, fatty acids, alcohols and also mixtures of terpenes, lignans, flavonoids and other aromatic substances, which form resins (Rayner and Boddy 1988). The presence of inorganic substances depends on the tree's growth conditions. Higher concentrations are detected in woods from tropical and subtropical areas. Around 50 % of ash is represented by calcium, followed by potassium, magnesium, manganese, sodium, phosphorus and chlorine (Fengel and Wegener 1989).

Cellulose represents the main organic structural component of the wood cell wall and its share in both coniferous and non-coniferous adult woods is around 42 ± 2 % (Torelli 1986). Cellulose is a homopolymer of β-D-glucopyranose, linked by β(1–4) linkages. Inter- and intramolecular hydrogen bonds between cellulose molecules result in formation of cellulose microfibrils (Eaton and Hale 1993). Hemicelluloses are short and branched homo- and heteropolymers and represent from 25 to 40 % of wood cell walls. Hemicelluloses are covalently linked to lignin and together with their noncovalent interactions to cellulose represent a coat for cellulose fibrils, protecting them from degradation and maintaining their integrity (Eaton and Hale 1993; Uffen 1997). Lignin is a complex, amorphous three-dimensional polymer of phenylpropane units and represents from 18 to 33 % of wood cell walls (Eaton and Hale 1993; Donaldson 2001). Lignin has a structural function in wood, causes impermeability and protects wood from wood-degrading organisms and oxidative stress (Pérez et al. 2002; Hatfield and Fukushima 2005).

The structures and organisation of the main components of wood cell walls and an extremely high carbon to nitrogen ratio (300–1,200:1) promote wood stability and cause resistance to deterioration and degradation, but on the other hand, wood can be an excellent source of organic matter. The idea of wood as an organic amendment to enrich organically poor soils or wood material as an accelerator of wood decay fungi in soil ecosystems is not new. Lignocellulosic residues generated from logging industry, agriculture, pulp and paper industry and construction or demolition industry have been suggested as organic supplements (e.g. Morgan et al. 1993; Boyle 1995; Lang et al. 2000; Jordan 2004; Dunisch et al. 2007; Baldrian 2008; McMahon et al. 2008; Tahboub et al. 2008; Belyaeva and Haynes 2009; Piškur et al. 2009; Barthes et al. 2010; Huang et al. 2010; Valentín et al. 2010). Barks of different tree species turned out to be a promising soil improver and growth medium (Solbraa 1979a, b; Valentín et al. 2010). In the 1970s the capacity of branch chips as soil fertilisers was envisaged and later a technique "Ramial Chipped Wood Technique (RCW)" was described. It was characterised as a "technique of the forest", imitating the way forest generates soils (Lemieux 1993; Caron et al. 1998; Germain 2007). Branches of non-coniferous species increase the fertility of soil to a greater extent than those of conifers due to their chemical composition, lignin structure and lack of inhibiting resins. Branches are compared to the trunk richer with minerals, amino acids and proteins, with more suitable C/N ratio, and are consequently a biostimulator of the natural cycle of nutrients through the promotion of humus formation (Germain 2007). RCW amendments have a positive effect on crop yield, except for the crops that immediately follow the first burying of RCW. Effects of RCW applications are affected by different factors, such as tree species, amount of RCW addition and wood chip size (Barthes et al. 2010). Favourable influences of pruned wood chips like improved soil organic matter content and aggregate stability are reported by Tahboub et al. (2008) also. They disproved the hypothesis about nitrogen immobilisation in the wood chip-amended soil and demonstrated that there was no influence on nitrogen content in leaves of trees, planted in wood chip-amended soils.

As summarised by Briceno et al. (2007), around 85 % of organic wastes originated by humans are forestry-farming wastes, and application of some organic residues has been carried out throughout human history due to their advantageous contribution to the physiochemical and biological properties of such amended soils. Addition of organic wastes was reported to promote soil structure, improve aeration, retain moisture, enhance humified components of soils and stimulate biological activity in soils (Graber et al. 2001; Namkoong et al. 2002; Dungan et al. 2003; Forge et al. 2003; Marschner et al. 2003; Briceno et al. 2007; Park et al. 2011). However, addition of organic wastes can have phytotoxic effects on seed germination or can destroy roots, suppress plant growth and cause nutrient loses. Additionally, organic wastes could represent a source of toxic compounds, such as heavy metals and pesticides (Hernandez-Apaolaza et al. 2000; Ceotto 2005; Briceno et al. 2007). Enrichment of soils with organic wastes can lead to rapid immobilisation of inorganic nitrogen as a consequence of high C/N ratio and thus inhibit nitrogen mineralisation (Amlinger et al. 2003). But Bridge and Prior (2007) summarised findings of studies about the influence of wood amendments on soil communities and nutrient availability (Kokalisburelle and Rodriguezkabana 1994; Visser et al. 1998)—wood supplements reduce fungal pathogens, alter soil communities and influence the establishment of mycorrhizal associations. Some studies indicate that wood application initially reduces soil nutrients due to increased microbial activities in wood material (Larsson et al. 1997; Bridge and Prior 2007). Wood amendments were also reported to influence the mobility of organic pollutants in soils (Grenni et al. 2009).

Table 3.1 Mass losses of the control and CCB-impregnated Norway spruce (*Picea abies*) specimens[a], immersed into 1 or 5 % aqueous solution of corn steep liquor (CSL), after 8 weeks of exposure to different wood decay fungi (adopted after Humar and Pohleven 2005)

		Wood decay fungi					
		Mass loss (%)					
c_{CCB} (%)	c_{CSL} (%)	Av[b]	Lp[b]	Pp[b]	Gt[b]	Tv[c]	Hf[c]
–	–	18.4 (3.4)	17.3 (3.5)	45.8 (4.0)	56.6 (5.4)	21.8 (4.2)	31.6 (5.8)
	1	18.5 (4.6)	9.4 (2.1)	50.4 (5.1)	19.0 (6.3)	18.8 (3.7)	20.7 (3.5)
	5	24.3 (3.0)	6.8 (2.4)	26.8 (4.4)	4.6 (3.9)	12.3 (2.1)	19.4 (3.4)
1	–	31.7 (3.0)	20.8 (3.2)	30.9 (6.8)	0.9 (0.2)	0.5 (0.1)	0.1 (0.3)
	1	29.1 (3.1)	17.4 (2.9)	34.7 (3.2)	0.5 (0.2)	0.5 (0.1)	0.2 (0.2)
	5	29.3 (2.9)	13.9 (1.8)	33.6 (2.6)	0.0 (0.3)	2.1 (0.1)	4.1 (1.0)
5	–	5.5 (1.0)	2.9 (0.9)	0.0 (0.3)	0.8 (0.2)	0.6 (0.2)	0.6 (0.2)
	1	1.1 (0.9)	1.5 (0.5)	0.0 (0.4)	0.3 (0.0)	3.1 (0.3)	5.2 (2.1)
	5	1.0 (0.4)	1.1 (0.6)	0.0 (0.4)	0.0 (1.0)	1.6 (0.6)	1.9 (0.1)

Standard deviations are given in parenthesis
[a]Wood preservative made of Cu, Cr and B
[b]Brown-rot fungi *Antrodia vaillantii* (Av), *Leucogyrophana pinastri* (Lp), *Postia placenta* (Pp) and *Gloeophyllum trabeum* (Gt)
[c]White-rot fungi *Trametes versicolor* (Tv) and *Hypoxylon fragiforme* (Hf)

3.2.1 Ecology of Wood Degradation

Wood is a biologically active material that in life-supporting conditions undergoes constant change and harbours a constantly changing community of different organisms (Garstang et al. 2002; Piškur et al. 2009). Wood degradation is conditioned by succession of fungal activities, by interactions among fungi and other organisms and by environmental factors. For example, Harris and Boddy (2005) assessed that the addition of wood material, overgrown with different fungal species, influenced the development of biomass and the amount and distribution of phosphorus in soils, previously inoculated with *Phanerochaete velutina*. Elissetche et al. (2006) ascertained that, compared to vermiculite, fungal species *Ganoderma australe* grows and degrades wood more intensively when wood material is in contact with soil. Nutrients from soil were translocated into wood, where they influenced on the fungal growth and wood degradation, but no proven impact on delignification was assessed. Higher concentrations of nitrogen enable fungal mycelium to grow at high concentrations of copper (Table 3.1) (Humar and Pohleven 2005) and promote wood degradation (van der Wal et al. 2007), even though lignin degradation can be inhibited by the addition of simple nitrogen sources (Boyle 1998). Wood degradation is also influenced by the soil type, which is a source of fungi and other microorganisms, and inhibitory substances that diffuse into wood and direct the course of wood degradation (Van der Wal et al. 2007). Additionally, the size of wood material is an important factor for the course of degradation. Van der Wal et al. (2007) noticed that, compared to bigger wood

Table 3.2 pH values of degraded Norway spruce wood (*P. abies*) after 2, 4 and 12 weeks of exposure to brown- and white-rot fungi (adopted after Humar et al. 2001)

Wood decay fungi	Weeks of exposure			
	0	2	4	12
Antrodia vaillantii	5.1	4.0	3.7	3.2
Trametes versicolor	5.1	5.2	5.1	3.9
Schizophyllum commune	5.1	5.4	6.2	5.5

chips, mass loss of sawdust is lower. Sawdust is rapidly colonised by bacteria, which exploit sugars and other available nutrients and at the same time excrete antifungal substances and lytic enzymes, which hinder the development of fungal colonisation and consecutively wood degradation.

Woody detritus is an important ecosystem component affecting soil development and water storage, providing major sources of energy and nutrients and representing an environment for various organisms (Harmon 2001; Kuffer et al. 2008). The relation between woody particles and thickness and portion of moss overgrowth on wood influences its degradation (Grebenc et al. 2004). Woody debris augments the diversity of saprophytic fungi in forest ecosystems (Lindhe et al. 2004; Ódor et al. 2006). A variability of wood species, mixture of twigs, branches or logs of different rates of decomposition provide a broad range of niches for wood decay fungi (Kuffer et al. 2008). For example, monitoring of macrofungi on beech woody debris revealed 244 different fungal species in Slovenian unmanaged forest sites (Piltaver et al. 2002).

Coniferous wood chips are more resistant and more durable than non-coniferous wood chips. During wood chip storage chemical deterioration, discolouration and fungal colonisation appear. At the beginning, due to the respiration of still-living parenchymal cells, rapid bacterial colonisation and exploitation of available nutrients, the temperature increases. Furthermore, pH decreases as a result of deacetylation of cellulose and microbial activities (Table 3.2). Acetic and other volatile organic acids are products of natural degradation and deterioration processes and as such they are important indicators of changes in wood material (Buggeln 1999; Garstang et al. 2002; Ashton et al. 2007; Piškur 2009; Piškur et al. 2009). Beside acetic acid, some wood decay fungi, particularly the copper-tolerant ones, excrete copious amounts of oxalic acid (Green and Clausen 2005). In parallel, mechanical properties of wood decrease very fast as well. Loss of stiffness is one of the first signs of decay (Wilcox 1993). Humar et al. (2006) reported that, after 1 week of exposure, *Gloeophyllum trabeum* reduced modulus of elasticity (MOE) by 7.4 % and *Antrodia vaillantii* by 8.3 % (Table 3.3). Another important sign of wood degradation is colour change, as already mentioned. Brown-rot fungi predominately degrade cellulose, leaving oxidised brown lignin. In contrary, white-rot fungi predominately degrade lignin, leaving bleached, white cellulose residues.

In the research made by Jirjis (2005), degradation processes of willow wood chips stored in a pile were monitored. At the beginning 10^9 cultivable fungal spores

Table 3.3 Losses of mass and modulus of elasticity (MOE) of Norway spruce wood (*P. abies*) specimens exposed to brown-rot fungi

	Wood decay fungi			
	Gloeophyllum trabeum		*Antrodia vaillantii*	
Weeks of exposure	Mass loss (%)	MOE loss (%)	Mass loss (%)	MOE loss (%)
0	0.0 (0.0)	2.3 (1.0)	0.0 (0.0)	2.3 (1.0)
1	−0.1 (0.2)	7.4 (1.8)	0.1 (0.0)	8.3 (3.4)
2	2.6 (1.2)	12.0 (3.6)	4.5 (1.1)	16.4 (6.2)
4	13.0 (3.6)	47.5 (7.4)	10.2 (1.8)	38.6 (5.9)
8	28.0 (8.8)	76.0 (10.9)	14.3 (3.2)	40.5 (8.3)

Standard deviations ($n = 5$) are given in parenthesis

were determined per kilogram of wood dry mass. Fungal growth was more intense on smaller wood chips (7–16 mm) and in the first months the number of spores increased by 8–20 times but remained constant afterwards. The number of spores on larger chips increased after 3 months. The most abundant fungal groups that are usually isolated from the wood chips are *Acremonium* sp., *Mortierella* sp., *Trichoderma* sp., *Mucor* sp., *Bjerkandera adusta*, *Gloeophyllum sepiarium*, *P. chrysosporium*, *Phlebiopsis gigantea*, *Trametes versicolor*, *Ceratocystis* sp., *Ophiostoma* sp., *Graphium* sp., *Aureobasidium* sp., *Alternaria* sp., *Trichoderma* sp., *Aspergillus* sp., *Gliocladium* sp. and *Penicillium* sp. (Zabel and Morrell 1992; Jirjis 2005; Schmidt 2006).

3.2.1.1   Succession

Fungal succession is a sequence of occurrence or presence and space arrangement of fungal species, which form a community in a defined ecological niche (adapted from Frankland 1998; Hyde and Jones 2002). Simultaneous effect of biotic and abiotic factors importantly influences the course of fungal appearance. Succession processes alter nutrients' proportions and characteristics of ecological niches. Primary colonisers exploit easily accessible nutrients and are followed by species, involved in cellulose, hemicelluloses and lignin degradation. At the same time, fungal and bacterial biomass is formed, which represents a new nutritive component in the ecological niche (Boddy 2001; Ponge 2005). Time pattern of a common succession involves bacteria, moulds and blue stain fungi, followed by white- and brown-rot fungi (Eaton and Hale 1993; Schmidt 2006). Nitrogen, metals, water and fungal metabolites (organic acids, toxins, antibiotics) increase with time. Intermediates of degradation processes influence the fungal community and are mineralised or integrated into soil organic components during decomposition processes (Boddy 2001; Ponge 2005). Various fungal species and populations and other organisms, competing for available nutrients, may be present in the substratum. Fungal interactions influence the dynamics of the complete fungal community and the course and extent of wood degradation (Vesel Tratnik and

Pohleven 1995; Woodward and Boddy 2008). Competition is the usual interaction among wood-degrading fungi. Capture of not yet colonised substratum depends on various factors, such as the capacity of fungal inoculum to be established, spore germination, rate of the mycelium spread, efficiency of substratum exploitation, etc. But, when the substratum is already colonised, the antagonistic mechanisms (volatile or diffusible compounds, parasitism, antifungal substances at the contact of two mycelia, firm sclerotial plates) are crucial factors for fungal growth. The colonising fungus can occupy only a part or the whole territory previously infected by another fungus. A deadlock appears when none of the competing fungi gain a competing advantage (Boddy 2000; Woodward and Boddy 2008).

Bacteria are the early colonisers of wood, exploiting easy available and degradable nutrients, like sugars, organic acids, pectin and cellulose (Clausen 1996). Some bacteria are reported to partially degrade cellulose and a minor part of lignin, thus having only little effect on wood degradation. Nevertheless, they have an important indirect influence on wood degradation through their effects on wood-degrading fungi (de Boer and Van der Wal 2008).

3.2.1.2 Endophytic Fungi

Wood decay fungi have a salient role in colonisation and degradation of wood by attacking wood cell components through enzymatic and nonenzymatic systems (Kirk and Farrell 1987; Zabel and Morrell 1992). In the initiation phase of wood decay fungal endophytes could have an important influence as they were shown to be capable of producing wood decay enzymes (Baum et al. 2003; Oses et al. 2006). The existence of fungal endophytes has been known since 1866, when Anton de Bary coined the term endophyte to denote the microflora in the inner tissues of plants (de Bary 1866). Fungal endophytes are fungi that colonise tissues of living plants and cause unapparent and asymptomatic infections entirely within plant tissues and for all or part of their life cycle cause no symptoms of disease in the invaded plant (Wilson 1995; Oses et al. 2006). Latent infections, present in healthy tissues, are probably triggered into the active phase by water loss and increased oxygen concentration. Endophytes can spread through the tree via xylem sap. In the research performed by Baum et al. (2003), few fungi were isolated from wood immediately after felling, but a larger number of isolates was cultivated after few weeks of wood incubation under sterile conditions. Basidiomycetes had a longer incubation period for penetration from wood samples compared to ascomycetes. There is also considerable circumstantial evidence that development of these latent propagules into mycelia, which cause decay, occurs in the field when sapwood begins to dry, as a result of drought, root damage or loss of transpiring canopy (Boddy 1994). Nevertheless, mechanisms by which decay fungi invade and degrade wood in living and dead trees are still unresolved (Eaton 2000). Some endophytes are potentially pathogenic (e.g. *Fomes fomentarius*, *Nectria coccinea*, *Botryosphaeria dothidea*, *Cenangium ferruginosum*, *Biscogniauxia mediterranea*), but disease can only develop in combination with other, mostly unknown, triggering factors (oxygen ratios, nutrient availability, water stress, etc.) (Jurc et al. 2000; Jurc and Ogris 2006; Sieber 2007; Moricca and Ragazzi 2008; Piškur et al. 2011b). Within the lifecycle of at least some wood decay fungi, there is an endophytic phase in which fungi do not degrade xylem (Boddy and Rayner 1983; Petrini and Fisher 1990; Kowalski and Kehr 1992; Eaton 2000; Baum et al. 2003; Oses et al. 2006, 2008; Parfitt et al. 2010). Epiphytes and endophytes that frequently colonise tissues of living leaves can also have a leading role in leaf litter decomposition. These fungi are reported to be primarily saprobic, being specifically adapted to colonise and utilise dead host tissues. Some species have the ligninolytic activity, which most likely has an important role in decomposition processes. For example, xylariaceous endophytes with ligninolytic activity persist until the late stages of decomposition and have important roles in further decomposition processes of structural components, nutrient dynamics and accumulation of soil organic matter (Osono 2006). On the other hand it was revealed that xylareaceous fungi which are totally adapted to endophytic life and colonise the branches extensively are rare in dead branches (Kowalski and Kehr 1996). It seems that the succession processes are governed by subtle environmental differences and competition relations leading to extremely diversified end results.

Diversity of endophytic fungal species in trees is high, for example, in twigs of *Carpinus caroliniana* more than 120 species were detected (Bills and Polishook 1991; Sieber 2007). In healthy needles of *Pinus nigra*, 99 taxa of endophytes were reported and they could be assigned to parasitic, facultative parasitic or saprobic mode of nutrition (Jurc and Jurc 1995; Jurc 1997). Species density and diversity below species level are also high within small volumes of tissues (Sieber 2007). This diversity is largely not identified and Sieber (2007) projects that around 465,000 endophytic species are undescribed. This large diversity of endophytic fungi has an enormous but unexploited potential for bioremediation.

3.2.2 Intermediates and Products of Wood Degradation

Decomposition of plant debris and its conversion to smaller organic fragments, production of organic acids and creation of anaerobic conditions are examples of ways in which soil microflora can influence the soil formation (Elliot et al. 1986). Easily degradable substances, such as proteins, starch and cellulose, are rapidly consumed by fungi, bacteria and other organisms. Through these processes, numerous by-products appear, such as NH_3, H_2S, CO_2, organic acids and other incompletely oxidised substances. Organic acids are also constituents of mineral soils and are important in the mobilisation and transport of metals, rock weathering and solubilisation. The final stage in the succession of decomposition processes includes decomposition of more resistant organic material, like encrusted hemicelluloses molecules and lignin (Stevenson 1982; Gadd 2007; Piškur et al. 2009).

Soil organic matter is formed from a complex and amorphous mixture of organic colloids and partially decomposed plant, animal, fungal and other microbial biomass. Processes of wood degradation and type of plant biomass under degradation significantly influence the composition and quantity of humic substances in soils. Wood degradation products and fungal and other microbial activities together with their influence on succession have an important role in formation of organic soil components (Swift 1977; Anderson and Domsch 1986; Chefetz et al. 1996). Humus is a major organic component of soil, formed by relatively inert humic substances and nonhumic material, like amino acids, carbohydrates and lipids. Both components are connected with covalent linkages and are difficult to separate (Stevenson 1982). The crucial source of soil humus are plant residues and the humic structures reflect the respective ecosystem and the source of organic matter (Shevchenko and Bailey 1996). Humic substances have important environmental functions due to their role in redox reactions, sorption, complexation and transport of pollutants, minerals and trace elements, sustaining plant growth and soil structure and controlling the biogeochemistry of organic carbon (summarised by Stevenson 1982; Grinhut et al. 2007). Fungi most likely have a dominant role in the formation, transformation and degradation of humic substance, especially in forests and areas with large amount of organic matter. Degradation and mineralisation of humic substance are led mainly by white-rot fungi, while modification and polymerisation is carried out by ascomycetes (Grinhut et al. 2007). Environmental factors, like nitrogen deposits, global environmental changes and forest practices, can alter the rate of lignin degradation (Ishikawa et al. 2007; Osono 2007).

3.3 Wood Degradation Processes by Fungi

3.3.1 White-Rot Fungi

White-rot fungi are able to degrade lignin more efficiently than other microbial groups. The degradation of lignin by white-rot fungi is an aerobic oxidative process (Kirk and Farrell 1987; Hammel 1996; Sanchez 2009). Nonspecific white-rot fungi degrade lignin, cellulose and hemicelluloses simultaneously and others cause selective delignification; that is, they remove lignin before the degradation of other cell wall components occurs (Zabel and Morrell 1992; Tuomela et al. 2000; Martinez et al. 2005). Partial degradation of lignin is caused by brown-rot fungi as well. This type of wood decay decomposes the polysaccharides in plant cell walls and leaves behind a brown, modified lignin residue (Kirk and Farrell 1987; Hammel 1997; Tuomela et al. 2000). The lignin-degrading system is based on extracellular oxidative enzymes, among which the principal ones are lignin peroxidases (LiP), manganese peroxidases (MnP) and laccases (Lac). Not all white-rot fungi produce all three enzyme types and different combinations of these enzymes are found among fungal species. Ligninolytic enzymes catalyse the oxidation of lignin, during which lignin free radicals are formed. The latter are chemically instable and carry out spontaneous nonenyzmatical degradative reactions, including aromatic ring and carbon–carbon bond cleavage, hydroxylation, demethylation, etc. Finally, the lignin macromolecule is spontaneously broken into smaller subunits (Kirk and Farrell 1987; Zabel and Morrell 1992; Hammel 1996; Pointing 2001; Sanchez 2009).

Both LiPs and MnPs are haem-containing glycosylated peroxidases. Just like other peroxidases, hydrogen peroxide, which is produced by H_2O_2-generating oxidases, is required for their functioning (Bennet et al. 2002; Martinez 2007). After LiPs are oxidised by H_2O_2, they oxidise the substrate (lignin) themselves, producing free radicals. The oxidised LiPs are reduced back to their native state by two consecutive one-electron oxidations of the substrate. LiPs are stronger oxidising agents as other peroxidises and have the ability to degrade even non-phenolic units of lignin (Kirk and Farrell 1987; Aust 1995; Sanchez 2009).

The mode of action of MnPs is similar to that of LiPs. After being oxidised by H_2O_2, MnPs are reduced to the initial state by two oxidations of Mn (II) to Mn (III). The latter, chelated by various organic acids, may act as diffusible oxidisers, acting on phenolic and non-phenolic lignin units and generating cation radicals in lignin during the decay process (Aust 1995; Hammel 1996; Sanchez 2009).

Laccases belong to polyphenol oxidases, which contain copper atoms in their catalytic centre (Thurston 1994). These extracellular glycoproteins catalyse the reduction of oxygen to water, and one-electron oxidation of the substrate, usually a phenolic compound, occurs concomitantly. In fungi, laccases have several functions, including lignin degradation (Thurston 1994; Baldrian 2006).

Even white-rot fungi, despite their exceptional ability to degrade lignin, cannot use lignin as a sole source of carbon and energy directly (Hammel 1997; Sanchez 2009). Lignin biodegradation occurs in secondary metabolism under nutrient, mostly carbon and nitrogen, limitation. Such mechanism enables fungi to consume the more easily accessible substrates rather than spending valuable energy for the synthesis of metabolically expensive ligninolytic enzymes (Hammel 1997). Therefore, through lignin-degrading processes fungi gain access to otherwise well-protected cellulose and hemicelluloses (Hammel 1997; Pointing 2001).

Using wood decay fungi for bioremediation (mycoremediation) is possible on a wide range of pollutants. Above all, white-rot fungi show the capability of degrading a variety of organic pollutants. Degradation of xenobiotics is thought to involve lignin-degrading system with its extracellular enzymes (Lac, LiPs and MnPs). Expression of ligninolytic genes is in general repressed by the presence of easily available nutrients (e.g. glucose) and induced in the presence of substrate polymers or their derivatives. The production of these enzymes is triggered by nutrient deficiency rather than by the presence of the pollutant (Bumpus et al. 1985; Kirk and Farrell 1987). However, there have been reports on the production of ligninolytic enzymes even in nutrient-sufficient media (Moreira et al. 1997; Janse et al. 1998). Laccase genes are transcribed at low constitutive levels in many basidiomycetous species (Cullen 1997; Aro et al. 2005). The expression can be enhanced using inducers, such as aromatic compounds, including degradation products of lignin, which also induce expression of other ligninolytic genes (Muñoz et al. 1997; Scheel et al. 2000).

Fig. 3.1 Calcium oxalate
crystals in the cell lumina of
the Norway spruce (*P. abies*)
wood specimens exposed to
brown-rot fungus *Antrodia
vaillantii*

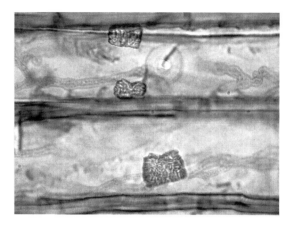

3.3.2   Brown-Rot Fungi

On the contrary, brown-rot fungi predominately degrade cellulose. One of the issues related to cellulose degradation is the fact that cellulose is surrounded by lignin preventing fungal enzymes to penetrate the cell wall, as they are too large. Therefore, the first degradation phases of brown-rot decay are nonenzymatic. Initial wood degradation is described by two mechanisms: Fenton's reaction and excretion of the organic acids. The first stages of decay are recognised by strength loss, accumulation of ions, decreased pH value of wood, etc. (e.g. Shortle 1990). pH changes of the infested wood are observed in vitro considerably before mass loss can be detected (Table 3.2).

Ions accumulating in wood infected by fungi can be associated with markedly increased acidity of wood (drop in pH) at all stages of decay (Table 3.2) (Shortle 1990). A substantial reduction of the pH values caused by fungi was observed both in liquid cultures and in solid wood substrates. Especially brown-rot fungi were able to cause a significant reduction of pH of wood. White-rot fungi were also associated with a pH decrease of degraded wood, though to a lesser extent (Jellison 1992). Oxalic acid has an important role in the production of acidic environment (Takao 1965; Shimada et al. 1994). Oxalic acid is a small organic acid with two low pK values ($pK1 = 1.27$ and $pK2 = 4.26$). It is often produced in great quantities by brown-rot fungi (Fig. 3.1) (Takao 1965; Green et al. 1991) and may be a very strong candidate for the central role in the production of the acid environment associated with brown-rot colonisation of wood (Jellison et al. 1997). Additionally, incipient brown-rot decay is similar to acid hydrolysis and is characterised by acid production (Green et al. 1991). Lastly, fungi produce and secrete many compounds containing acidic groups, such as phenolic acids and other simple organic acids, generated during metabolism (Jellison et al. 1997).

However, it should be considered that in the latter stages of cellulose degradation, enzymes are involved as well, predominately cellulases. Cellulase is an enzyme complex which breaks down cellulose to beta-glucose. Cellulases are widely distributed throughout the biosphere and are mostly manifested in fungal and microbial organisms. Cellulases refer to a family of enzymes which act in concert to hydrolyse cellulose. Three general types of enzymes make up the cellulase enzyme complex. Endocellulase breaks internal bonds to disrupt the crystalline structure of cellulose and expose individual cellulose polysaccharide chain. Exocellulase cleaves 2–4 units from the ends of the exposed chains produced by endocellulase, resulting in tetrasaccharides or disaccharides such as cellobiose. Cellobiase or beta-glucosidase hydrolyses the endocellulase products into individual monosaccharides (Eaton and Hale 1993).

The principle of the application of brown-rot fungi in bioremediation (bioleaching) processes is to convert the insoluble heavy metals found in waste wood into a soluble fraction through acidification with organic acids. Only then can the soluble heavy metal complex be leached from the wood. Thus, both the remediated wood fibre and metals can be reclaimed and recycled. The most important acid involved in this process is oxalic acid (Jellison et al. 1997). The most efficient oxalic-acid producers and consequently the most tolerant fungal isolates are found in the genus *Antrodia* (Shimada et al. 1994). Other wood-decaying fungi produce significantly lower concentrations of oxalic acid. Instead of oxalic acid, these fungi excrete other organic acids such as acetic and formic acids in order to optimise the pH value of the substrate (Shimada et al. 1994). Oxalic acid can react with insoluble chromium in wood to form chromium oxalate, which is soluble and can be leached out of wood. On the other hand, copper oxalate, which is formed between copper and oxalic acid, is insoluble and can only be leached with an ammonia solution (Humar et al. 2002a).

3.4 Establishment and Monitoring of Mycoremediation Systems

Natural microbial and fungal communities found in degraded and/or contaminated soils represent the heterogeneous potential for remediation. But ecological factors (nutrient availability, pH, unfavourable physiochemical properties, etc.) can hinder and prolong the restoration processes. The remediation potential of indigenous microflora can be enhanced with the addition of nutrients (biostimulation) or with the addition of living exogenous organisms into the remediated substrate (bioaugmentation). Referring to Sect. 3.2.1.2, the substrate used for biostimulation or as an organic amendment can also carry a variability of organisms that can express bioremediation potential.

Variable and complex microbial communities have a favourable effect on the stability of the ecosystem and stress resistance (Tilman 1996; McGradySteed et al. 1997; Bengtsson et al. 2000; Guimaraes et al. 2010). Supplements of saprophytic fungal species were noted to importantly influence on the structure of microbial communities (Andersson et al. 2003; Tornberg et al. 2003; Woodward and Boddy 2008; Piškur et al. 2011a). Some papers report inhibitory influence of the addition of white-rot fungi on the richness of microbial communities (Andersson et al. 2003; Piškur et al. 2011a), while others state the opposite (Federici et al. 2007).

The unique ligninolytic system of basidiomycetes inspired researchers to take advantage of white-rot fungi in various applications, including bioremediation of degraded or contaminated soils. However, when these fungi, normally living in wood, are inoculated into soil ecosystem, they encounter different problems, including soil-inhabiting fungi and bacteria, different nutrient status and characteristics of the environment and a much more heterogeneous environment compared to wood. To support the growth of inoculated white-rot fungi, addition of nitrogen and carbon sources is required. Different lignocellulosic substrates are usually applied, like corn cobs, straw, wood chips, bark and peat. Additionally, larger quantities of fungal inoculum reflect in faster and more successful colonisation (Leštan and Lamar 1996; Leštan et al. 1996; Baldrian 2008). As summarised by Ford et al. (2007), the amount of fungal addition into contaminated soils used in different studies ranges from 30 to 250 g kg^{-1} on a dry weight basis of colonised fungal inoculum to soil.

Selection of suitable fungal species, isolate and substrate for fungal growth is a fundamental step in the planning of mycoremediation technology. White-rot species have different abilities to survive and colonise soil. In the research by Tuomela et al. (1998), Šnajdr and Baldrian (2006) and as summarised by Baldrian (2008), species *P. chrysosporium*, *Trametes versicolor* and *Pleurotus ostreatus* were the most successful soil colonisers and strong competitors. Some wood-degrading fungi like *Gymnopilus penetrans* and *G. luteofolius* usually found on branches, stumps or roots and some other species like *Stropharia rugosoannulata* can also actively colonise soil ecosystems and can be used for bioremediation purposes (Valentin et al. 2009; Winquist et al. 2009; Steffen and Tuomela 2010). Nevertheless, this information should not be taken by itself. Fungal colonisation abilities and the success of remediation are dependent on numerous factors, especially in field experiments (Tornberg et al. 2003). Bioremediation of contaminated soils is usually treated with bioslurry reactors, biopiles, landfarming and restoration on site (Mougin et al. 2009; Piškur 2009).

Properties of fungal inocula, such as composition, age, methods of inoculation and moisture, affect the successfulness of bioremediation. As shown by Leštan et al. (1998), viability of mycelia fragments does not depend on the size of the fragments. Fungal spawn (inoculum) is usually produced on wood chips, bark, straw and some other agricultural waste products with the addition of nitrogen source, like wheat grain. Novel technologies use encapsulated inoculum, which is reported to improve survival and effectiveness of introduced fungi. As a mycelium carrier, various hydrogels and pelleted substrates made from carbon and nitrogen sources are used (Leštan and Lamar 1996; Leštan et al. 1998; Bennet et al. 2001; Schmidt et al. 2005; Ford et al. 2007; Giubilei et al. 2009; Piškur 2009; Piškur et al. 2009).

The selection of remediating fungal isolate should consider the following aspects: autochthonous fungal species, nonpathogenicity to plants and other organisms, noninvasiveness to adjacent areas, efficacy to degrade wood or pollutants, good competitiveness, etc. Additionally, the selection of biodegradation agents should also be considered beyond the species level, where further potential for biodegradation purposes could be revealed (Piškur et al. 2009). One should be aware of the fact that inoculated fungus represents a disturbing factor for normally occurring succession pathways. Application of selected fungal isolate into the environment should only be performed after careful considerations of all advantages and disadvantages.

3.4.1 Monitoring

Monitoring of established mycoremediation systems includes various methods, from the simple visual inspection for mycelium growth and formation of fruit bodies to methods measuring ergosterol contents and activities of ligninolytic enzymes. Products of wood degradation (e.g. volatile organic acids, mass loss, lignin degradation, etc.) and the amount of organic matter are also indicators of successfulness of the mycoremediation system. Respiration rates and temperature changes could also give an insight into successive processes after inoculation.

Evaluation of microbial communities in complex systems represents a unique challenge. A frequent estimation of all fungal species is around 1.5 million (Hawksworth 2001). Ecological studies of wood decay fungi are usually limited to the inventory of fungal fruiting bodies and to the cultivation techniques on different growth media (Vainio and Hantula 2000). In some species, formation of fruiting bodies is rare, variable and usually depends on growth factors. Isolation and cultivation of mycelia cultures under laboratory conditions accelerate the quickly growing species, and just around 17 % of all currently known fungi are cultivable on growth media (Vainio and Hantula 2000; Bridge and Spooner 2001; Hunt et al. 2004). The development of microbial communities and the influence of exogenously added substratum with or without fungal inoculum are nowadays usually monitored with the application of molecular methods, and analyses of DNA sequences, obtained from environmental samples, are becoming a valuable tool in the ecosystem research (Jellison and Jasalavich 2000). Molecular profiling techniques include extraction of the total genomic DNA from environmental samples (i.e. metagenomic DNA), amplification of a part of chromosomal DNA and separation of amplified products with agarose or polyacrylamide gel electrophoresis. Microbial diversity is usually studied using denaturing gradient gel electrophoresis (DGGE), temperature gradient gel electrophoresis (TGGE), terminal restriction fragment length polymorphism (T-RFLP) and single-strand conformation polymorphism (SSCP) (Kirk et al. 2004). The DGGE method was first used in 1993 for determining the bacterial community in biofilms (Muyzer et al. 1993). Afterwards, the method was also applied to the studies of fungi in different environments (Kowalchuk et al. 1997; van Elsas et al. 2000; Arenz et al. 2006). With DGGE method, dominant groups are detected and thus, the obtained fingerprints are actually a reflection of PCR products. DGGE is a suitable method for detection of shifts in microbial communities after, for example, addition of fungal inoculum (Nocker et al. 2007; Piškur et al. 2011a). Even though molecular methods are a popular tool, there are numerous deficiencies while performing them and analysing the obtained data. Critical accession to the evaluation of the results is therefore necessary (Muyzer and Smalla 1998).

3.5   Conclusions

The revitalisation of degraded and contaminated areas is a slow and long-term process, where a variety of different organisms and abiotic factors take part. Exogenously added organisms into such a sensitive system can hinder the development of indigenous community and repress the establishment of the ecosystem equilibrium.

When planning to establish remediation systems in the field, great attention should be paid to the impact of such strategies on the environment and the communities of different organisms. Laboratory experiments and experiments in mezo-conditions should be carefully planned and conducted. Even though, experiments in nature represent a unique and unpredictable research work and as such cannot be defined in an exact way.

Additionally, the knowledge about pathogenicity and dissemination of added organism is necessary. A thorough and careful selection for the final selection of remediating agents should be performed. Heterogeneous spectre of organisms and also different isolates should be tested "in vitro" in advance.

Monitoring of established mycoremediation system gives valuable information about the successfulness and the stability of such environments. Nevertheless, we should be aware of the pros and cons of the selected methods and also of the organisms involved. However, wood decay fungi and other organisms represent a research challenge and enormous possibilities for the salvation of environmental problems. Still, the application of mycoremediation strategies in nature should be executed with caution and scepticism.

Instead of conclusion, we would like to direct the reader towards some recent review papers and books dealing directly or indirectly with mycoremediation: Harms et al. (2011) described the metabolic and ecological features of fungi, important for bioremediation applications; Guimaraes et al. (2010) published a review paper about recent progresses in soil bioremediation technology; in their chapter, included in the Industrial Applications, Steffen and Tuomela (2010) represent developments in large-scale applications of fungal soil bioremediation; Baldrian (2008) argued the applicability of wood-inhabiting ligninolytic basidiomycetes in soils; Gadd (2007) emphasised the importance of fungi in geomycology. Several books on applicability of fungi in bioremediation technologies (e.g. Gadd 2001; Singh 2006) and about ecology of wood decay fungi (e.g. Jonsson and Kruys 2001; Dighton et al. 2005; Boddy et al. 2008) were published as well.

Acknowledgment The financial support from the Slovenian Research Agency through the research programmes P4-0107 and P4-0015 and the research project L4-3641 is acknowledged.

References

(2004) prCENT/TS 14961. Solid biofuels – Fuel specifications and classes. European Committee for Standardization, Brussels

Aitken MD, Venkatadri R, Irvine RL (1989) Oxidation of phenolic pollutants by a lignin degrading enzyme from the white-rot fungus *Phanerochaete chrysosporium*. Water Res 23:443–450

Amartey SA, Humar M, Ribeiro A, Helsen L, Ottosen L (2007) Remediation of CCA treated wood waste. In: Gallis C (ed) Management of recovered wood: reaching a higher technical, economic and environmental standard in Europe: proceedings, Klagenfurt, 2–4 May 2007. University studio press, Thessaloniki, pp 117–130

Amlinger F, Gotz B, Dreher P, Geszti J, Weissteiner C (2003) Nitrogen in biowaste and yard waste compost: dynamics of mobilisation and availability – a review. Eur J Soil Biol 39:107–116

Anderson TH, Domsch KH (1986) Carbon link between microbial biomass and soil organic matter. In: Megušar F, Gantar M (eds) Perspectives in microbial ecology. Fourth international symposium on microbial ecology, 24–29 August 1986. Slovene Society for Microbiology, Ljubljana, pp 467–471

Andersson BE, Lundstedt S, Tornberg K, Schnurer Y, Oberg LG, Mattiasson B (2003) Incomplete degradation of polycyclic aromatic hydrocarbons in soil inoculated with wood-rotting fungi and their effect on the indigenous soil bacteria. Environ Toxicol Chem 22:1238–1243

Arenz BE, Held BW, Jurgens JA, Farrell RL, Blanchette RA (2006) Fungal diversity in soils and historic wood from the Ross Sea Region of Antarctica. Soil Biol Biochem 38:3057–3064

Aro N, Pakula T, Penttilä M (2005) Transcriptional regulation of plant cell wall degradation by filamentous fungi. FEMS Microbiol Rev 29:719–739

Ashton S, Jackson B, Schroeder R (2007) Storing woody biomass. In: Hubbard W, Biles L, Mayfield C, Ashton S (eds) Sustainable forestry for bioenergy and bio-based products: trainers curriculum notebook. Forest Research Partnership, Athens, GA, pp 149–152

Atlas RM, Bartha R (1998) Microbial ecology: fundamentals and applications. Benjamin/Cumings, Menlo Park

Aust SD (1990) Degradation of environmental pollutants by *Phanerochaete chrysosporium*. Microb Ecol 20:197–209

Aust SD (1995) Mechanisms of degradation by white rot fungi. Environ Health Perspect 103:59–61

Baldrian P (2006) Fungal laccases – occurrence and properties. FEMS Microbiol Rev 30:215–242

Baldrian P (2008) Wood-inhabiting ligninolytic basidiomycetes in soils: ecology and constraints for applicability in bioremediation. Fungal Ecol 1:4–12

Barthes BG, Manlay RJ, Porte O (2010) Effects of ramial wood amendments on crops and soil: a synthesis of experimental results. Cah Agric 19:280–287

Baum S, Sieber T, Schwarze F, Fink S (2003) Latent infections of *Fomes fomentarius* in the xylem of European beech (*Fagus sylvatica*). Mycol Prog 2:141–148

Belyaeva ON, Haynes RJ (2009) Chemical, microbial and physical properties of manufactured soils produced by co-composting municipal green waste with coal fly ash. Bioresour Technol 100:5203–5209

Bengtsson J, Nilsson SG, Franc A, Menozzi P (2000) Biodiversity, disturbances, ecosystem function and management of European forests. Forest Ecol Manage 132:39–50

Bennet JW, Connick WJ, Daigle D, Wunch K (2001) Formulation of fungi for in situ bioremediation. In: Gadd GM (ed) Fungi in bioremediation. British Mycological Society, Cambridge, pp 97–112

Bennet JW, Wunch KG, Faison BD (2002) Use of fungi in biodegradation. In: Hurst CC (ed) Manual of environmental microbiology. ASM, Washington, DC, pp 960–971

Bills GF, Polishook JD (1991) Microfungi from *Carpinus caroliniana*. Can J Bot 69:1477–1482

Boddy L (1994) Latent decay fungi: the hidden foe? Arboric J 18:113–135

Boddy L (2000) Interspecific combative interactions between wood-decaying basidiomycetes. FEMS Microbiol Ecol 31:185–194

Boddy L (2001) Fungal community ecology and wood decomposition processes in angiosperms: from standing tree to complete decay of coarse woody debris. Ecol Bull 49:43–56

Boddy L, Rayner ADM (1983) Origins of decay in living deciduous trees: the role of moisture content and a re-appraisal of the expanded concept of tree decay. New Phytol 94:623–641

Boddy L, Frankland JC, van West P (eds) (2008) Ecology of saprotrophic basidiomycetes. Academic, Amsterdam

Boyle CD (1995) Development of a practical method for inducing white-rot fungi to grow into and degrade organopollutants in soil. Can J Microbiol 41:345–353

Boyle D (1998) Nutritional factors limiting the growth of *Lentinula edodes* and other white-rot fungi in wood. Soil Biol Biochem 30:817–823

Briceno G, Palma G, Duran N (2007) Influence of organic amendment on the biodegradation and movement of pesticides. Crit Rev Environ Sci Technol 37:233–271

Bridge PD, Prior C (2007) Introduction or stimulation? The association of *Stropharia aurantiaca* with bark and wood-chip mulches. Eur J Soil Biol 43:101–108

Bridge P, Spooner B (2001) Soil fungi: diversity and detection. Plant Soil 232:147–154

Buggeln R (1999) Outside storage of wood chips. Biocycle 40:32–34

Bumpus JA, Aust SD (1987) Biodegradation of environmental pollutants by the white rot fungus *Phanerochaete chrysosporium* – involvement of the lignin degrading system. Bioessays 6:166–170

Bumpus JA, Tien M, Wright D, Aust SD (1985) Oxidation of persistent environmental pollutants by a white rot fungus. Science 228:1434–1436

Caron C, Lemieux G, Lachance L (1998) Regenerating soils with ramial chipped wood. Publication No 83. Laval University, Quebec

Ceotto E (2005) The issues of energy and carbon cycle: new perspectives for assessing the environmental impact of animal waste utilization. Bioresour Technol 96:191–196

Chefetz B, Hatcher PG, Hadar Y, Chen YN (1996) Chemical and biological characterization of organic matter during composting of municipal solid waste. J Environ Qual 25:776–785

Clausen CA (1996) Bacterial associations with decaying wood: a review. Int Biodeterior Biodegrad 37:101–107

Cullen D (1997) Recent advances on the molecular genetics of ligninolytic fungi. J Biotechnol 53:273–289

de Bary A (1866) Morphologie und Physiologie der Pilze, Flechten und Myxomyceten. Engelman, Leipzig

de Boer W, Van der Wal A (2008) Interactions between saprotrophic basidiomycetes and bacteria. In: Boddy L, Frankland JC, van West P (eds) Ecology of saprotrophic basidiomycetes. Elsevier, Amsterdam, pp 143–153

de Boer W, Folman LB, Summerbell RC, Boddy L (2005) Living in a fungal world: impact of fungi on soil bacterial niche development. FEMS Microbiol Rev 29:795–811

Dighton J, White JF, Oudemans P (eds) (2005) The fungal community: its organization and role in the ecosystem, 3rd edn. CRC, Boca Raton

Donaldson LA (2001) Lignification and lignin topochemistry – an ultrastructural view. Phytochemistry 57:859–873

Dungan RS, Ibekwe AM, Yates SR (2003) Effect of propargyl bromide and 1,3-dichloropropene on microbial communities in an organically amended soil. FEMS Microbiol Ecol 43:75–87

Dunisch O, Lima VC, Seehann G, Donath J, Montoia VR, Schwarz T (2007) Retention properties of wood residues and their potential for soil amelioration. Wood Sci Technol 41:169–189

Eaton R (2000) A breakthrough for wood decay fungi. New Phytol 146:3–4

Eaton RA, Hale MDC (1993) Wood: decay, pests and protection. Chapman & Hall, London

Elissetche JP, Ferraz A, Freer J, Rodriguez J (2006) Influence of forest soil on biodegradation of *Drimys winteri* by *Ganoderma australe*. Int Biodeterior Biodegrad 57:174–178

Elliot ET, Hunt HW, Walter DE, Moore JC (1986) Microcosms, mesocosms and ecosystems: linking the laboratory to the field. In: Megušar F, Gantar M (eds) Perspectives in microbial ecology. Fourth international symposium on microbial ecology, 24–29 August 1986. Slovene Society for Microbiology, Ljubljana, pp 472–480

Federici E, Leonardi V, Giubilei MA, Quaratino D, Spaccapelo R, D'Annibale A, Petruccioli M (2007) Addition of allochthonous fungi to a historically contaminated soil affects both remediation efficiency and bacterial diversity. Appl Microbiol Biotechnol 77:203–211

Fengel D, Wegener G (1989) Wood: chemistry, ultrastructure, reactions. Walter de Gruyter, Berlin

Ford CI, Walter M, Northott GL, Di HJ, Cameron KC, Trower T (2007) Fungal inoculum properties: extracellular enzyme expression and pentachlorophenol removal in highly contaminated field soils. J Environ Qual 36:1599–1608

Forge TA, Hogue E, Neilsen G, Neilsen D (2003) Effects of organic mulches on soil microfauna in the root zone of apple: implications for nutrient fluxes and functional diversity of the soil food web. Appl Soil Ecol 22:39–54

Frankland JC (1998) Fungal succession – unraveling the unpredictable. Mycol Res 102:1–15

Gadd GM (ed) (2001) Fungi in bioremediation. British Mycological Society, Cambridge

Gadd GM (2007) Geomycology: biogeochemical transformations of rocks, minerals, metals and radionuclides by fungi, bioweathering and bioremediation. Mycol Res 111:3–49

Garstang J, Weekes A, Poulter R, Bartlett D (2002) Identification and characterisation of factors affecting losses in the large-scale, non-ventilated bulk storage of wood chips and development of best storage practices URN 02/1535. First Renewables, Leeds

Germain D (2007) Ramial chipped wood: the clue to a sustainable fertile soil. Universite Laval, Quebec

Giubilei MA, Leonardi V, Federici E, Covino S, Sasek V, Novotny C, Federici F, D'Annibale A, Petrucciola M (2009) Effect of mobilizing agents on mycoremediation and impact on the indigenous microbiota. J Chem Technol Biotechnol 84:836–844

Graber ER, Dror I, Bercovich FC, Rosner M (2001) Enhanced transport of pesticides in a field trial with treated sewage sludge. Chemosphere 44:805–811

Grebenc T, Piltaver A, Kraigher H (2004) Pomen velikih lesnih ostankov bukve (*Fagus sylvatica* L.) za ohranjanje pestrosti redkih in ogroženih vrst lignikolnih gliv. In: Brus R (ed) Staro in debelo drevje v gozdu. Biotehniška fakulteta, Oddelek za gozdarstvo in obnovljive vire, Ljubljana, pp 47–55

Green FI, Clausen CA (2005) Copper tolerance of brown-rot fungi: oxalic acid production in southern pine treated with arsenic-free preservatives. Int Biodeterior Biodegrad 56:75–79

Green F, Larsen MJ, Winandy JE, Highley TL (1991) Role of oxalic-acid in incipient brown-rot decay. Mater Org 26:191–213

Grenni P, Caracciolo AB, Rodriguez-Cruz MS, Sanchez-Martin MJ (2009) Changes in the microbial activity in a soil amended with oak and pine residues and treated with linuron herbicide. Appl Soil Ecol 41:2–7

Grinhut T, Hadar Y, Chen Y (2007) Degradation and transformation of humic substances by saprotrophic fungi: processes and mechanisms. Fungal Biol Rev 21:179–189

Guimaraes BCM, Arends JBA, van der Ha D, Van de Wiele T, Boon N, Verstraete W (2010) Microbial services and their management: recent progresses in soil bioremediation technology. Appl Soil Ecol 46:157–167

Hammel KE (1996) Extracellular free radical biochemistry of ligninolytic fungi. New J Chem 20:195–198

Hammel KE (1997) Fungal degradation of lignin. In: Cadisch G, Giller KE (eds) Plant litter quality and decomposition. CAB, Madison, pp 33–45

Harmon ME (2001) Moving towards a new paradigm for woody detritus management. Ecol Bull 49:269–278

Harms H, Schlosser D, Wick LY (2011) Untapped potential: exploiting fungi in bioremediation of hazardous chemicals. Nat Rev Microbiol 9:177–192

Harris MJ, Boddy L (2005) Nutrient movement and mycelial reorganization in established systems of *Phanerochaete velutina*, following arrival of colonized wood resources. Microb Ecol 50:141–151

Hatfield R, Fukushima RS (2005) Can lignin be accurately measured? Crop Sci 45:832–839

Hawksworth DL (2001) The magnitude of fungal diversity: the 1.5 million species estimate revisited. Mycol Res 105:1422–1432

Hawksworth DL, Colwell RR (1992) Microbial diversity 21: biodiversity amongst microorganisms and its relevance. Biodivers Conserv 1:221–226

Hernandez-Apaolaza L, Gasco JM, Guerrero F (2000) Initial organic matter transformation of soil amended with composted sewage sludge. Biol Fertil Soils 32:421–426

Huang DL, Zeng GM, Feng CL, Hu S, Lai C, Zhao MH, Su FF, Tang L, Liu HL (2010) Changes of microbial population structure related to lignin degradation during lignocellulosic waste composting. Bioresour Technol 101:4062–4067

Humar M, Pohleven F (2003a) Mikoremediacija s CCB (Cu/Cr/B) pripravki zaščitenega lesa. In: Glavič P, Brodnjak-Vončina D (eds) Slovenski kemijski dnevi, Maribor, 25. in 26. september 2003. Zbornik referatov s posvetovanja. FKKT, Maribor, pp 1–6

Humar M, Pohleven F (2003b) Razstrupljanje odpadnega s CCA ali CCB pripravki zaščitenega lesa z lesnimi glivami. Les 55:89–94

Humar M, Pohleven F (2005) Influence of a nitrogen supplement on the growth of wood decay fungi and decay of wood. Int Biodeterior Biodegrad 56:34–39

Humar M, Petrič M, Pohleven F (2001) Changes of the pH value of impregnated wood during exposure to wood-rotting fungi. Holz Roh Werkst 59:288–293

Humar M, Petrič M, Pohleven F, Šentjurc M, Kalan P (2002a) Changes in EPR spectra of wood impregnated with copper-based preservatives during exposure to several wood-rotting fungi. Holzforsch 56:229–238

Humar M, Pohleven F, Kalan P, Amartey S (2002b) Translokacija bakra iz zaščitenega lesa, izpostavljenega glivam razkrojevalkam lesa. Zbornik gozdarstva in lesarstva 67:159–171

Humar M, Bučar B, Pohleven F (2006) Brown-rot decay of copper-impregnated wood. Int Biodeterior Biodegrad 58:9–14

Hunt J, Boddy L, Randerson PF, Rogers HJ (2004) An evaluation of 18S rDNA approaches for the study of fungal diversity in grassland soils. Microb Ecol 47:385–395

Hyde KD, Jones EBG (2002) Introduction to fungal succession. Fungal Divers 10:1–4

Ishikawa H, Osono T, Takeda H (2007) Effects of clear-cutting on decomposition processes in leaf litter and the nitrogen and lignin dynamics in a temperate secondary forest. J Forest Res 12:247–254

Janse BJH, Gaskell J, Akhtar M, Cullen D (1998) Expression of *Phanerochaete chrysosporium* genes encoding lignin peroxidases, manganese peroxidases, and glyoxal oxidase in wood. Appl Environ Microbiol 64:3536–3538

Jellison J (1992) Cation analysis of wood degraded by white- and brown-rot fungi. The International Research Group on Wood Preservation, Document IRG/WP/1552-92:16

Jellison J, Jasalavich C (2000) A review of selected methods for the detection of degradative fungi. Int Biodeterior Biodegrad 46:241–244

Jellison J, Connolly J, Goodell B, Doyle B, Illman B, Fekete F, Ostrofsky A (1997) The role of cations in the biodegradation of wood by the brown rot fungi. Int Biodeterior Biodegrad 39:165–179

Jirjis R (2005) Effects of particle size and pile height on storage and fuel quality of comminuted *Salix viminalis*. Biomass Bioenergy 28:193–201

Jonsson BG, Kruys N (eds) (2001) Ecology of woody debris in boreal forests. Wallin & Dalholm, Lund

Jordan CF (2004) Organic farming and agroforestry: alleycropping for mulch production for organic farms of southeastern United States. Agroforest Syst 61–2:79–90

Jurc M (1997) Patogeni – simbionti – endofiti: sinonimi ali samostojne kategorije organizmov? In: Maček J (ed) Zbornik predavanj in referatov 3. slovenskega posvetovanja o varstvu rastlin v Portorožu od 4. do 5. marca 1997. Društvo za varstvo rastlin Slovenije, Ljubljana, pp 285–290

Jurc M, Jurc D (1995) Endophytic fungi in the needles of healthy-looking Austrian Pine (*Pinus nigra* Arn.). Acta Pharm 45:341–345

Jurc D, Ogris N (2006) First reported outbreak of charcoal disease caused by *Biscogniauxia mediterranea* on Turkey oak in Slovenia. Plant Pathol 55:299

Jurc D, Jurc M, Sieber TN, Bojović S (2000) Endophytic *Cenangium ferruginosum* (Ascomycota) as a reservoir for an epidemic of Cenangium dieback in Austrian Pine. Phyton (Horn) 40:103–108

Killham K (1995) Soil ecology. Cambridge University Press, Cambridge

Kirk TK, Farrell RL (1987) Enzymatic "combustion": the microbial degradation of lignin. Annu Rev Microbiol 41:465–505

Kirk JL, Beaudette LA, Hart M, Moutoglis P, Klironomos JN, Lee H, Trevors JT (2004) Methods of studying soil microbial diversity. J Microbiol Methods 58:169–188

Kokalisburelle N, Rodriguezkabana R (1994) Effects of pine bark extracts and pine bark powder on fungal pathogens, soil enzyme activity, and microbial populations. Biol Control 4:269–276

Kowalchuk GA, Gerards S, Woldendorp JW (1997) Detection and characterization of fungal infection of *Ammophila arenaria* (Marram Grass) roots by denaturing gradient gel electrophoresis of specifically amplified 18S rDNA. Appl Environ Microbiol 63:3858–3865

Kowalski T, Kehr RD (1992) Endophytic fungal colonization of branch bases in several forest tree species. Sydowia 44:137–168

Kowalski T, Kehr RD (1996) Fungal endophytes of living branch bases in several European tree species. In: Redlin SC, Carris LM (eds) Endophytic fungi in grasses and woody plants. APS, St. Paul, pp 67–86

Kuffer N, Gillet F, Senn-Irlet B, Aragno M, Job D (2008) Ecological determinants of fungal diversity on deadwood in European forests. Fungal Divers 30:83–95

Lang E, Kleeberg I, Zadrazil F (2000) Extractable organic carbon and counts of bacteria near the lignocellulose-soil interface during the interaction of soil microbiota and white rot fungi. Bioresour Technol 75:57–65

Larsson L, Stenberg B, Torstensson L (1997) Effects of mulching and cover cropping on soil microbial parameters in the organic growing of black currant. Commun Soil Sci Plant Anal 28:913–925

Lemieux G (1993) A universal pedogenesis upgrading processus: RCWs to enhance biodiversity and productivity. Food and Agriculture Organization (FAO), Rome

Leštan D, Lamar RT (1996) Development of fungal inocula for bioaugmentation of contaminated soils. Appl Environ Microbiol 62:2045–2052

Leštan D, Leštan M, Chapelle JA, Lamar RT (1996) Biological potential of fungal inocula for bioaugmentation of contaminated soils. J Ind Microbiol 16:286–294

Leštan D, Leštan M, Lamar RT (1998) Growth and viability of mycelial fragments of white-rot fungi on some hydrogels. J Ind Microbiol Biotechnol 20:244–250

Lindhe A, Asenblad N, Toresson H-G (2004) Cut logs and high stumps of spruce, birch, aspen and oak – nine years of saproxylic fungi succession. Biol Conserv 119:443–454

Marschner P, Kandeler E, Marschner B (2003) Structure and function of the soil microbial community in a long-term fertilizer experiment. Soil Biol Biochem 35:453–461

Martinez AT (2007) High redox potential peroxidases. In: Polaina J, MacCabe AP (eds) Industrial enzymes. Springer, Dordrecht, pp 477–488

Martinez AT, Speranza M, Ruiz-Duenas FJ, Ferreira P, Camarero S, Guillen F, Martinez MJ, Gutierrez A, del Rio JC (2005) Biodegradation of lignocellulosics: microbial chemical, and enzymatic aspects of the fungal attack of lignin. Int Microbiol 8:195–204

McGradySteed J, Harris PM, Morin PJ (1997) Biodiversity regulates ecosystem predictability. Nature 390:162–165

McMahon V, Garg A, Aldred D, Hobbs G, Smith R, Tothill IE (2008) Composting and bioremediation process evaluation of wood waste materials generated from the construction and
demolition industry. Chemosphere 71:1617–1628
Moreira MT, Feijoo G, SierraAlvarez R, Lema J, Field JA (1997) Biobleaching of oxygen
delignified kraft pulp by several white rot fungal strains. J Biotechnol 53:237–251
Morgan P, Lee SA, Lewis ST, Sheppard A, Watkinson RJ (1993) Growth and biodegradation by
white-rot fungi inoculated into soil. Soil Biol Biochem 25:279–287
Moricca S, Ragazzi A (2008) Fungal endophytes in Mediterranean oak forests: a lesson from
Discula quercina. Phytopathology 98:380–386
Mougin C, Boukcim H, Jolivalt C (2009) Soil bioremediation strategies based on the use of fungal
enzymes. In: Singh A, Kuhad RC, Ward OP (eds) Advances in applied bioremediation.
Springer, Heidelberg, pp 123–149
Muñoz C, Guillen F, Martinez AT, Martinez MJ (1997) Induction and characterization of laccase
in the ligninolytic fungus *Pleurotus eryngii*. Curr Microbiol 34:1–5
Muyzer G, Smalla K (1998) Application of denaturing gradient gel electrophoresis (DGGE) and
temperature gradient gel electrophoresis (TGGE) in microbial ecology. Antonie van
Leeuwenhoek 73:127–141
Muyzer G, de Wall EC, Uitterlinden AG (1993) Profiling of complex microbial populations by
denaturing gradient gel electrophoresis analysis of polymerase chain reaction-amplified genes
coding for 16S rRNA. Appl Environ Microbiol 59:695–700
Namkoong W, Hwang EY, Park JS, Choi JY (2002) Bioremediation of diesel-contaminated soil
with composting. Environ Pollut 119:23–31
Nocker A, Burr M, Camper AK (2007) Genotypic microbial community profiling: a critical
technical review. Microb Ecol 54:276–289
Ódor P, Heilmann-Clausen J, Christensen M, Aude E, van Dort KW, Piltaver A, Siller I,
Veerkamp MT, Walleyn R, Standovar T, van Hees AFM, Kosec J, Matočec N, Kraigher
H, Grebenc T (2006) Diversity of dead wood inhabiting fungi and bryophytes in semi-natural
beech forests in Europe. Biol Conserv 131:58–71
Oses R, Valenzuela S, Freer J, Baeza J, Rodríguez J (2006) Evaluation of fungal endophytes for
lignocellulolytic enzyme production and wood biodegradation. Int Biodeterior Biodegrad
57:129–135
Oses R, Valenzuela S, Freer J, Sanfuentes E, Rodriguez J (2008) Fungal endophytes in xylem of
healthy Chilean trees and their possible role in early wood decay. Fungal Divers 33:77–86
Osono T (2006) Role of phyllosphere fungi of forest trees in the development of decomposer
fungal communities and decomposition processes of leaf litter. Can J Microbiol 52:701–716
Osono T (2007) Ecology of ligninolytic fungi associated with leaf litter decomposition. Ecol Res
22:955–974
Parfitt D, Hunt J, Dockrell D, Rogers HJ, Boddy L (2010) Do all trees carry the seeds of their own
destruction? PCR reveals numerous wood decay fungi latently present in sapwood of a wide
range of angiosperm trees. Fungal Ecol 3:338–346
Park JH, Lamb D, Paneerselvam P, Choppala G, Bolan N, Chung J-W (2011) Role of organic
amendments on enhanced bioremediation of heavy metal(loid) contaminated soils. J Hazard
Mater 185:549–574
Pérez J, Muñoz-Dorado J, de la Rubia T, Martínez J (2002) Biodegradation and biological
treatments of cellulose, hemicellulose and lignin: an overview. Int Microbiol 5:53–63
Petrini O, Fisher PJ (1990) Occurrence of fungal endophytes in twigs of *Salix fragilis* and *Quercus
robur*. Mycol Res 94:1077–1080
Piltaver A, Matočec N, Kosec J, Jurc D (2002) Glive na odmrlem bukovem lesu v slovenskih
gozdnih rezervatih Rajhenavski Rog in Krokar. Zbornik gozdarstva in lesarstva 69:171–196
Piškur B (2009) Successive processes during the decomposition of wood, inoculated with fungus
Pleurotus ostreatus on degraded surfaces. PhD thesis, University of Ljubljana, Ljubljana,
Slovenia

Piškur B, Zule J, Piškur M, Jurc D, Pohleven F (2009) Fungal wood decay in the presence of fly ash as indicated by gravimetrics and by extractability of low molecular weight organic acids. Int Biodeterior Biodegrad 63:594–599

Piškur B, Bajc M, Robek R, Humar M, Sinjur I, Kadunc A, Oven P, Rep G, Al Sayegh Petkovšek S, Kraigher H, Jurc D, Pohleven F (2011a) Influence of *Pleurotus ostreatus* inoculation on wood degradation and fungal colonization. Bioresour Technol 102:10611–10617

Piškur B, Pavlic D, Slippers B, Ogris N, Maresi G, Wingfield M, Jurc D (2011b) Diversity and pathogenicity of Botryosphaeriaceae on declining *Ostrya carpinifolia* in Slovenia and Italy following extreme weather conditions. Eur J Forest Res 130:235–249

Pointing SB (2001) Feasibility of bioremediation by white-rot fungi. Appl Microbiol Biotechnol 57:20–33

Ponge JF (2005) Fungal communities: relation to resource succession. In: Dighton J, White JF, Oudemans P (eds) The fungal community. Its organization and role in the ecosystem. Taylor & Francis, Boca Raton, pp 169–180

Rayner ADM, Boddy L (1988) Fungal decomposition of wood. Its biology and ecology. Wiley, Chichester

Sanchez C (2009) Lignocellulosic residues: biodegradation and bioconversion by fungi. Biotechnol Adv 27:185–194

Šašek V, Cajthaml T, Bhatt M (2003) Use of fungal technology in soil remediation: a case study. Water Air Soil Pollut Focus 3:5–14

Scheel T, Hofer M, Ludwig S, Holker U (2000) Differential expression of manganese peroxidase and laccase in white-rot fungi in the presence of manganese or aromatic compounds. Appl Microbiol Biotechnol 54:686–691

Schmidt O (2006) Wood and tree fungi. Biology, damage, protection, and use. Springer, Berlin

Schmidt KR, Chand S, Gostomski PA, Boyd-Wilson KSH, Ford C, Walter M (2005) Fungal inoculum properties and its effect on growth and enzyme activity of *Trametes versicolor* in soil. Biotechnol Prog 21:377–385

Shevchenko SM, Bailey GW (1996) Life after death: lignin-humic relationships reexamined. Crit Rev Environ Sci Technol 26:95–153

Shimada M, Ma DB, Akamatsu Y, Hattori T (1994) A proposed role of oxalic-acid in wood decay systems of wood-rotting basidiomycetes. FEMS Microbiol Rev 13:285–296

Shortle WC (1990) Ionization of wood decay during previsual stages of wood decay. Biodeterior Res 3:333–348

Sieber TN (2007) Endophytic fungi in forest trees: are they mutualists? Fungal Biol Rev 21:75–89

Singh H (2006) Mycoremediation: fungal bioremediation. Wiley, Hoboken

Šnajdr J, Baldrian P (2006) Production of lignocellulose-degrading enzymes and changes in soil bacterial communities during the growth of *Pleurotus ostreatus* in soil with different carbon content. Folia Microbiol 51:579–590

Solbraa K (1979a) Composting of bark. I. Different bark qualities and their uses in plant production. Rep Norweg Forest Res Inst 34:285–328

Solbraa K (1979b) Composting of bark. II. Laboratory experiments. Rep Norweg Forest Res Inst 34:339–384

Steffen K, Tuomela M (2010) Fungal soil bioremediation: developments towards large-scale applications. In: Hofrichter M (ed) Industrial applications. Springer, Heidelberg, pp 451–467

Stevenson FJ (1982) Humus chemistry: genesis, composition, reactions. Wiley, New York

Swift MJ (1977) The ecology of wood decomposition. Sci Prog (Oxf) 64:175–199

Tahboub MB, Lindemann WC, Murray L (2008) Chemical and physical properties of soil amended with pecan wood chips. Hortscience 43:891–896

Takao S (1965) Organic acid production by Basidiomycetes. I. Screening of acid-producing strains. Appl Microbiol 13:732–737

Thurston CF (1994) The structure and function of fungal laccases. Microbiology 140:19–26

Tilman D (1996) Biodiversity: population versus ecosystem stability. Ecology 77:350–363

Torelli N (1986) Zgradba lesa. Biotehniška fakulteta, Oddelek za lesarstvo, Ljubljana

Tornberg K, Baath E, Olsson S (2003) Fungal growth and effects of different wood decomposing fungi on the indigenous bacterial community of polluted and unpolluted soils. Biol Fertil Soils 37:190–197

Tucker B, Radtke C, Kwon SI, Anderson AJ (1995) Supression of bioremediation by *Phaneorchaete chrysosporium* by soil factors. J Hazard Mater 41:251–265

Tuomela M, Lyytikäinen M, Oivanen P, Hatakka A (1998) Mineralization and conversion of pentachlorophenol (PCP) in soil inoculated with the white-rot fungus *Trametes versicolor*. Soil Biol Biochem 31:65–74

Tuomela M, Vikman M, Hatakka A, Itavaara M (2000) Biodegradation of lignin in a compost environment: a review. Bioresour Technol 72:169–183

Uffen RL (1997) Xylan degradation: a glimpse at microbial diversity. J Ind Microbiol Biotechnol 19:1–6

Vainio EJ, Hantula J (2000) Direct analysis of wood-inhabiting fungi using denaturing gradient gel electrophoresis of amplified ribosomal DNA. Mycol Res 104:927–936

Valentin L, Kluczek-Turpeinen B, Oivanen P, Hatakka A, Steffen K, Tuomela M (2009) Evaluation of basidiomycetous fungi for pretreatment of contaminated soil. J Chem Technol Biotechnol 84:851–858

Valentín L, Kluczek-Turpeinen B, Willför S, Hemming J, Hatakka A, Steffen K, Tuomela M (2010) Scots pine (*Pinus sylvestris*) bark composition and degradation by fungi: potential substrate for bioremediation. Bioresour Technol 101:2203–2209

van der Wal A, De Boer W, Smant W, van Venn JA (2007) Initial decay of woody fragments in soil is influenced by size, vertical position, nitrogen availability and soil origin. Plant Soil 301:189–201

van Elsas JD, Duarte GF, Keijzer-Wolters A, Smit E (2000) Analysis of the dynamics of fungal communities in soil via fungal-specific PCR of soil DNA followed by denaturing gradient gel electrophoresis. J Microbiol Methods 43:133–151

Vesel Tratnik N, Pohleven F (1995) Sukcesija in interakcija gliv povzročiteljic piravosti bukovine. Zbornik gozdarstva in lesarstva 46:163–176

Vidic I (2008) Degradation of chlorinated organic biocides by lignolytic fungi. MSc thesis, University of Ljubljana, Ljubljana, Slovenia

Visser S, Maynard D, Danielson RM (1998) Response of ecto- and arbuscular mycorrhizal fungi to clear-cutting and the application of chipped aspen wood in a mixedwood site in Alberta, Canada. Appl Soil Ecol 7:257–269

Wilcox WW (1993) Comparative morphology of early stages of brown-rot wood decay. IAWA J 14:127–138

Wilson D (1995) Endophyte – the evolution of a term, and clarification of its use and definition. Oikos 73:274–276

Winquist E, Valentin L, Moilanen U, Leisola M, Hatakka A, Tuomela M, Steffen KT (2009) Development of a fungal pre-treatment process for reduction of organic matter in contaminated soil. J Chem Technol Biotechnol 84:845–850

Woodward S, Boddy L (2008) Interactions between saprothropic fungi. In: Boddy L, Frankland JC, Van West P (eds) Ecology of saprotrophic basidiomycetes. Elsevier, Amsterdam, pp 125–141

Zabel RA, Morrell JJ (1992) Wood microbiology. Decay and its prevention. Academic, San Diego

Part II
Application of Mycoremediation Against Organic Pollutants

Chapter 4
Mycoremediation of Paper, Pulp and Cardboard Industrial Wastes and Pollutants

Shweta Kulshreshtha, Nupur Mathur, and Pradeep Bhatnagar

4.1 Introduction

The importance of paper is increasing day by day with the literacy and cultural development of countries. Demand of paper has dramatically increased in the past several years. This demand is satisfied by different types of paper and pulp industries existing in different countries. Paper and pulping operations and washout operations in paper and pulp industries have resulted in soil and water contamination with unused and discharged residues and effluents.

Improperly handled and disposed paper and pulp and cardboard industrial wastes imperil both human health and the environment. Paper and pulp (Klekowski et al. 2006) have proved toxic and mutagenic to human beings (Kulshreshtha et al. 2010a). Occupational and nonoccupational exposures of human beings to these industrial wastes have led to various health effects such as headaches, nausea, lung and skin irritations and congenital malformations (Shinka et al. 1991; Morikawa et al. 1997) and cancer (Felton et al. 2002).

Industrial development is pervasively connected with the disposal of number of toxic pollutants. Pollutants present in wastes are the major point of concern in many countries. Some of these pollutants require a high priority of treatment for protecting useful resources such as soil, water, air, etc. A growing concern for continuing deterioration of environment and public health has led to establishment

S. Kulshreshtha (✉)
Amity Institute of Biotechnology, Amity University of Rajasthan, Jaipur, Rajasthan, India
e-mail: shweta_kulshreshtha@rediffmail.com

N. Mathur
Department of Zoology, University of Rajasthan, Jaipur, Rajasthan, India
e-mail: nupurmathur123@rediffmail.com

P. Bhatnagar
Department of Life Sciences, The IIS University, Gurukul Marg, Mansarovar, Jaipur, Rajasthan, India
e-mail: pradeepbhatnagar1947@rediffmail.com

E.M. Goltapeh et al. (eds.), *Fungi as Bioremediators*, Soil Biology 32, DOI 10.1007/978-3-642-33811-3_4, © Springer-Verlag Berlin Heidelberg 2013 of legislation for control of quality of effluents and receiving water. Various government norms have also been made more and more stringent for the sake of nature. The environmental norms are made so strict in the countries like China that there has been a closure of almost 50 % of the papermaking units during the last 10 years (approximately 7,000 units in 1995 and only 3,500 in 2005) due to the use of inefficient procedures, outdated technologies, poor product quality and environmental ill effects. Similarly in India, in 1996, Supreme Court gave as many as 12 verdicts in pollution-related areas. Delhi Government had to shut down over 2,000 small-scale industries as a result of the Supreme Court's verdict in Nov. 2000 that no polluting industry must be allowed to operate in Delhi's residential areas. However, despite the regulations for toxic discharges into the environment, many situations exist in which discharging of paper and cardboard industrial pollutants and wastes to the water body is still a practice in many developing countries. For small-scale industries, environmental concerns are considered as just a good wish better to avoid and the pollution control measures are considered as a luxury which couldn't be afforded by the small-scale sector. However, now the situation is changing even for them, and the implementations of environment protection will have to proceed in a manner that sustains the economic health of the industries, yet ensures human health and environmental quality. Setting up a treatment plant at industrial level may be the ultimate solution for pollution caused by paper and cardboard industries. The common effluent treatment facilities are being created in the different industrial clusters at the small level.

Therefore, it becomes imperative to completely degrade paper and cardboard industrial pollutants which cannot be completely degraded by well-established techniques like conventional wastewater treatment methods. In recent years, there has been increasing interest in developing mycological techniques for remediation of paper and pulp industrial wastes which contaminate soil and water. Fungi can degrade paper and pulp industrial wastes by using their own enzymatic pathways. The main objective of bioremediation processes is to mineralise and degrade organic and inorganic contaminants.

4.2 Overview of Bioremediator Fungi for Remediation of Paper and Cardboard Industrial Wastes

To use fungi as agent of mycoremediation, it is necessary to know what makes paper—a good substrate for fungal growth. Paper is made up of cellulose- and lignin-containing fibres which are obtained from various sources such as agricultural residues, trees, cotton and linen rags, etc. Fungi can use lignin and cellulose for their growth and lead to the remediation of paper and cardboard industrial waste from the environment. This reveals the requirement of fungi that degrade lignin and cellulose by secreting nonspecific extracellular enzymes for remediating paper, cardboard and their industrial discharges. Generally, lignin and cellulose degrading fungi can be divided into two major categories, i.e. white-rot fungi and non-white-rot fungi.

Table 4.1 Comparison of the properties of MnP, LiP and Lac from white-rot fungi (Wesenberg et al. 2003)

Properties	Manganese peroxidase (MnP)	Lignin peroxidase (LiP)	Laccase (Lac)
E.C. no.	1.11.1.13	1.11.1.14	1.10.3.2
Prosthetic group	$Mn(II):H_2O_2$ oxidoreductases	Diarylpropan O_2, H_2O_2 oxidoreductases	p-benzenediol: O_2-oxidoreductases
	Heme	Heme	1 type-1 Cu, 1 type-2 Cu, 2-coupled type-3 Cu
MW(kDa)	32–62	38–47	59–110
Glycosylation	N-	N-	N-
Isoforms	Monomers, up to 11	Monomers, up to 15	Mono-, di-, tetramers; several
C–C cleavage	Present	Present	Absent
H_2O_2—regulated	Regulated	Regulated	Not regulated
Stability	Highly stable	Stable	Highly stable

White-rot fungi have ability to degrade lignin and chlorinated lignin derivatives effectively which are present in pulp, paper and cardboard industries (Eriksson 1991; Bajpai and Bajpai 1994) due to the production of a variety of extracellular ligninolytic enzymes such as lignin peroxidase, manganese peroxidase, laccase and H_2O_2-producing oxidases. Degradation of lignin is apparently a very energy-intensive process (Leisola et al. 1983; Boman et al. 1988) and requires specific cultivation conditions, necessary for the production of ligninolytic enzymes by white-rot fungi to degrade lignin (Boman et al. 1988). Various cultivation methods have been devised to overcome these difficulties. For instance, MyCoR process, MYCOPOR-system (Jaklin-Farcher et al. 1992), continuous-flow systems (Prasad and Joyce 1991) and immobilisation techniques employing white-rot fungi (Livernoche et al. 1983; Kirkpatrick et al. 1990) (Table 4.1).

Non-white-rot fungi also possess extracellular oxidative enzymes, in particular, lignin-degrading enzyme systems (LDS), for remediation of paper and pulp industrial wastes. These fungi may be present in the pulp and paper industrial waste indigenously and possess a well-developed lignin-degrading enzyme system which helps in bioremediation of lignin- and cellulose-containing waste. Autochthonous non-white-rot fungal strains are able to use lignin cellulose as carbon and energy source. Many fungal strains have been reported to be isolated from soils polluted by paper, pulp and cardboard industrial wastes (Table 4.3) belonging to the phycomycetes, zygomycetes and deuteromycetes. Phenotypic and biochemical assays revealed the ability of these filamentous fungi to synthesise extracellular oxidative enzymes and suggested a relationship between the LDS and paper and cardboard waste bioconversion.

Interest in mycoremediation of paper and pulp industrial waste logically led scientists to evaluate the potential of fungi for degrading cellulose, lignin, AOX, resins and polychlorobiphenyls (PCBs) present in paper and pulp industrial wastes. It is particularly of interest that fungi, applied for mycoremediation, is capable of producing extracellular enzymes in ligninlytic and non-ligninolytic conditions.

4.3 Overview of Fungal Enzymes Involves in Remediation of Paper and Cardboard Industrial Wastes

Fungal species, used for mycoremediation of paper and pulp mill waste and its constituents, are able to produce different enzymes depending on their genetic make-up and growth conditions. Ligninolytic fungal enzymes that degrade lignin also have potential to remediate the paper and pulp industrial waste. Due to non-specificity of enzymes, these can also transform a variety of organic and chloro-organic compounds. Key lignin degradation enzymes are oxidoreductases, i.e. two types of peroxidases, lignin peroxidase (LiP), manganese-dependent peroxidase (MnP), manganese-independent peroxidase (MIP) and a phenoloxidase, laccase. Lignin-degrading enzymes are applicable in the degradation of highly toxic environmental chemicals such as dioxins, polychlorinated biphenyls, various dyes and polyaromatic hydrocarbons and decolourisation of kraft bleach plant effluents.

Fungal enzyme treatments have the major advantage of requiring only a small stream of the process water to be cooled for generation of fungal culture filtrate with most of the degradation of detrimental organics being accomplished by enzyme (fungal culture filtrate) that function well at process water temperatures up to 75 °C. It is probable that enzymatic treatment will be relatively inexpensive and environmental friendly.

4.3.1 Lignin Peroxidase

LiPs are glycosylated heme proteins protoporphyrin IX which is differing in their catalytic mechanisms. LiPs act by abstracting single electron from aromatic rings of lignin and lignin model compounds, leading to the formation of cation radical and subsequent cleavage reactions.

These enzymes are potentially valuable in pulp, paper and cardboard industrial waste disposal because of their ability to degrade not only lignin but also various pollutants. These enzymes are produced by white-rot fungi (WRF) including edible and nonedible mushroom species, during their secondary metabolism. Since lignin oxidation provides no net energy to the fungus, synthesis and secretion of these enzymes are often induced in limited nutrient levels (mostly carbon and nitrogen) along with hydrogen peroxide. LiP interacts with lignin in the presence of a cofactor called veratryl alcohol (a secondary metabolite of white-rot fungi).

4.3.2 Manganese-Dependent Peroxidase

MnP are glycosylated glycoproteins (mol. wt. 32–62.5 kDa) with an iron porphyrin IX (heme prosthetic group) and are secreted in multiple isoforms. MnP preferentially oxidise Mn^{2+} into Mn^{3+} which is stabilised by chelators (oxalic acids) excreted by the fungi. Generated Mn^{3+}, a highly reactive intermediate, which, when stabilised by chelators, can diffuse from the enzyme active site to attack and oxidise the lignin structure in situ. Thus, MnP are able to oxidise and depolymerise lignin and other compounds present in paper and pulp mill effluent.

4.3.3 Laccases

Fungal laccases (Lacs) are produced by basidiomycetes as part of the ligninolytic enzyme system. This group of N-glycosylated extracellular blue oxidases with molecular masses of 60–390 kDa contains four copper atoms in the active site that are distributed among different binding sites. Laccases catalyse the oxidation of variety of aromatic hydrogen donors with the concomitant reduction of oxygen to water. Moreover, laccases do not only oxidise phenolic and methoxyphenolic acids but also decarboxylate them and attack their methoxy groups.

The most studied laccase-producing fungus is *Trametes versicolor*. *Phanerochaete chrysosporium* do not produce this enzyme (Hattaka 1994; Thurston 1994) or produce in defined culture medium containing cellulose and ammonium tartrate. Two laccase isozymes (I and II) of *Trametes versicolor* were purified by Bourbonnais et al. (1995). The same reactivities of two isozymes were found to be similar on most of the low-molecular-weight substances. However, significantly higher reactivity of laccase-I than laccase-II was exhibited with a polymeric substrate. The two isozymes had similar qualitative effects on kraft lignin and residual lignin in kraft pulp (Bourbonnais et al. 1995).

Laccases can oxidise many recalcitrant substances, such as chlorophenols, lignin-related structures, nonphenolic lignin models (Kawai et al. 1988) and dyes. Thus, these enzymes may be used for the degradation of pulp, paper and cardboard industrial wastes.

4.3.4 Versatile Peroxidase

A third group of peroxidases, versatile peroxidases (VPs), has been recently recognised that can be regarded as hybrid between MnP and LiP. VP has been found to be present in species of *Pleurotus* and *Bjerkandera*. Since they can oxidise not only Mn^{2+} but also phenolic and nonphenolic aromatic compounds including dyes, they may accelerate the bioremediation process of pulp and paper wastes.

4.3.5 H₂O₂-Producing Enzymes

White-rot fungi also have the ability to produce a number of oxidases, such as glucose oxidase (Kelley et al. 1986), glyoxal oxidase (Kersten and Kirk 1987), methanol oxidase (Eriksson and Nishida 1988), and veratryl alcohol oxidase (Bourbonnais and Paice 1988), that are capable of generating H_2O_2, presumably for utilisation by extracellular peroxidases during degradation of lignin.

In the first step of bioremediation, peroxidases and laccases degrade lignin and produce radical substances. These radical substances spontaneously re-polymerise in the absence of quinone oxidoreductase (CBQ) and cellobiose oxidase (CBO). It has also been shown that when *T. versicolor* laccase is incubated with glucose oxidase, improvement in lignin depolymerisation occurs. This is probably due to the action of glucose oxidase in reducing lignin and thereby limiting their repolymerisation. In this way, polymerisation/depolymerisation equilibrium is shifted towards degradation in the presence of other enzyme.

4.3.6 Other Enzymes

There is growing evidence for the participation of other important enzymes in the lignin degradation process. Two enzymes cellobiose, quinone oxidoreductase (CBQ) and cellobiose oxidase (CBO), play important role in the lignin degradation by reducing phenoxy radical compounds. MnP requires the supply of Mn(III)-complexing agent such as cellobionic acid, produced during the oxidation of cellobiose to cellobionic acid, for degradation of lignin. Mn (II) is also required by MnP during oxidation of lignin, since Mn (II) is converted to Mn (III) by this enzyme.

CBQ supply cellobionic acid reduces insoluble Mn (IV) to Mn (II), during the oxidation of cellobiose to cellobionic acid and hence recycling these cations. The small size of the organic acid Mn (III)-complexes makes them an important delignification agent due to their diffusability (Roy et al. 1994) in lignocellulosic wall of paper and cardboard. CBQ activity also inhibits the oxidation of veratryl alcohol (Ander et al. 1990).

The necessity of mycelial-bound ligninolytic enzyme or a hydrogen peroxide-producing system in the fungi plays an important role in dye decolourisation. Daniel et al. (1994) suggested a cooperative role between pyranose oxidase (POD) and MnP. Pyranose oxidase enzyme was reported to be a major source of hydrogen peroxide—an agent required for lignin degradation. Furthermore, both POD and hydrogen peroxide-dependent MnP occur in the periplasmic space of hyphae and extracellular medium.

4.4 Overview of Mycoremediation Process

"Myco" stands for fungi and "remediation" refers to removal of waste. Thus, "mycoremediation" refers to removal of waste by fungi. Mycoremediation is the form of bioremediation in which fungi is used to return polluted environment to less polluted state. Remediation of pollutants is possibly due to the presence of many highly active enzymes in the fungi, during which fungi are at their most metabolically active stage. Fungi inhabiting polluted environments are armed with various resistance and catabolic potentials. This catabolic potential of microbes in nature is enormous and is advantageous to mankind for a cleaner and healthier environment through bioremediation.

Mycoremediation is gaining attention as an alternative approach to presently available methods. It is important to understand the variables that control the rate and extent of fungal attack on paper, pulp and cardboard for meaningful exploitation of fungi. Paper and cardboard possess lignin and cellulose as main ingredients which are needed to be degraded by fungi and fungal enzymes. Ligninolytic enzymes degrade not only lignin but also several compounds present in the paper and cardboard industrial waste. The process of lignin degradation is proposed to be completed in two phases. In the initial primary phase, the ligninolytic enzyme system is synthesised, while during the secondary idiophasic metabolism, lignin is degraded. The white-rot fungi, in general, and *P. chrysosporium*, in particular, are by far the most active ligninolytic organisms described to date. These are widely used for the degradation of paper and pulp industrial waste. However, some indigenous fungi and known standard cultures can be used for remediation of paper and pulp industrial wastes.

4.4.1 Source of Fungal Cultures

When using fungi for bioremediation, availability of fungal inoculum is a practical consideration. Fungal cultures can be obtained in many ways for bioremediation of paper and cardboard industrial waste.

4.4.1.1 Standard Cultures

A lot of work has been reported in the literature for the treatment of pulp, paper and cardboard industrial effluents. Most of the work is reported to be based on the use of standard microbial strains. Cultures may be purchased from the well-known centres for fungal culture storage, given in Table 4.2. Isolates may also be purchased from the place reported in literature where the culture is deposited.

Table 4.2 Centres for getting fungal cultures

S. no.	Name of centre	Country
1	International Mycological Institute or Commonwealth Mycological Institute (CMI)	United kingdom
2	United Kingdom National Culture Collection (UKNCC)	United kingdom
3	American Type Collection Centre (ATCC)	United States of America
4	Central Bureau voor Schimmelcultures (CBS)	Netherlands
5	IHEM Culture collection	Belgium
6	MUCL Culture collection	Belgium
7	Herbarium Cryptogame Indiae Orientalis	India
8	Microbial Type Collection Centre, Chandigarh	India

4.4.1.2   Indigenous or Autochthonous Fungi

Indigenous fungi refer to the fungi present in waste itself. There are many fungi occurring naturally in conditions ecologically modified by industrial effluents. They degrade the surrounding substrates and wastes discharged by industries. These primary invaders change substratum conditions by their action making way to secondary invaders. Due to the continuous presence of microorganisms with industrial effluents and recalcitrant compounds, genetic potency develops to degrade these compounds. These isolates are present in low numbers and are not effective in bioremediation. To make them effective in mycoremediation process, these indigenous microbial communities can be enriched in the presence of intermediary metabolites of toxic compounds, and significant strains will be evolved with the process of adaptation which may possibly be employed under controlled conditions to degrade the industrial wastes. The metabolic activity leads to the change in the structure and function of the community. Based on this principle, indigenous fungi can be isolated from paper and cardboard industrial wastes. These can be isolated by using different techniques such as bait technique (using paper and pulp sludge), serial dilution method (using various fungal media) and blotter technique (pieces of rags and paper).

In bait and blotter technique, paper and cardboard pieces are used as bait/blotter which is wetted by industrial effluent and kept in petri dishes. These petri dishes are incubated at 25–35 °C for the particular period of time and observed daily for the growth of fungi. As fungal colonies appeared on bait, these are cultured on the suitable medium for obtaining pure culture which will be later screened for cellulolytic and ligninolytic ability.

In dilution method, initially, serial dilution of waste or contaminated soil/water is prepared in sterile normal saline. Appropriate dilution is inoculated on medium containing Mandel's salts solution with addition of 17.5 and 5.0 g/l phosphoric acid-swollen cellulose and 0.5 % L-sorbose as a colony restrictor and inducer of cellulose production (Wang et al. 1995). Further, all isolated fungi will be screened for their ability to degrade cellulose/lignin, dyes and other compound present in the paper and pulp industrial wastes.

After isolation of indigenous fungi, growth of the fungal strain is analysed for initially phenotypic characterisation of fungi. A scale based on colony growth is used for interpreting the results: nongrowth, residual growth, weak growth, moderate growth and good growth. Classic and molecular identification of fungal strain can be done by macro- and microscopic studies of morphological characters (hyphae, conidia, chlamydospores, conidiogenous cells and conidiophores). The information was compiled in a taxonomic description for comparison with specialised literature. For molecular identification of autochthonous paper and cardboard degrading fungal strains, PCR amplifications of 28S rRNA gene should be performed using the following specific primers: NL1 (forward), 5'-GCATATCAATAAGCGGAGGAAA AG-3', and NL4 (reverse), 5'-GGTCCG-TGTTTC AAGACGG-3'. PCR reactions can be performed in thermocycler using deoxyribonucleotide triphosphate and fungal genomic DNA as template. PCR conditions consisted of an initial denaturation at 95 °C for 5 min, 40 cycles of amplification at 95 °C for 35 s, annealing at 52 °C for 30 s, extension at 72 °C for 20 s and final extension at 72 °C for 10 min. The PCR product can be purified using the DNA purification kit and sequenced. Sequence analysis of nucleotide sequences can be performed using the Lasergene software package DNASTAR Programs, BLAST (Altschul et al. 1997) and FASTA (Pearson and Lipman 1988).

The ability of the fungal strains to produce extracellular oxidoreductases of lignindegrading system (LDS) can be performed by the 2, 2-azino-bis(3-ethylbenzothiazoline-6-sulphonic acid (ABTS) test as described previously by Saparrat et al. (2000). The chromogen ABTS is a very sensitive substrate that allows rapid screening of fungal strains producing the extracellular oxidative enzymes by means of a colourimetric assay at 420 nm (Saparrat et al. 2000). Each strain can be processed under controlled conditions at 30 °C in darkness, for definite period of time. A scale based on the colour intensity was used for interpreting the results: colourless, indicates no ABTS-oxidising activity; low colour intensity (+); moderate colour intensity (++); and high colour intensity (+++) indicating high ABTS-oxidising activity.

Mostly, indigenous fungi belong to phycomycetes, ascomycetes and deuteromycetes and are isolated from soil and water contaminated by paper and cardboard discharges. List of some of these fungi is given in Table 4.3.

4.4.1.3 Fungal Culture from Previously Inoculated Waste

It could be advantageous to use fungal culture procured from previously inoculated waste (Buswell 1994). Previously, inoculated waste is used here for waste treatment plant which is inoculated with fungal culture for treatment purpose. In the case of mushroom culture, culture can be collected from spent mushroom compost. Spent mushroom culture (i.e. *Pleurotus ostreatus*, oyster mushroom; *Lentinula edodes*, shiitake mushroom) is a by-product from commercial mushroom growers.

As with spent mushroom culture, pre-fruit body fungi can be obtained from commercial mushroom growers. It is an alternative to use colonised mushroom substrate before mushroom (fruit body) production.

Table 4.3 List of fungi distributed and isolated from water and soil samples of paper and pulp industrial effluent (Wahegaonkar and Sahasrabudhe 2005)

S. no.	Isolated fungi	Occurrence in water sample	Occurrence in soil sample
1	*Mucor* spp.	*	***
2	*Rhizopus* spp.	***	***
3	*Syncephalastrum racemosum*	***	***
4	*Chaetomium indicum*	*	*
5	*Corynascus spedonium*	00	*
6	*Eurotium herbariorum*	00	*
7	*Eupenicillium javanicum var. levitum*	00	*
8	*Pseudoeurotium multisporum*	00	*
9	*Acremonium* sp.	*	***
10	*Alternaria* spp.	*	***
11	*Aspergillus* spp.	00	***
12	*Aureobasidium indicum*	*	00
13	*Cephalosporium curtipes*	*	00
14	*Cephalosporium* sp.	*	00
15	*Cladosporium macrosporium*	*	00
16	*Curvularia* spp.	*	***
17	*Dendrostilbella indica*	00	*
18	*Drechslera* spp.	***	***
19	*Cunninghamella* spp.	*	***

"*" indicates the frequency of occurrence and "00" indicates non-occurring of fungi in the sample

4.4.2   Inoculation of Waste with Fungi

Another concern when using fungi for bioremediation of pulp and paper industrial wastes is how the inoculum is applied? Inocula of fungi can be applied in different ways depending upon the form of waste (liquid effluent/solid sludge) to be treated. Fungal inoculum can be applied by making the layers or in the form of homogenous mixture, if waste is present in sludge/solid form.

4.4.2.1   Inoculation of Sludge

Layering the Fungi with the Sludge/Soil

In this technique, fungi are applied in the form of layers in between the layers of sludge and soil. Layering the fungi with the waste would be easier and therefore more economical rather than mixing the fungi and waste.

Homogenised Mixture of Sludge or Soil with Fungi

In this technique, sludge, soil or solid waste is mixed with fungi in the form of homogenised mixture that degrades paper and pulp mill waste more effectively than soil and substrate layers.

4.4.2.2   Inoculation of Effluent

Fungus can be applied in five different ways in the effluent, if effluent is to be treated for remediation.

Agar Blocks

Fungi can be cultivated as monocultures on different media such as malt extract agar, potato dextrose agar, Czapek's agar and incubated for the definite period of time at the appropriate temperature. Then, agar blocks of equal size can be cut from the zone of active fungal growth and used for inoculation of paper, pulp and cardboard industrial effluent.

Spores Inoculation

Fungi can be grown on malt agar (2 %) medium in petri dishes for 72 h at the appropriate temperature. As sporulation occurs, spores can be separated from mycelium by filtration process and then suspended in sterile water. Paper and cardboard industrial effluent containing each flask can be inoculated with the equivalent of amount of spores/ml culture medium.

A novel method of using spores for bioremediation was developed by Childress et al. (1998). In this technique, spores can be encapsulated into alginate granules (1–3 mm in diameter) with Pyrex™ (non-nutritive filler such as saw dust, corn cob grits) for mycoremediation purpose. Encapsulated fungal spores can be stored in sterile petri dishes sealed with parafilm in refrigerator and should be checked regularly for the spore viability. The viability of spores is reported to be good in refrigerating condition as well as at room temperature. In future, this technique may be used for the mycoremediation of paper, pulp and cardboard industrial wastes contaminating soil and water.

Mycelium Inoculation

The strains can be grown in agitated cultures (200 rpm) in 100 ml flasks containing appropriate medium, for 3 days at 25–37 °C. Then the cultures can be centrifuged, and the mycelium can be ground in sterile distilled water. This mycelium can be used for the inoculation of flask containing pulp and paper mill effluent with equivalent amount of dry weight of mycelium per litre of effluent.

Enzymes Inoculation

Enzymes can be use directly, without immobilisation on different substrates, for bioremediation of pulp, paper and cardboard industrial effluent. However, these can treat a small quantity of waste effluents and sludge due to nonrecyclability of enzyme. In contrast to immobilised enzymes, these enzymes cannot be recycled and treat comparatively low amount of effluent.

The use of the purified enzymes showed limited activity in bioremediation. For example, MnP and LiP from *P. chrysosporium* can be used for bioremediation purpose. It was shown that only MnP had about 25 % decolourisation activity. Therefore, in vivo decolourisation, which attained more than 80 %, could depend on the production of other enzyme (Jaspers and Penninckx 1996). Due to this drawback, use of fungi is more beneficial for the mycoremediation purpose because the inoculated and adapted fungi can produce and secrete various enzymes as per the requirement for waste degradation.

Immobilised Mycelium or Fungal Enzymes

Fungal mycelium can be immobilised on various substrates such as polyurethane foam, agar gel, agarose gel, calcium alginate gel beads, magnetite, chitosan, etc. Thereafter, immobilised enzymes can be used for inoculation and remediation of pulp and paper mill effluents. Depending upon the requirement of bioremediation, substrate may vary. Immobilised enzymes could treat a large quantity of wastewater compared to soluble enzymes. Sometimes, degradation products absorb on the material used for immobilisation and accelerate the remediation process. For example, when phenols are oxidised by tyrosinases, quinones are obtained which adsorb rapidly and strongly onto chitosan beads (Muzarelli et al. 1994) and accelerate the degradation process.

Fungal enzyme such as laccase can be immobilised on carbon fibre electrodes using classical methods: physical adsorption, glutaraldehyde, carboimide and carbodiimde and glutaraldehyde for coupling laccase to carboxyl groups on carbon fibres. *P. chrysosporium* immobilised on cubes of polyurethane sponge of 1.5-cm sides or cylinders of polyurethane foam was found to produce ligninase under nitrogen sufficient condition (Chen et al. 1991).

Fungal enzymes LiP and MnP can be coated on nanoparticles for bioremediation purpose. Nano-assemblies of LiP and MnP are successfully fabricated and characterised on a flat surface as well as colloidal particles. During the assembly of enzymes on nanoparticles, a unique dynamic adsorption–desorption of enzyme layer occurs. Time, number of runs, nonaqueous media and drying of the enzyme layers have significant effect on the activity of assembled enzymes. A novel concept of using of silica nanoparticles can improve bio-catalysis. Formation of silica nanoparticles is based on electrostatic interaction between oppositely charged species. This may be successful strategy for treating the paper and cardboard industrial effluent in the coming future.

4.5 Bioreactors as Mycoremediation System

The pragmatic approach of biotreatment is analysed by way of using chemostat for enrichment and bioreactor for testing the relevance of the microorganisms in the treatment process. Various parameters and cultivation strategies need to be considered, for successful bioremediation of paper and pulp industrial wastes. In this context, various bioreactors have been tested for bioremediation of mill-made paper and pulp plants effluents and are discussed below.

4.5.1 Batch Treatment Process

Generally, for batch treatment, effluent treatment process is carried out in erlenmeyer flasks. Nagarathnamma and Bajpai (1999) conducted experiments in shake flasks and evaluated different fungi for bleach plant effluent treatment. They selected *Rhizopus oryzae* due to having high decolourisation efficiency and low sugar requirements during colour removal. During treatment, 92–95 % of the colour, 50 % of the COD and 72 % of the AOX were removed in 24 h at 25–45 °C and a pH of 3.5.

In a batch experiment, *Coriolus versicolor* was used to decolourise lignin-containing kraft E1-stage effluent, and it was found that both adsorption and oxidation proceeded best (Royer et al. 1985). Hence, the batch treatment was found to be successful in remediating the waste of paper, pulp and cardboard industries.

4.5.2 Continuous-Flow Systems

The results of using bench-scale bioreactor, in which aeration and mixing can be achieved with a diffuser that is placed at the side of the bottom of the reactor, provided a reasonable basis for setting out continuous-flow process consisting of a mixing tank, aeration basin and clarifier. In mixing tank, effluent would first be adjusted to the proper pH and mix with nutrient solution. The effluent would then go through a sedimentation basin where fungal solids would be removed and recycled to the head of the aeration unit. On the basis of this hypothesis, three continuous-flow laboratory-scale reactors have tested for bioremediation of pulp and paper mill waste (Prasad and Joyce 1991):

(a) System I consisted of a conventional oxidation basin for fungal cultivation. Mycelial mats of fungal species can be transferred to the reactor for decolourisation of paper mill effluent without aeration. This system could reduce effluent colour by at least 50 % for the first 6 days.
(b) System II consists of a vessel with increased height and decreased surface area. It possesses four baffles which divide the vessel into four compartments. Fungus culture, packed in wire bags, can be suspended in the middle of each zone with the supply of continuous aeration. Again, this system could reduce effluent colour by at least 50 % for the first 6 days.
(c) System III is similar to MyCoR reactor in which fungal mats can be clasped between circular wires, supported by outer central rings and securely fixed in a metallic frame, possessing the continuously rotating discs. In contrast to system I and II, system III has given the best results with a greater than 78 % total colour removal and 25 % COD reduction from extraction-stage paper mill effluent.

4.5.3 MyCoR Reactor

MyCoR reactor was described by Eaton et al. (1982), in which fungi are immobilised on the surface of rotating discs which are partially submerged during operation. These discs could be enclosed for additional oxygen supply and steam sterilisation.

In this reactor, at first growth of fungi has occurred till the nutrient is present, and then fungi become ligninolytic in nutrient-limiting condition. Simultaneously, decolourisation of the effluent occurs. The MyCoR process reduced colour in an alkaline stage spent liquor of pulp and paper industry by 80 % in less than 24 h in the laboratory.

In MyCoR process, only 40 % of the mycelium is in contact with substrate which seems to be the poor surface to volume ratio. Moreover, the mycelium is present in a thick layer that could result in deficient oxygen and nutrient supply and a lower general productivity.

4.5.4 MYCOPOR Process

In the MYCOPOR process, fungi are immobilised in polyurethane foam and used in a continuously trickling filter reactor to treat bleach plant effluent. This process lead to maximum decolouration, 50 % COD reduction and 80 % toxicity elimination of sulphate-based kraft pulp and paper mill effluent.

4.5.5   Fluidised-Bed Bioreactor

A fluidised-bed bioreactor operated in a continuous mode to treat and remediate bleach paper and pulp effluent. In this reactor, fungi can be immobilised on calcium alginate and other beads. This reactor has been tested and found to decolourise caustic stage effluent of pulp and paper industry by 69 % and reduce AOX of this effluent by 58 % at a retention time of 1 day.

4.5.6   Suspended Carrier Technology

This has been in use for a number of years for biological treatment of pulp and paper industrial waste and its bioremediation. Carrier particles such as polypropylene mats, polyester sponge-foam cubes, porous plastic foam cubes and ionically modified porous polyurethane granules have density close to that of water. These are used to minimise the energy required to keep the carriers suspended during treatment. Carrier particles should be engineered with more surface roughness and with surface charge as well as hydrophilicity to better accommodate fungi. Support matrixes are composed of particles that are kept in suspension by aeration. This has major advantages when compared to static biofilm systems that the risk for clogging of the stationary biomass support material with fines and fibres is eliminated.

This technology is used for treatment of bleached kraft pulp and paper mill effluent. At pH 7, 37 °C temperature and with hydraulic retention times of longer than 3.5 h, 55 % of COD removal was achieved. The suspended carrier treatment was also operated at pH 9 and 45 °C as well as at pH 7.0 and 50 °C, and 50 % reduction in COD in each case at a hydraulic retention time of 4 h was reported.

4.5.7   Moving Bed Biofilm

A full-scale treatment plant, based on moving bed biofilm system, is under construction in neutral sulphite paper mill for treating effluent. During a pilot-plant trial, 70 % COD reduction, 96 % BOD reduction and 98 % toxicity reduction were obtained at an organic load of about 25 kg COD/m^3 day. In this case, no clogging was found during the pilot-plant run.

4.5.8   Airlift Reactors

In airlift reactors, the fungi can be inoculated in the pelleted form. This strategy would facilitate recycling and the use of large amounts of fungal biomass. Mehna et al. (1995) used this reactor and a colour reduction of 92 % with a COD elimination of 69 % using pellets of *T. versicolor* strain in optimum conditions such as pH 4.5, temperature 30 °C and sucrose 7.5 g/l.

4.6 Factor Affecting Mycoremediation

There are specific conditions that need to be maintained for mycoremediation of paper and pulp waste and to operate the process at optimal efficiency. Any serious deviations could result in the complete shutdown of the system. In flask cultures, 80 % reduction in colour of paper and pulp mill effluent was observed after 6 days with *Trametes versicolor*. However, the same fungus, cultivated under optimal aeration in a laboratory fermentor with 0.8 % glucose plus 12 mM ammonium sulphate and at a controlled pH level of 5.0, reported to reduce 88 % colour units within 3 days. This emphasises the importance of culture conditions on the efficiency of decolourisation of effluent (Bergbauer et al. 1991) and simultaneously waste treatment. However, an effective biotreatment system must lead to the degradation of both high and low molecular mass compounds. White-rot fungi can achieve this effectively but possess complicated physiological demands for degradation of lignin-containing waste.

In shake flask, under optimal conditions (pH 4.0 and glucose as carbon source) *Trichoderma* sp. decreased the colour of kraft bleach plant effluent by 85 % and reduced the COD by 25 % after 3 days incubation in shake flasks. The maximum total decolourisation at pH 4.0 without an additional carbon source was 68.6 % after 3 days cultivation and therefore glucose stimulated decolourisation. Other carbon sources such as pulp and pith that are abundant and inexpensive increased the decolourisation after 6 days, since they were not metabolised immediately. The results of these studies again emphasised the importance of carrying out the treatments under strictly defined conditions (Prasad and Joyce 1991). However, there is not sufficient information about the nutritional, physiological and environmental parameters that influence the fungal degradation of lignin, cellulose, toxicants, dyes, etc.

Paper, pulp and cardboard industrial effluent possess highly coloured compounds, generally appeared due to the presence of lignin and lignin degradation compounds. Mycoremediation strategy requires the following factors that are well known to effect the delignification of paper and cardboard industrial effluent and hence decolourise the effluent.

4.6.1 Nitrogen

The onset of lignin degradation is triggered by nitrogen depletion in the medium. Lignin degradation is suppressed in nitrogen-rich medium possibly due to (1) high nitrogen content promotes rapid depletion of energy sources which are essential for lignin metabolism. (2) Nitrogen metabolism competes with lignin metabolism for fulfilling the requirement of same cofactors. (3) Nitrogen regulates the synthesis of one or more components of the lignin-degrading system. (4) Increased formation of biomass speeds up the rate of respiration and suppress lignin metabolism.

The increased nitrogen inhibits lignin degradation, while the increased carbohydrate supply stimulates. A medium containing an unlimited level of glucose and a limiting dose of nitrogen stopped primary growth and stimulated onset of ligninolytic activity during late stationary phase. In contrast to this, the ligninolytic activity considerably delayed in nitrogen-rich glucose-limited medium without adversely affecting the extent of fungal growth. The most effective nitrogenous suppressors of ligninolytic activity are ammonium chloride, glutamine, glutamate and histidine.

Mostly, white-rot fungi degrade lignin in better way in the presence of carbon to nitrogen ratio rather than the absolute levels of carbohydrates and nitrogen. Degradation of lignin by *Phanerochaete chrysosporium* and *Phlebia radiata* is dependent of C to N ratio. *Dichomitus squalens* and *Lentinus edodes* do not show any effect of C to N ratio. Thus, the effect of carbon nitrogen ratio is variable among the members of white-rot fungi.

In contrast to the physiological model proposed for *P. chrysosporium*, several commercially important and commonly occurring white-rot fungi produce higher ligninolytic enzyme activities in response to a nitrogen-rich medium. Moreover, laccase activities in *Lentinula edodes* and *Pleurotus ostreatus* and manganese-independent peroxidase (MiP) activities of *Bjerkandera* spp. were found to be enhanced by peptone (Kaal et al. 1995). Hence, mycoremediation of paper and pulp industrial waste requires appropriate amount of nitrogen depending on the type of fungi used.

4.6.2 Carbon Co-substrate

White-rot fungi cannot degrade lignin unless the co-substrate is supplemented simultaneously in the growth medium. Lignin is unable to serve as a growth substrate; the co-substrate is presumably the source of energy required for biosynthesis of individual components of ligninolytic system. Lignin degradation by white-rot fungi and *Aspergillus fumigatus* is dependent on the presence of readily metabolisable co-substrate such as glucose. The growth substrate is required because (1) energy recovered in lignin metabolism is too little to support growth and is at best only a marginal carbon and energy source for maintenance metabolism, (2) extent of ligninolytic activity is very low for fungal growth and (3) nature of carbon source also influences the ligninase activity by affecting the rate of H_2O_2 production.

An additional carbon source donates electrons which are cascaded down to the final electron acceptor. Sometimes the dye, present in the effluent, functions as final electron acceptor which helps in dye decolourisation of paper and cardboard industrial effluent. However, price of glucose may be a limiting factor in scaling-up projects. Addition of simple sugar to the wastewater source converts at least some of the biomass to glucose by brown-rot fungi. Simple sugar accelerates the production of phenoloxidases from the white-rot fungi which involve in breakdown a portion of biomass. As a result of the breaking down of biomass, thereby decreasing amount of phosphorous colour, odour, ammonia, suspended solids and sludge that are difficult to separate from paper and cardboard industrial effluent.

4.6.3   Oxygen or Air Tension

Lignin degradation is an oxidative process and needs the presence of oxygen at a partial pressure equal to that in the natural atmosphere. Increasing the O_2 levels in the culture has a strong activating effect on the rate of lignin degradation and production of LiP and MnP, which is generally optimal at high oxygen tension. Lignin is degraded much faster in the presence of oxygen than air. For example, *P. chrysosporium* reduced lignin of pulp when it is maintained in 100 % oxygen instead of air. The oxygen partial pressure has a profound effect on the rate and extent of lignin degradation by *P. chrysosporium*, but not on the growth of fungi.

4.6.4   Culture Agitation and Supplementation

Agitation is generally used to increase the rate of gas exchange between the atmosphere and culture medium. An initial period without agitation is needed to avoid severe inhibition of onset of degradation activities. Production of LiP and MnP is generally repressed by agitation in submerged liquid culture, while Lac production is often enhanced by agitation.

In agitated cultures, Tween 80 is essential for ligninase production. The supplementation of *P. chrysosporium* growth medium with emulsified vegetable oils, fatty acids and phospholipids sources enhanced the lignin peroxidase activity required for lignin, paper and pulp waste degradation. The extent of growth is good in both agitated and still fungal cultures.

Culture agitation results in pellet formation and strongly suppresses ligninolytic activity and slowing the waste bioremediation process. Agitation of cultures to increase the oxygen supply leads to increased formation of pellet due to the enclosing of fungi with pulp fibres which prevents optimal degradation possibly due to (1) the disturbance of the physiological state of cells on the pellet surface and (2) lower oxygen partial pressure which results in decomposition of pulp and paper fibres only on the exposed outer surface. Furthermore, a close physical association between fungi and paper, pulp and cardboard fibres is necessary. Fungi associated with paper, pulp and cardboard fibres are incubated in both agitated and still cultures during initial days. In contrast, if the pre-grown mycelial mat is agitated, it does not seem to affect lignin degradation.

In agitated submerged culture, the production of lignin peroxidases can be increased with various detergents and veratryl alcohol. This shows the effect of different supplementation on the production of LiP in agitated cultures.

4.6.5   Micronutrients

4.6.5.1   Manganese

The presence of manganese is necessary for degradation lignin-containing paper and cardboard fibres by white-rot fungi. Manganese is required for the functioning of an extracellular manganese peroxidase enzyme, first discovered in *P. chrysosporium*. Manganese peroxidase is present in numerous cultures of white-rot fungi. Mn regulates the production of LiPs and MnPs and ligninolytic activity of fungi. Ligninolytic activity increases with decreasing Mn(II) concentration.

4.6.5.2   Magnesium

Magnesium supplementation in the form of $MgCl_2 \cdot 6H_2O$ to bleach plant effluent increased the net consumption of chromophores by *C. versicolor* because these ions are known activators of many oxidases that play a role in delignification of paper, pulp and cardboard effluents by ligninases. Besides this, magnesium ions increased the degradation of chromophores indicating the involvement of Mg ions. For instance, addition of magnesium ions initiates the removal of paper mill effluent colour from its initial level.

4.6.5.3   Sulphur

Addition of sulphur also induces ligninolytic activity in *P. chrysosporium*. This shows the positive effect of sulphur on the remediation of paper and cardboard industrial waste.

4.6.6   Hydrogen Ion Concentration

Paper and cardboard waste degradation is based on lignin degradation which is quite sensitive to pH. Hence, adequate buffering is essentially required to control pH during lignin decomposition, consequently paper and cardboard waste degradation. The medium pH was found to be critical for decomposition of paper and cardboard industrial effluent. The optimum culture pH for delignification by *P. chrysosporium* is observed in the range of 4.0–4.5, with marked suppression above 5.5 and below 3.5. The pH requirement varies for growth and waste degradation. The optimum pH, however, for the growth was somewhat higher than that for lignin degradation. When pH is drifted from the optimum 3.5–5.5 range, the rate of CO_2 released is increased in *P. chrysosporium*.

4.6.7 Toxic Substances

The presence of toxic substances inhibits the mycoremediation process. Sometimes, toxic compound is generated during the treatment with fungi. For example, the toxic amines are generated when azo- and nitro compounds are reduced by fungi which are difficult to degrade. Further, these compounds create many difficulties in paper and pulp mill waste degradation.

4.7 Mycoremediation of Paper and Cardboard Industrial Wastes

Characteristics of waste must be considered for mycoremediation of different types of paper, pulp and cardboard industrial wastes. Wastes generated by these industries depend on the technique of making paper. Generally, all pulp and paper industries can be divided into three main categories on the basis of processing of raw material:

1. Handmade paper industries, process paper by hand or simple machinery
2. Kraft paper industries, process paper mechanical or chemical method
3. Cardboard industries, recycle wastepaper

These can be further divided into small-scale and large-scale categories on the basis of their production capacity.

4.7.1 Handmade Paper and Pulp Industries

Handmade paper industry is not the topic of discussion of this millennium but is an ancient art of making paper by hand still exist in many part of the world. According to American Paper and Pulp Association "Handmade Paper is a layer of entwined fibres, held together by the natural internal bonding properties of cellulose fibres lifted by hand, sheet by sheet on moulds in suspension of fibres in water with or without sizing."

Table 4.4 Difference in handmade paper and mill-made paper industries

S. no.	Characteristics	Handmade paper	Mill-made paper
1	Investment on plant on per ton per annum production	12,450	24,850
2	Employment generation on invest of each Rs. one crore	980	130
3	Employment generation of each 1,000 ton of production	1,050	25
4	Import of machinery	NIL	35 %
5	Resource input of production per each ton		
(a)	Forest based raw materials like bamboo and wood	NIL	2.5–3 ton
(b)	Coal	0.085 ton	1.50 ton
(c)	Electricity	150 kwh	2,000 kwh
(d)	Water	38 cu mts	300 cu mts
(e)	Chemicals	0.0051 ton	0.8 ton

Source: Khadi and Village Industries Commission, India

This ancient art of making paper by hand or simple machinery, i.e. handmade paper was discovered in China. Later it spread in Korea, Japan, Asia and Persia. In India, the first handmade paper was manufactured in the year 1159 AD. The first paper mill was set up at Serampur, West Bengal, in the year 1812 which used grass and jute as raw material for making paper. It still exists in India, many European and American companies, contributing in production of high-grade export quality paper.

As there has been phenomenal growth in the export market for Indian handmade paper and its products, especially in the developed countries like the United States of America, West Germany, European Countries, Australia, etc., India is being looked upon as the country with the maximum growth potential in handmade industry. In India, there are about 685 units of large-scale and small-scale handmade paper working all over the country. There are around 300 kagzi units still working in Sanganer (Jaipur) both large scale and small scale. These produce around Rs 21 crore worth of papers, providing full-time employment to 10,000 persons in the rural areas (Khadi Industries Village Commission report, Rajasthan, India). In spite of having competition in the global market, there are definite opportunities for smaller, local firms satisfying specific needs of paper. Differences in handmade paper and mill-made paper industries are depicted in Table 4.4.

The finest handmade papers are made from pure rag pulp, usually linen and cotton, which are washed, boiled and beaten to macerate the fibres. These fibres are then suspended in water where they can be lifted out by the papermaker using a mould and deckle. A mould is a screen of some sort, supported by a frame, which allows the surplus water to drain after dipping the fibres from the vat. A deckle is another frame on top of the mould which keeps the fibres from washing over the edges.

Handmade paper industries (HMP) utilise various lignocellulosic and cellulosic raw materials. On the basis of using raw material, these are further divided into two categories: (1) agro-residue-based HMPs (these industries use agroresidues for making paper such as hemp fibre, banana pulp, jute pulp, etc.) and (2) textile waste-based HMPs (these industries use cotton rags, hosiery rags, procured as tailor cuttings from textile industries for making paper). For processing of paper, these industries also use a large amount of water and discharge it into the nearby water bodies or in the agricultural fields.

4.7.1.1 Wastes of Handmade Paper Industries and Pollution

Depending on the type of raw material, handmade paper industries generate waste of different characteristics. These industries which use linen or cotton rags or mill broke as raw material having different characteristics compared to those that use agricultural residues and mill broke. Cotton rags are rich in cellulose; however, agro-residues were found to possess lignin and cellulose. Therefore, the processing of these raw materials differs from each other which lead to the generation of effluent with different characteristics. Water used in several steps of making paper is discharged as effluent-possessing chemicals used for the processing of raw materials.

Agro-based handmade paper and pulp industries adopt the procedure of processing the lignocellulosic pulp residues similar to mill-made paper industries. Hence, effluent characteristics are similar to the characteristics of kraft pulp and paper industries. However, the amount of effluent generated by these industries is low because all processes are done by hand or simple machinery.

Cotton- and hosiery rag-based handmade paper industries utilise various chemicals and huge amount of water which is discharged in the form of effluent. Earlier handmade paper used to come only in white colour, but now available in different colours and designs. Many small-scale industries are emerging in India, which are using various chemicals and chemical dyes for making paper.

Generally, these are considered as ecofriendly industries only on the basis of physicochemical characteristics. Although these industries use dyes and chemicals, there are few reports available on the effluent discharge and pollution-causing potential of these industries. These industrial discharges were found to possess effluent BOD, COD, pH and other parameters much above the discharging limit and having mutagenic potential possibly due to the use of dyes for colouring paper (Kulshreshtha et al. 2011).

These industries discharge two types of wastes: (1) effluent and (2) sludge. After making paper dyes, pulp residues containing effluent are discharged to nearby fields or in the water bodies which impart colour to the water body. The effluent colour may increase water temperature and decrease photosynthesis, both of which probably lead to a decreased concentration of dissolved oxygen.

Leftover or unused pulp residues are generated in the form of sludge and accumulated in the close proximity of industries or in the industrial drain and cause soil or water pollution.

4.7.1.2 Effluent Treatment

These are exempted from the list of pollution-causing industries, and therefore, effluent treatment of these industries is meagerly reported. Pulp fibres from handmade paper and pulp industries were separated out by filteration process and used for the cultivation of mushroom along with the sludge of unused pulp residues (Kulshreshtha et al. 2010b).

4.7.1.3 Sludge Treatment

Recently, sludge of these industries is utilised for mushroom cultivation, a macrofungi, for its bioremediation from the environment (Kulshreshtha et al. 2010b). Mushrooms are higher fungi capable of degrading lignin and cellulose efficiently due to the production of cellulolytic and ligninolytic enzymes. Mushroom is not only capable of bioremediation of waste but also provides a highly proteinaceous food. *Pleurotus florida* is a mushroom species which is recently used for the bioremediation of sludge generated by handmade paper and pulp industries.

The pulp and paper industrial sludges may mix either with effluent of same pulp and paper industry or with water. This mixing will provide thick slurry of composite waste (sludge and effluent). Generally, pulp and paper sludge is obtained in wet condition; therefore, no presoaking is required like wheat straw. This slurry of waste treats with 1.25 ml/l of 40 % formalin solution (v/v) and 0.75 g/l (w/v) of antifungal substance bavistin for 18–24 h to remove unwanted organisms. Pulp fibres are collected by filtering the slurry by a clean cloth and spread on a clean surface, to remove excessive water for maintaining 80 % humidity. Further, these pulp fibres sterilise by steam for 1 h to overcome the disturbance of other microorganisms. After cooling, wet pulp inoculates with spawn and pack in porous polythene bags for incubation. Incubation temperature varies per the requirement of mushroom species. When *Pleurotus* mycelium fully colonises the substrate, polythene bags can be removed and substrate bundles can be hanged. These bundles can be sprayed with water twice a day to maintain humidity, and time can be recorded for pinhead formation, immature fruiting body and mature fruiting body formation.

After obtaining the mature fruiting bodies (basidiocarps), mushroom fruiting bodies can be harvested and analysed in the laboratory for its nutritional characteristics, toxicity and genotoxicity. In our recent investigation, it was found that mushroom fruiting bodies of *P. florida* do not possess basepair and frameshift mutagens (Kulshreshtha et al. 2011). However, these basidiocarps cannot be considered safe for consumption until found free of other toxicants and metallic content. Preliminary results obtained with mushroom cultivation on handmade paper and pulp industrial wastes indicate the extended application of this mushroom species not only for treatment but also for proteinaceous food.

4.7.2   Mill-made Paper and Pulp Industries

About 1800–1860, all work sequences previously performed by hand were mechanised for making paper. This included the rag preparation, the use of fillers, pulp beating, the paper machine with its various parts and the machines required for finishing the paper (the headbox, wire section, press section, dryer section and units for reeling, smoothing and packaging). The paper industry has came a long way, and evolution of new sheet-forming principles (with fluid boundaries between paper and nonwoven fabrics) and chemical pulp processes have been the main process improvements.

The heavy demand for the paper helps in steady expansion of paper industries. When India became independent in 1947, there were 17 paper mills with an installed capacity of less than 0.14 million TPA (tonnes per annum). The organised growth of this sector began in 1951, with many small and medium paper mills being set up due to the government policy. The annual increase in paper production since independence has been in the range of 0–13 % with a current dead capacity of 1.1 million tonnes. At present, there are an estimated 525 pulp and paper mills with a total installed capacity of around 6.25 million TPA. However, the situation on the global market, i.e. increased demand, above all in the Third World, trends in chemical pulp prices and problems of location, is again raising capital intensity and encouraging the formation of big company groups with international operations. On the basis of pulping procedures, these are of various types as mentioned in Table 4.5.

4.7.2.1   Wastes of Mill-made Paper Industries and Pollution

Mill-made paper industries are also producing two types of wastes: (1) effluent and (2) sludge. Generally, these industries are involved in making paper by delignification of wood. Besides this, these industries consume large amount of water in different steps of papermaking process which reappear in the form of effluent. The most significant sources of pollution among various process stages are wood preparation, pulping, pulp washing, screening, washing, bleaching and paper machine and coating operations. Among the processes, pulping generates a high-strength wastewater especially by chemical pulping containing wood debris and soluble wood materials. Pulp bleaching generates most toxic substances as it utilises chlorine for brightening the pulp. Moreover, depending upon the type of the pulping process, various toxic chemicals such as resin acids, unsaturated fatty acids, diterpene alcohols, juvaniones, chlorinated resin acids and others are generated during pulp and papermaking process (Pokhrel and Viraraghavan 2004). All these chemicals are present in the effluent generated by these industries which is discharge to the adjacent water bodies or soil after use, and therefore, these industries pollute not only water but also soil. These pulp and paper industries are one of the 17 most polluting industries listed by the Central Pollution Control Board

Table 4.5 Different pulping method-based paper and pulp industries and risks associated with them

Pulping process	Details	Risks
Sulphate pulping	Alkaline sulphate process independent of wood species yields high-fibre properties	Bleaching with chlorine-containing chemicals such as (elemental or gaseous) chlorine, hypochlorite or with chlorine dioxide Dissolved lignin extracted with alkali
Sulphite pulping	Several special paper qualities, e.g. tissue, wood-free printing and writing papers, grease proof papers, etc., are prepared Suitable raw material for this is spruce Pine, birch and other hardwood species are not good Sulphite cooking is possible using Ca, Mg, Na or NH4 as a base chemical in cooking the liquor Two methods: acid sulphite process or bisulphite cooking process	The acid sulphite pulping process waste liquor is normally burned in a recovery boiler in an oxidative atmosphere with about dry solid content of 55–57 %. Except for when using a sodium-based waste liquor, there is not a chemical smelt layer on the bottom of the boiler and no risk for smelt/water explosions like in the case of a black liquor recovery boiler
Recycled pulping (RCF) and deinked (DIP) pulps	Releasing of ink during pulping is the cause of colour and toxicity in the effluent	Exposure to some hazardous chemicals (e.g. peroxide) No risks stemming from a recovery boiler
Mechanical pulping (groundwood (GW), thermomechanical (TMP), chemithermomechanical (CTMP) and bleached chemithermomechanical (BCTMP) pulps	Adjacent sulphate pulp mill recovery boiler may be used in cross recovery for impregnated chemicals	Exposure to some hazardous chemicals, e.g. peroxide No recovery boiler risks

(CPCB). Most mills probably release suspended solids, such as fibre and bark, organic matter, which increase the biological oxygen demand, chemical oxygen demand and colour and suspended solids.

Most of the solid waste from the pulp and paper industry consists of "mill residues", which comprises bark, wood residues, refuse (pulp, paper and cardboard), ash from combustion facilities and sludge from treatment of process water and deinking. The major contributors of solid wastes in pulp and paper mill effluents are bamboos and wood dust, line sludge, coal ash and fly ash and other rejects. Discharging of sludge and effluent of mill-made paper industries lead to the pollution of soil and water.

4.7.2.2 Effluent Treatment

Effluent treatment of different types of mill-made paper industries is reported by various scientists till date. Two-step bioreactor is designed for the treatment of pulp and paper industrial effluent by Singh and Thakur (2006) in which at first effluent was treated in anaerobic condition and then treated by fungi *Paecilomyces* sp. in the second bioreactor. This technique leads to reduction in colour (95 %), AOX (67 %), lignin (69 %), COD (75 %) and phenol (93 %) by third day when 7-day anaerobically treated effluent was further treated by fungi. Fungi used for treatment of effluent of different types of pulp and paper industries and effect of fungal treatment on different parameters are given in Table 4.6.

4.7.2.3 Sludge Treatment

Generally, the sludge of this type of industries is used for landfilling purpose as well as composting purpose. However, there is scarcity of literature on the treatment of mill-made paper industrial sludge and its bioremediation with fungi.

The use of chemithermomechanical pulp for the cultivation of mushroom is reported (Sivrikaya et al. 2002). Chemithermomechanical pulp is used as a substrate by few mushroom cultivators, who used to cultivate mushroom in the mushroom farm. During the cultivation period, both phytohormones 2.4-D and PS A6 can be used. Mature fruiting bodies of *Pleurotus sajor-caju* can be collected and used for examination metals, respectively. Postharvesting mature fruiting bodies of basidiocarp can be cleaned, dried in a drying oven at 103 °C and homogenised by grinding. All the edible parts of the mushroom basidiocarp can be stored in closed containers at room temperature after pretreatment.

Mushroom basidiocarps can be tested for the presence of toxic metals. Metal analysis of basidiocarp can be done by mixing 2 g of basidiocarp with 20 ml concentrated HNO_3, and this mixture is allowed to stand overnight. The mixture is heated carefully on a hot plate until the production of red NO_2 fumes ceased, and then samples are analysed for metallic content. These metals can be most conveniently determined by either the flame emission (FE) or atomic absorption (AA) methods.

Metal analysis of postharvesting mature fruiting bodies revealed the presence of trace element contents in mushroom *Pleurotus sajor-caju* cultivated on chemithermomechanical pulp with phytohormones PS A6 and 2.4-D. Fruiting bodies were found to possess many metals above the acceptance level which makes it unsuitable for consumption (Sivrikaya et al. 2002). Hence, fruiting bodies were found to possess many metals above the acceptance level which makes it unsuitable for consumption.

Table 4.6 Mycoremediation of paper and pulp industries effluent and parameters measured

Type of paper and pulp industry	Effluent type	Fungi used	Parameters reduced	References
Bagasse-based paper and pulp mill	Dissolved bagasse lignin, raw black and alkali stage liquors	Aspergillus foetidus	90–95 % of initial colour, COD removal	Sumathi and Phatak (1999)
	Dissolved bagasse lignin, raw black and alkali stage liquors	Schizophyllum commune	90 % colour reduction, 70 % BOD and 72 % COD reduction	Belsare and Prasad (1988)
Bagasse pith-based pulp and paper mill	Black liquor	Ceriporiopsis subvermispora	90 % colour, 45 % COD, 62 % lignin, 32 % AOX, 36 % EOX (extractable organic halide)	Nagarathnamma and Bajpai (1999)
Paper and pulp mill	Coloured effluent	Gliocladium virens, Phanerochaete chrysosporium, Coriolus versicolor, Trametes versicolor	42 % initial colour removal	Murugesan (2003)
	Black liquor contain colour, highly toxic chlorinated lignin-degradation products	Fomes lividus and Trametes versicolor	COD reduced to 1,984 mg/l (59.3 %), 68 % decolourisation with T. versicolor, 103 % inorganic chloride with T. versicolor	Selvam et al. (2002)
	Black liquor	Marine-derived fungus NIOCC #312	Better than NIOCC #2a strain of marine fungus in decolourisation of bleach plant effluent due to production of peroxidases	Raghukumar et al. (2008)
		Marine-derived fungus NIOCC #2a	Decolourisation of bleach plant effluent due to production of laccases	Raghukumar et al. (2008)
		Aspergillus niger	81 % decolourisation	Kannan et al. (1990)
		Trametes versicolor	93 % decolourisation	Bajpai et al. (1993) and Martin and Manzanares (1994)
		Trichoderma spp.	85 % decolourisation	Prasad and Joyce (1991)

(continued)

Table 4.6 (continued)

Type of paper and pulp industry	Effluent type	Fungi used	Parameters reduced	References
		Pleurotus sajor-caju, P. platypus, P. citrinopileatus	66.7 % decolourisation, 61.3 % COD reduction, inorganic chloride reduction	Raghunathan and Swaminathan (2004)
		Merulius aureus syn. *Phlebia* sp., *Fusarium sambucinum* Fuckel MTCC 3788	78.6 % colour, 79.0 % lignin and 89.4 % COD reduction	Malviya and Rathore (2007)
Bleached kraft pulp mill	Processing Eucalyptus globulus	*Trametes versicolor*	74–81 % COD	Freitas et al. (2009)
		Phanerochaete chrysosporium	Reduction in BOD and COD	Freitas et al. (2009)
		Rhizopus oryzae	Biodegradation of organic compounds	Freitas et al. (2009)
Bleach plant effluent	Black liquor contain colour, highly toxic chlorinated lignin–degradation products	Marine fungi *Sordaria fimicola* (NIOCC #298)	65–75 % decolourisation due to producing MnP	Raghukumar et al. (1996)
		Marine fungi *Halosarpheia ratnagiriensis* (NIOCC #321)	65–75 % decolourisation due to producing MnP	Raghukumar et al. (1996)
Eucalyptus paper pulp	Pitch-causing lipophilic compounds present	Laccase from the basidiomycete *Pycnoporus cinnabarinus*	Free and conjugated sitosterol removed	Gutiérrez et al. (2006)
Spruce pulp	Pitch-causing lipophilic compounds present	Laccase from the basidiomycete *Pycnoporus cinnabarinus*	Resin acids, sterol esters and triglycerides removed	Gutiérrez et al. (2006)
Flax pulp	Pitch-causing lipophilic compounds present	Laccase from the basidiomycete *Pycnoporus cinnabarinus*	Sterols and fatty alcohols removed	Gutiérrez et al. (2006)
Pine kraft black liquor	Black liquor	*Polyporus versicolor*	70 % decolourisation	Marton et al. (1969)
Kraft pulp and paper mill	Black liquor	*Aspergillus* spp.	Decolourisation	Dutta et al. (1985)
		Phanerochaete chrysosporium	Decolourisation	Thomas et al. (1981)
		Tinctoporia sp.	Decolourisation	Fukuzumi (1980)
Kraft pulp mill using chlorine bleaching	Black liquor	*Rhizopus oryzae*	Decolourisation and detoxification	Nagarathnamma and Bajpai (1999)

4.7.3   Cardboard Industries

Cardboard is a thick layer of paper made by shortened cellulosic fibres produced due to repeated recycling. Cardboard is a heavy wood-based type of paper notable for its stiffness and durability. It is used for a wide variety of purposes generally for packaging purposes. The growing demand of paper led to a greater mechanisation processes and materials. This produces a thick sheet of paper of much poorer quality. Due to shortage of rags, other papermaking materials are using nowadays which are available in plentiful amount and economic. In this context, various types of printed, nonprinted, coloured, laminated, non-laminated, paper and paperboards are collected for removing the waste from environment and recycle it into cardboard. These industries are based on the recycling of wastepaper and cardboards and laminated board. Recycling of newspaper and wastepaper and boards saves trees.

It is first invented in China in the fifteenth century. The first commercial cardboard box was produced in England in 1817, and first machine for producing large quantities of corrugated cardboard was built in 1874 by G. Smyth, and in the same year, Oliver Long improved upon Jones' design by inventing corrugated cardboard with liner sheets on both sides.

4.7.3.1   Wastes of Cardboard Industries and Pollution

These industries recycle the wastepaper, however, pollute the environment by discharging two types of wastes: (1) effluent and (2) sludge. Mostly, cardboard is prepared from unused paper and pulp fibres without using any type of chemicals for delignification of pulp fibres. However, these are considered as pollution-causing industries due to having low pH, high BOD and COD. Effluent of cardboard industries is highly coloured and generally possesses brown colour due to release of natural colour of lignin. Another possibility of brown-coloured effluent is the release of inks, dyes and chemicals in the pulping liquor during the pulp-making process. These wastepaper-based mills consume a large amount of raw water which is discharged as effluent. These industrial effluents are characterised by a high organic load in the form of shortened pulp fibres. The brown colour of the effluent is due to the release of ink, dyes and chemicals in the water during pulping used for making papers. The discharge of untreated effluent from these industries into water bodies causes poor water quality, and the brown colour of untreated effluent is detectable over long distances. Discharging of untreated effluent can cause eutrophication due to having organic nature. Due to economical and environmental reasons, consumption of water has decreased overtime, and today, there exist several zero-discharge recycled paper and cardboard mills. However, closing a water system leads to several different problems, such as increased demand for retention aids, decreased product quality and reduced felt life and corrosion.

Cardboard industries also cause pollution due to repeated recycling which results in shortening of pulp fibres due to repeated recycling. In the last stage of making paper, cardboard and pulp fibres become too short to be recycled which accumulate in adjacent area of industries. In this case, accumulation of nonrecycled left over pulp residues is the main cause of soil pollution, and biodegradable organic pulp residue waste will produce enormous odour in the area.

4.7.3.2 Effluent Treatment

Closing of drainage system of cardboard industries gives rise to problem of pungent odour, blockage, etc. Hence, it is necessary to find out the solutions to overcome these problems. One method is to reduce the compounds in the wastewater, on which the fungi feed on. As fungi used all the pollutants as source of energy and carbon, these are removed from the effluents.

An anaerobic/aerobic biological process for treating wastewater from a recycled paper mill was evaluated in both mesophilic and thermophilic conditions in laboratory-scale experiments. The pilot plant trials were carried out at Munksjo Lagamill AB, Sweden. Effluent can also be treated in an in-mill biological treatment of wastewater by using fungi, and it has been shown to be an attractive alternative for mycoremediation of wastewater from recycled paper/board production. Presently, there is no mycoremediation technology for the wastewater of paperboard or cardboard mill due to which Indian cardboard mills discharge brown-coloured water, treatment of which is meagerly reported. This untouched problem opens a good opportunity of research in the field of mycoremediation of waste.

4.7.3.3 Sludge Treatment

Sludge of these industries contains nonrecyclable pulp residues and is reported to be used for the cultivation of mushroom. Sludge is processed and inoculated as mentioned earlier for handmade paper industries. The results of this study show that when waste is used alone, mushroom cultivation required very long time as compared to straw. However, the same sludge is processed and mixed with wheat straw in equal amount; mushroom fruiting bodies can be obtained in short time with increased biological efficiencies (Kulshreshtha et al. 2010b).

4.8 Degradation of Paper, Pulp and Cardboard Industrial Pollutants

New treatment technologies must be designed to degrade those chemicals that pose the greatest threat to human health that are toxic and/or mutagenic, have a tendency to bioaccumulate, and are difficult to degrade. Using these criteria, the degradation of all mutagenic compounds, resin acids, chlorinated phenols and chlorinated aliphatic hydrocarbons should be the focus of new treatment processes.

4.8.1 Lignosulphonics

Generally, pulp, paper and cardboard industrial effluents are brown in colour which is due to the presence of lignin and its degradation products formed by the action of chlorine on lignin. This is a serious cause of concern from aesthetic and biological point of view. In the kraft process of making paper, lignin is converted to thio- and alkali lignin due to ligno-sulphonate in the sulphate process.

4.8.1.1 Mycoremediation of Lignosulphonics

These compounds give brown colour to the effluents. Many fungi can be used for the treatment of black liquor and mentioned in Table 4.6.

4.8.2 Heavy Metals and Inorganic Compounds

The main sources of Al, Cu, Cr, Ti, Fe, Mg and Zn in pulp and paper mill effluents are the chemicals used in pulping, additives used in papermaking.

4.8.2.1 Mycoremediation of Heavy Metals

Many fungi have been used for the bioremediation of heavy metals for long times. These include several fungi (*Aspergillus* spp. and *Rhizopus* spp.) and mushroom species (*Agaricus* spp., *Pleurotus* spp.). However, there is scarcity of report on the mycoremediation of heavy metals from paper and pulp industrial effluents. Metals absorbing fungi provide us a good opportunity of research to use them for mycoremediation of heavy metal content of paper and cardboard industrial effluent and sludge.

4.8.3 Dioxin and Dibenzofurans

Dioxins are the family of halogenated dibenzo-*p*-dioxin congeners. Dioxins are toxic, and environmentally persistent compounds arise as trace pollutants in the manufacture of pulp and paper bleaching which is the most considerable pollutant for bioremediation. Dioxins and furans are persistent and tend to accumulate in sediments and in human and animal tissues.

4.8.3.1 Mycoremediation of Dioxins

The metabolism of 2,7-dichlorodibenzo-*p*-dioxin by *Phanerochaete sordida YK-624* has been reported (Takada et al. 1996). Moreover, the oxidation of dibenzo-*p*-dioxin by fungal enzyme lignin peroxidase is also reported by Joshi and Gold (1993).

4.8.4 Organochlorine Compounds (AOX)

Absorbable organic halogens (AOX), total organochlorine (TOCl) and tertachlorinated dibenzodioxins (TCDD) are the most hazardous compounds of pulp and paper mill effluents, which besides being toxic, mutagenic, persistent and bioaccumulating are also known to contribute much to the pollution load in the effluents. These compounds are produced mainly by the reactions of residual lignin present in the wood fibre and chlorine used for bleaching which is a matter of great concern in the pulp and paper industry.

4.8.4.1 Mycoremediation of AOX

Mycoremediation of AOX from paper and pulp industrial effluent and degradation of chlorinated compounds such as mono-, di-, tri-, penta- and chlorophenols has been done by fungal isolates that were obtained from different effluent streams of pulp and paper industry and soil irrigated with this wastewater.

4.8.5 Resin Acids

The wastewaters of the pulp mills are found to contain measurable concentrations of resin acids and resin acid biodegradation products. Resin acids are the group of deterpenoid carboxylic acids which are toxic to fishes in recipient waters. Commonly found resin acids can be classified into two types: abietanes (abietic, dehydroabietic, neoabietic, palustric and levopimaric acids) and pimaranes (pimaric, isopimaric and sandaracopimaric acids). The most abundant resin acids are two abietanes, dehydroabietic acid (DHA) and abietic acid (ABA). The toxicity of chlorinated compound increases with increasing number of chlorine atoms on organic compounds. Recent studies indicate that natural resin acids and transformation products may accumulate in sediments and pose acute and chronic toxicity to fish. Several resins and their biotransformation compounds have also been shown to bioaccumulate and to be resistant to biodegradation than the original material.

4.8.5.1 Mycoremediation of Resin Acids

Resins are considered to be readily biodegradable. However, their remediation has been shown to vary. Until recently, the microbiology of resin acid degradation has received only scant attention. There is no conclusive evidence that fungi can completely degrade these compounds. In contrast, a number of bacterial isolates and activated sludge (Kostamo and Kukkonen 2003) have recently been described which are able to utilise dehydroabietic and isopimaric acids as their sole carbon source.

4.8.6 Lignin Cellulose Degradation

Lignin is a highly complex, stable and irregular polymer which is structurally similar to the resins. The lignin-degrading system has broad substrate specificity, includes a large range of oxidoreductases and hydroxylases such as laccases and high redox potential LiP, MnP, VP and others (see Sect. 4.3). Cellulose is a polysaccharide consisting of a linear chain of several hundred to over ten thousand $\beta(1 \rightarrow 4)$-linked D-glucose units. Cellulose is also an organic compound with the formula $(C6H10O5)_n$, possessing a complex structure, but it is easily degradable by cellulolytic fungi. When it is present in complex form with lignin, then its degradation occurs only after the removal of lignin.

4.8.6.1 Mycoremediation of Lignin and Cellulose

Sporotrichum pulverulentum was found to degrade lignin in the presence of phenol oxidase enzyme. The lignin degradation abilities of wild type, a phenol oxidase-less mutant and a phenol oxidase-positive revertant were compared by Ander and Eriksson (1976) to determine if phenol oxidase activity is necessary for lignin degradation by white-rot fungi. They found that the phenol oxidase-less mutant was unable to degrade kraft lignin. The phenol oxidase-positive revertant, however, regained the ability of the wild type to degrade kraft lignin.

Besides white-rot fungi, some non-white rots are also able to degrade lignin. For example, *Aspergillus foetidus*, an isolated fungus, reduced lignin content along with decolourisation, indicating strong correlation between the decolourisation and lignolytic processes. *Paecilomyces* sp. exhibited significant reduction in lignin (66 %) in sequential bioreactor supplemented with 1 % carbon and 0.2 % nitrogen source in 6 h of retention time (Singh and Thakur 2004).

4.8.7   Phenolic and Chlorophenolic Compounds

Chlorinated phenolic compounds are the most abundant recalcitrant wastes produced by the paper and pulp industry which accumulated in the effluents after secondary treatments. These compounds are produced upon the partial degradation of lignin during bleaching process. These compounds pose a big concern to human and environmental health due to their high toxicity to a wide range of organisms.

4.8.7.1   Mycoremediation of Phenolic and Chlorophenolic Compounds

White-rot fungi are a group of organisms very suitable for the removal of chlorinated phenolic compounds from the environment. Chlorophenols act as substrates for laccase and Mn-peroxidase that requires hydrogen peroxide and Mn (II) for reaction, which can react with all the chlorophenols including the most recalcitrant pentachlorophenol (PCP). The free-cell cultures of white-rot fungus *Panus tigrinus*, when adapting to high concentrations (up to 2,000 mg/l) of 2,4, 6-trichlorophenol (2,4,6-TCP), transformation of a mixture of 2,4-DCP, 2,4,6-TCP and PCP achieved (Leontievsky et al. 2000). The lifetime and reactivity of the biomass can be achieved by immobilisation of free cells on to a support matrix, which produces higher cell densities, higher and time-extended enzyme activities, greater enzyme stability and subsequently a reduction in treatment cost.

Exposure of *Trametes versicolor* to increasing amounts of pentachlorophenol (PCP) from 200 to 2,000 ppm leads to acclimatisation of the fungus to these toxic pollutants. Free-cell cultures of acclimatised *Trametes versicolor* were compared with cultures immobilised on nylon mesh for transformation of PCP and 2,4-dichlorophenol (2,4-DCP). A total addition of 2,000 ppm of 2,4-DCP and 3,400 ppm PCP were removed from the immobilised cultures with 85 % of 2,4-DCP and 70 % of PCP transformed by enzymes (laccase and Mn-peroxidase), 5 % 2,4-DCP and 28 % PCP adsorbed by the biomass and 10 % 2,4-DCP and 2 % PCP retained in the medium at the termination of the fermentation. In contrast free-cell cultures in the same medium with the same addition regime of PCP and 2,4-DCP transformed 20 % 2,4-DCP and 12 % PCP by enzyme action, adsorbed 58 % 2,4-DCP and 80 % PCP by the biomass and retained 22 % 2,4-DCP and 8 % PCP in the medium. This shows that immobilised acclimatised fungus can remove chlorophenols more efficiently than free acclimatised fungal culture (Selvam et al. 2002).

Lamar et al. (1993) compared the abilities of *P. chrysosporium*, *P. sordida* and *Trametes hirsuta* to degrade PCP and found that *P. chrysosporium* is best in degrading PCP. Extracellular peroxidases play an important role in the degradation of chlorophenols by *P. chrysosporium*. Extracellular laccases and peroxidases carry out the first productive step in the oxidation of chlorophenols, forming *para*-quinones and consequently releasing a chlorine atom. Further degradative steps involving several enzymes and highly reactive, nonspecific redox mediators produced by the fungus render it capable of efficiently degrading several toxic compounds. *Paecilomyces* sp. exhibited significant reduction in phenol (40 %) in sequential bioreactor supplemented with 1 % carbon and 0.2 % nitrogen source in 6 h of retention time (Singh and Thakur 2004).

Recently, the degradation of 2-chlorophenol (2-CP), 2,4-dichlorophenol (2,4-DCP), 2,4,6-trichlorophenol (2,4,6-TCP) and pentachlorophenol (PCP) was found to be bioremediated by white-rot fungus *Trametes pubescens* (González et al. 2010). These fungi may be used possibly for the bioremediation of paper and pulp industrial wastes in the near future.

4.8.8 Colour

Paper and pulp industries discharge highly coloured effluents. The brown colour of the effluent can be noticed many kilometres away from discharging point. The reason of this colour is possibly the lignin and its derivatives.

4.8.8.1 Mycoremediation of Colour

Terrestrial white-rot basidiomycetous fungi and their lignin-degrading enzymes laccase, manganese peroxidase and lignin peroxidases are useful in the treatment of brown-coloured effluent of paper and pulp industries. Fungi involved in decolourisation of paper and pulp are given in Table 4.6. Sometimes decolourised effluent was found to regain colour after prolonged incubation which is possibly due to high molecular weight colour components. To avoid decolourisation instability, fungal bioreactor can be coupled with ultrafilteration.

4.8.9 Lipophilic Extractives

Primary clarifier effluent (PE) and a secondary clarifier effluent (SE) from a treatment plant of a Finnish elementally chlorine-free (ECF)-bleached kraft pulp and paper mill was found to possess lipophilic organic compounds (Koistinen et al. 1998) originating from kraft pulping and papermaking and identified by straight gas chromatography/mass spectrometry (GC/MS) analyses. Lipophilic extractives also exert a negative impact in pulp and paper manufacturing causing the pitch problems and poor product quality (Back and Allen 2000).

4.8.9.1 Mycoremediation Lipophilic Extractives

Phlebia radiata, *Funalia trogii*, *Bjerkandera adusta* and *Poria subvermispora* strains are the most promising organisms for the degradation of both free and esterified sterols present in the effluent of eucalypt wood-based paper and pulp industries. *Pycnoporus cinnabarinus* and its laccase enzyme are used to treat different model lipids—alkanes, fatty alcohols, fatty acids, resin acids, free sterols, sterol esters and triglycerides—in the presence of 1-hydroxybenzotriazole as mediator. Unsaturated lipids attack via the corresponding hydroperoxides. The enzymatic reaction on sterol esters largely depended on the nature of the fatty acyl moiety, i.e. oxidation of saturated fatty acid esters initiated at the sterol moiety, whereas oxidation of unsaturated fatty acid esters initiated at double bonds of the fatty acid. In contrast, saturated lipids decreased when the laccase-mediated reactions carried out in the presence of unsaturated lipids suggesting participation of lipid peroxidation radicals. These results suggested the remediation of lipid mixtures containing paper and pulp industrial wastes using laccase-mediator system of different fungi.

4.8.10 Dissolved and Colloidal Substances

These substances are present in thermomechanical pulp and newsprint mill process waters. These are derived from the woody component and can be degraded by microbes in natural condition.

4.8.10.1 Mycoremediation of DCS

White-rot fungi *Trametes versicolor* possessed the highest growth on white water and highest activity against the DCS components present in thermomechanical white water. It shows significant decrease in the total dissolved and colloidal substances (DCS), carbohydrates and extractives in 2 days and at 30 °C.

4.9 Conclusion

Insufficient and inappropriate disposal of sludge and effluent wastes of paper and cardboard industries leads to various environmental and health problems. Therefore, management of the increasing quantities of solid and effluent of paper and cardboard industries is a global environmental issue. There is a lack of knowledge, organisation and planning in waste management due to insufficient information about regulations and due to financial restrictions in both large-scale and small-scale paper and cardboard industries. The issue of waste mycoremediation is necessary due to the increasing quantities of waste and an inadequate management system. Mycoremediation technology is a great boon which will definitely help us in improving the existing situation and to minimise environmental and health problems associated with paper and cardboard industrial wastes and pollutants. Therefore, in this chapter, we have discussed about the fungi, their enzymes and the role of both in mycoremediation process and factors affecting the process. Besides this, we have also discussed about the types of wastes and pollutants generated by handmade paper, mill-made paper and cardboard industries and strategies adopted for their mycoremediation. It was found that fungi have great capability to degrade not only lignin or cellulose but also a variety of pollutants. Therefore, these can be used successfully for remediating the paper and cardboard industrial waste in free or immobilised form. However, there is a great lacuna in mycoremediation technology of paper pulp and cardboard industrial waste which will provide us good opportunity of research. In brief, it is suggested that every industry should adopt mycoremediation technology and established treatment plant at industrial level before discharging the waste into environment for contributing in sustainable development without affecting the environment and production.

References

Altschul SF, Madden TL, Schaffer AA, Zhang J, Zhang Z, Miller W, Lipman DJ (1997) Gapped BLAST and PSI-BLAST: a new generation of protein database search programs. Nucleic Acids Res 25:338–340

Ander P, Eriksson KE (1976) The importance of phenol oxidase activity in lignin degradation by the white-rot fungus *Sporotrichum pulverulentum*. Arch Microbiol 109:1–8

Ander P, Mishra C, Farrell RL, Eriksson K-EL (1990) Redox reactions in lignin degradation: interactions between laccase, different peroxidases and cellobiose: quinone oxidoreductase. J Biotechnol 13:189–198. doi:10.1016/0168-1656(90)90104-J

Back EL, Allen LH (2000) Pitch control, wood resin and de-resination. Tappi, Atlanta, GA

Bajpai P, Bajpai PK (1994) Biological colour removal of pulp and paper mill wastewaters. J Biotechnol 33:211–220

Bajpai P, Mehna A, Bajpai PK (1993) Decolorization of kraft bleach effluent with white rot fungus *Trametes versicolor*. Process Biochem 28:377–384

Belsare DK, Prasad DY (1988) Decolorization of effluent from the bagasse-based pulp mills by white-rot fungus, *Schizophyllum commune*. Appl Microbiol Biotechnol 28:301–304

Bergbauer M, Eggert C, Kraepelin G (1991) Degradation of chlorinated lignin compounds in a bleach plant effluent by the white-rot fungus *Trametes versicolor*. Appl Microbiol Biotechnol 35:105–109

Boman B, Ek M, Eriksson K-EL, Frostell B (1988) Some aspects on biological treatment of bleached pulp effluents. Nordic Pulp Paper Res J 3:13–18

Bourbonnais R, Paice MG (1988) Veratryl alcohol oxidases from the lignin-degrading basidiomycete *Pleurotus sajor-caju*. Biochem J 255:445–450

Bourbonnais R, Paice MG, Reid ID, Lantheir P, Yaguchi M (1995) Lignin oxidation by laccase isozymes from *Trametes versicolor* and role of mediator 2,2′-azinobis (3-ethyl-benzthiazoline-6-sulfonate) in kraft lignin depolymerization. Appl Environ Microbiol 61:1876–1880

Buswell JA (1994) Potential of spent substrate for bioremediation purposes. Compost Sci Util 2:31–36

Chen AHC, Dosoretz CG, Grethlein HE (1991) Ligninase production by immobilized cultures of *Phanerochaete chrysosporium* grown under nitrogen sufficient conditions. Enzyme Microb Technol 13:404–407

Childress AM, Bennett JW, Connick WJ, Daigle DJ (1998) Formulation of filamentous fungi for bioremediation. Biotechnol Tech 12:211–214

Daniel G, Volc J, Kubatova E (1994) Pyranose oxidase, a major source of H2O2 during wood degradation by *Phanerochaete chrysosporium*, *Trametes versicolor*, and *Oudemansiella mucida*. Appl Environ Microbiol 60:2524–2532

Dutta SA, Parhad NM, Joshi SR (1985) Decolorization of lignin bearing waste by *Aspergillus sp.* IAWPC Technol Annu 12:32–37

Eaton DC, Chang HM, Joyce TW, Jeffries TW, Kirk TK (1982) Method obtains fungal reduction of the color of extraction-stage kraft bleach effluents. Tappi J 65:89–92

Eriksson K-EL (1991) Biotechnology: three approaches to reduce the environmental impact of the pulp and paper industry. Sci Prog 75:175–189

Eriksson KE, Nishida A (1988) Methanol oxidase of *Phanerochaete chrysosporium*. Methods Enzymol 161:322–326

Felton JS, Mark G, Cynthia K, Salmon P, Michael A, Kristen M, Kulhuman S (2002) Exposure to heterocyclic amine food mutagens/carcinogens: relevance to breast cancer. Environ Mol Mutagen 39:112–118

Freitas AC, Ferreira F, Costa AM, Pereira R, Antunes SC, Gonçalves F, Rocha-Santos TA, Diniz MS, Castro L, Peres I, Duarte AC (2009) Biological treatment of the effluent from a bleached kraft pulp mill using basidiomycete and zygomycete fungi. Sci Total Environ 407:3282–3289

Fukuzumi T (1980) Microbial decolorization and defoaming of pulping waste liquor. In: Kirk TK, Chang HM, Higuchi T (eds) Lignin biodegradation: microbiology, chemistry and potential applications, vol 2. CRC, Boca Raton, FL, pp 161–171

González LF, Sarria V, Sánchez OF (2010) Degradation of chlorophenols by sequential biological-advanced oxidative process using *Trametes pubescens* and TiO2/UV. Bioresour Technol 101:3493–3499

Gutiérrez A, del Río JC, Rencoret J, Ibarra D, Martínez ÁT (2006) Main lipophilic extractives in different paper pulp types can be removed using the laccase-mediator system. Appl Microbiol Biotechnol 72:845–851

Hattaka A (1994) Lignin modifying enzymes from selected white rot fungi: production and role in lignin degradation. FEMS Microbiol Rev 13:125–135

Jaklin-Farcher S, Szeker E, Stifler U, Messner K (1992) Scale up of the MYCOPOR reactor. In: 5th International conference on biotechnology in pulp and paper industry, Tokyo, pp 81–85

Jaspers CJ, Penninckx MJ (1996) Adsorption effects in the decolorization of kraft bleach plant effluent by *Phanerochaete chrysosporium*. Biotechnol Lett 18:1257–1260

Joshi DK, Gold MH (1993) Degradation of 2,4,5-trichlorophenol by the lignin-degrading basidiomycete *Phanerochaete chrysosporium*. Appl Environ Microbiol 59:1779–1785

Kaal EEJ, Field JA, Joyce TW (1995) Increasing ligninolytic enzyme activities in several white-rot Basidiomycetes by nitrogen-sufficient media. Bioresour Technol 53:133–139

Kannan K, Oblisami G, Loganathan BG (1990) Enzymology of lignocellulose degradation by *Pleurotus sajor-caju* during growth on paper mill sludge. Biol Waste 33:1–8

Kawai S, Umezawa T, Higuchi T (1988) Degradation mechanisms of phenolic b-1 lignin substructure model compounds by laccase of *Coriolus versicolor*. Arch Biochem Biophys 262:99–110

Kelley RL, Ramasamy K, Reddy CA (1986) Characterization of glucose oxidase-negative mutants of a lignin degrading basidiomycete *Phanerochaete chrysosporium*. Arch Microbiol 144:254–257

Kersten PJ, Kirk TK (1987) Involvement of a new enzyme, glyoxal oxidase, in extracellular H_2O_2 production by *Phanerochaete chrysosporium*. J Bacteriol 169:2195–2201

Kirkpatrick N, Reid ID, Ziomek E, Paice MG (1990) Biological bleaching of hardwood kraft pulp using *Trametes* (*Coriolus*) *versicolor* immobilized in polyurethane foam. Appl Microbiol Biotechnol 33:105–108

Klekowski E, David D, Levin E (2006) Mutagens in a river heavily polluted with paper recycling wastes: results of field and laboratory mutagen assays. Environ Mutagen 1:209–219

Koistinen J, Lehtonen M, Tukia K, Soimasuo M, Lahtiperä M, Oikari A (1998) Identification of lipophilic pollutants discharged from a Finnish pulp and paper mill. Chemosphere 37:219–235

Kostamo A, Kukkonen JVK (2003) Removal of resin acids and sterols from pulp mill effluents by activated sludge treatment. Water Res 37:2813–2820

Kulshreshtha S, Mathur N, Bhatnagar P (2010a) Genotoxic evaluation of handmade paper industrial effluent – a case study. Int J Chem Sci 8:2519–2528

Kulshreshtha S, Mathur N, Bhatnagar P, Jain BL (2010b) Bioremediation of industrial wastes through mushroom cultivation. J Environ Biol 31:441–444

Kulshreshtha S, Mathur N, Bhatnagar P (2011) Handmade paper and cardboard industries in health perspectives. Toxicol Ind Health 27:515–521

Lamar RT, Glaser JA, Evans JW (1993) Solid phase treatment of pentachlorophenol contaminated soil using lignin degrading fungi. Environ Sci Technol 27:2566–2571

Leisola M, Ulmer D, Haltmeier T, Fiechter A (1983) Rapid solubilization and depolymerization of purified Kraft lignin by thin layers of *Phanerochaete chrysosporium*. Eur J Appl Microbiol Biotechnol 17:117–120

Leontievsky AA, Myasoedova NM, Baskunov BP, Golovleva LA, Evans CS (2000) Transformation of 2,4,6-trichlorophenol by the white rot fungi *Panus tigrinus* and *Coriolus versicolor*. Biodegradation 11:331–340

Livernoche D, Jurasek L, Desrochers M, Doric J, Veliky IA (1983) Removal of color from kraft mill wastewaters with cultures of white-rot fungi and with immobilized mycelium of *Coriolus versicolor*. Biotechnol Bioeng 25:2055–2065

Malviya P, Rathore VS (2007) Bioremediation of pulp and paper mill effluent by a novel fungal consortium isolated from polluted soil. Bioresour Technol 98:3647–3651

Martin C, Manzanares P (1994) A study of the decolourization of straw soda- pulping effluents by *Trametes versicolor*. Bioresour Technol 47:209–214

Marton J, Stern AM, Marton T (1969) Decolourization of kraft black liquor with *Polyporus versicolor*, a white-rot fungus. Tappi J 53:1975–1981

Mehna A, Bajpai P, Bajpai PK (1995) Studies on decolorization of effluent from a small pulp mill utilizing agroresidues with *Trametes versicolor*. Enzyme Microb Technol 17:18–22

Morikawa Y, Shiomi K, Ishihara Y, Matsuura N (1997) Triple primary cancers involving Kidney, Urinary Bladder and Liver in a dye worker. Am J Ind Med 31:44–49

Murugesan K (2003) Bioremediation of paper and pulp mill effluents. Indian J Exp Biol 41:1239–1248

Muzarelli AAR, Ilari P, Xia W, Pinotti M, Tomasetti M (1994) Tyrosinase-mediated quinone tanning of chitinous materials. Carbohydr Polym 24:295–300

Nagarathnamma R, Bajpai P (1999) Decolorization and detoxification of extraction-stage effluent from chlorine bleaching of kraft pulp by *Rhizopus oryzae*. Appl Environ Microbiol 65:1078–1082

Pearson W, Lipman D (1988) Improved tools for biological sequence comparison. Proc Natl Acad Sci USA 85:2444–2448

Pokhrel D, Viraraghavan T (2004) Treatment of pulp and paper mill wastewater-a review. Sci Total Environ 333:37–58

Prasad DY, Joyce TW (1991) Color removal from kraft bleach plant effluents by *Trichoderma sp.* Tappi J 74:165–169

Raghukumar C, Chandramohan D, Michel FC Jr, Reddy CA (1996) Degradation of lignin and decolorization of paper mill bleach plant effluent by marine fungi. Biotechnol Lett 18:105–106

Raghukumar C, D D'S-T, Verma AK (2008) Treatment of colored effluents with lignin-degrading enzymes: an emerging role of marine-derived fungi. Crit Rev Microbiol 34:189–206

Raghunathan R, Swaminathan K (2004) Biological treatment of a pulp and paper industry effluent by *Pleurotus spp*. World J Microbiol Biotechnol 20:389–393

Roy BP, Paice MG, Archibald FS, Misra SK, Misiak JE (1994) Creation of metalcomplexing agents, reduction of manganese dioxide, and promotion of manganese peroxidase-mediated Mn(III) production by cellobiose:quinone oxidoreductase from *Trametes versicolor*. J Biol Chem 269:19745–19750

Royer G, Desrochers M, Jurasek L, Rouleau D, Mayer RC (1985) Batch and continuous decolorisation of bleached kraft effluent by a white-rot fungus. J Chem Technol Biotechnol 35B:14–22

Saparrat MCN, Margarita AM, Tournier HA, Cabello MN, Arambarri AM (2000) Extracellular ABTS-oxidizing activity of autochthonous fungal strains from Argentina in solid medium. Rev Iberoam Micol 17:64–68

Selvam K, Swaminathan K, Song MH, Chae KS (2002) Biological treatment of a pulp and paper industry effluent by *Fomes lividus* and *Trametes versicolor*. World J Microbiol Biotechnol 18:523–526

Shinka TY, Sawada Y, Morimoto S, Fujinaga T, Nakamura J, Ohkawa T (1991) Clinical study on urothelial tumors of dye workers in Wakayama City. J Urol 146:1504–1507

Singh P, Thakur IS (2004) Removal of colour and detoxification of pulp and paper mill effluent by microorganisms in two step bioreactor. J Sci Ind Res 63:944–948

Singh P, Thakur IS (2006) Colour removal of anaerobically treated pulp and paper mill effluent by microorganism in two step bioreactor. Bioresour Technol 97:218–223

Sivrikaya H, Bacak L, Saraçbaşı A, Toroğlu İ, Eroğlu H (2002) Trace elements in *Pleurotus sajor-caju* cultivated on chemithermomechanical pulp for bio-bleaching. Food Chem 79:173–176

Sumathi S, Phatak V (1999) Fungal treatment of bagasse based pulp and paper mill wastes. Environ Technol 20:93–98

Takada S, Nakamura M, Matsueda T, Kondo R, Sakai K (1996) Degradation of polychlorinated dibenzo-p-dioxins and polychlorinated dibenzofurans by the white rot fungus *Phanerochaete sordida* YK-624. Appl Environ Microbiol 62:4323–4328

Thomas W, Chang HM, Alton G, Eaton D, Kirk K (1981) Removal of Kraft bleach plant colour by ligninolytic fungus. In: Proc Appl Environ Conf, USA, pp 225–228

Thurston CF (1994) The structure and function of fungal laccases. Microbiology 140:19–21

Wahegaonkar N, Sahasrabudhe M (2005) Fungal diversity of paper industry and soils irrigated by the effluent. Nat Environ Pollut Technol 4:49–52

Wang D, Qu YB, Gao PJ (1995) The mechanism of increasing cellulase biosynthesis rate in *Trichoderma* by L-sorbose. Acta Mycol Sin 14:143–147

Wesenberg D, Kyriakides I, Agathos SN (2003) White-rot fungi and their enzymes for the treatment of industrial dye effluents. Biotechnol Adv 22:161–187

Chapter 5
Degradation of Petroleum Pollutant Materials by Fungi

Eri Hara and Hiroo Uchiyama

5.1 Introduction

Petroleum is one of the most important resources in today's industrial economy. However, the petroleum pollution to the soil and the aquatic environment is a problem worldwide; cleanup of the polluted site was imposed in the United States in 1980 by Superfund. Although physical and chemical techniques are mainly used, recently, bioremediation has been used to cleanup the petroleum-contaminated soil.

Although there is abundant fungal biomass in soil, bacteria forms the predominant group of microorganisms. They are involved in mineralization and humification of organic matter, and in biogeochemical cycles, and pollutant degradation. In particular, degradation of lignin, a constituent of a wide variety of environmental pollutants, by white-rot fungi, has been reported (Bumpus et al. 1985; Bumpus and Aust 1986; Aust 1990).

In addition, fungi play an important role in the utilization of petroleum pollutant materials, and several fungi such as *Aspergillus fumigatus* and *Penicillium* sp. exhibit greater ability to degrade petroleum pollutant materials than bacteria such as *Arthrobacter*, *Brevibacterium*, *Flavobacterium*, *Micrococcus*, and *Pseudomonas* (Cerniglia and Perry 1973). Fungi are able to degrade petroleum pollutant materials into much smaller, environmentally nonhazardous fractions as compared to bacteria, as demonstrated by Bartha and Atlas (1973).

In this chapter, we will describe bioremediation of petroleum pollutant materials, with focus on fungi.

E. Hara • H. Uchiyama (✉)
Graduate School of Life and Environmental Sciences, University of Tsukuba,
1-1-1 Tennodai, Tsukuba, Ibaraki 305-8572, Japan
e-mail: s1030300@u.tsukuba.ac.jp; uchiyama.hiroo.fw@u.tsukuba.ac.jp

E.M. Goltapeh et al. (eds.), *Fungi as Bioremediators*, Soil Biology 32,
DOI 10.1007/978-3-642-33811-3_5, © Springer-Verlag Berlin Heidelberg 2013

5.2 Principles in Bioremediation of Contaminated Soil

5.2.1 Definitions

The environmental microbial community is composed of a wide variety of microorganisms. Bioremediation is a technique that uses various metabolic activities of these microorganisms to degrade pollutants. Therefore, no secondary environmental pollution is created, if the microorganisms' metabolic activities are in the range of the natural degradation; moreover, this technique can be performed at a low cost. Bioremediation is divided into two areas: biostimulation, which activates indigenous microorganisms via addition of nutrients and improvement of various soil conditions; and bioaugmentation, which promotes pollutant degradation via inoculation of the microorganisms that can metabolize the target materials. Biostimulation and bioaugmentation are often used in combination to enhance microbial activity and promote pollutant degradation (Whiteley and Lee 2006).

5.2.2 Bioremediation Strategies

Bioremediation is further divided according to whether the pollutant has moved (ex situ) or not (in situ) from the pollution site.

5.2.3 In Situ Bioremediation

In situ bioremediation involves injection of an oxygen supply source, nutrients, etc., into the soil without excavation to activate indigenous microorganisms. Degradation efficiency is influenced by soil properties, difficulty of material supply, existence of groundwater, and fluidity or soil conditions such as coexisting materials. Various techniques have been adopted in response to such environmental conditions.

Bioventing involves supplying oxygen to the unsaturated zone above the groundwater level to activate the microorganisms. This activation can be amplified by supplying nutrients when the pollution is dense.

Biosparging involves increasing biodegradability and the volatilization effect of low-molecular-weight components via high-pressure injection of oxygen, air, and nutrients into the saturated zone.

The most widely used technique is the injection method that promotes biodegradation by supplying oxygen and nutrients into the groundwater. The material supply is increased by the combined use of a pumping well.

5.2.3.1 Ex Situ Bioremediation

Land farming, which expedites aeration because of soil excavation, and biopiling, which supplies water and nutrients to the soil, have been used to efficiently perform bioremediation (Mohn et al. 2001; Picado et al. 2001; Schoefs et al. 2003). Ex situ bioremediation is often considered to have the sure treatment effect since admixture of the uniform nutrients is possible, although processing yard is required. In addition, only ex situ bioremediation allows efficient optimization of incubation parameters, including pH, aeration, and moistening.

5.2.4 Bioremediation Designing Systems

The general workflow from preliminary investigation to actual bioremediation construction of the contaminated soil is as follows.

The pollution situation must first be accurately grasped. Planar distribution is generally examined on the pollution status of surface soil using a 10 m mesh, and the vertical distribution is clarified at $1,000\text{-m}^2$ intervals by boring. The groundwater level and pollution status must also be investigated.

Once pollution is confirmed in the preliminary investigation, the cleanup technique is selected accordingly. The treatment and cleanup countermeasures for petroleum-contaminated soil can be classified into physicochemical processes and bioremediation, although the choice is ultimately made according to the components and status of pollution. The criteria for selecting bioremediation are as follows:

– The pollutant content and components are in the biodegradable range.
– The soil condition and physicochemical properties are suitable.
– There is an area of cleanup yard within the polluted site.
– A general cleanup period is promised.
– The cleanup achievement level is not extremely severe.

To efficiently perform bioremediation, it is necessary to adjust soil conditions, including moisture, oxygen, nutrient concentration, and pH, in which the oil-degrading microorganisms are likely to be active. The concentration and method of adding the nutrients is especially important, as they influence the success or failure of bioremediation. It is important to confirm the biodegradability in the laboratory at the same time as determining effective mixture conditions. The goal is to use the correct nutrient concentration in order to maximize the effect of bioremediation while minimizing the amount of residual nutrients.

After laboratory testing, degradability is confirmed considering the weather conditions in the field. In addition to analyzing the soil's oil concentration and the constituents in a time-course study, it is also important to monitor the microorganisms that directly participate in the degradation cleanup.

The purpose of the actual cleanup is to reduce oil, oil stench, and oil film, effects of which are controlled by chemical analysis (oil concentration test, oil dissolution test, and stench concentration measurement). It is also necessary to control temperature and respiration activity as well as the behavior of each oil-degrading microorganism in order to maximize cleanup effect and safety. It is ideal to collect samples from at least 5 places and 3 cross sections for a total of 15 places in order to achieve uniform cleanup and measure the effect in replications of 3. Based on the analysis results, it is necessary to thoroughly regulate aeration and other factors to ensure favorable cleanup conditions.

5.3 Microbial Bioremediation

5.3.1 Bacteria

Bacteria that degrade the petroleum pollutant materials are reported in great numbers, and the degradation pathway has also been clarified. Here, the degradation pathways of alkanes and aromatics that are the easily biodegradable fraction of petroleum are shown.

The degradation pathway of n-octane is shown as a representative alkane in Fig. 5.1. The first step of alkane metabolism is catalyzed by alkane hydroxylases. To be specific, $alkM$ catalyzes the metabolism of long-chain alkanes (C_{12}–) via terminal oxidation with hydroxylase (monooxygenase) systems or with dioxygenase systems (Finnerty 1977), and $alkB$ catalyzes the metabolism of medium-chain-length (C_6–C_{12}) n-alkanes via a terminal oxidation pathway with monooxygenase systems (May and Katoposis 1990) and includes unknown substrate specificity, n-alkane oxidation pathways, and oxidation systems (Smits et al. 1999). The alkane terminal is oxidized by the alkane-degrading bacteria, is converted into alcohol, aldehyde, and higher fatty acids, and is finally metabolized in the β-oxidation pathway.

The degradation pathway of benzene is shown as a representative aromatic in Fig. 5.2. It is known that most aromatics are metabolized to a common intermediate, catechol, which is further oxidized via 2 ring-cleavage pathways, *ortho* and *meta*, which are catalyzed by catechol 1,2-dioxygenase (C12O) and catechol 2,3-dioxygenase (C23O), respectively (Sei et al. 1999). Finally, the metabolized aromatic enters the tricarboxylic acid cycle.

It has been reported that denitrifying microorganisms degrade alkane and alkylbenzenes under anaerobic conditions (Ehrenreich et al. 2000; Rabus et al. 1999). In addition, involvement of sulfate-reducing bacteria in anaerobic biotransformation pathways of fluorine and phenanthrene has been reported recently (Tsai et al. 2009).

Fig. 5.1 Degradation
pathway of *n*-octane. *AlkB*
works as an oxygenase of the
first departure in alkane
metabolism

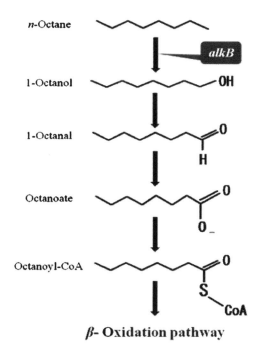

β- Oxidation pathway

5.3.2   Fungi

Petroleum pollutant material degradation is also reported by white-rot fungi such as
Pleurotus ostreatus and *Trametes versicolor* (Bogan et al. 1999; Rama et al. 2001). It is
possible that many white-rot fungi degrade lignin that is a persistent phenolic polymer.
Lignin peroxidase (LiP), manganese peroxidase (MnP), and laccases are well-known
enzymes that control lignin degradation. More recently, the enzyme versatile peroxidase (VP), found in *Pleurotus* (Camarero et al. 1999) and other fungi (Pogni et al.
2005), has been described as a new type of ligninolytic peroxidase that combines the
catalytic properties of LiP and MnP (Heinfling et al. 1998) and is able to oxidize typical
LiP and MnP substrates. LiP and MnP can catalyze recalcitrant non-phenolic lignin
oxidation to form a high redox potential oxo-ferryl intermediate during reaction of the
heme co-factor with H_2O_2 (Jensen et al. 1996; Gold et al. 2000; Perez et al. 2002;
Fig. 5.3). Laccases concomitantly catalyze 1-electron oxidation. At the same time, a
range of phenolic compounds are used as hydrogen donors during the 4-electron
reduction of molecular oxygen to H_2O (Fig. 5.3). The basis of these enzymatic
reactions, which pull out one electron from the substrate, is the electron oxidation
reaction. This reaction provides the radical and induces lignin to advance the degradation. These enzymes lack selectivity, resulting in a variety of intermonomer linkages.
This is because lignin contains a variety of aromatic structures that are formed in plant

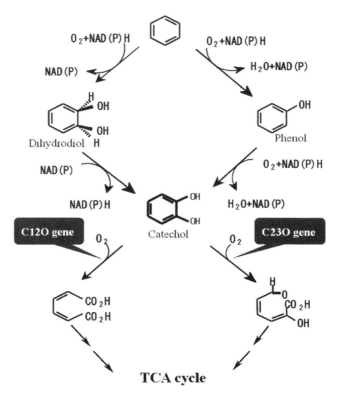

Fig. 5.2 Degradation pathway of benzene. C12O gene and C23O gene take part in *ortho* and *meta* ring cleavage of catechol that is common intermediate, respectively

$$LiP + H_2O_2 \rightarrow LiP \text{ compound } I + H_2O$$

$$LiP \text{ compound } I + AH2 \rightarrow LiP \text{ compound } II + AH$$

$$LiP \text{ compound } II + AH2 \rightarrow LiP + AH$$

$$MnP + H_2O_2 \rightarrow MnP \text{ compound } I + H_2O$$

$$MnP \text{ compound } I + Mn^{II} \rightarrow MnP \text{ compound } II + Mn^{III}$$

$$MnP \text{ compound } II + Mn^{II} \rightarrow MnP + Mn^{III}$$

$$Mn^{III} + AH2 \rightarrow Mn^{II} + AH$$

$$O_2 + 4AH2 \xrightarrow{\text{laccase}} 4AH + 2H_2O$$

Fig. 5.3 Catalysis reaction by LiP, MnP, and laccase. The mechanism of reaction is described in equation. AH2 is the reducing substrate. AH is the reducing substrate after one-electron oxidation

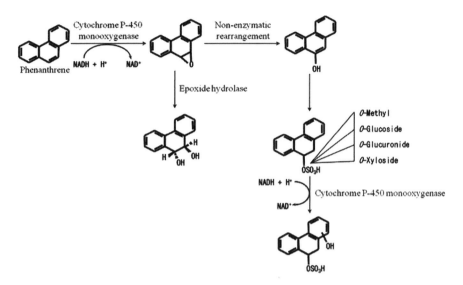

Fig. 5.4 Degradation pathway of phenanthrene by P-450 monooxygenase. A metabolic pathway is similar to that of mammals, which is considered to be the formation of diol-epoxides by P-450 monooxygenases and epoxide hydrolase. Another metabolic pathway involving hydroxylation by a monooxygenase, conjugation with sulfate ion, followed by a further hydroxylation to hydroxylarylsulfates compounds

cell walls by oxygen-radical coupling reactions of 4-hydroxycinnamyl, 3-methoxy-4-hydroxycinnamyl, and 3,5-dimethoxo-4-hydroxycinnamyl (Hammel 1992). Therefore, these enzymes are able to oxidize a wide variety of organic compounds.

A P-450 system for degradation of petroleum pollutant materials has also been reported (Tortella and Diez 2005; Bazalel et al. 1996). The biological functions of cytochrome P-450 monooxygenases, such as xenobiotic and steroidogenesis detoxification, are based on its ability to catalyze the insertion of oxygen into a wide variety of compounds. As a representative example, the degradation pathway of phenanthrene by P-450 monooxygenase is shown in Fig. 5.4. Numerous polyaromatic hydrocarbons (PAHs) can also be oxidized to phenols that are subsequently conjugated with sulfate, glucuronic acid, or glucose by cytochrome P-450 (Cerniglia et al. 1982, 1986; Casillas et al. 1996; Pothuluri et al. 1996).

5.4 Microbial Diversity and Biodegradation

Numerous microorganisms can degrade petroleum pollutant materials such as hydrocarbons. The elements (C, N, P, S, etc.) that function as the bases of biological materials and the energy necessary for constructing them must be acquired for microorganisms' growth. Many petroleum-degrading microorganisms utilize hydrocarbons as the carbon and energy source (Van et al. 2003). While many

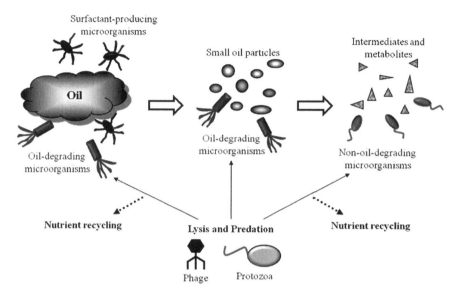

Fig. 5.5 Microbial degradation network based on Head et al. (2006). Many microorganisms directly or indirectly take part in the oil-degradation

microorganisms can decompose petroleum pollutant materials into CO_2 and H_2O under aerobic conditions, microorganisms that degrade into CH_4 under anaerobic condition are still being discovered. Numerous microorganisms can degrade and utilize hydrocarbons as a growth carbon source, including 79 genera of bacteria, 9 genera of cyanobacteria, 103 genera of fungi, and 14 genera of algae (Roger 2005). Bacteria and fungi belonging to specific taxonomic groups tend to have the ability of degrade oil. For example, *Achromobacter*, *Acinetobacter*, *Alcaligenes*, *Arthrobacter*, *Bacillus*, *Flavobacterium*, *Nocardia*, and *Pseudomonas* are well-known oil-degrading bacteria, while *Aspergillus*, *Mortierella*, *Penicillium*, and *Trichoderma* are well-known oil-degrading fungi (Leahy and Colwell 1990).

Many microorganisms appear to be directly or indirectly petroleum decomposition because individual microorganisms can metabolize only a limited range of petroleum pollutant materials (Britton 1984). For example, oil-degrading microorganisms that utilize petroleum pollutant materials by direct attack, non-oil-degrading microorganisms that use intermediates, biosurfactant-producing microorganisms that promote contact between oil-degrading microorganisms and oil, protozoa that prey on microorganisms, and phages that lyse microorganisms are related: the network is formed, and oil decomposition is systematically completed (Fig. 5.5). Understanding and controlling the mechanisms of such complex systems in bioremediation are important.

5.5   Factors Affecting Microbial Bioremediation

5.5.1   Environmental Factors

5.5.1.1   Temperature

Petroleum biodegradation is influenced by temperature, which affects the physical nature and chemical composition of the oil, the rate of hydrocarbon metabolism by microorganisms, and the composition of the microbial community (Atlas 1981). The optimal temperature range for petroleum biodegradation is 30–40 °C, as reported by Bossert and Bartha (1984). The onset of biodegradation is delayed at low temperatures. In addition, the viscosity of the oil increases, volatilization of toxic short-chain alkanes decreases, and water solubility increases (Atlas and Bartha 1972). In soils, even at temperatures as low as −1.1 °C, biodegradation on hydrocarbon has been observed (Leahy and Colwell 1990), but biodegradation rates are expected to reduce considerably below 15 °C (Dibble and Bartha 1979).

5.5.1.2   Moisture

Soil microorganisms can biodegrade petroleum hydrocarbons only within a limited range of favorable soil moisture conditions. The optimum moisture content for petroleum hydrocarbon biodegradation is 50–80 % (Bossert and Bartha 1984). Because the field capacity moisture content differs widely among soil types, it is important to experimentally determine the field capacity for the soil sample under investigation. A beneficial method is to add organic agents such as straw, woodchips, and rice hulls to the soil in order to increase water retention when the water-holding capacity of the soil is low.

5.5.1.3   Oxygen

Bacteria and fungi utilizing substrate oxidation pathway require molecular oxygen for the initial steps in the catabolism of petroleum pollutant materials. Oxygen availability in soil is dependent on the rate of microbial oxygen consumption, soil type, whether the soil is waterlogged, and the presence of utilizable substrates that can lead to oxygen depletion (Bossert and Bartha 1984). It has been observed that hydrocarbon biodegradation reduces considerably when the gaseous oxygen concentration in the soil drops below 2–5 % (v/v) (Huesemann and Truex 1996; Hurst et al. 1997; Leeson and Hinchee 1997; Hupe et al. 1999). To sustain the metabolic activities of microbial cells, the oxygen supply rate must match the overall oxygen consumption rate under equilibrium conditions (Anderson and Helder 1987; Huesemann and Truex 1996). Anaerobic bioremediation of petroleum pollutant materials has also been reported, but the degradation rates were too slow for a practical bioremediation system (Ward and Brock 1978). Bolliger et al. (1999) showed that mineralization of petroleum hydrocarbons in anaerobic aquifers is linked to the consumption of oxidants such as O_2, NO_3^-, or others, with subsequent production of the corresponding reduced species.

5.5.1.4 pH

In contrast to most aquatic ecosystems, soil pH can be highly variable, ranging from 2.5 in mine spoils to 11.0 in alkaline deserts (Bossert and Bartha 1984). Most heterotrophic bacteria and fungi favor a near-neutral pH, with fungi being more tolerant of acidic conditions (Atlas 1988). pH extremes, as observed in some soils, would, therefore, be expected to have a negative influence on the ability of microbial populations to degrade hydrocarbons. In general, it is assumed that the optimum pH for petroleum biodegradation in soils is 6.0–8.0. If the soil is overly acidic (pH < 6.0), lime or calcium carbonate may be added to increase the pH to the desired range, whereas if the soil is overly alkaline (pH > 8.0), elemental sulfur, ammonium sulfate, or aluminum sulfate may be added to lower the pH.

5.5.2 Material Factors

5.5.2.1 Nutrients

Nutrient availability is largely related to the adaptation of the concerned microorganisms and their growth by utilizing hydrocarbons. Nitrogen and phosphorus are considered the most important nutrients, because they are required for the incorporation of carbon into the biomass. A wide range of optimal C:N and C:P ratios have been reported in the literature. While Frankenberger (1992) recommends a C:N:P ratio of 100:10:1, Dibble and Bartha (1979) found optimal petroleum biodegradation with C:N and C:P ratios of 60:1 and 800:1, respectively. In contrast, Brown et al. (1983) reported that a C:N ratio of 9:1 was optimal for refinery sludge biodegradation, while Huddleston et al. (1984) suggested that the C:N ratio should be maintained between 25:1 and 38:1. Morgan and Watkinson (1989) carefully reviewed numerous biodegradation studies and found that optimal C:N ratios between 9:1 and 200:1 had been reported for waste oil and sludge. The best actual ratio is governed by the variable and complex composition of soils and other factors such as nitrogen reserves and the presence of nitrogen-fixing bacteria (Bossert and Bartha 1984).

5.5.2.2 Salts

KCl and NaCl contained in the soil influence the activity of oil-degrading microorganisms. For example, Shiaris (1989) reported a generally positive correlation between salinity and rates of mineralization of phenanthrene and naph-thalene in estuarine sediments. On the other hand, Ward and Brock (1978) showed that hydrocarbon metabolism rates decreased with increasing salinity in the range of 3.3–28.4 % and attributed the results to a general reduction in microbial metabolic rates. Excessive salts can be removed from the soil by water flushing or calcium treatment (Pollard et al. 1994).

5.5.2.3 Surfactants

The bioavailability and subsequent degradation of petroleum pollutant materials can be improved by adding chemicals and biosurfactants. In particular, using biosurfactants is expected since their persistence in the environment is low.

Many oil-degrading microorganisms synthesize and secrete biosurfactants. The oil is separated from the surface of the soil particle by the biosurfactant and is easily utilized by microorganisms, promoting decomposition. Ueno et al. (2006) inoculated *Pseudomonas aeruginosa* strain WatG, the oil-degrading bacterium in the diesel oil-contaminated soil, and confirmed that dirhamnolipid was produced in the soil during alkane decomposition. It is assumed that the decomposition was promoted by emulsification by the biosurfactant that *P. aeruginosa* WatG produced in diesel oil. In addition, the report stated that phenanthrene degradation was promoted by co-inoculation of phenanthrene-degrading and biosurfactant-producing bacteria (Dean et al. 2001). On the other hand, a case has been reported in which the indigenous microorganism was affected while the surfactants changed from in a concentration-dependent manner (Colores et al. 2000).

The utilization of biosurfactants in future bioremediation is expected to be very effective. Addition of 1–2 % biosurfactant to the soil is generally considered ideal. However, if a quantity greater than the critical micelle concentration is added, the pollutant does not come into contact with the degrading microorganisms. In such cases, it is necessary to dilute the soil to increase the opportunities for contact between the degrading microorganisms and pollutants (Billinqsley et al. 1999).

5.6 Bioremediation of Petroleum Pollutant Materials

5.6.1 Polyaromatic Hydrocarbons

PAH is the generic name for aromatics containing two or more benzene rings; they are constituents of crude oil. Therefore, during oil spills, the environment becomes highly polluted. Microbial degradation of PAHs became popular after bioremediation was adopted for the removal of pollutants from the Prestige ship spill on the north coast of Spain in November 2002 (Alcalde et al. 2006). Numerous microorganisms, including bacteria, yeasts, and fungi, are known to be able to degrade PAHs

(Mougin 2002; Aitken and Long 2004). White-rot fungi, including *Pleurotus ostreatus* and *Trametes versicolor*, are well known for their PAH-degrading abilities. Garon et al. (2004) assessed the potential of bioaugmentation by *Absidia cylindrospora*, as a method to accelerate fluorene biodegradation in soil slurries. In the presence of *A. cylindrospora*, $\geq$90 % of the fluorene was removed in 288 h, whereas in the absence of fungal bioaugmentation, the process required 576 h. In a recent study, *Aspergillus fumigatus* showed favorable anthracene degradation ability. The molecular structure of anthracene changed with the action of *A. fumigatus*, generating a series of intermediate compounds including phthalic anhydride, anthrone, and anthraquinone by ring-cleavage reactions (Ye et al. 2011).

A novel bioremediation technology using a combination of biological surfactant soil washing followed by PAH biological oxidation in the soil wash water by *Phanerochaete chrysosporium* was studied by Zheng and Obbard (2002). The removal efficiency for all the PAHs used was found to be >90 % by using a combination of surfactant soil washing and *P. chrysosporium*.

5.6.2 Benzene–Toluene–Ethylbenzene–Xylenes

As the main component of gasoline, benzene–toluene–ethylbenzene–xylenes (BTEX) is strongly related to environmental pollution around gas stations. Various studies have tried to advance the bioremediation of BTEX by fungi (Oh et al. 1998; Yadav and Reddy 1993; Prenafeta-Boldu et al. 2002, 2004). Chemostat experiments were conducted with four pure cultures to determine the effects of ethanol on BTEX biodegradation kinetics (Lovanh et al. 2002). A low ethanol concentration increased benzene removal efficiency, but a high ethanol concentration induced a negative effect. Advanced oxidation of BTEX by the extracellular hydroxyl radicals (*OH) generated by *Trametes versicolor* was demonstrated in a recent study (Aranda et al. 2010). The production of *OH was induced by incubation of *T. versicolor* with 2,6-dimethoxy-1,4-benzoquinone and Fe^{3+}-EDTA. All BTEX compounds (500 μM) were oxidized at a similar rate, reaching an average of 71 % degradation in 6 h.

A study provided details of the coupled biological and chemical (CBC) model for representing in situ bioremediation of BTEX (Maurer and Rittmann 2004). This model contains novel features that allow it to comprehensively track the footprints of BTEX bioremediation, even when the fate of those footprints is confounded by abiotic reactions and by the complex interactions among different kinds of microorganisms.

5.7 Conclusions and Perspectives

Bioremediation using fungi is expected because fungal enzymes (LiP, MnP, laccases, and P-450 monooxygenase) can oxidize a wide variety of organic compounds. In the natural environment, several species of white-rot fungi can co-metabolize petroleum pollutant materials with bacteria by improving compound bioavailability. A study carried out with anthracene confirmed that all the detectable oxidation products of the breakdown of this compound by white-rot fungi can be mineralized by indigenous bacteria (e.g., activated sludge) more rapidly than the mineralization of anthracene (Meulenberg et al. 1997). Inoculation of fungal-bacterial co-cultures into PAH-contaminated soil resulted in a significant improvement in the degradation of some high-molecular-weight PAHs such as chrysene, benzo [α] anthracene, and dibenzo [α, h] anthracene (Boonchan et al. 2000). Thus, it was suggested that the degradation of petroleum pollutant materials in nature is a consequence of sequential breakdown by fungi and bacteria, with the fungi performing the initial oxidation step (Meulenberg et al. 1997; Sack et al. 1997; Kotterman et al. 1998). In bioremediation, it is very important to understand the microbial network to achieve efficient degradation. However, both culturable and non-culturable microorganisms take part in degradation. To grasp the microbial network system including non-culturable microorganisms, use of molecular biological techniques such as stable isotope probing (SIP), denaturing gradient gel electrophoresis (DGGE), and terminal restriction fragment length polymorphism (T-RFLP) is effective. Furthermore, because bioremediation effectiveness differs from site to site, construction of a database of field experiments is needed. Consequently, in the future, bioremediation is expected to become an established scientific technology.

References

Aitken MD, Long TC (2004) Biotransformation, biodegradation, and bioremediation of polycyclic hydrocarbons. In: Singh A, Ward OP (eds) Biodegradation and bioremediation. Springer, Berlin, pp 83–124

Alcalde M, Ferrer M, Plou FJ, Ballesteros A (2006) Environmental biocatalysis: from remediation with enzymes to novel green processes. Trends Biotechnol 24:281–287

Anderson FO, Helder H (1987) Comparison of oxygen micro-gradients, oxygen flux rates and electron system activity in coastal marine sediments. Mar Ecol Prog Ser 37:259–264

Aranda E, Marco-Urrea E, Caminal G, Arias ME, Garcia-Romera I, Guillen F (2010) Advanced oxidation of benzene, toluene, ethylbenzene and xylene isomers (BTEX) by *Trametes versicolor*. J Hazard Mater 181:181–186

Atlas RM (1981) Microbial degradation of petroleum hydrocarbons: an environmental perspective. Microbiol Rev 45:180–209

Atlas RM (1988) Microbiology-fundamentals and applications, 2nd edn. Macmillan, New York, pp 352–353

Atlas RM, Bartha R (1972) Biodegradation of petroleum in seawater at low temperatures. Can J Microbiol 18:1851–1855

Aust SD (1990) Degradation of environmental pollutants. Microb Ecol 20:197–209

Bartha R, Atlas RM (1973) Biodegradation of oil in seawater: limiting factors and artificial stimulation. In: Ahern DG, Meyers SP (eds) The microbial degradation of oil pollutants. Centre for wetland resources, Louisiana State University, Baton Rouge, LA, pp 147–152

Bazalel L, Hadar Y, Fu PP, Freeman JP, Cerniglia CE (1996) Initial oxidation products in the metabolism of pyrene, anthracene, fluorene, and dibenzothiophene by the white rot fungus *Pleurotus ostreatus*. Apple Environ Microbiol 62:2554–2559

Billinqsley KA, Backus SM, Ward OP (1999) Effect of surfactant solubilization on biodegradation of polychlorinated biphenyl congeners by *Pseudomonas* LB400. Appl Microbiol Biotechnol 52:255–260

Bogan BW, Lamar RT, Burgos WD, Tien M (1999) Extent of humification of anthracene, fluoranthene, and benzo (alpha) pyrene by Pleurotus ostreatus during growth in PAH-contaminated soils. Lett Appl Microbiol 28:250–254

Bolliger C, Hohener P, Hekeler D, Haberli K, Zeyer J (1999) Intrinsic bioremediation of a petroleum hydrocarbon-contaminated aquifer and assessment of mineralization based on stable carbon isotopes. Biodegradation 10:201–217

Boonchan S, Britz ML, Stanley GA (2000) Degradation and mineralization of high-molecular-weight polycyclic aromatic hydrocarbons by defined fungal-bacterial cocultures. Appl Environ Microbiol 66:1007–1019

Bossert I, Bartha R (1984) The fate of petroleum in soil ecosystems. In: Atlas RM (ed) Petroleum microbiology. Macmillan, New York, pp 434–476

Britton LN (1984) Microbial degradation of aliphatic hydrocarbons. In: Gibson DT (ed) Microbial degradation of organic compounds. Marcel Dekker, New York, pp 89–129

Brown KW, Donnelly KC, Deuel LE (1983) Effects of mineral nutrients, sludge application rate, and application frequency on biodegradation of two oily sludges. Microb Ecol 9:363–373

Bumpus JA, Aust SD (1986) Biodegradation of environmental pollutants by the white rot fungus *Phanerochaete chrysosporium*: involvement of the lignin degrading system. Bioessays 6:166–170

Bumpus JA, Tien M, Wright DS, Aust AD (1985) Oxidation of persistent environmental pollutants by white rot fungus. Science 228:1434–1436

Camarero S, Sarkar S, Ruiz-Duenas FJ, Martinez MJ, Martinez AT (1999) Description of a versatile peroxidase involved in the natural degradation of lignin that has both manganese peroxidase and lignin peroxidase substrate interaction sites. J Biol Chem 274:10324–10330

Casillas RP, Crow SA, Heinze TM, Deck J, Cerniglia CE (1996) Initial oxidative and subsequent conjugative metabolites produced during the metabolism of phenanthrene by fungi. J Ind Microbiol 16:205–215

Cerniglia CE, Perry JJ (1973) Hydrocarbon utilization by *Cladosporium resinae*. In: Ahern DG, Meyers SP (eds) The microbial degradation of oil pollutants. Centre for Wetland Resources, Louisiana State University, Baton Rouge, pp 25–32

Cerniglia CE, Freeman JP, Mitchum RK (1982) Glucuronide and sulfate conjugation in the fungal metabolism of aromatic hydrocarbons. Appl Environ Microbiol 43:1070–1075

Cerniglia CE, Kelly DW, Freeman JP, Miller DW (1986) Microbiol metabolism of pyrene. Chem Biol Interact 57:203–216

Colores GM, Macur RE, Ward DM, Inskeep WP (2000) Molecular analysis of surfactant-driven microbial population shifts in hydrocarbon-contaminated soil. Appl Environ Microbiol 66:2959–2964

Dean SM, Jin Y, Cha DK, Wilson SV, Radosevich M (2001) Phenanthrene degradation in soils co-inoculated with phenanthrene-degrading and biosurfactant-producing bacteria. J Environ Qual 30:1126–1133

Dibble JT, Bartha R (1979) Effect of environmental parameters on the biodegradation of oil sludge. Appl Environ Microbiol 37:729–739

Ehrenreich P, Behrends A, Harder J, Widdel F (2000) Anaerobic oxidation of alkanes by newly isolated denitrifying bacteria. Arch Microbiol 173:58–64

Finnerty WR (1977) The biochemistry of microbial alkane oxidation: new insights and perspectives. Trends Biochem Sci 2:73–75

Frankenberger WT (1992) The need for a laboratory feasibility study in bioremediation of petroleum hydrocarbons. In: Calabrese EJ, Kostecki PT (eds) Hydrocarbon contaminated soils and groundwater, vol 2. Lewis, Boca Raton, FL, pp 237–293

Garon D, Sage L, Seigle-Murandi F (2004) Effect of fungal bioaugmentation and cyclodextrin amendment on fluorene degradation in soil slurry. Biodegradation 15:1–8

Gold MH, Youngs HL, Gelpke MD (2000) Manganese peroxidase. Met Ions Biol Syst 37:559–586

Hammel KE (1992) Oxidation of aromatic pollutants by lignindegrading fungi and their extracellular peroxidases. Met Ions Biol Sys 28:41–60

Head IM, Jones DM, Roling WF (2006) Marine microorganisms make a meal of oil. Nat Rev Microbiol 4:173–182

Heinfling A, Ruiz-Duenas FJ, Martinez MJ, Bergbauer M, Szewzyk U, Martinez AT (1998) A study on reducing substrates of manganese-oxidizing peroxidases from Pleurotus eryngii and Bjerkandera adusta. FEBS Lett 428:141–146

Huddleston RL, Bleckmann CA, Wolfe JR (1984) Land treatment biological degradation processes. In: Loehr RC, Malina JF (eds) Land treatment: a hazardous waste management alternative. Water resources symposium, November 13, University of Texas at Austin, pp 41–61

Huesemann MH, Truex MJ (1996) The role of oxygen diffusion in passive bioremediation of petroleum contaminated soil. J Hazard Mater 51:93–113

Hupe K, Heerenklage J, Stegmann R (1999) Influence of oxygen on the degradation of TPH contaminated soils. In: Alleman BC, Leeson A (eds) Bioreactor and ex situ biological treatment technologies. Battelle, Columbus, OH, pp 31–36

Hurst CJ, Sims RC, Sims JL, Sorensen DL, McLean JE, Huling S (1997) Soil gas oxygen tension and pentachlorophenol biodegradation. J Environ Eng 123(4):364–370

Jensen KAJ, Bao W, Kawai S, Srebotnik E, Hammel KE (1996) Manganese-dependent cleavage of nonphenolic lignin structures by Ceriporiopsis subvermispora in the absence of lignin peroxidase. Appl Environ Microbiol 62:3679–3686

Kotterman MJJ, Vis EH, Field JA (1998) Successive mineralization and detoxification of benzo [α] pyrene by the white rot fungus Bjerkandera sp. strain BOS55 and indigenous microflora. Appl Environ Microbiol 64:2853–2858

Leahy JG, Colwell RR (1990) Microbial degradation of hydrocarbons in the environment. Microbiol Rev 54:305–315

Leeson A, Hinchee RE (1997) Soil bioventing: principles and practice. Lewis, Boca Raton, FL

Lovanh N, Hunt CS, Alvarez PJ (2002) Effect of ethanol on BTEX biodegradation kinetics: aerobic continuous culture experiments. Water Res 36:3739–3746

Maurer M, Rittmann BE (2004) Formulation of the CBC-model for modeling the contaminants and footprints in natural attenuation of BTEX. Biodegradation 15:419–434

May SW, Katoposis AG (1990) Hydrocarbon monooxygenase system of Pseudomonas oleovorans. Methods Enzymol 188:3–9

Meulenberg R, Rijnaarts HHM, Doddema HJ, Field JA (1997) Partially oxidized polycyclic aromatic hydrocarbons shown an increased bioavailability and biodegradability. FEMS Microbiol Lett 152:45–49

Mohn WW, Radziminski CZ, Fortin MC, Reimer KJ (2001) On site bioremediation of hydrocarbon-contaminated Arctic tundra soils in inoculated biopiles. Appl Microbiol Biotechnol 57:242–247

Morgan P, Watkinson RJ (1989) Hydrocarbon degradation in soils and method for soil biotreatment. Crit Rev Microbiol 8:305–333

Mougin C (2002) Bioremediation and phytoremediation of industrial PAH-polluted soils. Polycycl Aromat Comp 22:1–33

Oh YS, Choi SC, Kim YK (1998) Degradation of gaseous BTEX biofiltration with Phanerochaete chrysosporium. J Microbiol 36:34–38

Perez J, Munoz-Dorado J, De La Rubia RT, Martinez J (2002) Biodegradation and biological treatments of cellulose, hemicelluloses and lignin: an overview. Int Microbiol 5:53–63

Picado A, Nogueira A, Baeta-Hall L, Mendonca E, de Fatima Rodrigues M, do Ceu Saagua M, Martis A, Anselmo AM (2001) Landfarming in a PAH-contaminated soil. J Environ Sci Health A Tox Hazard Subst Environ Eng 36:1579–1588

Pogni R, Baratto MC, Giansanti S, Teutloff C, Verdin J, Valderrama B, Lendzian F, Lubitz W, Vazquez-Duhalt R, Basosi R (2005) Tryptophan-based radical in the catalytic mechanism of versatile peroxidase from *Bjerkandera adusta*. Biochemistry 44:4267–4274

Pollard SJT, Hrudey SE, Fredorak PM (1994) Bioremediation of petroleum- and creosote-contaminated soils: a review of constraints. Waste Manage Res 12:173–194

Pothuluri JV, Evans FE, Heinze TM, Cerniglia CE (1996) Formation of sulfate and glucoside conjugates of benzo [e] pyrene by *Cunninghamella elegans*. Appl Microbiol Biotechnol 45:677–683

Prenafeta-Boldu FX, Vervoort J, Grotenhuis JTC, van Groenestijn JW (2002) Substrate interactions during the biodegradation of benzene, toluene, ethylbenzene, and xylene (BTEX) hydrocarbons by the fungus *Cladophialophora* sp. strain T1. Apple Environ Microbiol 68:2660–2665

Prenafeta-Boldu FX, Ballerstedt H, Gerritse J, Grotenhuis JTC (2004) Bioremediation of BTEX hydrocarbons: effect of soil inoculation with toluene-growing fungus *Cladophialophora* sp. strain T1. Biodegradation 15:59–65

Rabus R, Wilkes H, Schramm A, Harms G, Behrends A, Amann R, Widdle F (1999) Anaerobic utilization of alkylbenzenes and n-alkanes from crude oil in an enrichment culture of denitrifying bacteria affiliating with the beta-subclass of proteobacteria. Environ Microbiol 1:145–157

Rama R, Sigoillot JC, Chaplain V, Asther M, Jolivalt C, Mougin C (2001) Inoculation of filamentous fungi in manufactured gas plant site soils and PAH transformation. Polycycl Aromat Comp 18:397–414

Roger CP (2005) Metabolic indicators of anaerobic hydrocarbon biodegradation in petroleum-laden environments. In: Bernard O, Michel M (eds) Petroleum microbiology. American Society for Microbiology Press, Washington, DC, pp 317–336

Sack U, Heinze T, Deck MJ, Cerniglia CE, Martens R, Zadrazil F, Fritsche W (1997) Comparison of phenanthrene and pyrene degradation by different wood-decaying fungi. Appl Environ Microbiol 63:3919–3925

Schoefs O, Perrier M, Dochain D, Samson R (2003) On-line estimation of biodegradation in an unsaturated soil. Bioprocess Biosyst Eng 26:37–48

Sei K, Asano K, Tateishi N, Mori K, Ike M, Fujita M (1999) Design of PCR primers and gene probes for the general detection of bacterial populations capable of degrading aromatic compounds via catechol cleavage pathways. J Biosci Bioeng 88:542–550

Shiaris MP (1989) Seasonal biotransformation of naphthalene, phenanthrene, and benzo [a] pyrene in surficial estuarine sediments. Appl Environ Microbiol 55:1391–1399

Smits TH, Rothlisberger M, Witholt B, van Beilen JB (1999) Molecular screening for alkane hydroxylase genes in Gram-negative and Gram-positive strains. Environ Microbiol 1:307–317

Tortella GR, Diez MC (2005) Fungal diversity and use in decomposition of environmental pollutants. Crit Rev Microbiol 31:197–212

Tsai JC, Kumar M, Lin JG (2009) Anaerobic biotransformation of fluorine and phenanthrene by sulfate-reducing bacteria and identification of biotransformation pathway. J Hazard Mater 164:847–855

Ueno A, Hasanuzzaman M, Yumoto I, Okuyama H (2006) Verification of degradation of *n*-alkanes in diesel oil by *Pseudomonas aeruginosa* strain WatG in soil microcosms. Curr Microbiol 52:182–185

Van HJ, Singh A, Ward OP (2003) Recent advances in petroleum microbiology. Microbiol Mol Biol Rev 67:503–549

Ward DM, Brock TD (1978) Hydrocarbon biodegradation in hypersaline environments. Appl Environ Microbiol 35:353–359

Whiteley CG, Lee DJ (2006) Enzyme technology and biological remediation. Enzyme Microb Technol 38:291–316

Yadav JS, Reddy CA (1993) Degradation of benzene, toluene, ethylbenzene, and xylenes (BTEX) by lignin-degrading basidiomycetes *Phanerochaete chrysosporium*. Appl Environ Microbiol 59:756–762

Ye JS, Yin H, Qianq J, Penq H, Qin HM, Zhanq N, He BY (2011) Biodegradation of anthracene by *Aspergillus fumigates*. J Hazad Mater 15:174–181

Zheng Z, Obbard JP (2002) Polycyclic aromatic hydrocarbon removal from soil by surfactant solubilization and *Phanerochaete chrysosporium* oxidation. J Environ Qual 31:1842–1847

Chapter 6
Bioremediation of Organic Pollutants Using *Phanerochaete chrysosporium*

M.H. Fulekar, Bhawana Pathak, Jyoti Fulekar, and Tanvi Godambe

6.1 Introduction

Rapid industrialisation and urbanisation in the last century due to enormous technological innovations have led to the problem of environmental pollution and ecological concerns. Bioremediation is an option that offers the possibility to destroy or render harmless various contaminants using natural biological activity. As such, it uses relatively low-cost, low-technology techniques, which generally have a high public acceptance and can often be carried out on site. Bioremediation can take place under aerobic or anaerobic conditions. Under aerobic conditions, microorganisms consume atmospheric oxygen to function. Under anaerobic conditions, no oxygen is present. In this case, the microorganisms break down chemical compounds in the soil to release the energy they need. Ideally, bioremediation results in the complete mineralisation of contaminants to H_2O and CO_2 without the build-up of intermediates. Bioremediation processes can be broadly categorised into two groups: ex situ and in situ. Ex situ bioremediation technologies include bioreactors, biofilters, land farming, and some composting methods. In situ bioremediation technologies include bioventing, biosparging, biostimulation, liquid delivery systems, and some composting methods.

At present, bioremediation is considered as a less expensive alternative to physical and chemical means of degradation of organic pollutants. It deals with substances that are anthropogenic, distributed in nature, and recalcitrant (Fulekar 2010). Most research within the field of bioremediation has focused on bacteria,

M.H. Fulekar (✉) • B. Pathak • J. Fulekar
School of Environment and Sustainable Development, Central University of Gujarat,
Gandhinagar, Gujarat, India
e-mail: mhfulekar@yahoo.com; bhawanasp@hotmail.com; jyoti.mumbai406@gmail.com

T. Godambe
Department of Life Sciences, University of Mumbai, Mumbai, Maharashtra, India
e-mail: godambetanvi@yahoo.co.in

E.M. Goltapeh et al. (eds.), *Fungi as Bioremediators*, Soil Biology 32,
DOI 10.1007/978-3-642-33811-3_6, © Springer-Verlag Berlin Heidelberg 2013

with fungal bioremediation (mycoremediation) attracting interest just within the past two decades. The toxicity of many of the above-named pollutants limits natural attenuation by bacteria, but white rot fungi can withstand toxic levels of most organopollutants.

Fungal remediation refers to the use of fungi to remediate organic soil contaminants, primarily hydrocarbons. One group of fungi, *Phanerochaete chrysosporium* or white rot fungus, produces a family of enzymes called lignin peroxidases, or ligninases, which have extensive biodegrative properties. Remediation of soil using white rot fungus has been studied in both in situ and reactor-based systems.

6.2 White Rot Fungus

6.2.1 Taxonomic and Morphological Characteristics

The kingdom Fungi is divided into many subunits, each more restrictive than the next higher level. The white rot fungi are in the division Eumycota (true fungi), subdivision Basidiomycotina, class Hymenomycetes, subclass Holobasidio-mycetidae (Hawksworth et al. 1995). This subclass contains nearly all of the wood-decay fungi, as well as the mycorrhizal, litter, and decomposer fungi. It contains the mushrooms, puffballs, conks, and crustlike fungi. At the ordinal level (one step below the subclass), the biological variation among representatives is more reduced, especially as to function, and at the family level, there are only rare cases of functional diversity (mycorrhizal vs. wood decay, brown rot vs. white rot).

White rot fungi are a physiological rather than taxonomic grouping of fungi, comprising those fungi that are capable of extensively degrading lignin (a heterogenous polyphenolic polymer) within lignocellulosic substrates. The name white rot derives from the appearance of wood attacked by these fungi, in which lignin removal results in a bleached appearance to the substrate.

6.2.2 Life Cycle

The genetic incompatibility systems found operating in the white rot fungi are just as variable as those in the rest of the Basidiomycotina. There are homothallic species in which the life cycle of the organism can be completed by growth of a spore or other propagule containing a single haploid nucleus; that is, no anastomosis with elements from an individual with a compatible nucleus is needed. These species are uncommon, and no species known to be industrially important white rotters have been unqualifiedly demonstrated to be homothallic. Heterothallic species are more prevalent in the fungus kingdom. These species require that two compatible nuclei

Fig. 6.1 Life cycle of *Phanerochaete chrysosporium* showing the nuclear status of each state, assuming a heterothallic incompatibility system. *Source*: Young and Akhtar (1998)

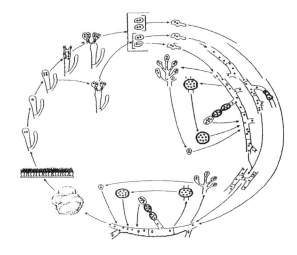

be associated through hyphal anastomosis in order for the life cycle to continue to completion, resulting in the formation of basidia and haploid basidiospores. Most of the white rot fungi possess a heterothallic incompatibility system (Fig. 6.1), which may be one of two types. The bipolar (unifactorial) system possesses a single incompatibility locus, which must differ in the associated nuclei, in order for completion of the life cycle. However, other fungi possess two incompatibility loci, both of which must differ in associated nuclei in order for a completed cycle to occur. This is called a tetrapolar (bifactorial) incompatibility system.

6.2.3   Role in Bioremediation

The white rot fungi that cause white rot of wood have recently become the object of increasing attention from workers in the hazardous waste field. These fungi normally grow on decaying wood and forest litter and appear to be unique among microorganisms in that they can rapidly depolymerise and mineralise lignin to carbon dioxide.

White rot fungi produce unique extracellular oxidative enzymes that degrade lignin, as well as related compounds found in explosive-contaminated materials, pesticides, and toxic wastes. Lignin plays a key role in the carbon cycle as the most abundant aromatic compound in nature, providing the protective matrix surrounding the cellulose microfibrils of plant cell walls. This amorphous and insoluble polymer lacks stereoregularity, and in contrast to cellulose and hemicellulose, it is not susceptible to hydrolytic attack. Although lignin is a formidable substrate, its degradation by certain fungi was recognised and described nearly 125 years ago. Collectively referred to as white rot fungi (since they degrade brown lignin and leave behind white cellulose), these are the only microbes capable of efficient depolymerisation and mineralisation of lignin. All are basidiomycetes, a fungal group that includes both edible mushrooms as well as plant pathogens such as smuts and rust. White rot fungi are also able to degrade a wide variety of environmental pollutants to carbon dioxide, including a number of chlorinated pollutants such as DDT [1,1,1-trichloro-2,2-bis(4-chlorophenyl) ethane], lindane (1,2,3,4,5, 6-hexachlorocyclohexane), chlordane (1,2,4,5,6,7,8,8-octachloro-3a,4,7,7a-tetrahydro-4-7-methanoindan), polychlorinated biphenyls-2,3,7,8-TCDD(2,3,7,8-tetrachlorodibenzo-p-dioxin), and 3,4-dichloroaniline. Biodegradation of the pollutants was observed only during secondary metabolism, occurring at high rates only under conditions of nutrient limitation, and is cometabolic, that is, a primary growth substrate such as cellulose or glucose is required (Bumpus et al. 1988).

White rot fungus is known to degrade polyaromatic hydrocarbons (PAHs), chlorinated aromatic hydrocarbons (CAHs), polycyclic aromatics, polychlorinated biphenyls, polychlorinated dibenzo(p)dioxins, the pesticides DDT and lindane, and some azo dyes (Cookson 1995). Degradation of Polyaromatic Hydrocarbons, which include benzo(a)pyrene, pyrene, fluorene, and phenanthrene, is favoured at nitrogen-limiting conditions and low pH (about 4.5) (Suthersan 1997). It has been documented that white rot fungus is able to mineralise tri-, tera-, and pentachlorophenol (PCP), and a group of microbes in soil can completely mineralise PCP (Suthersan 1997). Degradation of cyclodiene insecticides, including chlordane, by white rot fungus has been demonstrated by Suthersan (1997). White rot fungus has been observed to degrade TNT in laboratory-scale studies (using pure cultures); however, factors that limit their effectiveness (see below) may retard widespread use in the field. Results from bench-scale studies of mixed fungal and bacterial systems indicate that most of the degradation of TNT is attributable to bacteria and that most of the losses of TNT are due to adsorption onto the fungus and soil amendments.

Most known white rot fungi are basidiomycetes, although a few ascomycete genera within the *Xylariaceae* are also capable of white rot decay Eaton and Hale (1993). Four main genera of white rot fungi have shown potential for bioremediation: *Phanerochaete*, *Trametes*, *Bjerkandera*, and *Pleurotus* (Hestbjerg et al. 2003).

Recently, white rot fungus, the only organism that degrades wood, was shown to exhibit unique biodegradation capabilities. White rot fungi variously secrete one or more of three extracellular enzymes that are essential for lignin degradation and which combine with other processes to effect lignin mineralisation. They are often referred to as lignin-modifying enzymes or LMEs. The three enzymes comprise two glycosylated heme-containing peroxidases, lignin peroxidase (LiP, E.C. 1.11.1.14) and Mn-dependent peroxidase (MnP, E.C. 1.11.1.13) (Orth and Tien 1995), and a copper-containing phenoloxidase, laccase (Lac, E.C. 1.10.3.2) (Thurston 1994). The lignolytic enzymes of white rot fungi have a broad substrate specificity and have been implicated in the transformation and mineralisation of organopollutants with structural similarities to lignin (Aust et al. 2003).

6.3 Role of Lignin Peroxidase of White Rot Fungi in Biodegradation

The lignin peroxidases, which appear to be intimately involved in the mineralisation of chemicals, have several properties that make them particularly unique. For example, they are produced in response to nutrient limitation, not by derepression of enzyme synthesis. Therefore, the organism does not have to be adapted to the chemical. Lignocellulosic materials are an ideal nutrient source since they are natural substrates for the organism and are naturally low in nutrient nitrogen yet contain the necessary nutrients for growth and a carbon source. The enzymes are peroxidases so that the true substrate is H_2O_2, which is also naturally produced by the fungi, given a suitable carbon source. A logical choice is the fungus' natural nutrient cellulose. The t-I2O2 activates the enzymes which then react with the chemicals to initiate oxidation. The chemicals might react with the heme iron of the enzyme, the cation radical of the heme porphyrin, or other portions of the enzyme that might exist as a stable radical. The reaction can proceed until the concentration of chemical is essentially as low as the concentration of activated enzyme, as the reaction is a biomolecular free radical mechanism. This can result in essentially non-detectable levels of chemicals even though the rate of degradation may be slow at low chemical concentration. The lignin peroxidases and H_2O_2 are secreted by the fungi since lignin degradation must be at least initiated extracellularly. This has significance for the degradation of environmental pollutants. Often these chemicals are persistent in the environment because of their insolubility in water, yet this would not be a deterrent for the white rot fungi. The lignin peroxidases are like many peroxidases except that they have a very high oxidation potential and thus are potentially able to oxidise chemicals not oxidised by other peroxidises (Aust 1990).

6.4 Case Studies

White rot fungi have been demonstrated to be capable of transforming and/or mineralising a wide range of organopollutants.

6.4.1 *Biodegradation of Polycyclic Aromatic Hydrocarbons by White Rot Fungus*

Polycyclic aromatic hydrocarbons (PAHs) are an important class of persistent organic pollutants (POPs) in the environment which originate mainly from the incomplete combustion of fossil fuels, volcanic eruptions, and forest fires and are released into the environment in the form of exhaust and solid residues. In recent decades, with the continuing development of industry (particularly the annually increasing exploitation of petroleum and coal), the aromatic substances constantly emitted into the environment have increased, resulting in an ongoing increase in PAHs levels. This pattern of increasing environmental PAHs comprises a serious risk not only for humans but for whole ecosystems. Gaspare et al. (2009) analysed surface sediment and oyster samples from the inter-tidal areas of Dar es Salaam for 23 PAHs including the 16 compounds prioritised by US-EPA using GC/MS, and as a result, the total concentration of PAHs in the sediment was found to range from 78 to 25,000 ng/g dry weight, while the oyster concentrations ranged from 170 to 650 ng/g dry weight. It is thus evident that PAHs pollution has become ubiquitous in the region and that remediation of environmental PAHs requires our immediate attention.

White rot fungus is a species of mycelial fungus which is saprophytic on trees and leads to a decaying xylon white sponge-like mass. It can produce extracellular oxidase in the cell lumen. In addition to the decomposition of lignin, white rot fungus, with its capacity for high efficiency, low consumption, and broad-spectrum degradation of pollutants, is widely applied in wastewater treatments and bioremediation of polluted soil (and other urgent environmental applications), which include the degradation of dye, TNT, PAHs, and other toxic organic pollutants. Laccase (Lac), manganese peroxidase (MnP), and lignin peroxidase (LiP) are the major lignin degradation enzymes of white rot fungus (Wu et al. 2008). Laccase is a multicopper oxidase which can degrade a variety of complex structures of xenobiotics (Sealey and Ragauskas 1998; Couto 2007; Pozdnyakova et al. 2006). Knapp and Newby (1999) selected the five most effective strains for the decolouration study of a chemical industry effluent. All five yielded 70–80 % decolouration. Of these, the best were the strains of *Coriolus versicolor*. Vahabzadeh et al. (2004) studied the white rot fungus *Phanerochaete chrysosporium* treatment of molasses wastewater from an ethanolic fermentation plant. Under the condition of diluting and adding a certain quantity of the spore to the wastewater, the removal efficiency reached 75 % on the fifth day. Gomaa et al. (2008) studied the ability of the white rot fungus *P. chrysosporium* to decolourise Victoria blue B (VB) in textile dyes. The inhibition of laccase production (by adding various inhibitors to the shaken cultures) exerts a potent influence on the decolourisation of VB. When adding sodium azide and aminotriazole to inhibit the activities of endogenous catalase and cytochrome P-450 oxygenase, the decolourisation efficiency decreased by 100 and 70 %, respectively, while benzoate resulted in only a 50 % decrease. Chupungars et al. (2008) studied PAH degradation by *Agrocybe* sp. CU-43. At 100 ppm, fluorene was 99 % degraded within 6 days, while at the same concentration, 99 and 92 % degradation of phenanthrene and anthracene, respectively, occurred within 21 days, and fluoranthene and pyrene were reduced by 80 and 75 %, respectively, within 30 days. In a soil model, *Agrocybe* sp. CU-43 completely degraded 250 ppm fluorene at room temperature within 4 weeks. This investigation focused on the PAH degradation effect of aboriginal white rot fungus *Pseudotrametes gibbosa* (found in the northeast forested area of China) and *Pleurotus ostreatus* (which has been studied both domestically in China and overseas) with the goals of identifying a potent degradative aboriginal fungus and providing technical support for bioremediation utilising aboriginal white rot fungus.

6.4.2   Bioremediation of Contaminated Soils

Bioremediation of contaminated soils using white rot fungi has been investigated for many years (Aust 1990). Andersson and Henrysson (1996) studied PAH degradation by five white rot fungi (*T. versicolor* PRL 572, *T. versicolor* MUCL 28407, *P. ostreatus* MUCL 29527, *Pleurotus sajor-caju* MUCL 29757, and *P. chrysosporium* DSM 1556) by adding anthracene, benz[*a*]anthracene, and dibenz[*a,h*]anthracene to soil. The white rot fungi were cultivated in the soil contaminated with wheat straw and these pollutants. In a heterogenous soil environment, the fungi displayed different degradation abilities. *Trametes* showed poor degradation performance, while anthracene was completely transformed by *Phanerochaete* and *Pleurotus*. Marquez-Rocha et al. (2000) studied the degradation of PAHs adsorbed by the white rot fungus, *P. ostreatus*. After 21 days, 50 % of pyrene, 68 % of anthracene, and 63 % of phenanthrene were mineralised. The respective biodegradation percentage was increased to 75 %, 80 %, and 75 % when 0.15 % Tween 40 was added. Biodegradation of pyrene in the presence of a surfactant and H_2O_2 was 90 %. Eggen and Sveum (1999) investigated the effect of inoculation by the white rot fungus *P. ostreatus* for PAHs degradation in aged creosote-contaminated soil. *P. ostreatus* had an overall positive effect on the degradation of PAHs, and under the preconditions, the degradation increased with the increasing temperature. Chen et al. (2005) studied that temperature, medium composition, dissolved oxygen, and the moisture content in the soil influenced the degradation of PAHs.

The explosives TNT, HMX, and RDX are integral components of many munitions. Degradation of TNT was investigated using four different strains of white rot fungi (*Phanerochaete chrysosporium*, *Phanerochaete sordida*, *Phlebia brevispora*, and *Cyathus stercoreus*) in the liquid medium (Donnelly et al. 1997). The data indicated that TNT concentration (from 90 mg/L) in the liquid medium to below detection limits. *P. sordida* showed a relatively high growth rate and the fastest rate of TNT degradation. Chemical analysis revealed that the major metabolites in the initial transformation of TNT were the monoamino-dinitrotoluenes, which were also that white rot fungi are capable of metabolising and detoxifying TNT under aerobic conditions in a nonlignolytic liquid medium. The degradation of TNT by white rot fungi involved two distinct steps: the first step was degradation to OHADNT and ADNT, and the second step was a DANT (Aken et al. 1999). Axtell et al. (2000) reported that strains of *P. chrysosporium* and *P. ostreatus*, adapted to grow on high concentrations of TNT, were able to cause extensive degradation of TNT, HMX, and RDX.

Much research has been conducted on the degradation of chlorinated hydrocarbons. Zou and Zhang (1998) investigated the degradation of chlorinated pesticide by over 90 %. Ruiz-Aguilar et al. (2002) reported that three white rot fungi and achieved degradation of chlorinated pesticide by over 90 %. Ruiz-Aguilar et al. (2002) reported that three white rot fungi were used to degrade a mixture of PCBs at high initial concentrations from 600 to 3,000 mg/L, in the presence of a nonionic surfactant (Tween 80). The PCBs were extracted from historically PCB-contaminated soil. Preliminary experiments showed that Tween 80 exhibited the highest emulsification index of the three surfactants tested (Tergitol NP-10, Triton X-100, and Tween 80). Tween 80 had no inhibitory effect on fungal radial growth, whereas the other surfactants inhibited the growth rate by 75–95 %. PCB degradation ranged from 29 % to 70 %, 34 % to 73 %, and 0 % to 33 % for *T. versicolor* and from 29 % to 70 %, 34 % to 73 %, and 0 % to 33 % for *T. versicolor*, *Phanerochaete chrysosporium*, and *Lentinus edodes*, respectively, in 10-day incubation tests. The highest PCB transformation (70 %) was obtained with *T. versicolor* at an initial PCB concentration of 1,800 mg/L, whereas *P. chrysosporium* could modify 73 % at 600 mg/L. Zou and Zhang (1998) studied on the mechanism of degradation of chlorinated pesticides by white rot fungi by analysing degradation products with GC-MS. They found that the 1,1-Cl on the atom was more than the Cl atoms directly linked to the benzene ring during DDT degradation and that dechlorination of the chloride atom on benzene was the rate-limiting step of the reaction. Kamei et al. (2006) studied the degradation of 4,4′-DCB by the white rot fungi *P. chrysosporium*, and *Phanerochaete* sp. MZ142 was better than that of *P. chrysosporium*. Hydroxylation of 4,4′-DCB by *Phanerochaete* sp. MZ142 was different from hydroxylation by *Phanerochaete* sp. Although 2-OH-4,4′-DCB was not methylated, the metabolic pathway of 3-OH-4,4′-DCB was branched acid, 4-chlorobenzyl alcohol, and 4-chlorobenzaldehyde. Degradation of PCP by the fungus was reported previously (Ullah and Evans 1999). Degradation of lindane by *P. ostreatus* was studied by several researchers in relation to degradation conditions (Rigas et al. 2005). The degradation of TCE by the white rot fungus *T. versicolor* produces 2,2,2-trichloroethanol and CO_2 with chloral as an intermediate (Marco-Urrea et al. 2008). This pathway is different from the aerobic metabolic degradation of TCE.

6.4.3   Wastewater Treatment

In recent years, many researchers have indicated that white rot fungi are promising microorganisms in wastewater treatment. *P. chrysosporium* is the most investigated species. Fan et al. (2001) developed a process for dye wastewater which consisted of three parts—trifling electrolysis, biodegradation by white rot fungi, and flocculation precipitation. COD_{Cr} removal reached 90 %, and chroma declined from 12,800 to 80 and met the wastewater with white rot fungi containing peat, and the discharged water reached the standards in China. Huang and Zhou (1999) used white rot fungi to treat the TNT wastewater and achieved degradation of TNT by more than 99 %. Fang and Huang (2002) treated bleaching wastewater of paper pulp factory by a white rot fungus coagulation process, and COD_{Cr} and OD_{465} in the effluent were 185.1 mg/L and 0.0042 under optical conditions. The removal of COD and OD_{465} was 99.4 % and 86.5 %, respectively. The authors used polyurethane foam as carriers to immobilise the white rot fungus *P. chrysosporium* for the biodecolourisation of reactive dyes (Liang and Gao 2008). Results showed that stable decolourisation was as high as 95 % in an immobilised reactor system after incubating *P. chrysosporium* for only 2 days in comparison with 15 % in a suspended culture for 5 days. The maximum activity of MnP was 915.62 U/L in the immobilised system as compared with 324.90 U/L in the suspended culture along with the consumption of the carbon and nitrogen substrates (Liang and Gao 2008). Vahabzadeh et al. (2004) studied the white rot fungus *P. chrysosporium* for treating molasses wastewater from an ethanol fermentation plant. In diluted wastewater, addition of spores caused a fading rate of up to 75 % on the fifth day. The decolourisation was found to be correlated to the activity of the lignolytic enzyme system. The lignin peroxidise (LiP) activity was 185 U/L, while manganese peroxidise (MnP) activity was 25 U/L. Gomaa et al. (2008) studied the ability of the white rot fungus *P. chrysosporium* to decolourise Victoria blue B (VB) in textile dyes. Inhibition of laccase production by adding various inhibitors to the shaken cultures had a great negative influence on decolourisation. When sodium azide and aminotriazole were added to inhibit the activities of the endogenous catalase and cytochrome P-450 oxygenase, the decolourisation rate decreased by 100 % and 70 %, respectively. Addition of benzoate resulted in a decrease of 50 %. Ergul et al. (2009) studied treatment of olive mill wastewater and the operational conditions of *T. versicolor* FPRL 28A INI. The results showed removal of phenolics by 78 % in shake flasks and 39 % under static condition. In continuously stirred tank reactors (CSTR), the removal of total phenolics reached 70 %.

6.5 Mechanisms of Biodegradation

The mechanism of biodegradation by white rot fungus depends on the compound being degraded, but there are some consistent steps in the process regardless of the substrate. The lignolytic enzymes in white rot fungi catalyse the degradation of pollutants by using a non-specific free radical mechanism (Pointing 2001; Law et al. 2003). When an electron is added or removed from the ground state of a chemical, it becomes highly reactive, allowing it to give or take electrons from other chemicals. This provides the basis for the non-specificity of the enzymes and the ability of the enzymes to degrade xenobiotics, chemicals that have never been encountered in nature. The main reactions that are catalysed by the lignolytic enzymes include depolymerisation, demethoxylation, decarboxylation, hydroxylation, and aromatic ring opening. Many of these reactions result in oxygen activation, creating radicals that perpetuate oxidation of the organopollutants

(Reddy and Mathew 2001). Once the peroxidases have opened the aromatic ring structures by way of introducing oxygen, other more common species of fungi and bacteria can mineralise the products intracellularly into products such as CO_2 and other benign compounds (Hamman 2004).

Bioremediation using microorganisms is already an established technology, although almost all currently employed treatments use prokaryotes. Treatments employing white rot fungi offer the possibility to expand the substrate range of existing treatments via biodegradation of pollutants that cannot be removed by prokaryotes (or by chemical means). White rot fungal bioremediation treatments may be particularly appropriate for in situ remediation of soils, where recalcitrant compounds (e.g. the larger PAH) and bioavailability are problematic. A further application may lie in the operation of bioreactors for certain compounds (e.g. synthetic dyes) in liquid waste, where near-100 % degradation efficiencies have been achieved using white rot fungi.

6.6   Principal White Rot Fungi in Bioremediation: *Phanerochaete chrysosporium*

Phanerochaete chrysosporium as shown in Fig. 6.2, a lignin-degrading white rot fungus, is known to mineralise a wide range of chloroaromatic environmental pollutants to CO_2. It produces different extracellular enzymes involved in lignin degradation. *P. chrysosporium* has also been shown to mineralise a variety of recalcitrant aromatic pollutants. Ligninase has been shown to catalyse limited oxidation of benzo(*a*)pyrene and other polycyclic aromatics, as well as a number of phenolic pollutants. *P. chrysosporium* has been the most intensively studied white rot fungus. White rot fungi secrete an array of peroxidases and oxidases that act non-specifically via the generation of lignin free radicals, which then undergo spontaneous cleavage reactions. The non-specific nature and exceptional oxidation potential of the enzymes have attracted considerable interest for application in bioprocesses such as organopollutant degradation and fibre bleaching.

P. chrysosporium has several features that might make it very useful. First, unlike some white rot fungi, it leaves the cellulose of the wood virtually untouched. Second, it has a very high optimum temperature (about 40 °C), which means it can grow on wood chips in compost piles, which attain a very high temperature. These characteristics point to some possible roles in biotechnology.

In recent years, this white rot fungus *P. chrysosporium* has shown promise as an organism suitable for the breakdown of a broad spectrum of environmental pollutants, including polynuclear aromatic hydrocarbons (PAHs), 2,4,6-trinitrotoluene (TNT), and polychlorinated hydrocarbons (PCBs). *P. chrysosporium* is also able to degrade a wide variety of environmentally persistent xenobiotics to carbon dioxide, including a number of chlorinated hydrocarbons such as DDT [1,1,1-trichloro-2,2-bis (4-chlorophenyl)ethane], lindane (1,2,3,4,5,6-hexachlorocyclohexane), chloroanilines, and polychlorinated biphenyls.

Fig. 6.2 Picture showing white rot fungus— *Phanerochaete chrysosporium*

6.7   Case Studies

6.7.1   *Degradation of Pentachlorophenol by White Rot Fungus* (Phanerochaete chrysosporium)

Microorganisms are known to utilise phenolic substances, recalcitrant molecules, and even xenobiotic compounds as carbon source for their growth. *A. pseudomonas* sp. strain isolated from a consortium could be used very effectively for in situ bioremediation in an environment which is highly contaminated with PCP, other chlorinated phenols, and hexadecane (Murialdo et al. 2003). White rot fungi efficiently degrade lignin, a complex aromatic polymer in wood that is among the most abundant natural materials on earth (Martinez et al. 2004). The white rot fungi have been widely studied for their ability to degrade variety of environmental soil pollutants, including pentachlorophenol (Aiken and Logan 1996). Joyce et al. (1987) reported that white rot fungi were able to degrade pentachlorophenol and 2,4,6-trichlorophenol at concentrations up to 250 mg/L and it could be reduced to less than 5 mg/L in 96 h. If white rot fungi are to be used for large-scale bioremediation of PCP-contaminated soils and wastewaters, however, glucose media will need to be replaced by other less expensive carbon sources. The fungus can degrade various other xenobiotics such as polyaromatic hydrocarbons and chlorinated aromatic compounds and also pollutants, which are covalently bound to humic substances (Leung and Ponting 2002). The white rot fungus *P. chrysosporium* can be used for bioremediation of phenolic, xenobiotic compounds, and decolourisation of textile effluents. The siderophores detected from the culture of the organism have been found useful in the decolourisation and remediation of the effluent (Asamudo et al. 2005).

In this report, the results were examined by using lignosulphonate (LS) as a fungal growth medium. LS is a waste product of paper mill industry generated during the bleaching and pulping process. These solutions are highly coloured (dark brown) due to their lignin content and therefore usually disposed of by incineration. However, LSs contain large concentrations of wood sugars, nitrogen (17.6 g/L as NH_3-N), and other trace minerals needed for growth of fungi making LS a logical choice as an alternative growth substrate. The persistence of the dark colour of LS could pose a problem for its use in soil bioremediation or wastewater treatment processes. However, *P. chrysosporium* can decolour sulphonised azo dye compounds which are structurally similar to the substructures of LS and chlorolignins in paper mill waste streams.

6.7.1.1 Pentachlorophenol Degradation

P. chrysosporium has been demonstrated to mineralise up to 50.5 % of 14C-radiolabelled PCP when grown under lignolytic culture conditions (Mileski et al. 1988), although no confirmation that LMEs were involved was obtained. Growth of seven species belonging to the white rot fungal genus *Phanerochaete* was severely reduced in the presence of even low levels (5 ppm) of PCP (Lamar et al. 1993).

The extent of PCP degradation in static flask cultures was evaluated based on the disappearance of the PCP. PCP was added by injecting 10 µL of a 15 mg PCP/mL ethanol solution into five Erlenmeyer flask 3 days after inoculation with fungi. On days 5, 6, 7, and 8, the contents of each flask including mycelium were poured into a stainless steel blender. To remove mycelium and PCP that was attached to the sides of the flask was rinsed with 3 mL ultra pure water, 1 mL 95 % ethanol, and 3 mL ultra pure water. In order to ensure both the liquid and mycelia samples were homogenous, this mixture was blended on high speed for 2 min. This suspension was then centrifuged at 4 °C for 30 min to separate mycelium from the liquid medium. The total PCP in liquid was measured directly by injecting 20 µL into a high-performance liquid chromatography (HPLC, Varian) equipped with a C-18 reverse phase column. The mobile phase was a 75:25:0.125 mixture of acetonitrile/water/acetic acid. PCP was monitored at 238 nm and peak counts compared to a standard calibration curve.

6.7.1.2 Assays of Extracellular Enzyme

The laccase (*p*-diphenol oxygen oxidoreductase) enzyme was assayed during PCP degradation, and the assay was performed based on monitoring the rate of oxidation of syringaldazine (Sealey and Ragauskas 1998) in a spectrophotometer (ECIL, Hyderabad). The laccase plate assay was done as per the method of Srinivasan et al. (1995).

6.7.1.3 Pentachlorophenol Removal and Degradation

Scientist compared the ability of fungus grown in a LS medium to degrade PCP with fungus grown in a nitrogen-limited 2 % glucose medium by measuring the disappearance of PCP from solution. Pentachlorophenol was successfully degraded by *Phanerochaete chrysosporium* (TL 1) grown in media containing LS but lesser extent than cultures grown in glucose medium. Cultures of *Phanerochaete chrysosporium* (TL 1) which are grown on nitrogen-limited glucose medium removed 93 % of initial PCP during 5 days of incubation. When 2 % LS was used as the nitrogen source and carbon source, PCP removal was 85 %. Similarly, when LS was used as a nitrogen source, PCP removal was 86 %, and the overall PCP degradation in the two media containing LS (2 and 0.23 %) was more or less similar. Recovery of PCP from heat-killed control cultures appeared to be unaffected. It is reasonable to assume that mineralisation would follow dehalogenation since aromatic compounds such as phenol have been shown to serve as sole carbon sources for growth of *P. chrysosporium* (Krivobok 1994). Other investigators have generally observed more rapid removals of PCP by *P. chrysosporium* in liquid cultures (Lamar et al. 1993) than observed here. Since the LSs inhibit laccase enzyme activity, the percent PCP degradation was reduced for fungi grown in LS media compared to nitrogen-deficient glucose medium.

6.7.2   Biodegradation of 2,4,6-Trinitrotoluene

A number of nitroaromatic compounds such as nitrobenzenes, *o*- and *p*-nitrotoluene, 2,4-dinitrotoluene, and nitrobenzoates are mineralised at a relatively slow pace by microorganisms. These chemicals have multiple applications in the synthesis of polyurethane foams, herbicides, insecticides, pharmaceuticals, and explosives, all of which are more recalcitrant than the raw material from which they are synthesised. The most widely used nitroaromatic compound is 2,4,6-trinitrotoluene (TNT). TNT is more recalcitrant than mono- and dinitrotoluenes mainly because of the symmetric location of the nitro groups on the aromatic ring, an arrangement that limits attack by classic dioxygenase enzymes involved in the microbial metabolism of aromatic compound (Fig. 6.3).

The wood white rot fungus *Phanerochaete chrysosporium* has been the subject of intensive study, and more recently other white rot fungi and litter decay fungi have been investigated for their ability to transform TNT. This system contains lignin peroxidase, manganese peroxidise (MnP), oxidases, reductases, hydrogen peroxidase, veratryl alcohol, oxalate, and quinol oxidases. As in other organisms, the initial steps in the fungal degradation of TNT involve the reduction of nitro groups. Mycelia of *P. chrysosporium* reduce TNT to a mixture of 4-ADNT, 2-ADNT, 4-hydroxylamino-2,6-dinitrotoluene, and azoxytetranitrotoluenes.

Under lignolytic conditions, the aromatic compounds and azoxytetranitrotoluenes disappear, and mineralisation can be fairly extensive. Aust and Stahl provided evidence that TNT is reduced by a plasma membrane redox system in *P. chrysosporium* that requires live and intact mycelia. Any conditions that disrupt the integrity of the plasma membrane destroy the reductase activity. The presence of compounds known to inhibit the membrane redox systems also inhibits TNT reductase. The reduction coupled to the proton export system would be used by the fungus to maintain the extracellular physiological pH at approximately 4.5

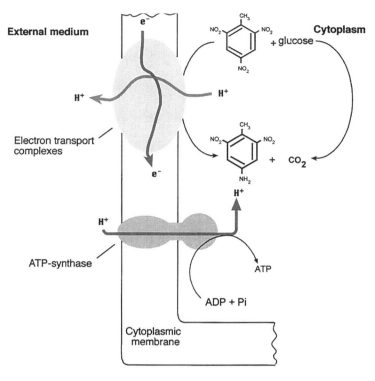

Fig. 6.3 Scheme showing the coupling of electron donor compounds, TNT oxidoreduction, and ATP synthesis, Pi, inorganic phosphate

(Esteve-Nunez et al. 2001). Rieble et al. (1994) had reported a membrane-associated TNT reductase activity that required NADPH as a co-substrate and the absence of molecular oxygen. Because activity was detected in the detergent-solubilised form, where a membrane potential could not be maintained, a membrane potential redox system was assumed to be inessential for this reaction. On the other hand, NAD(P) H-dependent intracellular TNT reductase activity has also been described (Michels and Gottschalk 1994). Regardless of the site where of TNT is reduced, further degradation of these compounds and eventual mineralisation of TNT by *P. chrysosporium* occur only when cultures are lignolytic, implying that lignin peroxidase, MnP, and/or other enzymes of the lignolytic system further transform the reduced products of TNT. The mechanism by which the fungi mineralise the explosive is not known, but preliminary information about the process has been reported (Hodgson et al. 2000; Michels and Gottschalk 1994). When lignolytic cultures of *P. chrysosporium* were incubated with 4-ADNT, a compound identified as 4-formamide-2,6-DNT was detected. This intermediate was transformed into 2-amino-4-formamide-6-nitrotoluene, a compound that disappeared rapidly under lignolytic conditions but not under nonlignolytic ones.

Most mineralisation studies have employed relatively low TNT concentrations, but one study has shown that *P. chrysosporium* can germinate from spores, grow vegetatively, and transform TNT at 20 ppm (Spiker et al. 1992). It was also shown that when TNT was added to established liquid-grown cultures, the fungus was able to transform TNT at levels of up to 100 ppm, although further germination and vegetative growth were inhibited. Fernando et al. (1990) demonstrated TNT mineralisation by *P. chrysosporium* in soils when present at levels of up to: 10,000 ppm. Bioremediation systems using *P. chrysosporium* for treating water, soils, sediments, and other materials that are contaminated with toxic or recalcitrant organopollutants or both have been found successful.

6.7.3 Biodegradation of Polynuclear Aromatic Hydrocarbons by Phanerochaete chrysosporium

Polycyclic aromatic hydrocarbons (PAHs) are benzene homologues formed from the fusion of four or more benzene rings. A considerable number of individual PAH are found, and these arise from natural oil deposits and vegetation decomposition, in addition to considerable anthropogenic production from the use of fossil fuels in heating and power production, wood burning, vehicular transport, run-off from bitumen roads, waste incineration, and industrial processes (Alloway and Ayres 1993). These compounds present huge problems of toxicity and persistence in the environment. Significantly, the white rot fungi are the only organisms capable of significant PAH mineralisation.

Phanerochaete chrysosporium has shown promise as an organism suitable for the breakdown of a broad spectrum of environmental pollutants, including polynuclear aromatic hydrocarbons (PAHs). It has also been shown that *P. chrysosporium* can degrade the more recalcitrant PAHs containing four or more rings. It appears that this degradation is due to the structural similarities of these pollutants to portions of the lignin substructure and the low-level specificity of the complement of ligninases produced by *P. chrysosporium*.

It was reported by Toby and Raymond 1992 that *P. chrysosporium* can be used as supplement for enhanced in situ biodegradation of PAHs in oil tar-contaminated soil obtained from a former oil gasification plant. Phenanthrene was one of the principal PAHs in the soil and was therefore used as a tracer to monitor mineralisation (i.e. complete degradation to CO_2 and water) and metabolism under various conditions.

6.7.4 Biodegradation of Gaseous Chlorobenzene by White Rot Fungus: Phanerochaete chrysosporium

Evaluation of the effect of white rot fungus *Phanerochaete chrysosporium* on removal of gaseous chlorobenzene was carried out by Wang et al. (2008). Fungal mycelium mixed with a liquid medium was placed into airtight bottles. A certain amount of chlorobenzene was injected into the headspace of the bottles under different conditions. At a certain interval, the concentrations in the headspace were analysed to evaluate the degradation of chlorobenzene by *P. chrysosporium*. The degradation effects of *P. chrysosporium* on chlorobenzene under different conditions were investigated. The difference in the optimum temperature for the growth of the fungi and chlorobenzene degradation was observed. The data indicated that a lower temperature (28 °C) would promote the degradation of chlorobenzene than the optimum temperature for the growth of the fungi (37 °C). A low nitrogen source concentration (30 mg N/L) had a better effect on degrading chlorobenzene than a high nitrogen source concentration (higher than 100 mg N/L). A high initial concentration (over 1,100 mg/m^3) of chlorobenzene showed an inhibiting effect on degradation by *P. chrysosporium*. A maximum removal efficiency of 95 % was achieved at the initial concentration of 550 mg/m^3. *P. chrysosporium* was found to have a rather good ability to remove gaseous chlorobenzene. A low nitrogen source concentration and a low temperature promote the removal of chlorobenzene by *P. chrysosporium*.

6.7.5   Biodegradation of Polychlorinated Biphenyls

Polychlorinated biphenyls (PCBs) are a family of compounds with a wide range of industrial applications in heat transfer fluids, dielectric fluids, hydraulic fluids, flame retardants, plasticisers, solvent extenders, and organic diluents. *Phanerochaete chrysosporium*, a lignin-degrading white rot fungus, is known to mineralise a wide range of chloroaromatic environmental pollutants to CO_2 (Hammel et al. 1986). Degradation of many of these pollutants was shown to be mediated by the lignin-degrading enzyme system of this organism. Major components of the lignin-degrading enzyme system include lignin peroxidises (LIPs), Mn(II)-dependent peroxidases (MNPs), and the H_2O_2-producing system, which are induced during secondary metabolism, under nutrient-limiting culture conditions but not under nutrient-rich conditions.

In the United States and the United Kingdom, complex PCB mixtures were manufactured under the trade name Aroclor. Three of the commonly used Aroclors are 1242, 1254, and 1260, which contain 42, 54, and 60 % chlorine by weight, with an average of 3, 5, and 6 chlorines per biphenyl molecule, respectively. Aroclors consist of a number of congeners which differ in the number and distribution of chlorines on the biphenyl nucleus. About 150 congeners have been reported in the environment. PCBs have entered into soil and sediment environments as a result of improper disposal of industrial PCB wastes and leakage of PCBs from electric transformers. Evidence for substantial degradation of polychlorinated biphenyl mixtures Aroclor 1242, 1254, and 1260 by the white rot fungus *Phanerochaete chrysosporium* based on congener-specific gas chromatographic analysis has been reported by Yadav et al. (1995). Maximal degradation (percent by weight) of

Aroclors 1242, 1254, and 1260 was 60.9, 30.5, and 17.6 %, respectively. Most of the congeners in Aroclors 1242 and 1254 were degraded extensively both in low-N (lignolytic) as well as high-N (nonlignolytic) defined media. Even more extensive degradation of the congeners was observed in malt extract medium. Congeners with varying numbers of *ortho*, *meta*, and *para* chlorines were extensively degraded. Aroclor 1260 was shown to undergo substantial net degradation by *P. chrysosporium*.

6.7.6   Other Compounds

6.7.6.1   Organochlorines

Organochlorines like DDT are highly persistent and are extracellular LMEs of *P. chrysosporium* were incapable of DDT degradation (Kohler et al. 1988). Biodegradation of DDT can result in toxic and persistent metabolites. One study has shown that 14C-radiolabelled 1,1-dichloro-2,2-bis(4-chlorophenyl)ethylene (DDE), an extremely toxic and persistent DDT breakdown product, is mineralised to $14CO_2$ by *P. chrysosporium* (Bumpus et al. 1993) degraded slowly in the environment. The mineralisation of the dioxin 2,7-dichlorodibenzo-*p*-dioxin by *P. chrysosporium* has been demonstrated (Valli et al. 1992). In this study purified LiP and MnP were capable of mineralisation in a multistep pathway involving sequential oxidation, reduction, and methylation reactions to remove the two Cl atoms and carry out ring cleavage.

6.7.6.2   Synthetic Dyes

Synthetic dyes are chemically diverse, with those commonly used in industry divided into those of azo, triphenylmethane, or heterocyclic/polymeric structure (Gregory 1993). Early studies showed that polymeric dyes were decolourised by lignolytic cultures of *P. chrysosporium* and that inhibitors of lignin degradation also inhibited dye decolourisation (Glenn and Gold 1983). LiP of *P. chrysosporium* has been shown to decolourise azo, triphenylmethane, and heterocyclic dyes in the presence of veratryl alcohol and H_2O_2 (Cripps et al. 1990; Ollikka et al. 1993).

6.7.6.3   BTEX

Benzene, toluene, ethylbenzene, and xylene used in the gasoline and aviation fuels are found to be degraded by *P. chrysosporium* (Yadav and Reddy 1993).

6.8   Bioremediation Mechanism of *Phanerochaete chrysosporium*

During the past several years, one organism which has been examined for its ability to degrade recalcitrant pollutants is the white rot fungus *P. chrysosporium*. Although the natural substrate degraded by this organism is lignin, the enzyme complement secreted by *P. chrysosporium* can degrade a variety of recalcitrant pollutants. This white rot fungi *P. chrysosporium* technology is very different from other well-established methods of bioremediation (e.g. bacterial systems). The differences are primarily due to the unusual mechanisms which nature has provided them with and several advantages for pollutant degradation. One distinct advantage these fungi have over bacterial systems is that they do not require preconditioning to the particular pollutant. Bacteria usually must be preexposed to a pollutant to allow the enzymes that degrade the pollutant to be induced. The pollutant also must be in a significant concentration; otherwise, induction of enzyme synthesis cannot occur. Thus, there is a finite level to which bacteria can degrade pollutants. Also because the induction of the degrading enzyme is not dependent on the pollutant in the fungi, the pollutant can be degraded to a near non-detectable level. In contrast to the bacterial system, the degradative enzymes of white rot fungi are induced by nutrient limitation. Thus, cultivation of the white rot fungi on a nutrient-limited substrate will initiate the process (Tuomela et al. 1999).

P. chrysosporium has shown to mineralise a variety of recalcitrant aromatic pollutants. Ligninase has been shown to catalyse limited oxidation of benzo(*a*)pyrene and other polycyclic aromatics, as well as a number of phenolic pollutants. The fungus can degrade various other xenobiotics such as polyaromatic hydrocarbons and chlorinated aromatic compounds, and also pollutants, which are covalently bound to humic substances (Pointing 2001). Humic substances consist of aromatic rings connected by flexible and rather long aliphatic chains (Tuomela et al. 1999). This structure is formed by oxidative ring opening lignin, loss of phenolic and methoxyl groups, and an increase in carboxyl and carbonyl group. Humic substances are thus less aromatic and have fewer methoxyl and more carboxyl groups than lignin. During the degradation of xenobiotics, the white rot fungi often polymerise or convert substantial amount of compounds to humic bound products.

It is clear that the biodegradative activity of *P. chrysosporium* is a complex one. Understanding the mechanisms of the biodegradation role of this fungus is very important if one must explore the unique enzyme system in it for remediation of coloured and complex, toxic effluents. The development of biotechnologies using white rot fungi has been implemented to treat various refractory wastes and to bioremediate contaminated soils. Degradation of many hazardous chemicals and wastes has been demonstrated on a laboratory scale, especially under sterile conditions. The technical challenge remains for the application including bacterial contamination and for the scaleup of the process. The white rot fungus *P. chrysosporium* has been applied for scaled-up bioremediation in the field.

6.9   The Requirements for Successful Bioremediation in Order to Better Understand Research Progress on White Rot Fungi

(a) Microorganisms must exist that have the needed catabolic activity.
(b) Those organisms must have the ability to transform the compound at reasonable rates and bring the concentration to levels that meet regulatory standards.
(c) They must not generate products that are toxic at the concentrations likely to be achieved during the remediation.
(d) The site must not contain concentrations or combinations of chemicals that are markedly inhibitory to the biodegrading species, or means must exist to dilute or otherwise render innocuous the inhibitors.
(e) The target compound(s) must be available to the microorganisms.
(f) Conditions at the site or in a bioreactor must be made conducive to microbial growth or activity, for example, an adequate supply of inorganic nutrients, sufficient O_2, or some other electron acceptor, favourable moisture content, suitable temperature, and a source of C and energy for growth if the pollutant is to be cometabolised.
(g) The cost of the technology must be less or, at worst, no more expensive than that of other (Alexander 1994).

More research and development is still needed for cost-effective and sustainable application. The stability of the enzymes in relation to the physicochemical nature of the effluents is an important factor in evaluating both technical and economic feasibility of using this organism commercially in bioremediation projects. Thus, the rate of enzyme inactivation is an important component of the overall kinetics of any proposed enzymatic process. Continuous research will eventually close the present gap in knowledge about the use of this organism.

6.10   Conclusion and Future Perspective

Industrial processes and operations generate the hazardous wastes such as petroleum hydrocarbons and pesticides. In spite of the present treatment technology, residual compounds have been persisting in soil, water, and environment. Research studies on bioremediation with special reference to the use of *Phanerochaete chrysosporium* for biodegradation of hazardous compounds have been reported. *Phanerochaete chrysosporium* —lignin degrader, white rot fungus—is known to degrade pentachlorophenol, polychlorinated biphenyls, and chlorinated aromatic compound pollution into environmentally friendly compounds.

The growth of *Phanerochaete chrysosporium* and the conditions required for the biodegradation of compounds with special reference to research studies are highlighted so that white rot fungus can be used as effective measure for the biodegradation of hazardous waste compounds such as polycyclic aromatic hydrocarbons (PAHs), chlorobenzene, organochlorines, and synthetic dyes. *Phanerochaete chrysosporium* will be effective and efficient microbial source for the hazardous waste compounds to clean up the environment.

References

Aiken BS, Logan BE (1996) Degradation of pentachlorophenol by the white rot fungus Phanerochaete chrysosporium grown in ammonium lignosulphonate media. Biodegradation 7(3):175–182

Aken BV, Hofrichter M, Scheibner K, Hatakka AI, Naveau H, Agathos SN (1999) Transformation and mineralization of 2,4,6-trinitrotoluene (TNT) by manganese peroxidase from the white rot basidiomycete *Phlebia radiata*. Biodegradation 10:83–91

Alexander M (1994) Biodegradation and bioremediation. Academic, New York

Alloway JB, Ayres DC (1993) Chemical principles of environmental pollution. Chapman and Hall, London

Andersson BE, Henrysson T (1996) Accumulation and degradation of dead-end metabolites during treatment of soil contaminated with polycyclic aromatic hydrocarbons with five strains of white rot fungi. Appl Microbiol Biotechnol 46:647–652

Asamudo NU, Daba AS, Ezeronye OU (2005) Bioremediation of textile effluent using *Phanerochaete chrysosporium*. Afr J Biotechnol 4:1548–1553

Aust SD (1990) Degradation of environmental pollutants by *Phanerochaete chrysosporium*. Microb Ecol 20:197–204

Aust SD, Swaner PR, Stahl JD (2003) Detoxification and metabolism of chemicals by white rot fungi. *Pesticide decontamination and detoxification*. In: Zhu JJPC, Aust SD, Lemley Gan AT (eds) Pesticide decontamination and detoxification. Oxford University Press, Washington, DC, pp 3–14

Axtell C, Johnston CG, Bumpus JA (2000) Bioremediation of soil contaminated with explosives at the Naval Weapons Station Yorktown. Soil Sediment Contam Int J 9:537–548

Bumpus JA, Mileski G, Brock B, Ashbaugh W, Aust SD (1988) Biological oxidations of organic compounds by enzymes from a white rot fungus. In: Land disposal, remedial action, incineration and treatment of hazardous waste. Proceedings of the fourteenth annual research symposium, May 1988, Cincinnati, OH

Bumpus JA, Kakar SN, Coleman RD (1993) Fungal degradation of organophosphorous insecticides. Appl Biochem Biotechnol 39:715–726

Chen J, Hu JD, Wang XJ, Tao S (2005) Degradation of polycyclic aromatic hydrocarbons from soil by white rot fungi. Environ Chem 24:270–274 (in Chinese)

Chupungars K, Rerngsamran P, Thaniyavarn S (2008) Polycyclic aromatic hydrocarbons degradation by *Agrocybe sp.* CU-43 and its fluorene transformation. Int Biodeter Biodegrad 1:1–7

Cookson JT (1995) Bioremediation engineering: design and application. McGraw Hill, New York

Couto SR (2007) Decolouration of industrial azo dyes by crude laccase from *Trametes hirsute*. J Hazard Mater 148:768–770

Cripps C, Bumpus JA, Aust SD (1990) Biodegradation of azo and heterocyclic dyes by *Phanerochaete chrysosporium*. Appl Environ Microbiol 56:1114–1118

Donnelly KC, Chen JC, Huebner HJ, Brown KW, Autenrieth RL, Bonner JS (1997) Utility of four strains of white rot fungi for the detoxification of 2,4,6-trinitrotoluene in liquid culture. Environ Toxicol Chem 16:1105–1110

Eaton RA, Hale MDC (1993) Wood, decay, pests and prevention. Chapman and Hall, London

Eggen T, Sveum P (1999) Decontamination of aged creosote polluted soil: the influence of temperature, white rot fungus *Pleurotus ostreatus*, and pretreatment. Int Biodeter Biodegrad 43:125–133

Ergul FE, Sargin S, Ongen G, Sukan FV (2009) Dephenolisation of olive mill wastewater using adapted *Trametes versicolor*. Int Biodeter Biodegrad 63:1–6

Esteve-Nunez A, Caballero A, Ramos J (2001) Biological degradation of 2,4,6-trinitotoluene. Microbiol Mol Biol Rev 65(3):335–352

Fan WP, Cao HJ, Zhang J, Wei H (2001) Study on the treatment of dyeing effluents by using rice straw powder immobilized mycelium of white rot fungus. Ind Water Treat 21:19–21 (in Chinese)

Fang JZ, Huang SB (2002) Process of oxidation by white rot-fungi and coagulation for treating bleaching effluents. Environ Sci Technol 25:12–13 (in Chinese)

Fernando T, Bumpus JA, Aust SD (1990) Biodegradation of TNT (2,4,6-trinitrotoluene) by *Phanerochaete chrysosporium*. Appl Environ Microbiol 56:1666–1671

Fulekar MH (2010) Environmental biotechnology: recent advances. Springer, New York

Gaspare L, Machiwa JF, Mdachi SIM, Streck G, Brack W (2009) Polycyclic aromatic hydrocarbon (PAH) contamination of surface sediments and oysters from the inter-tidal areas of Dares Salaam. Tanzania Environ Pollut 157:24–34

Glenn JK, Gold MH (1983) Decolorization of several polymeric dyes by the lignin-degrading basidiomycete *Phanerochaete chrysosporium*. Appl Environ Microbiol 45:1741–1747

Gomaa OM, Linz JE, Reddy CA (2008) Decolorization of Victoria blue by the white rot fungus, *Phanerochaete chrysosporium*. World J Microbiol Biotechnol 24:2349–2356

Gregory P (1993) Dyes and dye intermediates. In: Kroschwitz JI (ed) Encyclopedia of chemical technology, vol 8. Wiley, New York, pp 544–545

Hamman S (2004) Bioremediation capabilities of white rot fungi. Review article, Springer

Hammel KE, Kalyanaraman B, Kirk TK (1986) Oxidation of polycyclic aromatic hydrocarbons and dibenzo[p]dioxins by *Phanerochaete chrysosporium* ligninase. J Biol Chem 261:16948–16952

Hawksworth DL, Kirk PM, Sutton BC, Peggler DN (1995) Ainsworth and Bisby's dictionary of the fungi, 8th edn. CAB, Oxon, xii+616 pp, 195

Hestbjerg H, Willumsen PA, Christensen M, Andersen O, Jacobsen CS (2003) Bioaugmentation of tar-contaminated soils under field conditions using *Pleurotus ostreatus* refuse from commercial mushroom production. Environ Toxicol Chem 22(4):692–698

Hodgson J, Rho D, Guiot SR, Ampleman G, Thiboutot S, Hawari J (2000) Tween 80 enhanced TNT mineralization by *Phanerochaete chrysosporium*. Can J Microbiol 46:110–118

http://www.worldoffungi.org/Mostly_Mycology/Lucy_GoodeveDocker_bioremediation_website/whiterotfungi.htm

Huang J, Zhou SF (1999) Study on the biodegradation of TNT packing wastewater by white rot fungi. Environ Sci Technol (3):17–19 (in Chinese)

Joyce TW, Chang HM, Vasudevan B, Taneda H (1987). Degradation of hazardous organics by one white rot fungus-*Phanerochaete chrysosporium*. In Proceedings of the 184th American Chemical Society National Meeting, New Orleans, LA, Aug. 30–Sept.4, American Chemical, pp 217–217

Kamei I, Kogura R, Kondo R (2006) Metabolism of 4,4′-dichlorobiphenyl by white rot fungi *Phanerochaete chrysosporium* and *Phanerochaete sp.* MZ142. Appl Microbiol Biotechnol 72:566–575

Knapp JS, Newby PS (1999) The decolourisation of a chemical industry effluent by white rot fungi. Water Res 33:575–577

Kohler A, Jager A, Willeshansen H, Graf H (1988) Extracellular ligninase of *Phanerochaete chrysosporium* Burdsall has no role in degradation of DDT. Appl Microbiol Biotechnol 29:618–620

Krivobok S (1994) Diversity in phenol metabolizing capabilities in 809 strains of micromycetes. Microbiology 17:51–60

Lamar RT, Glaser JA, Evans JW (1993) Solid phase treatment of pentachlorophenol contaminated soil using lignin degrading fungi. Environ Sci Technol 27:2566–2571

Law WM, Lau WN, Lo KL, Wai LM, Chiu SW (2003) Removal of biocide pentachlorophenol in water system by the spent mushroom compost of *Pleurotus pulmonarius*. Chemosphere 52:1531–1537

Leung PC, Ponting SB (2002) Effect of different carbon and nitrogen regimes on poly R decolourisation by white rot fungi. Mycol Res 72:219–226

Liang H, Gao DW (2008) Enhanced biodecolorization of reactive dyes by immobilized *Phanerochaete chrysosporium*. J Biotechnol 136:S676

Marco-Urrea E, Parella T, Gabarrell X, Caminal G, Vicent T, Adinarayana Reddy C (2008) Mechanistics of trichloroethylene mineralization by the white rot fungus *Trametes versicolor*. Chemosphere 70:404–410

Marquez-Rocha FJ, Hernandez-Rodriguez VZ, Vazquez-Duhalt R (2000) Biodegradation of soil-adsorbed polycyclic aromatic hydrocarbons by the white rot fungus *Pleurotus ostreatus*. Biotechnol Lett 22:469–472

Martinez D, Larrondo LF, Putnam N, Maarten D, Sollewijn G, Katherine J (2004) Genome sequence of the lignocellulose degrading fungus *Phanerochaete chrysosporium* strainRP78. Nat Biotechnol 22:695–700

Michels J, Gottschalk G (1994) Inhibition of the lignin peroxidase of *Phanerochaete chrysosporium* by hydroxylamino-dinitrotoluene, an early intermediate in the degradation of 2,4,6-TNT. Appl Environ Microbiol 60:187–194

Mileski GJ, Bumpus JA, Jurek MA, Aust SD (1988) Biodegradation of pentachlorophenol by the white rot fungus *Phanerochaete chrysosporium*. Appl Environ Microbiol 54:2885–2889

Murialdo SE, Fenoglio R, Haure PM, Gonzalez JF (2003) Degradation of phenol and chlorophenols by mixed and pure cultures. Water SA 29:457–463

Ollikka P, Alhonmaki K, Leppanen V-M, Glumoff T, Raijola T, Suominen I (1993) Decolorization of azo, triphenylmethane, heterocyclic, and polymeric dyes by lignin peroxidase isoenzymes from *Phanerochaete chrysosporium*. Appl Environ Microbiol 59:4010–4016

Orth AB, Tien M (1995) Biotechnology of lignin degradation. In: Esser K, Lemke PA (eds) The mycota. II. Genetics and biotechnology. Springer, Berlin, pp 287–302

Pointing SB (2001) Feasibility of bioremediation by white rot fungi. Appl Microbiol Biotechnol 57:20–33

Pozdnyakova NN, Nowak JR, Turkovskaya OV, Haber J (2006) Oxidative degradation of polyaromatic hydrocarbons catalyzed by blue laccase from *Pleurotus ostreatus* D1 in the presence of synthetic mediators. Enzyme Microb Technol 39:1242–1249

Reddy CA, Mathew Z (2001) Bioremediation potential of white rot fungi. In: Gadd GM (ed) Fungi in bioremediation. Cambridge University Press, Cambridge

Rieble S, Joshi DK, Gold M (1994) Aromatic nitroreductase from the basidiomycete *Phanerochaete chrysosporium*. Biochem Biophys Res Commun 205:298–304

Rigas F, Dritsa V, Marchant R, Papadopoulou K, Avramides EI, Hatzianestis I (2005) Biodegradation of lindane by Pleurotus ostreatus via central composite design. Environ Int 31:191–196

Ruiz-Aguilar GML, Fernandez-Sanchez JM, Rodriguez-Vazquez R, Poggi-Varaldo H (2002) Degradation by white rot fungi of high concentrations of PCB extracted from a contaminated soil. Adv Environ Res 6:559–568

Sealey J, Ragauskas AJ (1998) Residual lignin studies of laccase delignified kraft pulps. Enzyme Microb Technol 23:422–426

Spiker JK, Crawford DL, Crawford RL (1992) Influence of 2,4,6-trinitrotoluene (TNT) concentration on the degradation of TNT in explosive-contaminated soils by the white rot fungus *Phanerochaete chrysosporium*. Appl Environ Microbiol 58:3199–3202

Srinivasan C, D'Souza TM, Boominathan K, Reddy CA (1995) Demonstration of laccase in the white rot basidiomycete *Phanerochaete chrysosporium* BKM-F1767. Appl Environ Microbiol 61:4274–4277

Suthersan S (1997) Remediation engineering design concepts. CRC Press, Boca Raton, FL

Thurston CF (1994) The structure and function of fungal laccases. Microbiology 140:19–26

Toby SB, Raymond LL (1992) Enhanced Bioremediation of Phenanthrene in Oil Tar-Contaminated Soils Supplemented with *Phanerochaete chrysosporium*. Appl Environ Microbiol 58(9):3117–3121

Tuomela M, Lyytikainen M, Oivanen P, Hatakka A (1999) Mineralization and conversion of pentachlorophenol (PCP) in soil inoculated with the white rot fungus *Trametes versicolor*. Soil Biol Biochem 31:65–74

Ullah MA, Evans CS (1999) Bioremediation of pentachlorophenol pollution by the fungus Coriolus versicolor. Land Contam Reclamat 7:255–260

Vahabzadeh F, Mehranian M, Saatari AR (2004) Color removal ability of *Phanerochaete chrysosporium* in relation to lignin peroxidase and manganese peroxidase produced in molasses wastewater. World J Microbiol Biotechnol 20:859–864

Valli K, Wariishi H, Gold MH (1992) Degradation of 2,7-dichlorodibenzo-*p*-dioxin by the lignin-degrading basidiomycete *Phanerochaete chrysosporium*. J Bacteriol 174:2131–2137

Wang C, Xi JY, Hu HY, Wen XH (2008) Biodegradation of gaseous chlorobenzene by white rot fungus *Phanerochaete chrysosporium*. Biomed Environ Sci 21:474–478

Wu YC, Luo YM, Zou DX, Ni JZ, Liu WX, Teng Y, Li ZG (2008) Bioremediation of polycyclic aromatic hydrocarbons contaminated soil with *Monilinia sp.*: degradation and microbial community analysis. Biodegradation 19:247–257

Yadav JS, Reddy CA (1993) Degradation of benzene, toluene, ethylbenzene and xylenes (BTEX) by the lignin-degrading basidiomycete *Phanerochaete chrysosporium*. Appl Environ Microbiol 59:756–762

Yadav JS, Quensen JF III, Tiedje JM, Reddy CA (1995) Degradation of polychlorinated biphenyl mixtures (Aroclors 1242, 1254 and 1260) by the white rot fungus *Phanerochaete chrysosporium* as evidenced by congener specific analysis. Appl Environ Microbiol 61 (7):2560–2565

Young RA, Akhtar M (eds) (1998) Environmentally friendly technologies for the pulp and paper industry. Wiley, New York. ISBN 0-471-15770-8

Zou SC, Zhang ZX (1998) The biodegradation of organochlorinated pesticides by *P. chrysosporium* fungi. Acta Scientiarum Naturalium Universitatis Sunyatseni 37:112–115 (in Chinese)

Chapter 7
Bioremediation of PAH-Contaminated Soil by Fungi

Irma Susana Morelli, Mario Carlos Nazareno Saparrat, María Teresa Del Panno, Bibiana Marina Coppotelli, and Angélica Arrambari

7.1   Introduction

Polycyclic aromatic hydrocarbons (PAHs) constitute a class of hazardous organic chemicals consisting of three or more fused benzene rings in linear, angular, and cluster arrangements (Cerniglia 1992). PAHs are unique contaminants in the

I.S. Morelli (✉)
Centro de Investigación y Desarrollo en Fermentaciones Industriales, CINDEFI
(UNLP, CCT-La Plata, CONICET), La Plata 1900, Argentina

Cátedra de Microbiología, Facultad de Ciencias Exactas, UNLP, La Plata, Argentina

Comisión de Investigaciones Científicas de la Provincia de Buenos Aires (CIC-PBA),
La Plata, Argentina
e-mail: guri@biol.unlp.edu.ar

M.C.N. Saparrat
Instituto de Botánica Spegazzini, Facultad de Ciencias Naturales y Museo, UNLP,
La Plata, Argentina

Instituto de Fisiología Vegetal, INFIVE (UNLP, CCT-La Plata, CONICET), La Plata, Argentina

Cátedra de Microbiología Agrícola, Facultad de Ciencias Agrarias y Forestales, UNLP,
La Plata, Argentina
e-mail: masaparrat@yahoo.com.ar

M.T. Del Panno
Centro de Investigación y Desarrollo en Fermentaciones Industriales, CINDEFI
(UNLP, CCT-La Plata, CONICET), La Plata 1900, Argentina

Cátedra de Microbiología, Facultad de Ciencias Exactas, UNLP, La Plata, Argentina

B.M. Coppotelli
Centro de Investigación y Desarrollo en Fermentaciones Industriales, CINDEFI
(UNLP, CCT-La Plata, CONICET), La Plata 1900, Argentina

A. Arrambari
Instituto de Botánica Spegazzini, Facultad de Ciencias Naturales y Museo, UNLP,
La Plata, Argentina

E.M. Goltapeh et al. (eds.), *Fungi as Bioremediators*, Soil Biology 32,   159
DOI 10.1007/978-3-642-33811-3_7, © Springer-Verlag Berlin Heidelberg 2013

environment because they are generated continuously by the inadvertently incomplete combustion of organic matter, for instance, in forest fires, home heating, traffic, and waste incineration (Johnsen et al. 2005). It is estimated that more than 90 % of the total burden of PAHs resides in the surface soils, where most of these compounds accumulate (Wild and Jones 1995). These ubiquitous organic pollutants exhibit strong carcinogenic and toxic properties (Berthe-Corti et al. 2007).

In soil, PAHs may undergo adsorption, volatilization, photolysis, and chemical oxidation, although microbial transformation is the major degradation process. The bioremediation of soil contaminated with PAHs should be a more efficient, financially affordable, and adaptable choice than physicochemical treatment because it presents potential advantages such as the complete degradation of the pollutants, lower treatment cost, greater safety, and less soil disturbance (Habe and Omori 2003).

Several microorganisms, such as bacteria, yeasts, and filamentous fungi, are capable of using and mineralizing different types of PAHs. Low-molecular-weight (LMW) PAHs, such as naphthalene, phenanthrene, and anthracene, are usually readily degraded by bacteria and fungi in soil and under laboratory conditions (Peng et al. 2008). However, high-molecular-weight (HMW) PAHs (four and more rings) are more persistent, in part because of their low bioavailability, due to their strong adsorption onto the soil organic matter. Of the microorganisms identified to have the capability to degrade PAHs in the environment, fungi have been shown to be relatively more successful than bacteria in breaking down HMW PAHs (Potin et al. 2004). Furthermore, filamentous fungi have an advantage over unicellular forms since the fungal mycelium could grow into the soil and distribute itself through the solid matrix to degrade PAHs (Cerniglia 1997).

This chapter summarizes the recent information on the metabolic pathway of the fungal transformation of PAHs and provides a critical review of previous work about fungal bioremediation of PAH-contaminated soil. In addition, this chapter discusses some of the most recently used fungal technology to enhance PAHs bioremediation process.

7.2 PAHs as Soil Contaminants

7.2.1 Physicochemical and Toxicology Properties of PAHs

It is well established that the fate of PAHs in the environment is primarily controlled by their physicochemical properties (Ferreira 2001). PAHs are nonpolar compounds with general physicochemical properties such as (1) low water solubility; (2) low Henry's law constant, a coefficient which represents the air/water partitioning; (3) high hydrophobicity or lipophilicity, represented by the n-octanol/water partition coefficient (Log K_{ow}); (4) low organic carbon partition coefficient; and (5) high bioconcentration factor (Berthe-Corti et al. 2007). The physicochemical properties of some commonly studied PAHs are given in the Table 7.1.

Table 7.1 Chemical structure, physicochemical characteristics, and carcinogenicity of 16 PAHs considered priority pollutant (Berthe-Corti et al. 2007; Pazos et al. 2010; http://www.sigmaaldrich.com)

Substance	CAS number	Mol wt (g mol⁻¹) structure	Boiling point (°C)	Vapor pressure (Pa, 25 °C)	Water solubility (mg l⁻¹, 25 °C)	Log K_{OW}	Henry's law constant (atm-m³ mol⁻¹)	IARC carcinogenicity group[a]
Naphthalene	91-20-3	128.17	218	11.3	32	3.37	–	–
Acenaphthylene	208-96-8	152.20	92–93	0.12	3.93	4.07	1.45×10^{-3}	–
Acenaphthene	82-32-9	154.21	95	0.30	1.93	3.92	7.91×10^{-5}	–
Fluorene	86-73-7	166.22	295	0.10	1.89	4.90	6.50×10^{-6}	3
Anthracene	120-12-7	178.20	340–342	0.97×10^{-3}	0.015	4.45	1.77×10^{-5}	3
Phenanthrene	85-01-8	178.24	340	2×10^{-2}	1.20	4.55	2.56×10^{-5}	3
Fluoranthene	206-44-0	202.26	375	1.2×10^{-3}	0.20–0.26	4.90	6.50×10^{-6}	3
Pyrene	129-00-0	202.26	393	0.82×10^{-3}	0.0077	4.88	1.14×10^{-5}	3
Benz[a]anthracene	56-55-3	228.29	435	2.5×10^{-5}	0.010	5.61	1.00×10^{-6}	2A

(continued)

Table 7.1 (continued)

Substance	CAS number	Mol wt (g mol⁻¹) structure	Boiling point (°C)	Vapor pressure (Pa, 25 °C)	Water solubility (mg l⁻¹, 25 °C)	Log K_{ow}	Henry's law constant (atm-m³ mol⁻¹)	IARC carcinogenicity group[a]
Chrysene	218-01-9	228.29	448	0.07×10^{-2} (20 °C)	0.0028	5.16	1.05×10^{-6}	3
Benzo[b] fluoranthene	205-99-2	252.32	481	6.7×10^{-5}	0.0013×10^{-7}	6.04	1.22×10^{-5}	2B
Benzo[k] fluoranthene	207-08-9	252.32	481	5.2×10^{-8}	–	6.11	–	2B
Benzo[a]pyrene	50-32-8	252.32	495	7×10^{-5}	0.0023	6.06	4.90×10^{-7}	2A
Benzo[g,h,i] perylene	191-24-2	276.34	550	1.3×10^{-8}	0.00026	6.50	1.44×10^{-7}	3
Indeno[1,2,3-cd] pyrene	193-39-5	276.24	530	1×10^{-5}	0.062	6.58	6.95×10^{-8}	2B
Dibenz[a,h] anthracene	53-70-3	278.35	524	1.3×10^{-8}	0.0005×10^{-10}	6.84	7.30×10^{-8}	2A

[a]Carcinogenicity group according to IARC classification 2004: http://monographs.iarc.fr/ENG/Classification/index.php. (1) Carcinogenic to humans, (2A) probably carcinogenic to humans, (2B) possibly carcinogenic to humans, (3) not classifiable as to carcinogenicity to humans, (4) probably not carcinogenic to humans

Generally, LMW PAHs are more volatile, more water soluble, and less lipophilic than HMW PAH (Table 7.1). An increase in the size and angularity of a PAH molecule results in a concomitant increase in hydrophobicity and electrochemical stability, two of the primary factors which contribute to PAH persistence in the environment (Cerniglia 1992).

Many PAH compounds are known or suspected to be toxic, mutagenic, and, in some cases, carcinogenic (Table 7.1). PAHs were, perhaps, the first recognized environmental carcinogens (Haritash and Kaushik 2009). In addition to the increase in environmental persistence with the increase in PAH molecule size, evidence suggests that in some cases, PAH genotoxicity also increases with size, up to at least four or five fused benzene rings (Cerniglia 1992). Besides human and animal health risks, PAHs are also a potential risk to the soil ecosystems (Berthe-Corti et al. 2007).

On the basis of their abundance and toxicity, the US Environmental Protection Agency (EPA), the World Health Organization (WHO), and the Communautée Economique Européenne (CEE) selected a list of 16 PAH as priority contaminants (Table 7.1).

7.2.2 Origen of Soil Contamination with PAHs

The PAHs are formed during the thermal decomposition of organic molecules and their subsequent recombination. Incomplete combustion at high temperature (500–800 °C) or subjection of organic material at low temperature (100–300 °C) for long periods results in PAH production (Haritash and Kaushik 2009).

PAHs released into the environment may originate from many sources that include natural sources (volcanoes activity, forest fires, etc.) and anthropogenic sources (tobacco smoke, burning of wood, fossil fuel, application of pesticides in agriculture or for wood protection, municipal solid waste incineration, and industrial activity related with petroleum refining and transport).

Though PAH are the chief pollutants of air, soil acts as the ultimate depository of these chemicals. In consequence and as a result of anthropogenic activities, the concentration of PAHs in soils has increased considerably since the nineteenth century. However, the concentration of PAHs in soils varies widely depending on the level of industrial development of the area. For example, an analysis of different municipal areas of Ensenada City (Argentina) revealed concentrations of PAHs between 0.10 and 56.90 mg kg^{-1} (dry soil) of the upper soil layer (Pessacq et al. 2010).

7.2.3 Fate of PAHs in Soil

Soil is one of the major sinks for organic pollutants due to their strong affinity to naturally occurring organic matter (Yang et al. 2010). Once in the soil, PAHs undergo many loss processes: volatilization, leaching, uptake by biota, and physical

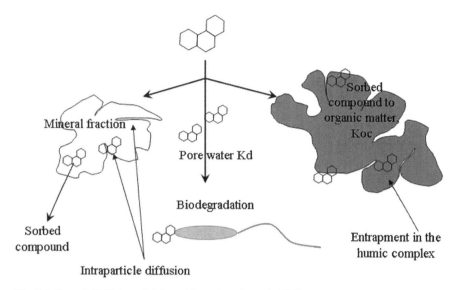

Fig. 7.1 Fate of PAHs in soil (adapted from Semple et al. 2003)

mass transport. However, the principal PAH loss process from soil is through microbial degradation (Doick et al. 2005). Rate and extension of a PAH biodegradation process depends on many factors such as the physicochemical characteristic of PAHs, soil composition and texture, weather and climate, and composition and activity of the microbial community.

The fate of PAHs in soil is largely governed by their bioavailability. Reduced bioavailability results from the interaction of pollutants with both organic and inorganic components of the soil matrix: diffusion limitation due to sequestration of the pollutant into nanoscale pores, binding to soil minerals by ionic or electrostatic interactions, oxidative covalent coupling of the pollutant with soil organic matter, and partition/dissolution of the pollutants into the soil organic matter (Head 1998) (Fig. 7.1). Hence, the rate of biodegradation may be limited by the rates of diffusion from the micropores, partitioning out of the organic matter, or desorption from surfaces.

7.3 Fungal Transformation of PAHs

The use of fungi and their enzymatic systems to decontaminate and detoxify soils and waters constitutes a promising management in bioremediation strategies. Yeasts and/or filamentous fungi, which can be either naturally occurring or deliberately introduced into a polluted site, can remediate the environment mediating by

Transformation of PAHs

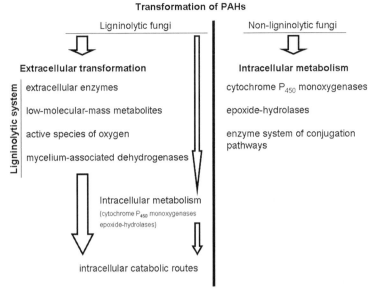

Ligninolytic fungi | Non-ligninolytic fungi

Extracellular transformation | **Intracellular metabolism**

extracellular enzymes | cytochrome P_{450} monoxygenases

low-molecular-mass metabolites | epoxide-hydrolases

active species of oxygen | enzyme system of conjugation pathways

mycelium-associated dehydrogenases

Ligninolytic system

Intracellular metabolism
(cytochrome P_{450} monoxygenases
epoxide-hydrolases)

intracellular catabolic routes

Fig. 7.2 Main fungal groups involved in the transformation of PAHs: enzymatic systems

several mechanisms. In this sense, the transformation of PAHs and other toxic organic compounds by fungi can play three roles in fungal biology:

1. An assimilative role that consumes and breaks down pollutants (being it used as sole carbon source) and thus yields carbon and energy for the degrading fungus and goes along with the conversion into harmless by-products such as simpler organic compounds and the mineralization (complete transformation into CO_2 and H_2O) of the compound or part of it.

2. Detoxification mechanisms which involve chemical alterations in the pollutants and the consequent conjugation, polymerization or immobilization of them into soil humic substances (humification and stabilization) (Cerniglia 1984) through covalent links to macromolecular soil organic matter (Richnow et al. 1997) where only the turnover of the organic matter would lead to the availability of the pollutant carbon. PAH conjugates are generally less toxic and more soluble than their respective parent compounds.

3. Co-metabolism, which is the degradation of PAHs or other pollutants without generation of energy and carbon for the fungal metabolism. A contaminated soil could be detoxified by fungi naturally growing on wood such as white-rot ones mainly *via* co-metabolism, obtained its source of carbon for mycelial growth from degradation of available and assimilable substrates such as cellulose or glucose. In this sense, the attack of pollutants by those fungi can be stimulated in the soil by the addition of straw or wood chips.

7.3.1 Metabolic Pathways for Fungal Transformation of Pollutants

Lignin is a complex aromatic heteropolymer with a large array of interunit linkages, being one of the most recalcitrant natural compounds towards both chemical and biological transformation. Therefore, its removal represents a key step in the carbon cycle (Ruiz-Dueñas and Martínez 2009). Based on the ability to degrade and mineralize it, the fungi can be classified in two main groups: ligninolytic fungi or also known as the white-rot fungi (WRF), which grow on wood and forest-leaf litter (Martínez et al. 2005; Saparrat et al. 2008a), and non-ligninolytic ones, which have differential metabolic strategies to transform PAHs and other pollutants (Fig. 7.2). The ability to open and mineralize the aromatic ring of lignin and PAHs is characteristic of ligninolytic fungi (Bezalel et al. 1997). While the enzymatic systems involved in the intracellular metabolism of PAHs by the fungi, either ligninolytic ones or non-ligninolytic ones, include cytochrome P_{450} monooxygenases and epoxide hydrolases, WRF transform, degrade, and mineralize lignin significantly, as well as several pollutants (including PAHs), by means of an oxidative and unspecific complex system that includes extracellular oxidoreductases. Many ligninolytic fungi such as *Phanerochaete chrysosporium* and *Pleurotus ostreatus* produce both non-ligninolytic and ligninolytic type enzymes, but it is unclear to what extent each enzyme contributes to the transformation of the PAH molecule (Bamforth and Singleton 2005).

7.3.2 Ligninolytic System from WRF

Lignin is a complex three-dimensional bulky polymer, being the result of reactions among three *p*-hydroxycinnamyl alcohols, *p*-coumaryl, coniferyl, and sinapyl alcohols, and their acylated forms (Martínez et al. 2005). Lignin is probably one of the most recalcitrant compounds synthesized by plants, which is mostly abundant in trees and the main contributor to wood strength (Saparrat et al. 2008b). Recently, lignin has gained increasing attention, in parallel with global population growth and industrial development in the world, as its degradation is the key factor in lignocellulose decay in nature, being it only degraded by a few organisms such as WRF (mainly from the phylum Basidiomycota; Martínez et al. 2005) as well as in the industrial application of plant biomass (Jurado et al. 2011). Due to its high molecular size and heterogeneous structure, lignin can only be attacked outside the fungal cell through a complex oxidative and unspecific system that involves extracellular enzymes, low-molecular-mass metabolites, which are mainly redox mediators, and active species of oxygen (Saparrat et al. 2002, 2008b, 2010). In this sense, the degradation of lignin by WRF was defined by Kirk and Farrell (1987) as enzymatic combustion and the concomitant action of those extracellular components and reductive reactions carried out by cell-bound systems seem to determine the effectiveness of WRF to degrade and mineralize lignin as well as other recalcitrant compounds such as single aromatic molecules, humic acids, and other xenobiotics

Table 7.2 Synthesis and roles of laccases in fungi (adapted from Saparrat and Balatti 2005)

Fungus	Ecological and/or taxonomical status	Putative physiological role
Agaricus bisporus	*Basidiomycota*	Oxidation of phenolics
Armillaria mellea	Root pathogen white-rot *Basidiomycota*	Formation of rhizomorphs
Armillaria mellea	Root pathogen white-rot *Basidiomycota*	Oxidation of phenolics
Aspergillus nidulans	Soil anamorph of *Ascomycota*	Synthesis of pigments
Aureobasidium pullulans	Black yeast	Synthesis of pigments
Botrytis cinerea	Soft rot anamorph of *Ascomycota*	Oxidation of phytoalexins
Chaetomium thermophilium	Cellulolytic *Ascomycota*	Humification
Chalara paradoxa	Anamorph of *Ascomycota*	Oxidation of phenolics
Cochliobolus heterostrophus	Plant pathogen *Ascomycota*	Synthesis of melanin
Colletotrichum graminicola	Plant pathogenic anamorph of *Ascomycota*	Fungal survival
Coriolopsis rigida	White-rot *Basidiomycota*	Degradation of aromatic compounds
Cryphonectria parasitica	Chestnut blight *Ascomycota*	Oxidation of polyphenolics
Cryptococcus neoformans	AIDS-related opportunistic pathogen anamorph of *Basidiomycota*	Oxidation of catecholamines and melanization
Fusarium proliferatum	Soil-inhabiting anamorph of *Ascomycota*	Oxidation of phenolics and humification
Gaeumannomyces graminis var. *tritici*	Phytopathogenic *Ascomycota*	Melanin polymerization and lignin depolymerization
Heterobasidion annosum	Root pathogen white-rot *Basidiomycota*	Oxidation of lignans and lignin degradation
Lentinus edodes	White rot *Basidiomycota*	Fruiting body formation
Lentinula (Lentinus) edodes	White-rot *Basidiomycota*	Mechanism of defense
Leptosphaerulina briosiana	*Ascomycota*	Synthesis of melanin
Marasmius quercophilus	Leaf litter-degrading white-rot *Basidiomycota*	Oxidation of aromatic compounds
Ophiostoma novo-ulmi	Dutch elm disease pathogen *Ascomycota*	Oxidation of phenols or resistance
Penicillium chrysogenum	Soil-inhabiting anamorph of *Ascomycota*	Lignin mineralization
Pleurotus eryngii	White-rot *Basidiomycota*	Lignin degradation
Pycnoporus cinnabarinus	White-rot *Basidiomycota*	Ligninolysis and pigment synthesis
Rhizoctonia praticola	Soil anamorph of *Basidiomycota*	Oxidation of phenolics
Rhizoctonia solani	Soil–living plant–pathogenic anamorph of *Basidiomycota*	Detoxification
Schizophyllum commune	White-rot *Basidiomycota*	Fruit body development

(continued)

Table 7.2 (continued)

Fungus	Ecological and/or taxonomical status	Putative physiological role
Tetraploa aristata	Litter-associated anamorph of *Ascomycota*	Melanization process
Thelephora terrestris	Ectomycorrhizal *Hymenomycetes* (*Basidiomycota*)	Oxidation of phenolics and pigmentation
Verticillium dahliae	Anamorph of *Ascomycota*	Synthesis of 1,8-dihydroxynaphthalene melanin

including PAHs (Jurado et al. 2011). Extracellular nature of ligninolytic system enables to their enzymes to be able to diffuse into the soil/sediment matrix and potentially oxidize PAHs with low bioavailability (Rodríguez et al. 2004). Likewise, simple products from degradation can enter the fungal hyphae and to be incorporated into intracellular catabolic routes.

The extracellular ligninolytic enzymatic systems have low substrate specificity and include ligninolytic peroxidases, laccases, oxidases responsible for the production of extracellular hydrogen peroxide (H_2O_2), and reductases (Martínez et al. 2005). These enzyme systems exhibit differential characteristics depending on the fungal species, strains, and culture conditions, being also regulated by chemical agents and several nutrients such as nitrogen, as well as by their concentration level (Martínez et al. 2005; Saparrat et al. 2008b). In this sense, several patterns of lignocellulose decay by WRF and ligninolytic enzyme synthesis either involving a single enzyme type or various associated enzyme complexes acting sinergically may equally (alternately) participate in an active degradation of lignin (Saparrat and Balatti 2005). Among peroxidases, which require H_2O_2 as co-substrate for enzyme catalysis, there are according to Martínez et al. (2005) and Ruiz-Dueñas and Martínez (2009)

(i) Lignin peroxidases (LiP, E.C.1.11.1.14) which are able to oxidize phenolic units as well as directly non-phenolic lignin units (which are up to 90 % of the polymer) because of its high redox potential.

(ii) Manganese peroxidases (MnP, E.C.1.11.1.13) acting preferentially on phenolic units, but also on non-phenolic units in the presence of mediators; this enzyme generates Mn^{3+}, which acts as a diffusible oxidizer on phenolic or non-phenolic lignin units via lipid peroxidation reactions.

(iii) Versatile peroxidases (VP, E.C.1.11.1.16), a third type of ligninolytic peroxidase that combines the catalytic properties of LiP, MnP, and plant/microbial peroxidases oxidizing phenolic compounds; it is also able to oxidize azo dyes not oxidized by MnP or LiP.

Laccases (*p*-diphenol: oxygen oxidoreductase; EC 1.10.3.2), which also participate in other physiology processes in different fungal groups to be or not ligninolytic ones (Table 7.2), oxidize preferentially lignin phenolic units, similarly to the ligninolytic peroxidases, but they also act on non-phenolic units in the presence of mediators (Saparrat et al. 2008b; Cañas and Camarero 2010). Laccases oxidize aromatic amines, a wide number of phenolic compounds including chlorophenols, secondary aliphatic polyalcohols, anthraquinone dyes, and to a certain extent, some PAHs, such as anthracene, as well as some inorganic ions (like Mn^{2+}), inorganic and organic metal ion complexes such as ferrocyanide, ferrocenes, and cytochrome c or an electron transferred directly from an electrode (Saparrat et al. 2008b). These substrates also are oxidized by peroxidases. However, laccases, unlike peroxidases, do not require hydrogen peroxide (utilize molecular oxygen as oxidant), raising the interest of biotech companies. Furthermore, laccases can participate in the production of active oxygen species and the oxidation of azo and indigo dyes, and other PAHs, compounds that cannot be oxidized by laccases on their own (Cañas and Camarero 2010). The laccase-substrate couple constitutes the laccase-mediator system (LMS), which oxidizes compounds by the laccase generated free radicals, suggesting that LMS might have a more powerful catalytic activity than peroxidases. In this sense, LMS allows the degradation of xenobiotic compounds and chlorine-free bleaching of paper pulp (Jurado et al. 2011).

Both laccases and peroxidases catalyze the one-electron oxidation of lignin as well as other aromatic compounds such as PAHs, thereby generating radicals such as phenoxy or cationic ones from either phenolic aromatic units or on non-phenolic units, respectively. These evolve in different nonenzymatic reactions, including bond cleavage and the reaction with O_2 or water and active oxygen species to give hydroxy- or keto-derivatives, generating a selection of quinones and acids (Rodríguez et al. 2004).

On the other hand, two extracellular oxidases, glyoxal (GLOX, E.C.1.2.3.5) and aryl-alcohol (AAO, E.C.1.1.3.7) oxidases, have been reported for extracellular H_2O_2 production. H_2O_2 is required by ligninolytic peroxidases, but it can also be involved in lignocellulose degradation as the precursor of hydroxyl radical ($^{\bullet}OH$), the strongest oxidizing agent produced by fungi through the iron-catalyzed Haber–Weiss reaction (Guillén et al. 2000). On a hypothetical basis, it was previously proposed that this free radical may initiate the attack on lignin in the initial stages of wood decay, when the small size of pores in the still-intact cell wall prevents the penetration of ligninolytic enzymes (Gómez-Toribio et al. 2009). Then, lignin degradation proceeds by oxidative attack of the enzymes described above. Guillén et al. (2000) described the synergistic action of laccase and AAO in $^{\bullet}OH$ production. AAO oxidizes aromatic and aliphatic polyunsaturated primary alcohols and is involved, together with intracellular dehydrogenases (such as aryl-alcohol dehydrogenases), in extracellular H_2O_2 production through a redox cycle. Oxygen activation in redox cycling reactions involving quinones (either from breakdown of lignin or synthesized *de novo* by fungi), laccases or peroxidases, and quinone reductases have been also reported for extracellular hydroxyl radical production (Gómez-Toribio et al. 2009). Furthermore, extracellular reductases such as cellobiose dehydrogenase (CDH, E.C.1.1.99.18) and cellobiose–quinone oxido-reductase (CQO, E.C.1.1.5.1), which is a proteolytic product of CDH, may catalyze the reduction of phenolic products derived from lignin degradation, and thus avoiding their posterior repolymerizing and acting as a nexus between lignolysis and cellulolysis (Temp and Eggert 1999). For a deep analysis and discussion of the catalytic, structural characteristics and other molecular aspects of the ligninolytic enzymes as well as their degradation mechanism, it is recommend to consult other reviews (Martínez et al. 2005; Cañas and Camarero 2010; Jurado et al. 2011).

7.3.3 Ligninolytic Fungi as Potential Tools in Bioremediation Strategies

WRF are promising bioremediation agents because of their ability to degrade or transform PAHs and a large variety of other aromatic pollutants (such as the BTEX group, an acronym that stands for benzene, toluene, ethylbenzene, and xylenes; chlorophenols; polychlorinated biphenyls; and dyes with different chromophore groups) that can reach soil and water, thus reducing their toxicity. They can use lignocellulosic substrates for growth, making them suitable for inoculation into contaminated matrices. In addition, WRF can exert a beneficial effect on the autochthonous microbiota, improving porosity and water-holding capacity of soil and thus making the total degradation of recalcitrant pollutants easier. The biotrans-formation of pollutants by WRF involves several types of processes initiated either by ligninolytic enzymes or mycelial-bound redox systems, which generate free radicals, which can then either undergo another enzyme-catalyzed oxidation, or other nonenzymatic transformations via enzymatic combustion process.

The first studies on pollutant degradation by WRF were carried out with *Phanerochaete chrysosporium*, an organism capable of degrading a broad spectrum of pollutants. As this well-studied fungus often demands specific conditions for growth and expression of its ligninolytic system, more WRF species have been screened for their ability to transform and detoxify soil pollutants under different environmental conditions. In this sense, *Pleurotus ostreatus*, *Trametes villosa*, and *Coriolopsis rigida* showed themselves to be highly effective at degrading both the aliphatic and aromatic fractions of crude oil from contaminated soils when com-pared in a 90-day laboratory experiment using a natural, not-fertilized soil contaminated with 10 % crude oil (Colombo et al. 1996). Most of the WRF belonging to phylum Basidiomycota, including several species from *Bjerkandera*, *Coriolopsis*, *Phlebia*, and *Trametes*, are able to degrade PAHs and other pollutants, although their transformation capability greatly depends on environmental conditions. Thus, the particular conditions leading to detoxification should be analyzed in each case, paying special attention to nitrogen and carbon contents, as well as to the supplementation with the enzyme cofactors and inducers, such as copper for laccase, which can greatly enhance enzyme production (Saparrat et al. 2010; Jurado et al. 2011). Therefore, although the applicability of these fungi seems very promising, further research is still needed to achieve maximum efficiency in the removal of pollutants and shift our experimental work from artificially contaminated soils to contaminated soils present in the environment due to human activities.

7.3.4   Non-ligninolytic Fungi with Ability for Transforming PAHs and Other Hazardous Organic Compounds

Non-ligninolytic fungi (generally known also as micromycetes) conform a fungal group that is widely distributed in the ecosystems. Taxonomically, non-ligninolytic fungi belong to phylum Ascomycota, mainly as their anamorph forms, and also to Zygomycota one. They can grow on a broad spectrum of carbon compounds such as glucose and other monosaccharides, sucrose, cellulose, starch, proteins, and lipids, as well as hydrocarbons, including PAHs, which are used as single C source. Among non-ligninolytic fungi, *Aspergillus* and *Penicillium* species, which are the dominant genera in soils and sediments contaminated with petroleum or PAHs, show an outstanding ability to degrade HMW PAHs in pure culture (Launen et al. 1995; April et al. 2000; Chaillan et al. 2004; Romero et al. 2010). *Penicillium janthinellum* was previously reported for its ability to transform pyrene and benzo [a]pyrene into phenols, diols, and quinones, a reaction catalyzed by cytochrome P_{450} monooxygenases (Launen et al. 1995). *Penicillium ochrochloron* is also capable to degrade and use pyrene as sole carbon source (Saraswathy and Hallberg 2005). Romero et al. (2010) reported recently the mineralization of ^{14}C-benzo(a) pyrene by non-ligninolytic filamentous fungi isolated from industrial polluted sediments, such as *A. flavus* and *Paecilomyces farinosus*, with and without C_{16} as co-substrate, though the presence of this latter C-compound enhanced the biomass production and benzo(a)pyrene uptake as C-source, confirming increasing bio-degradation of xenobiotics by co-metabolism. All these features make these non-ligninolytic fungi another promising biological alternative in the course of detoxification processes for the bioconversion of PAHs. In this sense, non-ligninolytic fungi have high ecological plasticity and higher competitive abilities as well as being rather tolerant to stress situations when compared to basidiomyce-tous fungi (Saparrat and Hammer 2006; Saparrat et al. 2008a). However, their degradative abilities are more limited and/or slower when compared to those from ligninolytic fungi. Schmidt et al. (2010) reported the ability of several soil micromycetes to metabolize PAHs such as phenanthrene, fluoranthene, and pyrene as single carbon source and *Cunninghamella elegans* (*Zygomycota*) was by far the most efficient degrader. This latter fungal species has been reported broadly by its ability to oxidize numerous PAHs to phenols by oxidation at Cyt P_{450} (Cerniglia 1992). Although it is considered that the produced epoxides are hydrolyzed to transdihydrodiols or to phenols which are subsequently conjugated with sulfate, glucuronic acid, or glucose (including those from *C. elegans* such as glucoside-conjugated phase-two metabolites from phenanthrene, fluoranthene, and pyrene), the metabolites reported by Schmidt et al. (2010) were mono-, di-, or trihydroxylated and conjugated only with sulfate. Sulfate-conjugated PAH metabolites have also been previously reported for *Aspergillus niger*, *A. terreus*, *Penicillium glabrum*, and *Syncephalastrum racemosum* (Casillas et al. 1996; Capotorti et al. 2004; Wunder et al. 1997). Furthermore, the mineralization of ^{14}C-metabolites produced by *C. elegans* was in all cases extremely slow compared to that of the parent ^{14}C-PAHs in soil slurries (Schmidt et al. 2010). High water solubility, low lipophilicity, and slow mineralization of the metabolites such as PAH conjugates indicate a potential problem of leaching of fungal PAH metabolites from polluted soil to the groundwater aquifers. However, micromycetes (as previously demonstrated for ligninolytic fungi) might also detoxify PAHs such as pyrene and their metabolites through the immobilization into soil humic substances (formation of bound residues or humification) as an alternative pathway for bioremediation of PAH-contaminated soils. The formation of bound residues of organic pollutants (or humification of xenobiotic compounds) is under discussion as a complement to complete mineralization (Sack and Fritsche 1997). In this sense, the covalently binding to PAH metabolites (such as from the transformation of pyrene and benzo(*a*)pyrene by *Penicillium janthinellum* SFU403 and *A. flavus* and *Paecilomyces farinosus*, respectively) with cell molecules forming inextractable cell-associated products (ICAP) is a theoretically feasible mechanism for the formation of bound residues by micromycetes (Launen et al. 2000; Romero et al. 2010). However, as far as we know, no data is available about their contribution in bioremediation of soils polluted with PAHs and about if ICAP are formed by fungi *in situ* and remain stable over time in the presence of the soil microbial community, having this process high potential in bioremediation as well as its fate as parent compound for humification reactions that occurs in the soil. Therefore, mineralization and formation of bound residues of added PAHs might together be considered as detoxification processes since the tested fungi destroy the pollutant or have the potential to render it unavailable to soil organisms. However, until stable immobilization and complete mineralization of PAHs and their metabolites, from transformation by non-ligninolytic fungi that might be in the soil and in its groundwaters, are reached, the presence of compounds derived from PAHs with certain level of toxicity, activity, mutagenicity, and/or carcinogenicity makes the environmental monitory necessary.

7.4 Fungal Remediation of PAH-Contaminated Soil

Bioremediation is the technological process whereby biological systems are harnessed to effect the cleanup of environmental pollutants (Head 1998). An effective bioremediation strategy depends on the ability of microorganism to degrade the target compounds towards mineralization or to a very low level with production of nontoxic metabolites (Silva et al. 2009).

Effective biodegradation of PAH in soil is a function of their physical and chemical properties, concentration, rate of diffusion and bioavailability. It also depends on soil type, moisture content, availability of carbon substrates, nitrogen or other nutrients, aeration conditions, pH, and temperature, as well as the number and types of PAH-degrading microorganisms present.

Fungal remediation, or mycoremediation, which means fungal treatment or fungal-based remediation, is a promising technique for the cleanup of PAH-contaminated soil (Wu et al. 2008a). Generally fungal rates of PAHs degradation are slow and inefficient compared to bacteria. However, fungi are capable of degrading HMW PAHs by different enzyme systems (see Sect. 7.3). This, together with their physiological versatility, their ability to grow in environments with low nutrient concentrations, low humidity and acidic pH, their filamentous way of growing that allows an efficient colonization and exploration of contaminated soil, and the fact that the fungal mycelia constitute a large fraction of the soil biomass, provide to the fungi a significant ecological role, contributing considerably to the transformation of PAHs in soil.

7.4.1 Role of Fungi in Bioremediation Process of PAH-Contaminated Soil

Microbial degradation represents the major route responsible for the ecological recovery of PAH-contaminated sites; however, the success of bioremediation projects has been limited by the failure to remove HMW PAHs (Boonchan et al. 2000).

The initial attack on HMW PAHs in soil by fungal exoenzymes appears to be more likely than attack by bacterial intracellular enzymes. Fungal exoenzymes have the advantage that they may diffuse to the highly immobile HMW PAHs. This is in contrast to bacterial PAH dioxygenases, which are generally cell bound because they require NADH as a cofactor (Johnsen et al. 2005). A positive correlation has been found between PAH degradation and ligninolytic enzyme activity in soil (Novotný et al. 2004). Also, the transformation of PAHs by non-ligninolytic fungi may mobilize PAHs for further degradation by bacteria.

Fungi usually do not mediate complete mineralization of PAH, producing metabolites with higher water solubility and enhanced chemical reactivity. In soil these metabolites are subject to further metabolic attack by other soil microorganisms, extracellular enzymes, chemical reactions, or sorption to soil surfaces.

Since fungal remediation is a promising strategy for bioremediation of PAH-contaminated soil, different innovative fungal technologies for improving the rate and extension of PAH biodegradation are being developed.

7.4.2 Fungal Technology in PAH-Contaminated Soil Remediation

7.4.2.1 Bioaugmentation

Bioaugmentation can be defined as the technique for improving the bioremoval capacity of a contaminated matrix by introducing specific competent microbial strains or communities.

Within mycoremediation technologies, most bioaugmentation studies have been carried out using filamentous fungi, in particular WRF (Bogan et al. 1999; in der Wiesche et al. 2003; Leonardi et al. 2008; Acevedo et al. 2011). The interest in WRF is due to their ability to synthesize relatively unspecific enzymes that can degrade HMW PAH, and their hyphal growth enables them to penetrate and to diffuse into soil, thus reaching pollutants and acting, at the same time, as effective dispersion vectors of indigenous pollutant-degrading bacteria (Leonardi et al. 2008). Also, soils contaminated with PAH often contain high levels of other pollutants such as heavy metals, which are often derived from the same sources as PAH (Maliszewska-Kordybach and Smreczak 2003). In such a difficult case, the use of WRF may give some advantages over bacterial bioaugmentation. Fungi display a high ability to immobilize toxic metals by insoluble metal oxalate formation, biosorption, or chelation onto melanin-like polymers (D'Annibale et al. 2005). However, the potential colonization of WRF in soil is reported to be limited, and PAHs depletion by WRF may be hindered by limiting environmental factors (Wu et al. 2008a).

Autochthonous bioaugmentation (ABA) is defined as a bioaugmentation technology that uses microorganisms indigenous to the site (soil, sand, and water) to be decontaminated (Ueno et al. 2007). The use of filamentous fungi isolated from PAH-contaminated soil presents the advantage that they are adapted not only to the presence of contaminants but also to the environmental conditions of the site.

Several authors have reported that indigenous filamentous non-white-rot fungi isolated from PAH-contaminated soil transform PAHs significantly (Potin et al. 2004; D'Annibale et al. 2006), suggesting that known PAH-degrading fungi are relatively less compared to the highly diverse fungi kingdom and there is a huge fungi pool in terrestrial ecosystem from which potent PAH-degrading strains remain to be explored (Wu et al. 2008a). D'Annibale et al. (2006) showed that the isolation of fungi from a contaminated soil (*Allescheriella* sp., *Stachybotrys* sp., and *Phlebia* sp.) followed by their reinoculation at the same site managed a significant removal of PAHs and the detoxification of the soil.

For use in bioaugmentation, the PAH degrader inoculum should ideally mineralize the PAHs, minimizing the production and accumulation of toxic metabolites. Fungi usually do not mediate complete mineralization of pollutants, so sequential fungal–bacterial degradation is necessary for soil detoxification (Wu et al. 2008a). Therefore, the effect of indigenous soil bacteria on fungal bioaugmentation efficiency has been widely studied (Kotterman et al. 1998; Wu et al. 2008a; Borràs et al. 2010). Borràs et al. (2010) found that the degradation capacity of *T. versicolor* was negatively affected in the presence of bacteria, and only weak fungal/bacterial synergistic effects were observed in the case of removal of HMW PAHs. In the same way, the results found by Kotterman et al. (1998) and Schmidt et al. (2010) suggest that not all of the fungus-oxidized metabolites were easily mineralized by indigenous microflora.

7.4.2.2 Enzymatic Treatment

Enzymatic treatment of PAH-contaminated soil is an alternative to conventional bioremediation (Wu et al. 2008b). The use of enzymes in the degradation of organic compounds presents several advantages compared to the use of microorganisms such as their high reaction activity towards recalcitrant pollutants, their capacity to act in the presence of many xenobiotic substances under a wide range of environmental conditions, and their low sensitivity or susceptibility to the presence of predators and inhibitors of microbial metabolism (Gianfreda and Rao 2004). Moreover, enzymes are able to reach substrates in pores with small dimensions, roughly 100 times smaller than bacteria (Quiquampoix et al. 2002). However, the number of published reports dealing with enzymatic remediation of soil is limited, and the cost of these biocatalysts is still too high to implement commercially attractive enzyme-based treatments in field applications (Torres et al. 2003).

Wu et al. (2008b) evaluated the potential of enzymatic remediation by free laccase of a long term PAH-contaminated soil and its effect on soil microbial biomass. They found that the treatment of soil by free laccase transformed several PAHs efficiently, decreased the genotoxicity, and showed a low impact on soil microbial community, suggesting that laccase can be an efficient and safe remedial agent.

Also, Acevedo et al. (2010) studied the performance of free and nanoclay-immobilized MnP from *Anthracophyllum discolor* in the PAHs degradation in sterile and nonsterile soil. The immobilized MnP enhanced the enzymatic transformation of anthracene of soil after 24 h, and the anthraquinone (more soluble and less toxic than the anthracene) was identified as the major by-product.

However, more studies must be performed to show that enzymatic in situ bioremediation may be an effective and desirable alternative to current remediation strategies (Acevedo et al. 2010).

7.5 Conclusions and Perspectives

Degradation of PAHs by fungi has been widely studied. As result, a remarkable understanding of catabolic pathways and the enzymes involved in the PAHs degradation have been obtained. However, the development of mycoremediation technology for PAH-contaminated soil treatment is still now a fruitful area for future researches. The stability and persistence of the inoculated fungi and/or their enzymes in the environment, their synergistic or competitive relationship with the indigenous soil microflora, and the production and accumulation of more soluble and mobile by-products are some of the subjects that require further investigation.

Acknowledgment Arambarri A. M. and Saparrat M. C. N. are research members of CONICET, and Morelli I.S. is research member of CIC-PBA. Coppotelli B. M. is postdoctoral fellow of CONICET. This review was partially supported by a grant from ANPCyT (PICT 884) and CONICET (PIP 1422) Argentina.

References

Acevedo F, Pizzul L, Castillo MP, González ME, Cea M, Gianfreda L, Diez MC (2010) Degradation of polycyclic aromatic hydrocarbons by free and nanoclay-immobilized manganese peroxidase from *Anthracophyllum discolor*. Chemosphere 80:271–278

Acevedo F, Pizzul L, Castillo MP, Cuevas R, Diez MC (2011) Degradation of polycyclic aromatic hydrocarbons by the Chilean white-rot fungus *Anthracophyllum discolor*. J Hazard Mater 185:212–219

April TM, Fought JM, Currah RS (2000) Hydrocarbon-degrading filamentous fungi isolated from flare pit soils in northern and western Canada. Can J Microbiol 46:38–49

Bamforth SM, Singleton I (2005) Bioremediation of polycyclic aromatic hydrocarbons: current knowledge and future directions. J Chem Technol Biotechnol 80:723–736

Berthe-Corti L, Del Panno MT, Hulsch R, Morelli IS (2007) Bioremediation and bioaugmentation of soils contaminated with polyaromatic hydrocarbons. Curr Trends Microbiol 3:1–30

Bezalel L, Hadar Y, Cerniglia CE (1997) Enzymatic mechanisms involved in phenanthrene degradation by the white rot fungus *Pleurotus ostreatus*. Appl Environ Microbiol 63:2495–2501

Bogan BW, Lamar RT, Burgos WD, Tien M (1999) Extent of humification of anthracene, fluoranthene, and benzo[alpha]pyrene by *Pleurotus ostreatus* during growth in PAH-contaminated soils. Lett Appl Microbiol 28:250–254

Boonchan S, Britz ML, Stanley GA (2000) Degradation and mineralization of high-molecular-weight polycyclic aromatic hydrocarbons by defined fungal bacterial cocultures. Appl Environ Microbiol 66:107–1019

Borràs E, Caminal G, Sarrà M, Novotný C (2010) Effect of soil bacteria on the ability of polycyclic aromatic hydrocarbons (PAHs) removal by *Trametes versicolor* and *Irpex lacteus* from contaminated soil. Soil Biol Biochem 42:2087–2093

Cañas AI, Camarero S (2010) Laccases and their natural mediators: biotechnological tools for sustainable eco-friendly processes. Biotechnol Adv 28:694–705

Capotorti G, Digianvincenzo P, Cesti P, Bernardi A, Guglielmetti G (2004) Pyrene and benzo(a) pyrene metabolism by an *Aspergillus terreus* strain isolated from a polycyclic aromatic hydrocarbons polluted soil. Biodegradation 15:79–85

Casillas RP, Crow SA, Heinze TM, Deck J, Cerniglia CE (1996) Initial oxidative and subsequent conjugative metabolites produced during the metabolism of phenanthrene by fungi. J Ind Microbiol 16:205–215

Cerniglia CE (1984) Microbial metabolism of polycyclic aromatic hydrocarbons. Adv Appl Microbiol 30:31–71

Cerniglia CE (1992) Biodegradation of polycyclic aromatic hydrocarbons. Biodegradation 3:351–368

Cerniglia CE (1997) Fungal metabolism of polycyclic aromatic hydrocarbons: past, present and future applications in bioremediation. J Ind Microbiol Biotechnol 19:324–333

Chaillan F, Le Fleche A, Bury E, Phantavong Y, Grimont P, Saliot A (2004) Identification and biodegradation potential of tropical aerobic hydrocarbon-degrading microorganisms. Res Microbiol 155:587–595

Colombo J, Cabello M, Arambarri AM (1996) Biodegradation of aliphatic and aromatic hydrocarbons by natural soil microflora and pure cultures of imperfected and lignolitic fungi. Environ Pollut 94:355–362

D'Annibale A, Ricci M, Leornadi V, Quaratino D, Micione E, Petruccioli M (2005) Degradation of aromatic hydrocarbons by white-rot fungi in a historically contaminated soil. Biotechnol Bioeng 90:723–731

D'Annibale A, Rosetto F, Leonardi V, Federici F, Petruccioli M (2006) Role of autochthonous filamentous fungi in bioremediation of a soil historically contaminated with aromatic hydrocarbons. Appl Environ Microbiol 72:28–36

Doick KJ, Klingelmann E, Burauel P, Jones KC, Semple KT (2005) Long-term fate of polychlorinated biphenyls and polycyclic aromatic hydrocarbons in an agricultural soil. Environ Sci Technol 39:3663–3670

Ferreira MMC (2001) Polycyclic aromatic hydrocarbons: a QSPR study. Chemosphere 44:124–146

Gianfreda L, Rao MA (2004) Potential of extracellular enzymes in relation of polluted soils: a review. Enzyme Microb Technol 33:339–354

Gómez-Toribio V, García-Martin AB, Martínez MJ, Martínez AT, Guillén F (2009) Induction of extracellular hydroxyl radical production by white-rot fungi through quinone redox cycling. Appl Environ Microbiol 75:3944–3953

Guillén F, Gómez-Toribio V, Martínez MJ, Martínez AT (2000) Production of hydroxyl radical by the synergistic action of fungal laccase and aryl alcohol oxidase. Arch Biochem Biophys 383:142–147

Habe H, Omori T (2003) Genetic of polycyclic aromatic hydrocarbon metabolism in diverse aerobic bacteria. Biosci Biotechnol Biochem 67:225–243

Haritash AK, Kaushik CP (2009) Biodegradation aspects of polycyclic aromatic hydrocarbons (PAHs): a review. J Hazard Mater 169:1–15

Head IM (1998) Bioremediation: towards a credible technology. Microbiology 144:599–608

in der Wiesche C, Martens R, Zadrazil F (2003) The effect of interaction between white-root fungi and indigenous microorganisms on degradation of polycyclic aromatic hydrocarbons in soil. Water Air Soil Pollut 3:73–79

Johnsen AR, Wick LY, Harms H (2005) Principles of microbiol PAH-degradation in soil. Environ Pollut 133:71–84

Jurado M, Martinez AT, Martinez MJ, Saparrat MCN (2011) Wastes from agriculture, forestry and food processing. Application of white-rot fungi in transformation, detoxification, or revalorization of agriculture wastes: role of laccase in the processes. In: Murray Moo-Young (ed.), Comprehensive Biotechnology, Second Edition, Elsevier, pp 595–603

Kirk TK, Farrell RL (1987) Enzymatic 'combustion': the microbial degradation of lignin. Annu Rev Microbiol 41:465–505

Kotterman MJJ, Vis EH, Field JA (1998) Successive mineralization and detoxification of benzo[a]pyrene by the white rot fungus *Bjerkandera* sp strain BOS55 and indigenous microflora. Appl Environ Microbiol 64:2853–2858

Launen L, Pinto LJ, Wiebe C, Kiehlmann E, Moore MM (1995) The oxidation of pyrene and benzo[a]pyrene by non-basidiomycete soil fungi. Can J Microbiol 41:477–488

Launen LA, Pinto LJ, Percival PW, Lam SFS, Moore MM (2000) Pyrene is metabolized to bound residues by *Penicillium janthinellum* SFU403. Biodegradation 11:305–312

Leonardi V, Giubilei MA, Federici E, Spaccapelo R, Šašek V, Novotný C, Petruccioli M, D'Annibale A (2008) Mobilizing agents enhance fungal degradation of polycyclic aromatic hydrocarbons and affect diversity of indigenous bacteria in soil. Biotechnol Bioeng 101:273–285

Maliszewska-Kordybach B, Smreczak B (2003) Habitual function of agricultural soils as affected by heavy metals and polycyclic aromatic hydrocarbons contamination. Environ Int 28:719–728

Martínez AT, Speranza M, Ruiz-Dueñas FJ, Ferreira P, Camarero S, Guillén F, Martínez MJ, Gutiérrez A, del Río JC (2005) Biodegradation of lignocellulosics: microbiological, chemical and enzymatic aspects of fungal attack to lignin. Int Microbiol 8:95–204

Novotný Č, Svobodová K, Erbanová P, Cajthaml T, Kasinath A, Lang E, Šašek V (2004) Ligninolytic fungi in bioremediation: extracellular enzyme production and degradation rate. Soil Biol Biochem 36:1545–1551

Pazos M, Rosales E, Alcántara T, Gómez J, Sanromán MA (2010) Decontamination of soils containing PAHs by electroremediation: a review. J Hazard Mater 177:1–11

Peng RH, Xiong AS, Xue Y, Fu XY, Gao F, Zhao W, Tian YS, Yao QH (2008) Microbial biodegradation of polyaromatic hydrocarbons. FEMS Microbiol Rev 32:927–955

Pessacq J, Bianchini FE, Terada C, Da Silva M, Morelli IS, Del Panno MT (2010) Effect of different stress treatments on microbial catabolic diversity of chronically hydrocarbon contaminated-soil in Buenos Aires, Argentina. In: Book of abstracts of 13th international symposium on microbial ecology. International Society for Microbial Ecology, Seattle

Potin O, Rafin C, Veignie E (2004) Bioremediation of an aged polycyclic aromatic hydrocarbons (PAHs)-contaminated soil by filamentous fungi isolated from the soil. Int Biodeterior Biodegrad 54:45–52

Quiquampoix H, Servagent-Noinville S, Baron MH (2002) Enzymes adsorption on soil mineral surfaces and consequences for the catalytic activity. In: Burns RG, Dick RP (eds) Enzymes in the environment: activity, ecology, and applications. Marcel Dekker, New York, pp 285–306

Richnow HH, Seifert R, Hefter J, Link M, Francke W, Schaefer G, Michaelis W (1997) Organic pollutants associated with macromolecular soil organic matter: mode of binding. Org Geochem 26:745–758

Rodríguez E, Nuero O, Guillén F et al (2004) Degradation of phenolic and non-phenolic aromatic pollutants by four *Pleurotus* species: the role of laccase and versatile peroxidase. Soil Biol Biochem 36:909–916

Romero MC, Urrutia MI, Reinoso HE, Moreno Kiernan M (2010) Benzo[a]pyrene degradation by soil filamentous fungi. J Yeast Fungal Res 1:25–29

Ruiz-Dueñas FJ, Martínez AT (2009) Microbial degradation of lignin: how a bulky recalcitrant polymer is efficiently recycled in nature and how we can take advantage of this. Microb Biotechnol 2:164–177

Sack U, Fritsche W (1997) Enhancement of pyrene mineralization in soil by wood-decaying fungi. FEMS Microbiol Ecol 22:77–83

Saparrat MCN, Balatti PA (2005) The biology of fungal laccases and their potential role in biotechnology (chapter 4). In: Thangadurai D, Pullaiah T, Tripathy L (eds) Genetic resources and biotechnology, vol 3. Regency, New Delhi, pp 94–120, 366 pp. ISBN 81-89233-28-9

Saparrat MCN, Hammer E (2006) Decolorization of synthetic dyes by the deuteromycete *Pestalotiopsis guepinii* CLPS no. 786 strain. J Basic Microbiol 46:28–33

Saparrat MCN, Guillén F, Arambarri AM, Martínez AT, Martínez MJ (2002) Induction, isolation, and characterization of two laccases from the white-rot basidiomycete *Coriolopsis rigida*. Appl Environ Microbiol 68:1534–1540

Saparrat MCN, Rocca M, Aulicino MB, Arambarri A, Balatti P (2008a) *Celtis tala* and *Scutia buxifolia* leaf litter decomposition by selected fungi in relation to their physical and chemical properties and the lignocellulolytic enzyme activity. Eur J Soil Biol 44:400–407

Saparrat MCN, Mocchiutti P, Liggieri CS, Aulicino M, Caffini N, Balatti PA, Martínez MJ (2008b) Ligninolytic enzyme ability and potential biotechnology applications of the white-rot fungus *Grammothele subargentea* LPSC no. 436 strain. Process Biochem 43:368–375

Saparrat MCN, Balatti PA, Martínez MJ, Jurado M (2010) Differential regulation of laccase gene expression in *Coriolopsis rigida* LPSC No. 232. Fungal Biol 114:999–1006

Saraswathy A, Hallberg R (2005) Mycelial pellet formation by Penicillium ochrochloron species due to exposure to pyrene. Microbiol Res 160:375–383

Schmidt S, Christensen J, Johnsen A (2010) Fungal PAH-metabolites resist mineralization by soil microorganisms. Environ Sci Technol 44:1677–1682

Semple KT, Morriss WJ, Paton GI (2003) Bioavailability of hydrophobic organic contaminants in soils, fundamental concepts and techniques for analysis. Eur J Soil Sci 54:809–818

Sigma–Aldrich Company (http://www.sigmaaldrich.com)

Silva IS, Santos ED, Menezes CR, Faria AF, Franciscon E, Grossman M, Durrant LR (2009) Bioremediation of a polyaromatic hydrocarbon contaminated soil by native soil microbiota and bioaugmentation with isolated microbial consortia. Bioresour Technol 100:4669–4675

Temp U, Eggert C (1999) Novel interaction between laccase and cellobiose dehydrogenase during pigment synthesis in the white rot fungus *Pycnoporus cinnabarinus*. Appl Environ Microbiol 65:389–395

Torres E, Bustos-Jaimes I, Le Borgne S (2003) Potential use of oxidative enzymes for the detoxification of organic pollutants. Appl Catal B 46:1–15

Ueno A, Ito Y, Yumoto I, Okuyama H (2007) Isolation and characterization of bacteria from soil contaminated with diesel oil and the possible use of these in autochthonous bioaugmentation. World J Microbiol Biotechnol 23:1739–1745

Wild SR, Jones KC (1995) Polynuclear aromatic hydrocarbons in the United Kingdom environment: a preliminary source inventory and budget. Environ Pollut 88:91–108

Wu YC, Luo YM, Zou DX, Ni JZ, Liu WX, Teng Y, Li ZG (2008a) Bioremediation of polycyclic aromatic hydrocarbons contaminated soil with *Monilinia* sp.: degradation and microbial community analysis. Biodegradation 19:247–257

Wu YC, Teng Y, Li ZG, Liao XW, Luo YM (2008b) Potential role of polycyclic aromatic hydrocarbons (PAHs) oxidation by fungal laccase in the remediation of an aged contaminated soil. Soil Biol Biochem 40:789–796

Wunder T, Marr J, Kremer S, Sterner O, Anke H (1997) 1-methoxypyrene and 1,6-dimethoxypyrene: two novel metabolites in fungal metabolism of polycyclic aromatic hydrocarbons. Arch Microbiol 167:310–316

Yang Y, Zhang N, Xue M, Tao S (2010) Impact of soil organic matter on the distribution of polycyclic aromatic hydrocarbons (PAHs) in soils. Environ Pollut 158:2170–2174

Chapter 8
Sequential Soil Vapor Extraction and Bioremediation Processes Applied to BTEX-Contaminated Soils

António Alves Soares, José Tomás Albergaria, Valentina F. Domingues, Paolo De Marco, and Cristina Delerue-Matos

8.1 Soil Contamination

Natural soils are highly diverse and heterogeneous media. They differ owing to their inorganic and organic composition. Several properties depend on climatic factors (precipitation and temperature) and on the richness of plant cover, which in turn is itself highly dependent on climate and soil characteristics (Chernyanskii and Gennadiyev 2005). Soils comprise a great quantity and variety of life forms: metazoans (i.e., multicellular animals, like countless forms of worms and arthropods), plant roots, and a plethora of types of microorganisms. Each of these living beings plays its part in the cycling of nutrients that keeps a soil healthy and fertile.

On average, half of a soil's volume is pore space, which is occupied by water and gases in varying proportions: layers below the water table level are said saturated as water fills up pores thus excluding gases. In the layers above the water table (vadose zone), more or less of the pore space is occupied by gases, with variable concentrations of O_2 and CO_2 depending on the greater or lesser rates of exchange with the atmosphere (Farrell and Elliott 2007).

The organic fraction, both dead and living, characterizes fertile soils: removing it leaves behind just sterile sand and dust. The source of organic matter in soil is, in most cases, material produced by plants aboveground. This material fuels almost

A.A. Soares • J.T. Albergaria • V.F. Domingues • C. Delerue-Matos (✉)
Requimte, Instituto Superior de Engenharia do Porto, Rua de S. Tomé,
4200-072 Porto, Portugal
e-mail: tucasalves@hotmail.com; jtva@isep.ipp.pt; vfd@isep.ipp.pt; cmm@isep.ipp.pt

P. De Marco
IBMC, Universidade do Porto, R. Campo Alegre, 823, Porto, Portugal

Centro de Investigação em Ciências da Saúde (CICS), Instituto Superior de Ciências
da Saúde – Norte, CESPU, Gandra PRD, Portugal
e-mail: pdmarco@clix.pt

E.M. Goltapeh et al. (eds.), *Fungi as Bioremediators*, Soil Biology 32,
DOI 10.1007/978-3-642-33811-3_8, © Springer-Verlag Berlin Heidelberg 2013

entirely the nutrient cycles in soil, representing food for herbivores, detritivores, and symbiotic, pathogenic, or saprophytic microbes, all of which in turn become nourishment for secondary consumers. A substantial part of the organic matter supply into soils comes in the form of plant root exudates: it was estimated that up to 40 % of photosynthates produced by a plant may, in some cases, be excreted at the level of the rhizosphere (Bertin et al. 2003). This organic carbon input is quite important: physicochemical and nutritional conditions in the proximity of roots are dramatically different, and microorganisms tend to be much more abundant there than in the rest of soil: this is called "the rhizosphere effect" (Gregory 2006). So, in essence, in soil too, like in any other compartment in the biosphere, sunlight is the ultimate source of energy that makes cycles "go round."

Soil performs manifold functions such as the support for terrestrial vegetation; storage, internal cycling, and processing of nutrients (like nitrogen and phosphorous); providing habitat for resident and transient soil macrofauna; retention of water; supporting plant growth; regulation of hydrological flows and atmospheric chemical composition; and, from a cultural point of view, providing a platform for recreational activities and opportunities for noncommercial activities (Morvan et al. 2008). This demonstrates the extreme importance that soils, and its quality, have in all societies. However, these same societies are the main culprits for the contamination and the deterioration of the quality of soils, because of the multiple human activities that introduce in soil a wide variety and large amounts of contaminants. This can occur intentionally (e.g., in the case of agriculture with pesticides and fertilizers) or accidentally due to spills, inadequate handling, or leaks. Contaminants that are released into the air too can be transferred and deposited onto soil by precipitation (Shaylor et al. 2009).

The identification and remediation of contaminated soils is done all over the world and it is well advanced in some regions: in the United States of America, since 1982, of the 3,240 priority contaminated sites identified, 1,620 of those were already remediated (USEPA 2010); in the South American continent, Uruguay identified 1,006 potentially contaminated sites based on the activities that operate at those sites. The metallurgic and chemical industries are responsible for two thirds of the potential contaminated sites, 41 % and 24 %, respectively (DINAMA 2005). In Europe, 1.8 million potentially contaminated sites have been identified, of which 246,000 were estimated to be really contaminated. Indeed, 81,000 remediation actions have been performed at some of those sites (EEA 2010). The contaminants that can be found at each location depend closely on the activities that operate at the site, to relate human activities. Aiming at site remediation, it is extremely important to know what contaminants are present in the soil, because this will determine which remediation technology is more appropriate.

The most common chemicals found at contaminated sites are petroleum hydrocarbons (Siddique et al. 2006), solvents, pesticides, and heavy metals (Sorensen and Holmstrup 2005). The occurrence of pollution is correlated with the degree of industrialization and intensity of chemical usage.

A broad range of industrial and commercial activities has impacts on soil through the release of a broad variety of chemicals: heavy metal-loaded fluids, mineral oil, polycyclic aromatic hydrocarbons (PAH), chlorinated hydrocarbons, and aromatic hydrocarbons. The relative contribution of each class may vary greatly from country to country. For example, in terms of occurrences of the specific contaminant at investigated sites, mineral oil is the most frequent pollutant in the Czech Republic, Italy (more than 45 %), and Latvia (25 %). Heavy metals are the pollutants most often found in Sweden (40 %) and Belgium-Flanders (35 %). Chlorinated hydrocarbons are the most frequent contaminants in Austria (25 %), while cyanide is the most widespread pollutant in the Netherlands (about 20 %) (European Agency: EIONET data flow progress report 2010).

National reports indicate that heavy metals and mineral oil are the most frequent soil contaminants at investigated sites in Europe, while mineral oil and chlorinated hydrocarbons are the most frequent contaminants found in groundwater (EEA 2010).

The risks to human health and the environment vary considerably with each specific contaminant, the specific conditions at the site, and the level of exposure of the public. In fact, the risks are determined by the physicochemical properties of the contaminants such as solubility, mobility, volatility, sorption capacity, and persistence, as well as the pathways from source to potential receivers (e.g., the existence of an impermeable layer and the permeability and thickness of the unsaturated zone) and the frequency and length of exposure. Therefore, the assessment of the impacts of contamination has to be assessed on a case-by-case basis (EEA 2010).

The protection of groundwater resources and the exposure of humans via drinking water from ground sources are reported by far as the most important motives for the application of risk-reduction measures. The protection of the soil per se deserves a relatively low importance as an objective (this is only reported in Hungary in 10 % of all remediated sites) (European Agency: EIONET data flow progress report 2010). This may be due to the lack of specific regulations covering the soil medium but also to the wider dispersion and farther-reaching effects of contaminants in groundwater compared to soil. Relatively low priority, with a few exceptions, is assigned to the remediation of sites where exposure to the contaminants arises from surface water, soil gases, and the food chain. Only in very few cases are remediation activities reported to be triggered by the protection of ecosystems, e.g., in Croatia and the Czech Republic, and the loss of biodiversity (Croatia, Malta) (European Agency: EIONET data flow progress report 2010).

The main contaminants identified in these cases include heavy metals, the BTEX group of hydrocarbons, PAHs, chlorinated hydrocarbons, and mineral oil, as well as inorganic compounds such as asbestos.

BTEX is an acronym for benzene, toluene, ethylbenzene, and the three xylene isomers. This group of volatile organic compounds (VOCs) contaminates the soil through spills involving petroleum products such as gasoline, diesel fuel, lubricating oil, and heating oil from leaking oil tanks. Benzene is used in the production of synthetic materials such as plastics, synthetic rubber, insecticides, and paints. Toluene is used as paint solvent, resins, and oils. Ethylbenzene is found in products like paints, plastics, and pesticides. Finally, xylenes are used in leather industries or as a printing solvent (Salanitro et al. 1997). These compounds will tend to be found as vapors in soil atmosphere as well as dissolved in the water phase because of their relatively high water solubility and low K_{ow} (octanol–water partition coefficient) values. These compounds are not retained very efficiently by soil particles or constituents because of their relative hydrophilic nature and can be transported several kilometers downstream of the source (Wolfe et al. 1997). Due to their water solubility (especially in the case of benzene), BTEX always migrate from soil to groundwater systems and contaminate drinking water supplies often far from their source. BTEX compounds have raised increasing concern; benzene in particular is a potent mutagen that can lead to hematological effects, which may ultimately lead to aplastic anemia and development of acute myelogenous leukemia (ATSDR 2004). Acute exposures to high-level BTEX have been associated with skin and sensory irritation, impact on the respiratory system, and central nervous system depression. More prolonged exposure to BTEX affects the liver, kidneys, and the blood systems (WHO 2000). According to USEPA, from 1982 to 2005, BTEX compounds were among the contaminants more often addressed at contaminated sites, 25 %, second only to halogenate volatile organic compounds, found in 42 % of the contaminated sites (USEPA 2010). For these reasons, remediation of BTEX-contaminated soils is paramount to avoid public health hazards. According to the same report, soil vapor extraction, incineration, and bioremediation were the most used technologies to remediate soils contaminated with BTEX.

8.2 Physicochemical Remediation Technologies

The remediation of contaminated soils can be performed through the utilization of technologies that involve physicochemical, thermal, or biological processes. The physicochemical processes employ physical, chemical, and/or electrical properties of pollutants to destroy, convert, separate, or even just immobilize them in the soil structure. Thermal treatments use heat to increase volatility, to burn, to decompose, to destroy, or to melt the contaminants. These processes are usually expensive due to the cost of energy, equipment, and maintenance. Finally, biological remediation uses microorganisms or plants to degrade contaminants to innocuous or less toxic compounds. These latter processes are typically inexpensive but slow and are sensitive to the presence of toxins and high concentrations of the contaminants. The choice of the most appropriate remediation technology to be applied to a specific case depends on soil type, environmental parameters, and contaminant properties. In this chapter, the most common remediation technologies will be described, emphasizing those that use biological processes, namely, those employing fungi.

Among the physicochemical remediation technologies, soil vapor extraction (SVE) and solidification/stabilization (S/S) can be highlighted because they are two of the most used technologies (USEPA 2010).

SVE, also known as soil venting or vacuum extraction, is the most used in situ remediation technology, and it is applicable to soils contaminated with volatile and semi-volatile organic compounds located in the unsaturated zone of the soil. This technology induces airflow in the subsurface of the soil through an applied vacuum, creating and enhancing the volatilization of contaminants (Suthersan 1996). SVE uses the volatility of the contaminants to allow mass transfer from the pollutant adsorbed, dissolved, and/or located in free phases in the soil to the vapor phase, which is then transported by the airstream to the exterior of the soil where it is properly treated (Fischer et al. 1996). SVE is efficient when the soils to be treated have high porosity (allowing the airflow to easily permeate through all the soil matrix and extract the contaminant), low contents of water and organic matter (Alvim-Ferraz et al. 2006) (if the contaminant is partitioned mostly in the gas phase of the soil, it will be more easily extracted), and the contaminant has a vapor pressure above 1.0 mmHg (it will volatilize easily), low water solubility (the contaminant will dissolve less in the aqueous phase of the soil and transfer more easily to the gas phase), and a low soil adsorption coefficient (adsorption to the organic matter of the soil too hinders the mass transfer to the gas phase) (Albergaria et al. 2006).

S/S is an ex situ or in situ technology that uses physical and chemical means to reduce the mobility of contaminants in the soil, mostly metals. *Solidification* refers to processes that encapsulate the soil aiming at the formation of a solid material and the reduction of the migration of the contaminant. This process is accomplished through chemical reactions between the contaminated soil and binding reagents or by mechanical processes (Jeffrey et al. 1995). Cement, proprietary additives, and phosphate are the most common binders (Sparrevik et al. 2009). When this technique is applied to fine soil particles, it is commonly designated as microencapsulation, whereas, when applied to large blocks of soil, it is designated as macroencapsulation. *Stabilization* involves chemical reactions with soil that immobilize the contaminants and reduce their solubility and, consequently, their leachability. Ex situ S/S operations require excavation of the contaminated soil, and the material obtained after the treatment must be disposed of at appropriate sites. In situ S/S uses auger/caisson systems and injector head systems to add binders to the contaminated soil without any need of excavation (Wiles 1987).

The remediation of contaminated soils by thermal treatment uses heat to destroy or extract, by volatilization and other mechanisms, the contaminants in soil. This technology is commonly used with soils contaminated with inorganic pollutants, radionuclides, explosives, and metals (Hinchee and Smith 1992). As for the other treatments, it can be performed in situ or ex situ. In situ treatments commonly include conductive heating, electrical resistive heating, hot air injection, hot water injection, steam-enhanced extraction, and radio frequency heating. The volatilization of the contaminants turns them more mobile, which allows them to be extracted through extraction wells to the exterior of the soil where they are properly treated in specific units (Hinchee and Smith 1992).

Within in situ technologies, the following can be pointed out:

– *Electric resistance heating*—Utilization of arrays of electrodes located around a central neutral electrode to create a concentrated flow of current toward the central point. The resistance to current flow in soil generates heat that increases soil temperature above 100 °C, leading to the volatilization of contaminants that are recovered via vacuum extraction and processed at the surface (Beyke and Smith 2009; Mori et al. 2005).

– *Injection of hot air*—In deep subsurface situations, the hot air is injected at high pressure through wells or soil fractures, while in surface soils, hot air is usually injected in combination with soil mixing or tilling (Park et al. 2005; Schmidt et al. 2002).

– *Radio frequency heating*—Uses electromagnetic energy to increase soil temperature (over 300 °C) and enhance SVE using rows of vertical electrodes embedded in soil. The heated soil is bounded by two rows of ground electrodes with energy applied to a third row midway between the ground rows (Daniel and Pearce 1994; Roland et al. 2007).

– Vitrification—Uses electric current to melt the contaminated soil at temperatures between 1,600 and 2,000 °C. After cooling, the obtained soil is chemically stable, leach-resistant, glassy, and crystalline material similar to obsidian or basalt rock. Organic contaminants are destroyed during the process while radionuclides and heavy metals will be retained within the vitrified product (USEPA 2006; Staley 1995).

– *Low-temperature thermal desorption*—It is one of the most used ex situ thermal technologies for the remediation of soils contaminated with organic compounds. The equipment used to perform this treatment, denominated thermal desorbers, are designed to heat soils to temperatures sufficiently high that the contaminants volatilize and desorb from the soil. The main objective of this equipments is the desorption of the contaminants but, depending on the contaminants that are present in soil and the temperature of operation, partial or complete decomposition of the contaminants can occur. To prevent atmospheric contamination, the vaporized emissions from the soil pass through treatment units before being released to the atmosphere (USEPA 1994; Abramovitch et al. 2003).

8.3 Biological Remediation

8.3.1 Soil Microbiology

Biological processes are mainly based on the utilization of soil microorganisms to degrade or detoxify pollutants. Decaying organic matter, and especially the remains of the slow-degrading polymer lignin, also called humus, are of the utmost importance in conferring to soil mechanical structure and some of their capacity to retain water and dissolved mineral ions (Piccolo 2002; Rasse et al. 2006) but also in allowing, directly or indirectly, countless forms of microbial life to thrive. The microbes most readily associated to soil are obviously bacteria and fungi. However, microalgae and photosynthetic bacteria (cyanobacteria) can be found in the top illuminated layers (Metting 1981; Whitton 2000); Archaea, formerly called

Archaebacteria, are also widely present in soils (Kent and Triplett 2002; Leininger et al. 2006); and unicellular predators, protozoans, can be found in all cases (Bonkowski 2004). It is worth observing that this sketchy list puts together structurally very different microorganisms: on the one hand Archaea and bacteria, with simple cells devoid of nucleus or internal compartments (prokaryotes), and on the other fungi, unicellular algae, and protozoans, which exhibit much more complex eukaryotic cells.

From the physiological point of view, microbes in soil can be divided into two main groups: autotrophs (organisms utilizing CO_2 as source of carbon) and heterotrophs (organisms using directly environmental organic matter). Within the former group, microalgae and cyanobacteria are active exclusively in the top layers where sunlight, their energy source, is available. Heterotrophs on the other hand take advantage of organic matter as source of both energy and carbon. Many soil bacteria fall into this group as do all fungi. Another important physiological distinction is between aerobic and anaerobic microbes. Strictly aerobic organisms depend on the presence of O_2 for their functioning because their respiratory machinery requires molecular oxygen as the final acceptor of electrons. Within anaerobic organisms, we can distinguish respiratory ones, which use other compounds as final electron acceptors of respiration (such as oxidized species of N, S, Mn, Mg, Fe, As, Se, or other inorganic and sometimes organic compounds), from fermentative ones, which manage to obtain energy from the interconversion of organic molecules. Nevertheless, this description is but a simplification of reality, since cases of intermediate and facultative alternative behavior in many species make the whole picture much more intricate (Andresen 1992; Vallini et al. 1997; Anandham et al. 2009).

Special consideration has to be given to an alternative lifestyle that many soil microorganisms entertain: symbiosis. Many soil bacteria and fungi live associated to plant roots exchanging services (like nutrient extraction or nitrogen fixation) for a constant supply of carbon source produced by photosynthesis aboveground (Franche et al. 2009; Dodd 2000). Many of such associations are extremely relevant to the survival of wild plants as well as to agriculture: it is renowned how important nodulating diazotrophs (rhizobia, *Frankia*) are to their hosts, and the notion that most known plants are mycorrhizal, harboring symbiotic fungi at the root level, is commonplace nowadays (Brundrett 2009).

Estimates produced in the last two decades stating that less than 1 % of natural microbial species had been cultivated or were by any means culturable (Amann et al. 1995) may be disproportionate. Nonetheless, data obtained in the past two decades by molecular biological techniques clearly show that in any environment, and in soils in particular, a huge untapped microbial biodiversity is present on which we still have scant knowledge (Da Rocha et al. 2009). In addition, microorganisms that have been cultivated and studied in detail will probably reflect that minority of wild microbes that adapt easily to artificial lab conditions, which means they may in fact correspond to a less competitive fraction in their real natural environment. Possibly an even bigger flaw in our understanding of the soil microflora derives from the scientific practice of simplifying systems for analysis, in this case isolating single species from the rest of the community: this obviously yields always a very partial view of a microorganism's actual lifestyle in the wild, where it is embedded in a complex physical medium and surrounded by myriads of cells belonging to scores of different species. Microbiologists have tried to overcome these problems by employing ever more sophisticated molecular tools, which circumvent the biases introduced by in vitro cultivation (Shannon et al. 2002; Tringe et al. 2005), and to study organisms in situ, within their complex natural settings (Yarwood et al. 2002; Mummey et al. 2006; Eickhorst and Tippkötter 2008). Nevertheless, we are still very far from grasping soil microbial ecology in all its complexity.

Soil particles in situ tend to form aggregates, technically called peds, held together by water surface tension and sticky organic matter (Adl 2008). Pores between and within peds may be filled with gases and/or water: microniches with extremely dissimilar conditions with respect to oxygen tension, humidity, pH, availability of nutrients, possible exposure to xenobiotics, proximity with a plant root or with antibiotic-producing organisms are created micrometers away from each other (Mummey et al. 2006; Becker et al. 2006; Or et al. 2007). Each microenvironment will allow just the microbes best adapted to the specific local conditions to thrive, thus fostering a huge variety of different strains and species.

8.3.2 Biological Technologies

In general terms, *bioremediation* is the elimination or immobilization or inactivation of contaminants from a polluted environment operated by biological action. In the light of this definition, it is clear that, for bioremediation to proceed successfully and sufficiently rapidly, two fundamental types of factors need to be available: first, adequate organisms that will be able to withstand the existing levels of pollution and transform it into nontoxic compounds—this is technically known as the genetic or functional potential (Zehr and Capone 1996; Martin-Laurent et al. 2004); second, physicochemical parameters that will allow these microorganisms to operate. These may include, depending on the particular process, presence or lack of O_2, presence of alternative electron acceptors (e.g., NO_3^- or SO_4^{2-}), presence of inorganic and/or organic nutrients, adequate humidity levels, pH, and temperature (Holden and Firestone 1997; Soares et al. 2010). When all these features are already present at the site, bioremediation will proceed spontaneously—in this case, we speak of natural or passive or intrinsic bioremediation (EPA, OSWER Directive 9200.4-17P 1999).

The utilization of biological processes to remediate contaminated soils is increasing due to its efficiency and low costs associated. There are several types of biological remediation technologies such as natural attenuation, bioventing, land farming, biopiles, phytotechnologies, and enhanced bioremediation.

The term *natural attenuation* refers to the reliance on natural processes to achieve site-specific remediation objectives through physical, chemical, or biological processes that, under favorable conditions, occur without any human intervention, and that reduce the mass, toxicity, or mobility of contaminants in soil. These processes can include phenomena such as biodegradation, dispersion, dilution, sorption, volatilization, chemical or biological stabilization, transformation, or destruction of contaminants (Mulligan and Yong 2004). Biological degradation (aerobic or anaerobic) of contaminants is one of the most important components of natural attenuation and is commonly active in soils contaminated with petroleum hydrocarbons (Kao and Prosser 2000) or chlorinated solvents (Davis et al. 2002).

However, conditions in situ will be optimal for attenuation just rarely: human intervention to adjust one or more of the parameters in order to allow or accelerate bioremediation is technically referred to as biostimulation or enhanced bioremediation. When physicochemical parameters are not optimal, they can be improved by aeration (increasing oxygenation), or sealing off with liners (inducing anaerobic conditions), or amendment with inorganic or organic nutrients, water, acids, or bases (Mulligan and Yong 2004).

If on the other hand the lacking factor is genetic potential, the problem can be solved by bioaugmentation—the addition of selected microorganisms or mixes of different species. Sometimes bioaugmentation is performed using undifferentiated microbe-rich matter (activated sludge, manure, etc.) (Siméon et al. 2008; Ros et al. 2010) or indirectly, by augmenting the local microflora through the addition of copious amounts of nutrients (Boopathy et al. 1994; Östberg et al. 2006). In some special cases of mixed inorganic/organic contamination (Roane et al. 2001; Fernandes et al. 2009), selected microbes of two specific types have been employed (dual bioaugmentation) so that one strain will somehow detoxify inhibitory heavy metal concentrations permitting to the other strain to degrade the organic pollutants present.

From the description of the rhizosphere effect and of symbioses provided above, an obvious consequence follows: soils with a full plant cover will be rich in active roots, and this will be accompanied by the presence of a thriving microbial community. In fact, plant cover and land management hugely affect soil microflora (Yao et al. 2000; Larkin 2003). This fact is exploited in the technique called rhizoremediation, whereby polluted soils are actively colonized with suitable plants which will lead to augmentation of the microflora and thus to an enhancement of bioremediation (Gerhardt et al. 2009).

Bioventing is an in situ remediation technology that aims at stimulating microorganisms of the unsaturated zone of soil to biodegrade organic pollutants by enhancing their metabolism by the introduction of air into soil and, in most cases, the addition of nutrients. Bioventing can be, in some cases, similar to SVE, but, while in SVE the main goal is the removal of contaminants primarily through volatilization, in bioventing systems, the objective is to promote in situ biodegradation of pollutants and minimize volatilization, through the utilization of lower airflow rates compared with SVE (Jørgensen 2007). Bioventing has proven to be extremely efficient in the remediation of soils contaminated with petroleum products, namely, gasoline, jet fuels, kerosene or diesel fuel (Van Eyk 1997), and benzene (Soares et al. 2010). This remediation technology is often used in soils contaminated with pollutants with vapor pressures lower than 0.5 mmHg (USEPA 1994).

Land farming and *biopiles* are ex situ remediation technologies that are used to reduce concentrations of pollutants (e.g., petroleum constituents) in excavated soils through the use of biodegradation enhanced by the addition of air, moisture, and nutrients. While soils treated by land farming are aerated by tilling or plowing, biopiles are aerated forcing air to move through the soil by injection or extraction through slotted or perforated piping placed throughout the pile (USEPA 1994; Lin et al. 2010).

Phytotechnologies use plants to extract, degrade, contain, or immobilize contaminants in soil when the contamination is shallow and in low or moderate levels, due to the fact that high levels of contaminants may be toxic to plants and inhibit their growth. These technologies are used to treat a wide range of contaminants, including metals, volatile organic compounds, PAHs, petroleum hydrocarbons, radionuclides, and explosives through several mechanisms such as phytoextraction, rhizodegradation, phytodegradation, and phytovolatilization (Kamath et al. 2004; Prasad et al. 2010). Phytoextraction uses the uptake of contaminants by the plant roots, with subsequent accumulation in plant tissue, which have to be harvested and properly treated. Rhizodegradation is the process whereby degradation occurs in the root zone. Phytodegradation, as phytoextraction, involves the uptake of contaminants but, in this case, they are degraded through metabolic processes within the plant. Phytovolatilization is the uptake of contaminants into the plant and their subsequent transpiration to the atmosphere. It is mainly used to remediate contaminated groundwater, but can also be applied to soluble soil contaminants (USEPA 2006).

8.4   Bioremediation by Fungi

Several authors also state that biodegradation is usually driven by a consortium of different species of microorganisms, including algae, bacteria, protozoan, and fungus (Mazzeo et al. 2010; Racke Kenneth 1990). Bacteria and fungi exhibit different abilities to metabolize hydrocarbons. Both yeasts (nonfilamentous fungi) and bacteria show decreasing abilities to metabolize alkanes with increasing chain length. Filamentous fungi instead do not exhibit preferential degradation for a particular chain length. Similar patterns of hydrocarbon catabolism are known in bacteria and fungi, but there is considerable variability among individual isolates. The hyphal structure of fungal mycelia and their larger surface area allow better penetration of the hydrocarbon drops and hydrocarbon-impregnated soil aggregates, the interior of which may be anoxic. Many species of fungi are xero- and osmotolerant. Most fungi secrete extracellular enzymes that may assist in the initial degradation of hydrocarbons. Soil yeast isolates produce alcohols and aldehydes as a result of oxidation of the hydrocarbons. This may be a potential candidate for seeding in the biodegradation of oil-polluted terrestrial ecosystems.

It has been proved that soil disruption caused by pollution with hydrocarbons decreases biodiversity by selecting for microbial species better adapted to survive in the changed environment (Hughes et al. 2007). Physicochemical changes due to this pollution affect soil structure and fertility, and therefore the fauna and flora. Hydrocarbon-polluted soils become relatively sterile with the death of all but resistant microbial life forms. Within aromatic hydrocarbons, BTEX constituents are among the most susceptible to elimination from the environment by indigenous microorganisms (Prenafeta-Boldú et al. 2004). Prior exposure of an ecosystem to a pollutant has been correlated with the possibility of that particular compound to be degraded. Successful biodegradation is a function of the presence of the activity of a fungal population capable of degrading target contaminants.

Indeed, the key components in bioremediation are the microbial enzymes involved in the degradative reactions leading to the elimination or detoxification of the chemical pollutant. Due to superior robustness and metabolic versatility, mixed cultures rather than pure cultures are currently being applied for the treatment of petroleum wastes in fermentor-based systems. In such a situation, a pre-acclimated hydrocarbon-degrading culture may be subsequently exposed to a variety of heterogeneous hydrocarbon-contaminated waste streams (Okoh and Trejo-Hernandez 2006).

Fungal degradation of monoaromatic compounds has clear implications for bioremediation, and the role of fungi in the removal of BTEX from the environment in studies published the last 6 years is subject of this revision.

The process of degradation of woody plant material has a central place in terrestrial carbon cycling using a nonspecific free radical mechanism (the overwhelming majority of carbon fixed by land plants is found in lignocellulose). A relatively large number of basidiomycete fungi are found naturally as wood or leaf-litter degraders, and the majority of these—the white rot fungi—are capable of breaking down lignin. However, the ability to degrade xenobiotics is not limited to basidiomycetes; it is also found in other xylotrophs (e.g., those causing brown rot and soft rot), as well as in a broad range of soil-dwelling fungi (Rabinovich et al. 2004). The initial steps in this process involve an oxidase (laccase) and a manganese peroxidase. Lignin peroxidase is a crucial component in some fungi, and there are some less well-studied peroxidases that have atypical substrate specificities.

One of the reports indicated that lignin-degrading white rot fungi, such as *Phanerochaete chrysosporium*, can degrade an extremely diverse group of environmental pollutants (Pinedo-Rivilla et al. 2009). Since then, there has been intense worldwide research to unravel the potential of white rot fungi in bioremediation. This ability of white rot fungi to degrade a wide spectrum of environmental pollutants sets them apart from many other microbes used in bioremediation.

White rot fungi offer a number of advantages for use in bioremediation. The key enzymes of the lignin degradation system (LDS) are extracellular, obviating the need to internalize the substrates and allowing substrates of low solubility to be oxidized. Furthermore, the extracellular enzyme system of the white rot fungi enables these organisms to tolerate relatively higher concentrations of toxic pollutants than would otherwise be possible. The constitutive nature of the key enzymes involved in the LDS obviates the need for these organisms to be adapted to the chemical being degraded. White rot fungi are also ubiquitous in nature.

Although they degrade lignin, they cannot utilize it as a source of energy for growth and, instead, require co-substrates such as cellulose or other carbon sources (Singh 2006). The preferred substances for growth of white rot fungi in nature are lignocellulosic substrates. Therefore, inexpensive lignocellulosics such as corncobs, straw, peanut shells, and sawdust can be added as nutrients at contaminated sites to obtain enhanced degradation of pollutants by these organisms. Finally, white rot fungi grow by hyphal extension and thus can reach pollutants in the soil in ways that other organisms cannot.

8.4.1　Fungal Enumeration

Preliminary analyses of the total heterotrophic and specific hydrocarbon-degrading fungal counts can provide a good understanding of soil biological conditions and the extent of acclimatization of indigenous fungal populations to site conditions. Viable counts indicate the viability of the indigenous fungal population capable of supporting bioremediation. In addition to the preliminary analyses, monitoring fungal populations during soil remediation is important to gauge fungal activity in hydrocarbon degradation. *Trametes gibbosa* has been measured by enzymatic activity and biomass dry weight (Zanaroli et al. 2010); other methods include direct inoculation of serial dilutions (Pozdnyakova et al. 2008), laccase activity (Pozdnyakova et al. 2008), hyphal dry weight measurements (Garcia-Pena et al. 2008; Hughes et al. 2007), ergosterol quantification (Ballaminut and Matheus 2007), mycorrhizal colonization (Volante et al. 2005), 18S rDNA-based PCR-TGGE (temperature gradient gel electrophoresis) (Prenafeta-Boldú et al. 2004), quantification of proteins (Vigueras et al. 2008), carbon dioxide and O_2 monitorization (Garcia-Pena et al. 2008), colonization of substrate, loss of organic matter, pH variation, and total nitrogen (Ballaminut and Matheus 2007). Yet other techniques such as epifluorescence direct counts, most probable number techniques, DNA probes, lipid assays, and metabolic indicator methods can be used (Singh 2006).

8.4.2　Soil Microcosm Test

Soil microcosm tests can be conducted in simple jars of contaminated soil (Fernandes et al. 2009; Soares et al. 2010) or at pilot scale with soil columns (Soares et al. 2010). Good results are expected when real environmental conditions are incorporated in the model of microcosm design. These tests include sterile treatments as appropriate controls to separate the effects of abiotic hydrocarbon loss from actual biodegradation (Soares et al. 2010).

Laboratory-based studies have shown that fungi are able to degrade a wide range of organic pollutants and have great potential for use as inoculants to remediate contaminated soil. However, soil is a heterogeneous environment, and it is to be expected that remediation experiments using fungal inocula will show varying degrees of success. For example, soil environmental conditions such as pH, nutrients, and oxygen levels may not be optimal for fungal growth or for activity of the fungal extracellular enzymes involved in pollutant transformation.

8.4.3 BTEX Application

Antarctic fungi showed a range of susceptibilities to high or saturated concentrations of hydrocarbons, but hyphal extension rates were reduced more by aromatic than aliphatic hydrocarbons. Fungi, such as *Mortierella* sp., were found in Antarctic soils that can degrade and metabolize hydrocarbons, with potential for bioremediative applications. However, as happens with bacteria, bioremediation was dependent on soil physical and chemical parameters such as temperature, nutrients, and moisture availability (Hughes et al. 2007).

Arbuscular mycorrhizal fungi have been studied as potential tools in the reclamation of sites contaminated by BTEX. Studies performed with *Gigaspora rosea*, *Gi. margarita*, and *Glomus mosseae* showed that plants colonized by arbuscular mycorrhizal fungi are able to reduce the persistence of BTEX in polluted soils (Volante et al. 2005).

A soil inoculation study with toluene-metabolizing fungus *Cladophialophora* sp. strain T1 indicated that fungal inoculation of soil might be a viable technique to improve biodegradation of BTEX pollutants during bioventing, especially when a low pH limits the activity of the indigenous bacteria (Prenafeta-Boldú et al. 2004). *Paecilomyces lilacinus* consumed toluene as the sole carbon source, although in a gas-phase biofilter, obtaining a removal efficiency of 53 % (Vigueras et al. 2008). Another gas-phase biofilter with *Paecilomyces variotii* showed complete degradation of toluene to CO_2 and biomass. Benzene and the xylenes were only partially degraded when fed individually (Garcia-Pena et al. 2008). *Trametes gibbosa* in microbial consortia showed degradation activity of diesel fuel compounds (Zanaroli et al. 2010). The application of the white rot fungus *Pleurotus ostreatus* in oil-polluted soils has shown that the fungus degraded mainly the simpler aromatic fraction (mono and di), whereas local soil microflora degraded paraffin and naphthene oil fractions (Pozdnyakova et al. 2008) (Table 8.1).

8.5 Remediation Techniques (SVE and Bioremediation on BTEX)

SVE is, as said before, extremely efficient for the extraction of volatile or semi-volatile contaminants from the saturated zone of the soil. However, under certain conditions, such as soils with high contents of organic matter, the efficiencies of the

Table 8.1 Fungus activity in BTEX bioremediation

Fungus	Benzene	Toluene	Ethylbenzene	Xylene	Ref.
Trichoderma sp.	–	Degradation activity			Hughes et al. (2007)
Mollisia sp.	–	Degradation activity			
Penicillium commune	–	Degradation activity			
Mortierella sp.	–	Degradation activity			
Cladophialophora sp.	Not metabolized	Degradation activity			Prenafeta-Boldú et al. (2004)
Gigaspora rosea	Degradation activity as arbuscular mycorrhizal fungi				Volante et al. (2005)
Gi. margarita					
Glomus mosseae					
Paecilomyces lilacinus	–	Degradation activity	–	–	Vigueras et al. (2008)
Paecilomyces variotii	Degradation activity	Complete degradation	–	Degradation activity	Garcia-Pena et al. (2008)
Pleurotus ostreatus	Degradation activity				Pozdnyakova et al. (2008)
Trametes gibbosa	Nonspecific activity				Zanaroli et al. (2010)

process decrease significantly endangering the achievement of the defined cleanup goals (Albergaria et al. 2006). In these cases, it is necessary to complement the remediation process with other technologies that allow the remediation to be completed. Bioremediation is one of the technologies that is used more frequently nowadays to complement SVE because of its proved efficiency and low cost (USEPA 2010). In other circumstances, the combination of these technologies is used in temporal succession: in an initial phase, SVE is applied to extract rapidly and efficiently high amounts of contaminant that are present in soil, reducing the toxicity that high levels of contaminant create to soil microorganisms, and then, in the final stage, bioremediation is used to complete the process. In many cases of bioremediation, the main organisms responsible for the degradation are not known.

Based on this, experimental work was developed in a laboratorial installation with different soils (sandy soils with four water contents and humic soils with three organic matter content levels): all the soils were contaminated separately with benzene, toluene, ethylbenzene, and xylene at an initial level of 250 mg kg^{-1} and remediated by SVE until the concentration of contaminant in the extracted emissions was below 1.0 mg L^{-1}. The results showed that in approximately 29 % of the cases, SVE was not efficient to achieve the remediation levels defined by Spanish legislation (Real Decreto 9/2005) because soil organic matter content influenced negatively the SVE process turning it less efficient and more time consuming (Soares et al. 2010). Those cases occurred with humic soils (with 4, 14, and 24 % of organic matter content) and more clearly when contaminants with lower vapor pressures were used (namely, xylene) (Albergaria 2010).

In those soils which through this treatment did not reach the cleanup goals, bioremediation was implemented.

The first stage of the bioremediation experiments was to verify if the native microorganisms of the soil were able, by themselves, to degrade the contaminants or if inoculation of BTEX-degrading microorganisms into the soil was necessary. Six tests were performed in flasks stoppered with Teflon valves (Mininert™, VICI®, Valco instruments) containing soil with an organic matter content of 14 %, nutrients (minimum medium obtained from mineral salts medium and prepared according to Kelly et al. (1994)), and water. A different soil preparation was used in each flask: one non-sterile, one sterilized by autoclaving, and four more sterilized and inoculated separately with one of four strains of bacteria (*Pseudomonas stutzeri* str. OX1, *Labrys portucalensis* str. F11, *Pseudomonas fluorescens* str. ST, and *Pseudomonas putida* str. KT2440) (De Marco et al. 2004). The tests showed that the native microorganisms present in soil (flask with non-sterilized soil) were capable to degrade the contaminants present in soil with yield identical or better than the best inoculated strain (Soares et al. 2010). Following this conclusion, bioremediation was performed based only on the native microorganisms (most probably mainly fungi) of the soil.

The bioremediation experiments were performed in the same stainless steel columns where SVE remediation had taken place and required complementary treatment. During the experiments, the degradation was studied by monitoring the concentration of contaminant in the soil gas phase by gas chromatographic analysis.

These tests showed that the native microorganisms of the soil were able to degrade the contaminants to the levels established by legislation; however, above certain levels of contamination (that depended on the specific BTEX contaminant present in soil), the soil environment became toxic to the microorganisms inhibiting their remediation activity. The highest degradation rates were obtained with soils contaminated with benzene, while the lowest rates were obtained when xylene was used. Other authors such as Amin et al. (2009), Cozzarelli et al. (2010), Ni et al. (2010), and Wolicka et al. (2009) also concluded that the biodegradation times for ethylbenzene were longer than for benzene. The soils with higher content of organic matter presented higher degradation rates which led to faster remediations (Soares et al. 2010) which corroborate the contention that organic matter acts as support for microbial communities.

Based on these results, it was possible to conclude that the resident soil microflora could degrade the contaminants to the pre-defined cleanup goals. Although no attempts have been carried out in order to identify which organisms were responsible for the degrading activities, the nature of the organic matter present in the soil plus the observation of mycelia growing on the surface of the soil and at different depths in test columns during the course of the experiments seem to suggest that fungi were involved. This may match with the level of contaminants that inhibited the degradation. The monitoring registered two distinct stages: an initial one that was characterized by a stabilization of the concentration of the contaminant in the gas phase and a final stage where the concentration began to decrease more rapidly indicating that the degradation was occurring.

Acknowledgment The authors are grateful to Fundação para a Ciência e Tecnologia (Project PTDC/ECM/68056/2006) for the material support for this work.

References

Abramovitch RA, Chang QL, Hicks E, Sinard J (2003) In situ remediation of soils contaminated with toxic metal ions using microwave energy. Chemosphere 53:1077–1085

Adl SM (2008) Setting the tempo in land remediation: short-term and long-term patterns in biodiversity recovery. Microbes Environ 23:13–19

Agency for Toxic Substances and Disease Registry (ATSDR) (2004) Interaction profile for benzene, toluene, ethylbenzene, and xylenes (BTEX). U.S. Department of Health and Human Services, Public Health Service, Atlanta, GA

Albergaria JT (2010) Previsão do Tempo de Remediação de Solos Contaminados Usando a Extracção de Vapor. PhD Thesis. Faculdade de Engenharia da Universidade do Porto, Porto

Albergaria JT, Delerue-Matos C, Alvim-Ferraz MCM (2006) Remediation efficiency of vapour extraction of sandy soils contaminated with cyclohexane: influence of air flow rate and of water and natural organic matter contents. Environ Pollut 143:146–152

Alvim-Ferraz MCM, Albergaria JT, Delerue-Matos C (2006) Soil remediation time to achieve clean-up goals: II: influence of natural organic matter and water contents. Chemosphere 64:817–825

Amann RI, Ludwig W, Heinzschleifer K (1995) Phylogenetic identification and in situ detection of individual microbial cells without cultivation. Microbiol Rev 59:143–169

Amin FE, Kee WK, Surif S (2009) Biodegradation of hydrocarbon benzene, toluene, ethylbenzene and xylene (BTEX) by consortium bacterial culture. In: Prosiding Seminar Kimia Bersama UKM-ITB VIII

Anandham R, Gandhi PI, Kwon SW, Sa TM, Kim YK, Jee HJ (2009) Mixotrophic metabolism in Burkholderia kururiensis subsp. thiooxydans subsp. nov., a facultative chemolithoautotrophic thiosulfate oxidizing bacterium isolated from rhizosphere soil and proposal for classification of the type strain of Burkholderia kururiensis as Burkholderia kururiensis subsp. kururiensis subsp. nov. Arch Microbiol 191:885–894

Andresen RA (1992) Diversity of eukaryotic algae. Biodivers Conserv 1:267–292

Ballaminut N, Matheus DR (2007) Characterization of fungal inoculum used in soil bioremediation. Braz J Microbiol 38:248–252

Becker JM, Parkin T, Nakatsu CH, Wilbur JD, Konopka A (2006) Bacterial activity, community structure, and centimeter-scale spatial heterogeneity in contaminated soil. Microb Ecol 51:220–231

Bertin C, Yang X, Weston LA (2003) The role of root exudates and allelochemicals in the rhizosphere. Plant Soil 256:67–83

Beyke G, Smith GJ (2009) Advances in the application of in situ electrical resistance heating. In: Abstracts of the 11th international conference on environmental remediation and radioactive waste management, pp 1009–1017

Bonkowski M (2004) Protozoa and plant growth: the microbial loop in soil revisited. New Phytol 162:617–631

Boopathy R, Kulpa CF, Manning J, Montemagno CD (1994) Biotransformation of 2,4,6-trinitrotoluene (TNT) by co-metabolism with various co-substrates: a laboratory-scale study. Bioresour Technol 47:205–208

Brundrett MC (2009) Mycorrhizal associations and other means of nutrition of vascular plants: understanding the global diversity of host plants by resolving conflicting information and developing reliable means of diagnosis. Plant Soil 320:37–77

Chernyanskii SS, Gennadiyev AN (2005) Climate. In: Lal R (ed) Encyclopedia of soil science, 2nd edn. CRC, Boca Raton, FL, pp 291–299

Cozzarelli IM, Bekins BA, Eganhouse RP, Warren E, Essaid HI (2010) In situ measurements of volatile aromatic hydrocarbon biodegradation rates in groundwater. J Contam Hydrol 111:48–64

Da Rocha UN, van Overbeek L, van Elsas JD (2009) Exploration of hitherto-uncultured bacteria from the rhizosphere. FEMS Microbiol Ecol 69:313–328

Daniel DE, Pearce JA (1994) Enhanced vapor extraction of organics from contaminated soil using radio frequency heating. Waste Manage 14:347

Davis JW, Odom JM, De Weerd KA, Stahl DA, Fishbain SS, West RJ, Klecka GM, DeCarolis JG (2002) Natural attenuation of chlorinated solvents at Area 6, Dover Air Force Base: characterization of microbial community structure. J Contam Hydrol 57:41–59

De Marco P, Pacheco CC, Figueiredo AR, Moradas-Ferreira P (2004) Novel pollutant-resistant methylotrophic bacteria for use in bioremediation. FEMS Microbiol Lett 234:75–80

Dirección Nacional de Medio Ambiente (DINAMA) (2005) Plan Nacional de Implementación del Convenio de Estocolmo (NIP). Retrieved January 11th 2011 from http://www.dinama.gub.uy

Dodd JC (2000) The role of arbuscular mycorrhizal fungi in agro- and natural ecosystems. Outlook Agric 29:55–62

Eickhorst T, Tippkötter R (2008) Detection of microorganisms in undisturbed soil by combining fluorescence in situ hybridization (FISH) and micropedological methods. Soil Biol Biochem 40:1284–1293

European Environment Information and Observation Network (EIONET) (2010) Eionet priority data flows, May 2009–April 2010. Retrieved January 11th 2011 from http://www.eea.europa.eu/publications/eionet-priority-data-flows-may/at_download/file

European Environmental Agency (EEA) (2010) Progress in management of contaminated sites (CSI 015). Retrieved September 26th 2009 from http://www.eea.europa.eu/dataandmaps/indicators/progress-in-management-of-contaminated-sites/progress-in-management-of-contaminated-1

Farrell RE, Elliott JE (2007) Soil air (chapter 64). In: Carter MR, Gregorich EG (eds) Soil sampling and methods of analysis, 2nd edn. CRC, Boca Raton, FL

Fernandes VC, Albergaria JT, Oliva-Teles T, Delerue-Matos C, De Marco P (2009) Dual augmentation for aerobic bioremediation of MTBE and TCE pollution in heavy metal-contaminated soil. Biodegradation 20:375–382

Fischer U, Schulin R, Keller M, Stauffer F (1996) Experimental and numerical investigation of soil vapor extraction. Water Resour Res 32:3413–3427

Franche S, Lindström K, Elmerich C (2009) Nitrogen-fixing bacteria associated with leguminous and non-leguminous plants. Plant Soil 321:35–59

Garcia-Pena I, Ortiz I, Hernandez S, Revah S (2008) Biofiltration of BTEX by the fungus Paecilomyces variotii. Int Biodeterior Biodegrad 62:442–447

Gerhardt KE, Huang X-D, Glick BR, Greenberg BM (2009) Phytoremediation and rhizoremediation of organic soil contaminants: potential and challenges. Plant Sci 176:20–30

Gregory PJ (2006) Roots, rhizosphere and soil: the route to a better understanding of soil science? Eur J Soil Sci 57:2–12

Hinchee RE, Smith LA (1992) In situ thermal technologies for site remediation, 1st edn. Lewis (CRC), Boca Raton, FL

Holden PA, Firestone MK (1997) Soil microorganisms in soil cleanup: how can we improve our understanding? J Environ Qual 26:32–40

Hughes KA, Bridge P, Clark MS (2007) Tolerance of Antarctic soil fungi to hydrocarbons. Sci Total Environ 372:539–548

Jeffrey LM, Smith LA, Nehring KW, Brauning SE, Gavaskar AR, Sass BM, Wiles CC, Mashni CI (1995) The application of solidification and stabilization to waste materials. Lewis (CRC), Boca Raton, FL

Jørgensen KS (2007) In situ bioremediation. Adv Appl Microbiol 61:285–305

Kamath R, Rentz JA, Schnoor JL, Alvarez PJJ (2004) Phytoremediation of hydrocarbon-contaminated soils: principles and applications. Stud Surf Sci Catal 151:447–478

Kao CM, Prosser J (2000) Evaluation of natural attenuation rate at a gasoline spill site. J Hazard Mater 82:275–289

Kelly DP, Baker SC, Trickett J, Davey M, Murrell JC (1994) Methanesulphonate utilization by a novel methylotrophic bacterium involves an unusual monooxygenase. Microbiology 140:1419–1426

Kent AD, Triplett EW (2002) Microbial communities and their interactions in soil and rhizosphere ecosystems. Annu Rev Microbiol 56:211–236

Larkin RP (2003) Characterization of soil microbial communities under different potato cropping systems by microbial population dynamics, substrate utilization, and fatty acid profiles. Soil Biol Biochem 35:1451–1466

Leininger S, Urich T, Schloter M, Schwark L, Qi J, Nicol GW, Prosser JI, Schuster SC, Schleper C (2006) Archaea predominate among ammonia-oxidizing prokaryotes in soils. Nature 442:806–809

Lin TC, Pan PT, Cheng SS (2010) Ex situ bioremediation of oil-contaminated soil. J Hazard Mater 176:27–34

Martin-Laurent F, Cornet L, Ranjard L, López-Gutiérrez J-C, Philippot L, Schwartz C, Chaussod R, Catroux G, Soulas G (2004) Estimation of atrazine-degrading genetic potential and activity in three French agricultural soils. FEMS Microbiol Ecol 48:425–435

Mazzeo DEC, Levy CE, de Angelis DD, Mann-Morales MA (2010) BTEX biodegradation by bacteria from effluents of petroleum refinery. Sci Total Environ 408:4334–4340

Metting B (1981) The systematics and ecology of soil algae. Bot Rev 47:195–312

Mori K, Maki S, Tanaka Y (2005) Warm and hot stamping of ultra high tensile strength steel sheets using resistance heating. Ann CIRP 54:209–212

Morvan X, Saby NPA, Arrouays D, Le Bas C, Jones RJA, Verheijen FGA, Bellamy PH, Stephens M, Kibblewhite MG (2008) Soil monitoring in Europe: a review of existing systems and requirements for harmonization. Sci Total Environ 391:1–12

Mulligan CN, Yong RN (2004) Natural attenuation of contaminated soils. Environ Int 30:587–601

Mummey D, Holben W, Six J, Stahl P (2006) Spatial stratification of soil bacterial populations in aggregates of diverse soils. Microb Ecol 51:404–411

Ni Y, Kim D, Chung M, Lee S, Park H, Rhee Y (2010) Biosynthesis of medium-chain-length poly (3-hydroxyalkanoates) by volatile aromatic hydrocarbons-degrading Pseudomonas fulva TY16. Bioresour Technol 101:8485–8488

Okoh AI, Trejo-Hernandez MR (2006) Remediation of petroleum hydrocarbon polluted systems: exploiting the bioremediation strategies. Afr J Biotechnol 5:2520–2525

Or D, Smets BF, Wraith JM, Dechesne A, Friedman SP (2007) Physical constraints affecting bacterial habitats and activity in unsaturated porous media – a review. Adv Water Resour 30:1505–1527

Östberg TL, Jonsson AP, Lundström US (2006) Accelerated biodegradation of n-alkanes in aqueous solution by the addition of fermented whey. Int Biodeterior Biodegrad 57:190–194

Park G, Shin HS, Ko SO (2005) A laboratory and pilot study of thermally enhanced soil vapor extraction method for the removal of semi-volatile organic contaminants. J Environ Sci Health A Eng Tox Hazard Subst Control 40:881–897

Piccolo A (2002) The supramolecular structure of humic substances: a novel understanding of humus chemistry and implications in soil science. Adv Agron 75:57–134

Pinedo-Rivilla C, Aleu J, Collado IG (2009) Pollutants biodegradation by fungi. Curr Org Chem 13:1194–1214

Pozdnyakova NN, Nikitina VE, Turovskaya OV (2008) Bioremediation of oil-polluted soil with an association including the fungus Pleurotus ostreatus and soil microflora. Appl Biochem Microbiol 44:60–65

Prasad MNV, Freitas H, Fraenzle S, Wuenschmann S, Markert B (2010) Knowledge explosion in phytotechnologies for environmental solutions. Environ Pollut 158:18–23

Prenafeta-Boldú FX, Ballerstedt H, Gerritse J, Grotenhuis JTC (2004) Bioremediation of BTEX hydrocarbons: effect of soil inoculation with the toluene-growing fungus Cladophialophora sp. strain T1. Biodegradation 15:59–65

Rabinovich ML, Bolobova AV, Vasil'chenko LG (2004) Fungal decomposition of natural aromatic structures and xenobiotics: a review. Appl Biochem Microbiol 40:1–17

Racke Kenneth D (1990) Pesticides in the soil microbial ecosystem. In: Enhanced biodegradation of pesticides in the environment, vol 426. American Chemical Society, pp 1–12

Rasse DP, Dignac MF, Bahri H, Rumpel C, Mariotti A, Chenu C (2006) Lignin turnover in an agricultural field: from plant residues to soil-protected fractions. Eur J Soil Sci 57:530–538

Real Decreto 9/2005 from the January 14th, Ministerio de La Presidencia, Boletín Oficial del Estado 15:1833–1843

Roane TM, Josephson KL, Pepper IL (2001) Dual-bioaugmentation strategy to enhance remediation of contaminated soil. Appl Environ Microbiol 67:3208–3215

Roland U, Holzer F, Buchenhorst D, Kopinke FD (2007) Results of field tests on radio-wave heating for soil remediation. Environ Sci Technol 41:8447–8452

Ros M, Rodríguez I, García C, Hernández T (2010) Microbial communities involved in the bioremediation of an aged recalcitrant hydrocarbon polluted soil by using organic amendments. Bioresour Technol 101:6916–6923

Salanitro JP, Dorn PB, Huesemann MH, Moore KO, Rhodes IA, Rice JLM, Vipond TE, Western MM, Wisniewski HL (1997) Crude oil hydrocarbon and soil ecotoxicity assessment. Environ Sci Technol 31:1769–1776

Schmidt R, Gudbjerg J, Sonnenborg TO, Jensen KH (2002) Removal of NAPLs from the unsaturated zone using steam: prevention of downward migration by injecting mixtures of steam and air. J Contam Hydrol 55:233–260

Shannon D, Sen AM, Johnson DB (2002) A comparative study of the microbiology of soils managed under organic and conventional regimes. Soil Use Manage 18:274–283

Shaylor H, McBride M, Harrison E (2009) Sources and impacts of contaminants in soil. Cornell University, Department of Crop and Soil Sciences, Waste Management Institute, Ithaca, NY

Siddique T, Rutherford PM, Arocena JM, Thring RW (2006) A proposed method for rapid and economical extraction of petroleum hydrocarbons from contaminated soils. Can J Soil Sci 86:725–728

Siméon N, Mercier G, Blais J-F, Ouvrard S, Cébron A, Leyval C, Goergen J-L, Guedon E (2008) Décontamination de sols pollués par les hydrocarbures aromatiques polycycliques par biodégradation en présence de substrats organiques supplémentaires. Environ Eng Sci 7:467–479

Singh H (2006) Mycoremediation: fungal bioremediation. Wiley, Hoboken, NJ

Soares AA, Albergaria JT, Domingues VF, Alvim-Ferraz MCM, Delerue-Matos C (2010) Remediation of soils combining soil vapor extraction and bioremediation: benzene. Chemosphere 80:823–828

Sorensen TS, Holmstrup M (2005) A comparative analysis of the toxicity of eight common soil contaminants and their effects on drought tolerance in the collembolan Folsomia candida. Ecotoxicol Environ Safe 60:132–139

Sparrevik M, Eek E, Grini RS (2009) The importance of sulphide binding for leaching of heavy metals from contaminated Norwegian marine sediments treated by stabilization/solidification. Environ Technol 30:831–840

Staley LJ (1995) Vitrification technologies for the treatment of contaminated soil. In: Tedder W, Pohland FG (eds) Emerging technologies in hazardous waste management V. American Chemical Society, pp 102–120

Suthersan SS (1996) Soil vapor extraction. In: Suthan S. Suthersan (ed) Remediation engineering: design concepts. Lewis, Boca Raton, FL, pp 27–88

Tringe S, von Mering C, Kobayashi A, Salamov AA, Chen K, Chang HW, Podar M, Short JM, Mathur EJ, Detter JC, Bork P, Hugenholtz P, Rubin EM (2005) Comparative metagenomics of microbial communities. Science 308:554–557

United States Environmental Protection Agency (USEPA) (1994) How to evaluate alternative cleanup technologies for underground storage tank sites: a guide for corrective action plan reviewers. Retrieved in January 18th 2010 from http://www.epa.gov/oust/pubs/tum_ch6.pdf

United States Environmental Protection Agency (USEPA) (2006) In situ treatment technologies for contaminated soil. Retrieved in January 18th 2010 from http://www.epa.gov/tio/tsp/download/542f06013.pdf

United States Environmental Protection Agency (USEPA) (2010) Superfund remedy report, 13th edn. Retrieved October 10th 2010 from http://clu-in.org/asr

United States Environmental Protection Agency (USEPA), OSWER Directive 9200.4-17P (1999) http://www.epa.gov/swerust1/directiv/d9200417.pdf

Vallini G, Pera A, Agnolucci M, Valdrighi MM (1997) Humic acids stimulate growth and activity of in vitro tested axenic cultures of soil autotrophic nitrifying bacteria. Biol Fertil Soils 24:243–248

Van Eyk J (1997) Petroleum hydrocarbon biodegradation. In: Balkema AA (ed) Petroleum bioventing, an introduction. Balkema, Rotterdam, pp 122–183

Vigueras G, Shirai K, Martins D, Franco TT, Fleuri LF, Revah S (2008) Toluene gas phase biofiltration by Paecilomyces lilacinus and isolation and identification of a hydrophobin protein produced thereof. Appl Microbiol Biotechnol 80:147–154

Volante A, Lingua G, Cesaro P, Cresta A, Puppo M, Ariati LM, Berta G (2005) Influence of three species of arbuscular mycorrhizal fungi on the persistence of aromatic hydrocarbons in contaminated substrates. Mycorrhiza 16:43–50

Whitton BA (2000) Soils and rice fields. In: Whitton BA, Potts M (eds) The ecology of cyanobacteria – their diversity in time and space. Kluwer, Dordrecht

Wiles CC (1987) A review of solidification/stabilization technology. J Hazard Mater 14:5–21

Wolfe W, Haugh C, Webber A, Diehl T (1997) Preliminary conceptual models of the occurrence, fate, and transport of chlorinated solvents in karst regions of Tennessee, Water-Resources Investigations Report 97-4097. U.S. Geological Survey

Wolicka D, Suszek A, Borkowski A, Bielecka A (2009) Application of aerobic microorganisms in bioremediation in situ of soil contaminated by petroleum products. Bioresour Technol 100:3221–3227

World Health Organization (WHO) (2000) Air quality guidelines, Chapter 5.2 Benzene. Retrieved November 12th 2010 from http://helid.desastres.net/en/d/Js13481e/4.1.9.html

Yao H, He Z, Wilson MJ, Campbell CD (2000) Microbial biomass and community structure in a sequence of soils with increasing fertility and changing land use. Microb Ecol 40:223–237

Yarwood RR, Rockhold ML, Niemet MR, Selker JS, Bottomley PJ (2002) Noninvasive quantitative measurement of bacterial growth in porous media under unsaturated-flow conditions. Appl Environ Microbiol 68:3597–3605

Zanaroli G, Di Toro S, Todaro D, Varese GC, Bertolotto A, Fava F (2010) Characterization of two diesel fuel degrading microbial consortia enriched from a non acclimated, complex source of microorganisms. Microb Cell Fact 9:10

Zehr JP, Capone DG (1996) Problems and promises of assaying the genetic potential for nitrogen fixation in the marine environment. Microb Ecol 32:263–281

Chapter 9
Biological Remediation of Petroleum Hydrocarbons in Soil: Suitability of Different Technologies Applied in Mesocosm and Microcosm Trials

M. Nazaré F. Couto, M. Clara P. Basto, and M. Teresa S. D. Vasconcelos

9.1 Introduction

Contamination with petroleum hydrocarbons (PHC) is a global problem with environmental implications, which leads to the need for environmental matrixes remediation, including soil.

Biological remediation, which relies on the capabilities of microorganisms and/or plants (Chaudhry et al. 2005) to eliminate or accumulate contaminants from a contaminated matrix, is an attractive approach to remediate contaminated areas as it is cost-effective and restores soil structure, providing beautiful landscapes.

9.1.1 Bioremediation

Biodegradation of PHC is characterised by their use as a carbon source by microorganisms, resulting in the formation of lower molecular weight compounds or in the transformation of PHC in more polar compounds with the same carbon number of the parent compound (Aldrett et al. 1997). The utilisation of these compounds by the microorganisms depends on environmental constraints and on the chemical nature of the contaminant in soil (Aldrett et al. 1997). To be considered an effective

M.N.F. Couto (✉) • M.C.P. Basto
CIIMAR/CIMAR, Centro Interdisciplinar de Investigação Marinha e Ambiental, Rua dos Bragas 289, 4050-123 Porto, Portugal

Departamento de Química e Bioquímica, Faculdade de Ciências, Universidade do Porto, Rua do Campo Alegre, 4169-007 Porto, Portugal
e-mail: maria.couto@fc.up.pt; mcpbasto@fc.up.pt

M.T.S.D. Vasconcelos
CIIMAR/CIMAR, Centro Interdisciplinar de Investigação Marinha e Ambiental, Rua dos Bragas 289, 4050-123 Porto, Portugal
e-mail: mtvascon@fc.up.pt

E.M. Goltapeh et al. (eds.), *Fungi as Bioremediators*, Soil Biology 32,
DOI 10.1007/978-3-642-33811-3_9, © Springer-Verlag Berlin Heidelberg 2013

application, the bioremediation process should reveal a higher efficiency than the natural decontamination rate in the nature (Bento et al. 2005).

Transformation of contaminant by microorganisms involves its destruction or immobilisation, avoiding its transfer between environmental compartments as occur in physicochemical treatments that can, for example, entail the movement of contaminants from soil to atmosphere (Sabaté et al. 2004). In most areas and in natural circumstances, PHC-degrading microorganisms embrace very few compared with the total number of microorganisms (Joo et al. 2008). Nevertheless, at PHC-contaminated sites, these populations may grow and increase their abundance as they use PHC as a carbon source (Gallego et al. 2001; Bento et al. 2005; Das and Mukherjee 2007).

A process to improve bioremediation is land farming. Soil tillage increases oxidative potential of the soil, promoting physical volatilisation as well as photochemical oxidation. It also improves conditions for microorganisms resulting in enhanced natural degradation activity (Huang et al. 2005). However, in aged oil sludge-contaminated soil, the efficiency of this technique may be limited (Huang et al. 2005).

9.1.1.1 Bioaugmentation

A strategy to improve the capacity of a contaminated matrix (soil or other biotope) to remove contamination uses soil amendment, for instance, by means of bioaugmentation, in order to improve the degradative capacity of contaminated areas by introduction of specific competent strains or consortia of microorganisms to the affected environment (El Fantroussi and Agathos 2005).

Indigenous microorganisms, capable of degrading the contaminant, are usually present in contaminated matrixes, especially if site was priorly exposed to the contaminant (Bento et al. 2005, Sabaté et al. 2004) having most of the times a crucial impact in remediation. However, if environmental conditions are unfavourable or if microbial community does not have the ability to degrade certain components (e.g. high molecular weight compounds) or to emulsify insoluble compounds (Mancera-López et al. 2008), soil might not present the sufficient metabolic potential to degrade contaminant (El Fantroussi and Agathos 2005). Bioaugmentation can also be applied in soils with absent or relatively low numbers of competent degraders (Bento et al. 2005; Mrozik and Piotrowska-Seget 2010), when PHC contamination is toxic to indigenous microorganisms leading to the inhibition of their biodegradative potential, in small-scale sites where cost of nonbiological methods exceeds cost for bioaugmentation (Mrozik and Piotrowska-Seget 2010), and in areas contaminated with compounds requiring long acclimation or adaptation period of time (Mrozik and Piotrowska-Seget 2010). Most of the reports are related with laboratory-scale (microcosm) studies with only a few well-documented field studies (El Fantroussi and Agathos 2005). Many reports reveal the effectiveness of bioaugmentation (Bento et al. 2005; El Fantroussi and Agathos

2005; Agarry et al. 2010); however, in other cases, it has failed (Bento et al. 2005; El Fantroussi and Agathos 2005).

Different possibilities can be carried out for bioaugmentation. In one of the options, indigenous microorganisms are isolated from the target contaminated soil, and after culturing under laboratory conditions, the preadapted bacterial strains return to the same soil—reinoculation of soil (Mrozik and Piotrowska-Seget 2010). This process of selection and culturing of organisms directly from the local site has shown to give the best results (Devinny and Chang 2000) either in laboratory (Bento et al. 2005) or in the field.

The relations of inoculated microorganisms with novel biotic and abiotic conditions (the ecological background), especially in complex biotopes as soil (El Fantroussi and Agathos 2005), can be critical in terms of activity, migration, survival, etc., possibly adversely affecting the outcome of the bioaugmentation strategy in a soil (Di Toro et al. 2008). Exogenous microorganisms can suffer a rapid decline in population size ("microbiostasis" or "obstinacy") (El Fantroussi and Agathos 2005) where predation by protists and competition with autochthonous microorganisms for poor nutrients and electron acceptors/donors are very often responsible for this process (Di Toro et al. 2008). For these reasons, there is some scepticism about commercial bioaugmentation products' efficiency (Aldrett et al. 1997), because when they are used, there is an application of microbial species that may be greatly different from those existent in the local environment (Bento et al. 2005) and, in consequence, the added microorganisms may not compete well with the indigenous microorganism in the soil to remain dominant or viable (Devinny and Chang 2000). In addition, the possible benefits of these products may not justify their cost (Bento et al. 2005; Frankenberger et al. 1989) sometimes with native microorganisms being more active (16–25 % of loss in crude oil weight) than commercial inocula (Thouand et al. 1999).

Another issue connected with soil inoculation is the delivery of suitable microorganisms to the desired sites. In one hand, it is simple to disperse inoculants into surface matrix, but in another hand, it can be very complicated to do it in subsurface environments (Mrozik and Piotrowska-Seget 2010).

9.1.1.2   Biostimulation

When an oil spill occurs, a considerable amount of PHC enters in the environment. Due to the high input of carbon compounds, biodegradation can be compromised by nutrient availability (Ayotamuno et al. 2009), namely nitrogen and phosphorus. The adjustment of the environment to stimulate indigenous microorganisms to degrade contaminants (Aldrett et al. 1997) can be carried out by adding N and P (Bento et al. 2005) causing positive effects in the attenuation of PHC (Gallego et al. 2001; Sarkar et al. 2005; Kogbara 2008). However, lower performances in biostimulation than in bioaugmentation or even natural attenuation have also been observed (Seklemova et al. 2001; Bento et al. 2005). One explanation for lower biostimulation performances may also be the limited nutrient bioavailability (Bento et al. 2005) as reported for phosphorus that can became unavailable to microorganisms because of their low solubility (Mills and Frankenberger 1994).

Biostimulation has been sometimes associated with tilling and/or with phytoremediation (Ayotamuno et al. 2009).

9.1.2 Phytoremediation

In certain cases, a strategy that uses plants for remediation purposes (Gerhardt et al. 2009; Glick 2010) can be carried out, with the action of plants including degradation, adsorption, accumulation and volatilisation of compounds or enhancement of soil rhizosphere activity (Newman and Reynolds 2004).

Phytoremediation technology is cost-effective, nonintrusive (Alkorta and Garbisu 2001), relatively easy to implement and environmentally friendly, and there is no need for disposal sites avoiding excavation and heavy traffic. Disadvantages are related with higher recovery times comparing with physical and chemical technologies, potential limitations related with climatic and geological determinants as well as extremes of environmental toxicity.

9.1.2.1 Rhizoremediation

Individual abilities of both plants and microorganisms may have certain limitations to remediate organics (Chaudhry et al. 2005), and a synergistic action between them can be required to achieve efficient degradation. The activity of microorganisms in the root zone stimulates root exudation which further stimulates microbial activity (Banks et al. 2003), as plants can supply a quantity of readily available organic substances that will serve for augmentation and long-term survival of diverse fungal and bacterial communities in the rhizosphere (Alkorta and Garbisu 2001). Additionally, root exudates constitute a continuously natural source of readily available C that may support co-metabolic processes (Fang et al. 2001) such as co-oxidation of soil-bound and recalcitrant chemicals (Chaudhry et al. 2005).

The composition of microbial community in rhizosphere is known to vary both qualitatively and quantitatively from that in a non-rhizospheric soil (Chaudhry et al. 2005), also having larger adaptability for growth on different carbon substrates (including contaminants) than non-rhizosphere microflora (Johnson et al. 2004). The microbial populations in the rhizosphere are composed of different and synergistic communities rather than a single strain having, consequently, distinct enzymatic capabilities and probably working in a sequential and synergistic order (Chaudhry et al. 2005).

When chemical stress occurs in soil, a plant may react by increasing/changing its exudation to the rhizosphere, which then alters the microbial composition or activity of the rhizosphere (Lee et al. 2008) by selectively enrichment for specific degraders or alternatively secretion of compounds that induce the necessary degradative pathways (Fang et al. 2001). It is assumed that, in stress conditions, plants could selectively increase a degrading population in its rhizosphere by altering exudation or simply causing a non-specific augment in microbial numbers, some of which may be degraders or have some other mechanism of protection against the toxic contaminant.

For a valuable microbe–plant synergy, the soil microbial population must contain species able of transforming soil-bound and recalcitrant organics by means of an adaptation and selection within the existent microhabitats. As a consequence, population characteristics will change significantly in response to the type and degree of a contamination (Chaudhry et al. 2005). For instance, the composition of a microbial community has changed in a soil contaminated with dibenzothiophene (DBT), with time favouring DBT-degrading microorganisms (Chaudhry et al. 2005).

Since the release of compounds or enzymes from roots is presumed to be associated with rhizosphere bioremediation and plant types vary in the nature and quantity of compounds exuded, it follows that both the plant species and soil type could be significant factors influencing the efficacy of phytoremediation (Lee et al. 2008). Some studies have concluded that plants stimulated bioremediation of PHC (Banks et al. 2003; Newman and Reynolds 2004; Gerhardt et al. 2009; Peng et al. 2009; Agamuthu et al. 2010; Cheema et al. 2010) whereas other investigations have led to the conclusion that vegetation have not been effective in the degradation of natural and synthetic compounds (Lalande et al. 2003; Liste and Prutz 2006) with degradative activities differing among species and varieties of plants (Wiltse et al. 1998; Liste and Alexander 2000). It was also reported (Euliss et al. 2008) that only three from five different plant species tested had higher PHC degradation than non-vegetated treatment, and it was explained that the increase in remediation affected by plants was coupled with a difference in microbial diversity.

Compared to bulk soil, soil adjacent to the roots contains, in many cases, increased microbial numbers and population (Macek et al. 2000; Alkorta and Garbisu 2001; Chaudhry et al. 2005; Kirk et al. 2005; Kaimi et al. 2006; Agamuthu et al. 2010), presenting a greater range of metabolic capacities (Alkorta and Garbisu 2001) with changes in the community structure (Muratova et al. 2003). However, sometimes, enhanced number of degraders did not have the highest degradation rates (Phillips et al. 2006). It was also reported (Bento et al. 2005) that an increase in microbial activity was not followed by an increase in the number of specific PHC-degrading microorganisms and/or heterotrophic population. This could indicate that specialised microorganisms were adjusting to the changing substrate conditions, increasing their metabolic activity in a stressed environment and thus limiting the growth of the microbial population (Devinny and Chang 2000; Bento et al. 2005). Above the controversy, it is well known that in PHC-contaminated sites, bacterial abundance and proportion of bacteria capable of hydrocarbon degradation are higher than in non-contaminated areas (Piehler et al. 2002).

Both chemical composition of root exudates and rates of exudation differ significantly between plant species (Alkorta and Garbisu 2001) with changes in soil microbial communities being plant-specific (including decrease or increase in microbial diversity), suggesting that different plants may enhance rhizosphere degradation by selection of a variety of microbial communities (Euliss et al. 2008).

Plants vary widely with respect to root parameters such as morphology, root exudation, fine root turnover, root decomposition and associated microbial communities (Mueller and Shann 2006). Some researchers screened different species characterised by the exudation of phenols that support polychlorinated byphenyl (PCBs)-degrading bacteria and found that not all plants produce and release the same types of phenolic compounds what will imply that some plants may preferentially harbour PCBs degrading in their rhizosphere (Alkorta and Garbisu 2001).

To perform an effective remediation effect, plants should establish in soil environment, and the velocity of this process is not the same for different individual or mixed treatments (Banks et al. 2003). Additionally, plants should proliferate in the presence of high concentrations of contaminants (Huang et al. 2005).

Association of plants and its effect in remediation has also been tested (Cheema et al. 2010). The primary hypothesis is that the effects of mixed plant populations could be cumulative, with the positive benefits of each individual species summing to a greater whole (Phillips et al. 2006). Combinations of root types allow combination of exudate patterns stimulating microbial communities, namely the catabolic potential (Cheema et al. 2010). However, some studies (Banks et al. 2003; Phillips et al. 2006) on the efficacy of phytoremediation for weathered hydrocarbon-contaminated soil have found that a combination of species resulted in a decrease in degradation potential compared to single plant. Allelopathic interactions between two species used in the mixed plant treatment can increase desorption of PHC probably by alteration of both microbial and plant inputs (Phillips et al. 2009). It seems that one of the species of the mixture can be dominant, selectively influencing degrader populations, situation that will not immediately favour increased overall degradation (Phillips et al. 2006).

9.1.3   Remediation of Aged Contamination

In weathered contaminated soils, the more volatile/soluble PHC will be the most susceptible to change by volatilisation/chemical reaction/leaching/biotransformation, occurring an enrichment in nonvolatile fractions (Banks et al. 2003), that can be entrapped in soil micropores. In fact, bioavailability will be limited by means of adsorption and/or precipitation phenomena (Torres et al. 2005) and/or entrapment of contaminants within soil micropores (Banks et al. 2003).

Therefore, actual environmentally contaminated and aged soils often behave differently from laboratory-spiked soils, with respect to remediation (Huang et al. 2005), contaminants being, in the second case, inaccessible to plant roots or microbial degraders (Gao et al. 2007) due to the formation of soil-bounded and insoluble compounds.

Plants may enhance biodegradation of entrapped contaminants (Banks et al. 2003) by different ways. On one hand, fine roots and root hairs penetrate micropores, disrupt soil aggregates concomitantly increasing exposed surface area and transport microorganisms that were attached to the root surface, allowing them to thrive in regions of the soil that were inaccessible without roots (Banks et al. 2003), also increasing oxygen contents in that part of the soil. On the other hand, root exudates may enhance desorption of contaminants from soil as plants have been shown to secrete enzymes and secondary metabolites that also have a surfactant activity, which may increase available contaminant concentration in soil (Chaudhry et al. 2005).

9.1.4   The Role of Surfactants

Nonpolar organic contaminants, like PHC, will bind to hydrophobic groups of the surfactants or will accumulate at the centre of micelles (aggregates of surfactant molecules), and thus, the hydrophobic contaminant solubility in aqueous phase will increase. Therefore, microbial degradation may be facilitated by the addition of a surfactant to the soil, as it will promote desorption of contaminant and facilitate their transfer from the solid to the water phase. In this case, surfactant could be helpful to enhance the efficacy of the *phytoremediation explanta* (Gao et al. 2007). However, there are conflicting views and evidence as to whether the presence of surfactants in soil enhances (Haigh 1996; Torres et al. 2005; Cheng et al. 2008), has no effect or even inhibits (Haigh 1996; Gao et al. 2007) the degradation of hydrophobic organic contaminants. However, in successful cases, degradation can be enhanced even in the presence of low surfactant concentrations (Haigh 1996) or in the absence of an effect on desorption.

9.2   Aims of the Work

It has been proved that biological remediation can, in certain cases, provide an efficient way to remediate contaminated matrixes. However, to be competitive in the market, its overall scope and rate of decontamination (Chaudhry et al. 2005) should be enhanced, in order to efficiently compete with physical and chemical methods. Field studies are still scarce in the literature. Field experiments are subject to unpredictable variables, as interactions among contaminant and chemicals, uncontrolled climatic conditions as well as adverse biotic factors (e.g. herbivore damage) that frequently occur and influence the results.

Here, the results of a work whose main purpose was to search for a suitable biological technology for removal PHC from a soil with weathered contamination were reported. It was executed, from 2007 to 2010, a pilot project, in mesocosms trials, which took place in the environment of a petroleum refinery (Porto Refinery from the *Galp Energia* group). Within the same period, shorter experiments were also carried out in outdoor microcosm scale but at laboratory environment. Although laboratory experiments are not as realistic as the field ones, they allow a better control of the experimental conditions. Different experiments were carried out, in parallel, in order to obtain comparable results, and the following specific aims were globally persecuted: (1) comparison of the effectiveness of phytoremediation (different vascular plants were tested for this purpose) with bioremediation (by means only of indigenous microorganisms—natural attenuation—and by using bioaugmentation) and (2) a possible synergistic effect resulting of simultaneous application of more than one remediation technology (phytoremediation, bioremediation and soil amendments). Therefore, it was compared the efficiency of PHC remediation in soil non-vegetated by the action of microorganisms and in soil vegetated, which may take advantage of joint action of the plant and of the microorganisms that grow around the roots.

9.3   Experimental

9.3.1   General Description of the Work

This work consisted, basically, of a mesocosm-scale experiment, which was carried out in the refinery environment (field study—Fig. 9.1a), and of a microcosm-scale experiment that took place outdoor in laboratory environment (lab study—Fig. 9.1b).

In both studies, it was compared the efficiency of PHC remediation in soil non-vegetated by the action of microorganisms and in soil vegetated, which may take advantage of joint action of the plant and of the microorganisms that grow around the roots. In the field study, the effectiveness of the remedial biological process in the absence and presence of soil amendments was systematically compared.

Plants tested in the field study were *Cortaderia selloana* (Poaceae family), *Scirpus maritimus* (Cyperaceae family) and *Juncus maritimus* (Juncacea family). The two last plant species were chosen because of their tolerance to salinity, which will make them potentially suitable for remediation of PHC in coastal environments, where accidental oil spills sometimes occur and where petroleum refineries are often installed.

In the study in microcosm scale, *S. maritimus* and *J. maritimus* were also used, alone and in association, as well as another plant species, *Halimione portulacoides* (Chenopodiaceae family), that also existed in the area where the other two species were collected for transplantation (see below).

The remediation potential of PHC of any of the plant species used had never been searched before, as far as we could ascertain.

In all studies, the potential of each tested technology for remediation of both recent and old contamination was also investigated, since the degree of bioavailability of PHC can be very different in the two cases, being usually lower in the second one, for reasons that were discussed above.

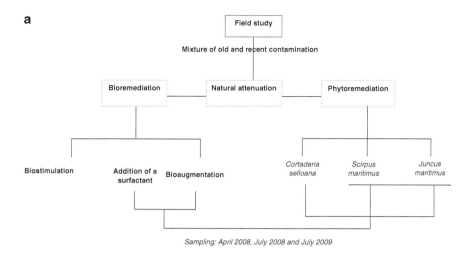

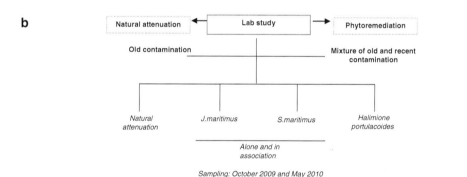

Fig. 9.1 (**a**) Experimental design of field study. (**b**) Experimental design of laboratory study

9.3.2 Field Work

9.3.2.1 Preparation of Containers with Contaminated Soil

Soil contaminated with crude oil ($ca.10$ mg g^{-1}) was excavated from a refinery area, in July 2007, near a tank container where an oil spill occurred in a previous unknown date. Just after excavation, the removed soil was recontaminated with *Galp Turbinoil* (turbine oil) in order to increase the amount of PHC (Couto et al. 2010), the total PHC (TPHC) concentration being $ca.$ 20 mg g^{-1}.

Just after recontamination, soil was homogenised and distributed uniformly to 21 containers of high-density polyethylene with 1.0 m^2 of base area and about 70 cm of height. Sixteen containers received transplants of different plants (see below), and the remaining were left without vegetation to be used both as control and in bioremediation experiments.

9.3.2.2 Plant Species for Transplantation and Bioremediation Experiments

Plants tested in the field study were *C. selloana*, which grew spontaneously in another area of the refinery (PHC from several sources in a TPHC concentration of 0.5 mg g^{-1} with contamination of other natures) from where the transplants were obtained for the study, as well as *S. maritimus* and *J. maritimus*, which grew spontaneously in a salt marsh of the low estuary of *Douro* river (TPHC concentration in the rhizosediment was of 0.2 mg g^{-1}).

All plants were transplanted to target containers without a previous washing of the roots with the purpose of diminish plant stress.

Five containers were used for bioremediation experiments. One container was left as it was throughout the study—natural attenuation (serving as control not only for bioremediation but also for phytoremediation experiments). The remaining received (1) biostimulation by adding (1.1) only nutrients and (1.2) nutrients and a nonionic surfactant and (2) bioaugmentation in the presence of added (2.1) nutrients or (2.2) nutrients and a nonionic surfactant amendments. All the amendments and aeration were performed weekly.

9.3.2.3 Regular Amendments

During the bioremediation study, the non-vegetated containers received, weekly, a nutrient amendment in order to attain a final proportion of C:N:P of about 62:5:2. For this purpose, a mixture of suitable amounts of NH_4NO_3 and $(NH_4)_2HPO_4$ of commercial grade was used. The vegetated containers were amended with the nutrient solution only once, at the commencement of the work.

For surfactant amendment, a 0.15 % (v/v) aqueous solution of a nonionic surfactant $[CH_3–(CH_2)_{8 \text{ or } 10}–(OCH_2CH_2)_6–OH]$ was added to the soil weekly (Couto et al. 2010), during the first 9 months of experiment (bioremediation study) or during the first year experiment (phytoremediation study).

A commercial bioaugmentation product, containing microorganisms that use PHC as a carbon source, was used for bioaugmentation. Despite normally used for wastewater treatment, the product can be applied also to soil remediation (supplier's information) (Couto et al. 2010). This product was added to the soil weekly, with a concentration of approximately 1.6×10^{-3} mg g^{-1} of soil (Couto et al. 2010), during the first 9 months, in bioremediation study, and during the first year in phytoremediation study.

Water was regularly added to all containers whenever required. Manual tillage was performed, only in non-vegetated soil, down to about 10 cm of depth with a garden hoe. This procedure was carried out during the first 9-month period of experiment that is up to the first analysis of the soil in non-vegetated containers.

9.3.3   Laboratory Experiment

One part of the soil that had been subjected to natural attenuation and nutrient amendment in the field study (about 40 kg in total) was collected after 21 months of exposure (in March 2009) and divided in two parts. One part was left as it was, that is, with less biodegradable fraction of PHC-*old contamination*, (10.4 $\pm$ 0.9 mg g^{-1}) and the remaining soil was recontaminated with turbine oil, by means of manual homogenisation (Couto et al. 2011a). Soil remained untreated for 2 days, and its TPHC concentration was determined as being 30 $\pm$ 2 mg g^{-1} (Couto et al. 2011a).

Twenty-four pots of 1.5 L of capacity were used for remediation trials. The pots (filled with about 660 g of contaminated soil) were placed in the exterior of the laboratory building in a non-covered area. Three pots with soil with *old contamination* and three other ones with a soil with a mixture *old and recent contamination* remained non-vegetated to be used as controls (natural attenuation) (Couto et al. 2011a). In this study, *S. maritimus* and *J. maritimus* (alone and in association) were also used, as well as another salt marsh plant, *H. portulacoides*.

Biweekly, all the vases were supplied with one quarter-strength-modified Hoagland nutrient solution (Hoagland and Arnon 1950). This procedure was carried out in the laboratory experiment and not in the field attending to the different quantity of soil involved in each experiment (660 g *vs.* 400 kg) and, in consequence, the possible quick ending of readily available inorganic nutrients in the first case.

9.3.4   Soil Sampling

In the mesocosm-scale experiment, soil cores with 20 cm depth were collected from five random different locations (around plants in vegetated containers) of each container. Each core was divided in four segments, 0–5 cm, 5–10 cm, 10–15 cm and 15–20 cm depths and manually homogenised to obtain a composite sample.

In the microcosm-scale experiment, the procedures were similar to those described above, except with regard to the depth of sampling, as only the soil fraction at the level of the root system was collected, and the number of cores per pot, which was just one in this case.

Due to the dimensions of the work in mesocosm-scale, sampling faced additional problems that usually do not occur in microcosm scale. For instance, in the present work, despite physical soil homogenisation, it was not possible to prevent the appearance of some aggregates of contamination in the soil. Therefore, additional precautions were taken during sampling, in order to prevent overestimation of PHC in the samples.

9.3.5 Experimental Materials

9.3.5.1 Soil Characterisation

Soil composition, in terms of grain size, was performed by dry sieving, and it presented a fraction of silt and clay lower than 9 %. Organic matter content (7.5 %) was determined by loss on ignition (4 h at 500 °C) (Couto et al. 2010). The pH value was 6.4 (determined with a Mettler Toledo International pH/mV meter equipped with a combined pH electrode Inlab 302, from the same brand). The same equipment with a combined redox electrode (Inlab 501) was used to estimate the soil redox potential (Couto et al. 2010) at the end of the study, which was ≥ 400 mV in all cases.

9.3.5.2 Microbial Analysis

Total cell count of microorganisms (TCC) in the soil sample was performed using a modification of the $4',6'$-diamidino-2-phenylindole (DAPI) direct count method (Kepner and Pratt 1994; Porter and Feig 1980).

The most probable number of culturable hydrocarbon degraders (MPN) determination was carried out adapting a protocol already developed (Wrenn and Venosa 1996). Crude oil and turbine oil (supplied by the refinery) were used as the selective carbon source for determination of specific degraders.

9.3.5.3 Determination of Total Petroleum Hydrocarbons

Two different methodologies were used for evaluating TPHC content in soil, in order to estimate remediation efficiency of the biological technologies tested.

Fourier transform infrared spectrophotometry (FTIR) was used to estimate the level of contamination (Couto et al. 2012a). Briefly, about 1 g of soil was mixed with 1 g of anhydrous sodium sulphate and with 10 mL of tetrachloroethylene, and an ultrasonic extraction was performed for 30 min, with a final cleaning step using deactivated silica gel. The absorbance of the extract was measured at the maximum near 2,926 cm^{-1}and compared with calibration curve.

Gas chromatography with flame ionisation detection (GC-FID) was used to provide information about the profile of the distribution of PHC between C_{10} and C_{40} and about the relative magnitude of different PHC. A Varian 3800 gas chromatograph provided with a column VF-5ht (Varian Factor Four) 30 m $\times$ 0.25 mm $\times$ 0.25 μm was used for this purpose as well as a protocol adapted from the literature (Saari et al. 2007) and from the previously described methodology. Again, 1 g of soil and 1 g of anhydrous sodium sulphate for drying and ultrasonic extraction (30 min) were used. After extraction, the six millilitres of extract (a mixture of acetone:n-hexane (1:1(v/v))) was solid phase extracted (Florisil) (Couto et al. 2011). The chromatographic programme was the one reported by Saari et al. 2007.

9.3.6   Statistics

Degradation of PHC and MPN (log transformed prior analysis) were examined. Means, based on three replicates for each treatment, were compared by analysis of variance (ANOVA) followed by a Tukey Test.

9.4   Interpretation of Data

9.4.1   Field Study

9.4.1.1   Bioremediation

Efficiency of bioremediation was evaluated after 9 and 24 months of permanence of the soil in containers. Only during the first 9-month period, soil was subject to several amendments, whose effects were also considered. Results of percentage of TPHC that remained in the soil after 9 (Couto et al. 2010) and 24 months of exposure are shown in Fig. 9.2.

Only results of non-aerated soil are presented in Fig. 9.2, attending that, in general, tillage did not improved remediation efficiency (Couto et al. 2010). Some advantages can be related with this practice (Gogoi et al. 2003; Vasudevan and Rajaram 2001), but in this work, it does not prove to be efficient, in part explained by soil texture as well as aerated conditions (pE > 400 mV) (periodic watering).

After 9 months of exposure, in general, it was observed that in surface soil, the percentage of TPHC degradation was between 20 and 50 %, decreasing with depth between 10 and 35 %. The enhanced remediation at surface soil corroborates the importance of photochemical oxidation, namely in atmospheres that are rich in volatile organic compounds (Couto et al. 2010). Nevertheless, even at surface, PHC degradation was not higher that 50 % of initial TPHC, which coincide with the recent contamination (added just before the beginning of the work).

Natural attenuation was as efficient as several soil treatments tested. In fact, PHC-degrading microorganisms, ubiquitous in most environments, are present in sites where contaminants may serve as organic carbon sources. In this case, they started to establish when the first oil spill occurred; this means some years ago.

Treatments with nutrients, nutrients and surfactant and bioaugmentation plus nutrients did not significantly increase degradation. However, the combination of bioaugmentation, nutrients, and surfactant-enhanced remediation, especially until 5 cm depth, is the maximum value obtained in this part of the field experiment (Couto et al. 2010). The use of commercial bioaugmentation products has some advantages over laboratorial augmentation, attending that they are ready to use and no laboratorial prework is necessary to carry out a remediation strategy. Nevertheless, the survival of the introduced microorganisms can be compromised especially

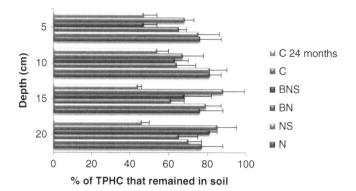

Fig. 9.2 Percentage of TPHC [mean value and confidence limits ($P < 0.05$)] that remained in non-vegetated soil subjected to different experimental conditions* after 9 (adapted from Couto et al. 2010) and 24 months of exposure (only for the control soil). *N, nutrients amendment; NS, nutrients and surfactant amendments; BN, bioaugmentation and nutrients amendments; BNS, bioaugmentation, nutrients and surfactant amendments; C, control soil (natural attenuation)

when a well-adapted autochthonous population is already degrading the contaminant, as it happened in the refinery soil. In this case, however, the use of a commercial bioaugmentation product together with inorganic nutrients and non-ionic surfactant amendments revealed a better efficiency than natural attenuation.

The enumeration of microbial cells was carried out only in soil collected at 5–10 cm depth. This layer was chosen because it corresponded to the layer that contained the highest density of roots in the case of vegetated soil to allow a comparison. After 9-month period of exposure, the number of total cell counts in 5–10 cm soil layer was similar to those found in the initial soil, with variations of 30 % or lower (Couto et al. 2010). This result corroborated the idea that changes in numbers of PHC degraders can occur, but values of total bacteria have little fluctuations, being not statistically influenced by the different biological remediation treatments tested.

After 24 months of exposure in the absence of amendments, percentages of TPHC removal were similar at all depths, being about 50 % of total contamination. As this value corresponds to 100 % of recent contamination, it seems that old contamination was refractory to bioremediation.

9.4.1.2 Plant Effects

After 12 months of exposure, none of the tested plant species could enhance remediation significantly (Couto and Vasconcelos 2009). This fact suggested that transplants in rhizoremediation strategies of PHC might require extended exposure periods for providing suitable results. This aspect may be more relevant when the soil that needs remediation presents weathered contamination and, therefore, a microbiological community well adapted to that carbon source. This will be because

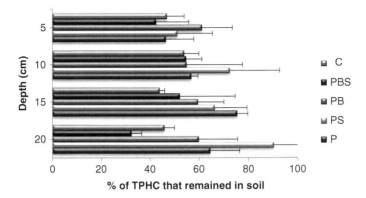

Fig. 9.3 Percentage of TPHC [mean value and standard deviation ($n = 3$)] that remained in *C. selloana* vegetated soil subjected to different experimental conditions* (adapted from Couto et al. 2012b) after 24 months of exposure. *C, control soil (natural attenuation); PBS, vegetated soil with bioaugmentation and surfactant amendments; PB, vegetated soil with bioaugmentation amendment; PS, vegetated soil with surfactant amendment; P, vegetated soil

both plants and the associated soil microorganisms need time for mutual adaptation before the system can act as an efficient degrader of contamination (Couto and Vasconcelos 2009).

In the case of *C. selloana*, a positive role on the degradation of PHC was not visible even after 24 months of exposure (Couto et al. 2012b) (Fig. 9.3). Indeed, the mitigation of the contamination levels was not significantly different in the presence and in the absence of the plant (control soil).

The MPN of culturable turbine oil and crude oil degraders present in the layer with highest root density (5–10 cm depth) was of the same magnitude order in the presence and in the absence of the plant (Couto et al. 2012b), which was compatible with absence of synergetic effects between plant and rhizosphere microorganisms. Therefore, the plant did not show to be able of either uptake or stimulated microbial degradation of PHC.

Additionally, the simultaneous use of surfactant and bioaugmention amendments in the presence of the plant slightly improved the removal of PHC, which suggested that an increase of PHC bioavailability favoured the use of the contaminant as carbon source by the inocula. The presence of *C. selloana* apparently facilitated the migration of additives and microorganisms into the deeper layers of soil, which can be considered a secondary but positive role of the plant on the remediation (Couto et al. 2012b). However, this effect, by itself, does not support the application of the spontaneously grown *C. selloana* as a refinery resource for remediation purposes.

Neither *J. maritimus* nor association of this plant species with *S. maritimus* displayed potentialities for PHC remediation in the present study (results not shown).

In contrast, after 24 months of exposure, *S. maritimus* revealed capability to improve the efficiency of PHC degradation, relative to natural attenuation (that

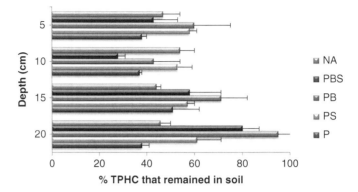

Fig. 9.4 Percentage of TPHC [mean value and standard deviation ($n = 3$)] that remained in *S. maritimus* vegetated soil subjected to different experimental conditions* after 24 months of exposure (adapted from Couto et al. 2012c). *C, control soil (natural attenuation); PBS, vegetated soil with bioaugmentation and surfactant amendments; PB, vegetated soil with bioaugmentation amendment; PS, vegetated soil with surfactant amendment; P, vegetated soil

took place in non-vegetated soil and removed *ca.* 50 % of TPHC), although statistically significant difference (15 % higher) was only achieved in 5–10 cm depth layer, where there was higher root density (Couto et al. 2012c), as can be seen in Fig. 9.4.

An analysis of GC-FID chromatogram of the initial soil revealed (data not shown) the presence of a mixture of turbine oil and crude oil, without the more volatile fractions of the last one, as expected due to the weathered character of this contamination.

GC-FID data (not shown) confirmed that the layer where plant most favoured PHC degradation was the one between 5 and 10 cm depth, being particularly efficient in the degradation of heavier ($C_{22}-C_{40}$) fractions (Couto et al. 2012c). With depth increasing, rhizospheric effect was becoming less pronounced, with plant in 10–15 cm layer, only promoting removal of $C_{22}-C_{40}$ fraction and with slight degradation inhibition between 15 and 20 cm depth (where root has a lower field of action attending to its surface and density) (Couto et al. 2012c).

Microbiological analyses, in 5–10 cm soil layer, corroborated "rhizospheric effect" attending that MPN of crude oil degraders (old contamination) was tendencially higher in the presence of plant. This fact indicated that synergistic interaction between root and microorganisms was, probably, more effective in weathered contamination than in the recent one. Additionally, a comparison of MPN values found at *S. maritimus* initial rhizosphere (in estuarine environment, with a contamination *ca.* 50 times lower than that present in refinery soil and with carbon sources apparently different from those found in the refinery environment) and after exposure in refinery soil showed that turbine oil degraders significantly increased and crude oil degraders tendencially increased (Couto et al. 2012c).

These facts, all together, allowed concluding that after a 20-month period, the system plant rhizosphere was working. The difficulty in adaptation, in this contaminated soil, can be partially explained by its weathered contamination. In one hand, the more labile PHC fractions had already disappeared with some PHC fractions, very probably, being sorbed to soil particles. On the other hand, when the oil spill happened, some years ago, PHC degraders started to establish since they are ubiquitous in soil environment. Twenty-four months before, when transplants occurred, that microflora acted as a buffer to incoming microorganisms, competing and thriving with them, fact that delayed the effectiveness of transplants.

Comparing with only the effect of *S. maritimus*, both addition of nonionic surfactant and bioaugmentation did not enhance remediation efficiency. It is worth of note that bioaugmentation had the lowest performance in this set of experiments, and once again, the possible explanation is the existence of physiologically adapted microorganisms that started to establish some years ago (Couto et al. 2012c). As observed with *C. selloana*, for *S. maritimus*, the simultaneous addition of surfactant and bioaugmentation product also increased TPHC degradation (remediation efficiency double that observed for the plant alone) but, in this case, in 5–10 cm soil layer, where the root density was higher. To note, this was also the layer where *C. selloana* has higher root density. However, the morphology of the roots of *C. selloana* favours the physical retention of PHC, which will upset the possible effect of the amendments. Any case, significant improvement of degraders was not found.

9.4.2 Laboratory Study

The field work was complemented with a less extensive (a 7-month period) laboratorial work. The percentages of TPHC that remained either in soil that initially contained only old contamination or in that containing a mixture of old and recent contamination (Couto et al. 2011) are shown in Fig. 9.5.

9.4.2.1 Old Contamination

Practically, PHC removal was inexistent in old contaminated soil, except for soil vegetated with *S. maritimus* and *H. portulacoides*. It is worthy of note that this aged contamination presented, mainly, recalcitrant PHC fractions. Nevertheless, the species *S. maritimus* could significantly enhance remediation, in *ca.* 13 %. *H. portulacoides* seemed also to favour remediation (a 10 % of TPHC decrease was observed, but this result was not significantly different from natural attenuation) (Couto et al. 2011). The positive plant effect can be explained either by enhanced contaminant's bioavailability, desorption or co-metabolism. However, the presence of *J. maritimus* and association of *S. maritimus* and *J. maritimus* did not cause any enhancement of PHC degradation, which demonstrated that the nature of the plant may condition the results.

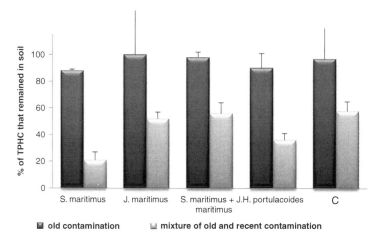

Fig. 9.5 Percentage of TPHC [mean value and standard deviation ($n = 3$)] that remained in vegetated and non-vegetated soil after 7 months of exposure in the laboratory experiment (adapted from Couto et al. 2011). C, control soil (natural attenuation)

9.4.2.2 Mixture of Old and Recent Contamination

S. maritimus could cause marked positive effects on PHC remediation in soil containing a mixture of recent and old contamination: *ca.* 80 % of TPHC removal and 40 % being inside the refractory category (Couto et al. 2012c). *H. portulacoides* also significantly favoured PHC degradation, although in a less extent than *S. maritimus*.

As was observed in the field study, neither *J. maritimus* alone nor the association of *S. maritimus* and *J. maritimus* in the same pots (to note that these two species were associated in the site where they were collected) was able to significantly increase PHC remediation relative to natural attenuation, even in the presence of recent contamination. It has been reported that mixed plant populations may cause synergetic effects, displaying higher total positive effect than individual one (Cheema et al. 2010). However, in the present case, the presence of *J. maritimus* seemed to neutralise the effectiveness of *S. maritimus* by itself.

In non-vegetated soil (natural attenuation), in 7 months of exposure, about 63 % of recent contamination (about 42 % of total contamination) was removed (Couto et al. 2011). This fact indicated that indigenous microorganisms were able to consume, as a carbon source, most of the available turbine oil in a relatively short period of time.

In the work carried out in the refinery environment, a slightly lower effectiveness of TPHC removal (*ca.* 50 %) was observed in the upper soil layer after 9 months of exposure (see Sect. 9.4.1.1). Any case, the results of this work demonstrated that indigenous microorganisms can have an important role in restoration of ecosystems disturbed by recent PHC contamination, turbine oil in the present case. Nevertheless, microorganisms alone were slower in degrading PHC than in the presence of the plants *S. maritimus* and *H. portulacoides*. In addition, the results of this work cannot be inferred that indigenous microorganisms by themselves were able to degrade old contamination.

At the end of the laboratorial work, MPN presented in soil with only old contamination was lower than those presented in soil with mixture of old and recent contamination, probably as a result of toxic effect of turbine oil. At the beginning of the experiment, crude oil degraders were higher than turbine oil degraders, as it was expected attending to different times of contamination (some years to crude oil and 21 months to turbine oil) (Couto et al. 2011).

9.5 Conclusions

Studies in both mesocosm- and microscale provided results coherent in substance, which gave consistence to the information obtained, and confidence in the final conclusions that it was possible to extract.

As a whole, the studies of bioremediation allowed to conclude that, considering a certain period of exposure, PHC degradation at soil surface layer (0–5 cm depth) was more extensive than at higher depth. As at depths down to 20 cm depth (the maximum attained in the field study) aerobic conditions (redox potential ≥ 400 mV) were also observed (caused by aeration and watering), it was admitted that, probably, photochemical oxidants, like O_3, present in sunny days (very common in Portugal) favoured PHC degradation. Indeed, petroleum refinery and urban atmospheres can present volatile organic compounds and nitrogen dioxide, which are precursors of secondary oxidants. For the deeper layers of soil, where the action of both sunlight and photochemical oxidants cannot be felt, it was concluded that bioremediation by indigenous microorganisms could only remove recent contamination, which was of turbine oil in the present work. Among the different amendments tested, only a combination of bioaugmentation with inorganic nutrients and surfactant amendments could make faster PHC degradation relatively to natural attenuation, thus reducing, therefore, the time required for soil recovery. Indeed, after a 9-month exposure in the field experiment, natural attenuation could remove *ca.* 33 % of initial TPHC, which correspond to 66 % of recent contamination, whereas in the presence of amendments, the percentage of TPHC removal was 50 % of total initial contamination or 100 % if only the recent one was considered. After a 24-month exposure, 100 % removal of recent contamination was observed either in the absence or presence of amendments.

Therefore, an important practical conclusion of this work is that a cost-effective way to reduce half-life for degradation of PHC of recently contaminated soil in the refinery (or maybe elsewhere) can be a periodic revolving of the soil, like tillage, in order to expose the different layers of contaminated soil to atmospheric oxygen. A combination of soil revolving with bioaugmentation, together with nutrients and surfactant amendments, may result in an additional improvement of PHC degradation rate. However, as this last procedure will raise markedly the price of the remediation treatment, a cost-benefit analysis should be performed before deciding to carry out amendments.

With regard to action of the plants, among those tested in the field study, *S. maritimus* was the only one that revealed capability to improve the efficiency of PHC degradation. In fact, besides the plant has tolerated, without apparent harm, the levels of soil contamination (20 mg g^{-1} in the field study and 30 mg g^{-1} in the laboratory study), in soil vegetated by *S. maritimus*, degradation of PHC was not only faster than the natural attenuation but also it was extensive to weathered contamination, being particularly efficient for the heavier PHC from weathered contamination. This seems to be an important advantage relatively to natural attenuation in aerobic conditions, when it is necessary to remediate soil with old contamination, to which bioremediation was not effective. Going into some detail, at the end of the field study (24-month exposure), the presence of the plant resulted in an increase of 15 % of remediation in the soil layer with higher root density (between 5 and 10 cm depth) and a marked improvement of remediation efficiency in that layer (28 % higher than natural attenuation) was observed when both nonionic surfactant amendment and bioaugmentation were combined with the presence of the plant. In the laboratory study, the results were still quite more expressive, given that, in only 7-month exposure, *S. maritimus* was capable of removing not only the recent contamination but also 40 % of the older contamination which was refractory to natural attenuation.

A more general conclusion that emerges from the fact that the salt marsh plant *S. maritimus* has demonstrated capability for PHC remediation of both recent and weathered PHC contamination is that this plant species has potential to be used, either in situ or ex situ, for recovering coastal sediments that have suffered accidental oil spills.

The laboratory study revealed furthermore that *H. portulacoides* also has potential to be used for enhancing remediation of PHC contamination in soil. Therefore, this plant species deserves a more extensive study in mesocosm-scale or directly in the field if possible.

In contrast, *J. maritimus* and association of *J. maritimus* and *S. maritimus* did not reveal capability to improve PHC remediation in soil, the presence of *J. maritimus* inhibiting anyway the capability of *S. maritimus* when the two plants were used in association. As the two plants are sometimes associated in the natural environment, the presence of *J. maritimus* might be a limitation for application of *S. maritimus* in PHC remediation in situ when the two plants species are associated there.

Future research work is necessary to complement and substantiate the information provided by the present study. For instance, as the results on the suitability of *S. maritimus* in terms of PHC remediation are very promissory, it will be very interesting to extend the study to recover contaminated sediments, either ex situ or in situ.

It is worthwhile to conduct similar studies for *H. portulacoides*, which despite having been the subject of a more limited study in this work, also showed potential for PHC remediation, and it is disseminated by several estuaries.

Since *J. maritimus* itself and association of this species and *S. maritimus* did not reveal capability to enhance PHC remediation in the tested soil and these two plant species are sometimes associated in the natural estuarine environment, there will be interest in investigating whether the presence of *J. maritimus* may be a limitation for application of *S. maritimus* in PHC remediation in situ, when the two plant species grow together.

In any case, at least the two following main purposes should be achieved: (a) the suitability for PHC remediation of the plant in the endogenous place and (b) the time that plant requires to act as an efficient remediator of PHC.

Acknowledgement The author acknowledges *Fundação para a Ciência e a Tecnologia* for the Ph.D. scholarship of M. N. Couto (SFRH/31816/2006) that was co-financed by POPH/FSE and *Refinaria do Porto* (GALP Energy) for financial support and logistical support by C. Santos.

References

Agamuthu P, Abioye OP, Aziz AA (2010) Phytoremediation of soil contaminated with used lubricating oil using *Jatropha curcas*. J Hazard Mater 179:891–894

Agarry SE, Owabor CN, Yusuf RO (2010) Studies on biodegradation of kerosene in soil under different bioremediation strategies. Bioremediat J 14:135–141

Aldrett S, Bonner JS, Mills MA, Autenrieth RL, Stephens FL (1997) Microbial degradation of crude oil in marine environments tested in a flask experiment. Water Res 31:2840–2848

Alkorta I, Garbisu C (2001) Phytoremediation of organic contaminants in soils. Bioresour Technol 79:273–276

Ayotamuno JM, Kogbara RB, Agoro OS (2009) Biostimulation supplemented with phytoremediation in the reclamation of a petroleum contaminated soil. World J Microbiol Biotechnol 25:1567–1572

Banks M, Schwab P, Liu B, Kulakow P, Smith J, Kim R (2003) The effect of plants on the degradation and toxicity of petroleum contaminants in soil: a field assessment. Adv Biochem Eng Biotechnol 78:75–96

Bento FM, Camargo FAO, Okeke BC, Frankenberger WT (2005) Comparative bioremediation of soils contaminated with diesel oil by natural attenuation, biostimulation and bioaugmentation. Bioresour Technol 96:1049–1055

Chaudhry Q, Blom-Zandstra M, Gupta SK, Joner E (2005) Utilising the synergy between plants and rhizosphere microorganisms to enhance breakdown of organic pollutants in the environment. Environ Sci Poll Res 12:34–48

Cheema SA, Imran Khan M, Shen C, Tang X, Farooq M, Chen L, Zhang C, Chen Y (2010) Degradation of phenanthrene and pyrene in spiked soils by single and combined plants cultivation. J Hazard Mater 177:384–389

Cheng KY, Lai KM, Wong JWC (2008) Effects of pig manure compost and nonionic-surfactant Tween 80 on phenanthrene and pyrene removal from soil vegetated with *Agropyron elongatum*. Chemosphere 73:791–797

Couto MNFS, Vasconcelos MTSD (2009) Phytoremediation of hydrocarbons from a refinery's soil – the role of rhizosphere. In: Proceedings of remediation technologies symposium, Milan

Couto MNPFS, Monteiro E, Vasconcelos MTSD (2010) Mesocosm trials of bioremediation of contaminated soil of a petroleum refinery: comparison of natural attenuation, biostimulation and bioaugmentation. Environ Sci Pollut Res 17:1339–1346

Couto MNPF, Basto MCP, Vasconcelos MTSD (2011) Suitability of different salt marsh plants for petroleum hydrocarbons remediation. Chemosphere 84(8):1052–1057

Couto MNPFS, Borges JR, Guedes P, Almeida R, Monteiro E, Almeida CM, Basto MCP, Vasconcelos MTSD (2012a) An improved method for determination of petroleum hydrocarbons from soil using a simple ultrasonic extraction and Fourier transform infrared spectrophotometry. Pet Sci Technol, in press

Couto MNPFS, Pinto D, Basto MCP, Vasconcelos MTSD (2012b) Role of natural attenuation, phytoremediation and hybrid technologies in the remediation of a refinery soil with old/recent petroleum hydrocarbons contamination. Environ Technol, 1–8

Couto MNPFS, Basto MCP, Vasconcelos MTSD (2012c) Suitability of Scirpus maritimus for petroleum hydrocarbons remediation in a refinery environment, Environ Sci Pollut Res Int, 19(1), 86–95

Das K, Mukherjee AK (2007) Crude petroleum-oil biodegradation efficiency of *Bacillus subtilis* and *Pseudomonas aeruginosa* strains isolated from a petroleum oil contaminated soil from North-East India. Bioresour Technol 98:1339–1345

Devinny JS, Chang S-H (2000) Bioaugmentation for soil remediation. In: Wise DL, Trantolo DJ, Cichon EJ, Inyang HI, Stottmeister U (eds) Bioremediation of contaminated soils. Marcel Dekker, New York, pp 465–488

Di Toro S, Zanaroli G, Varese GC, Marchisio VF, Fava F (2008) Role of Enzyveba in the aerobic bioremediation and detoxification of a soil freshly contaminated by two different diesel fuels. Int Biodeter Biodegr 62:153–161

El Fantroussi S, Agathos SN (2005) Is bioaugmentation a feasible strategy for pollutant removal and site remediation? Curr Opin Microbiol 8:268–275

Euliss K, C-h H, Schwab AP, Rock S, Banks MK (2008) Greenhouse and field assessment of phytoremediation for petroleum contaminants in a riparian zone. Bioresour Technol 99:1961–1971

Fang C, Radosevich M, Fuhrmann JJ (2001) Atrazine and phenanthrene degradation in grass rhizosphere soil. Soil Biol Biochem 33:671–678

Frankenberger WT, Emerson KD, Turner DW (1989) In situ bioremediation of an underground diesel fuel spill: a case history. Environ Manage 13:325–332

Gallego JLR, Loredo J, Llamas JF, Vázquez F, Sánchez J (2001) Bioremediation of diesel-contaminated soils: evaluation of potential techniques by study of bacterial degradation. Biodegradation 12:325–335

Gao Y-Z, Ling W-T, Zhu L-Z, Zhao B-W, Zheng Q-S (2007) Surfactant-enhanced phytoremediation of soils contaminated with hydrophobic organic contaminants: potential and assessment. Pedosphere 17:409–418

Gerhardt KE, Huang X-D, Glick BR, Greenberg BM (2009) Phytoremediation and rhizore-mediation of organic soil contaminants: potential and challenges. Plant Sci 176:20–30

Glick BR (2010) Using soil bacteria to facilitate phytoremediation. Biotechnol Adv 28:367–374

Gogoi BK, Dutta NN, Goswami P, Krishna Mohan TR (2003) A case study of bioremediation of petroleum-hydrocarbon contaminated soil at a crude oil spill site. Adv Environ Res 7:767–782

Haigh SD (1996) A review of the interaction of surfactants with organic contaminants in soil. Sci Total Environ 185:161–170

Hoagland DR, Arnon DI (1950) The water-culture method for growing plants without soil (CAES Circular 347). Agricultural Experiment Station, Davis, CA, p 32

Huang X-D, El-Alawi Y, Gurska J, Glick BR, Greenberg BMA (2005) A multiprocess phytoremediation system for decontamination of persistent total petroleum hydrocarbons (TPHs) from soils. Microchem J 81:139–147

Johnson DL, Maguire KL, Anderson DR, McGrath SP (2004) Enhanced dissipation of chrysene in planted soil: the impact of a rhizobial inoculums. Soil Biol Biochem 36:33–38

Joo H-S, Ndegwa PM, Shoda M, Phae C-G (2008) Bioremediation of oil contaminated soil using *Candida catenulata* and food waste. Environ Pollut 156:891–896

Kaimi E, Mukaidani T, Miyoshi S, Tamaki M (2006) Ryegrass enhancement of biodegradation in diesel-contaminated soil. Environ Exp Bot 55:110–119

Kepner RLJ, Pratt JR (1994) Use of fluorochromes for direct enumeration of total bacteria in environmental samples: past and present. Microbiol Rev 58:603–615

Kirk JL, Klironomos JN, Lee H, Trevors JT (2005) The effects of perennial ryegrass and alfalfa on microbial abundance and diversity in petroleum contaminated soil. Environ Pollut 33:455–465

Kogbara RB (2008) Ranking agro-technical methods and environmental parameters in the bio-degradation of petroleum-contaminated soils in Nigeria. Electron J Biotechnol 11(1):113–125

Lalande T, Skipper H, Wolf D, Reynolds C, Freedman D, Pinkerton B, Hartel P, Grimes L (2003) Phytoremediation of pyrene in a Cecil soil under field conditions. Int J Phytoremediation 5:1–12

Lee S-H, Lee W-S, Lee C-H, Kim J-G (2008) Degradation of phenanthrene and pyrene in rhizosphere of grasses and legumes. J Hazard Mater 153:892–898

Liste H-H, Alexander M (2000) Accumulation of phenanthrene and pyrene in rhizosphere soil. Chemosphere 40:11–14

Liste H-H, Prutz I (2006) Plant performance, dioxygenase-expressing rhizosphere bacteria, and biodegradation of weathered hydrocarbons in contaminated soil. Chemosphere 62:1411–1420

Macek T, Macková M, Kás J (2000) Exploitation of plants for the removal of organics in environmental remediation. Biotechnol Adv 18:23–34

Mancera-López ME, Esparza-García F, Chávez-Gómez B, Rodríguez-Vázquez R, Saucedo-Castañeda G, Barrera-Cortés J (2008) Bioremediation of an aged hydrocarbon-contaminated soil by a combined system of biostimulation bioaugmentation with filamentous fungi. Int Biodeter Biodegr 61:151–160

Mills SA, Frankenberger WT Jr (1994) Evaluation of phosphorus sources promoting bioremedia-tion of diesel fuel in soil. Environ Contam Tox 53:280–284

Mrozik A, Piotrowska-Seget Z (2010) Bioaugmentation as a strategy for cleaning up of soils contaminated with aromatic compounds. Microbiol Res 165:363–375

Mueller KE, Shann JR (2006) PAH dissipation in spiked soil: impacts of bioavailability, microbial activity, and trees. Chemosphere 64:1006–1014

Muratova A, Hübner T, Narula N, Wand H, Turkovskaya O, Kuschk P, Jahn R, Merbach W (2003) Rhizosphere microflora of plants used for the phytoremediation of bitumen-contaminated soil. Microbiol Res 158:151–161

Newman LA, Reynolds CM (2004) Phytodegradation of organic compounds. Curr Opin Biotechnol 15:225–230

Peng S, Zhou Q, Cai Z, Zhang Z (2009) Phytoremediation of petroleum contaminated soils by Mirabilis Jalapa L. in a greenhouse plot experiment. J Hazard Mater 168:1490–1496

Phillips LA, Greer CW, Germida JJ (2006) Culture-based and culture-independent assessment of the impact of mixed and single plant treatments on rhizosphere microbial communities in hydrocarbon contaminated flare-pit soil. Soil Biol Biochem 38:2823–2833

Phillips LA, Greer CW, Farrell RE, Germida JJ (2009) Field-scale assessment of weathered hydrocarbon degradation by mixed and single plant treatments. Appl Soil Ecol 42:9–17

Piehler MF, Maloney JS, Paerl HW (2002) Bacterioplanktonic abundance, productivity and petroleum hydrocarbon biodegradation in marinas and other coastal waters in North Carolina, USA. Mar Environ Res 54:157–168

Porter KG, Feig YS (1980) The use of DAPI for identifying and counting aquatic microflora. Limnol Oceanogr 25:943–948

Saari E, Peramaki P, Jalonen J (2007) A comparative study of solvent extraction of total petroleum hydrocarbons in soil. Microchim Acta 158:261–268

Sabaté J, Viñas M, Solanas AM (2004) Laboratory-scale bioremediation experiments on hydrocarbon-contaminated soils. Int Biodeter Biodegr 54:19–25

Sarkar D, Ferguson M, Datta R, Birnbaum S (2005) Bioremediation of petroleum hydrocarbons in contaminated soils: comparison of biosolids addition, carbon supplementation, and monitored natural attenuation. Environ Pollut 136:187–195

Seklemova E, Pavlova A, Kovacheva K (2001) Biostimulation-based bioremediation of diesel fuel: field demonstration. Biodegradation 12:311–316

Thouand G, Bauda P, Oudot J, Kirsch G, Sutton C, Vidalie J (1999) Laboratory evaluation of crude oil biodegradation with commercial or natural microbial inocula. Can J Microbiol 45:106–115

Torres LG, Rojas N, Bautista G, Iturbe R (2005) Effect of temperature, and surfactant's HLB and dose over the TPH-diesel biodegradation process in aged soils. Process Biochem 40:3296–3302

Vasudevan N, Rajaram P (2001) Bioremediation of oil sludge-contaminated soil. Environ Int 26:409–411

Wiltse CC, Rooney WL, Chen Z, Schwab AP, Banks MK (1998) Greenhouse evaluation of agronomic and crude oil-phytoremediation potential among alfalfa genotypes. J Environ Qual 27:169–173

Wrenn BA, Venosa AD (1996) Selective enumeration of aromatic and aliphatic hydrocarbon degrading bacteria by a most-probable number procedure. Can J Microbiol 42:252–258

Chapter 10
Pesticides Removal Using Actinomycetes and Plants

Analía Alvarez, María S. Fuentes, Claudia S. Benimeli, Sergio A. Cuozzo, Juliana M. Saez, and María J. Amoroso

10.1 Introduction

Organochlorine pesticides have been used extensively all over the world for public health and agricultural purposes. Currently, their use is being phased out because of their toxicity, environmental persistence, and accumulation in the food chain. Hexachlorocyclohexane (HCH) is one of the most extensively used organochlorine pesticides for both agriculture and medical purposes. Though the use of technical mixture containing eight stereoisomers was banned in several advanced countries in the 1970s, many developing countries continue to use lindane (γ-HCH) for

A. Alvarez • S.A. Cuozzo
Planta Piloto de Procesos Industriales y Microbiológicos (PROIMI-CONICET), Avenida Belgrano y Pasaje Caseros, 4000 Tucumán, Argentina

Universidad Nacional de Tucumán, Tucumán, Argentina
e-mail: alvanalia@gmail.com; sergio_cuozzo@yahoo.com

M.S. Fuentes • J.M. Saez
Planta Piloto de Procesos Industriales y Microbiológicos (PROIMI-CONICET), Avenida Belgrano y Pasaje Caseros, 4000 Tucumán, Argentina
e-mail: sfuentes@proimi.org.ar; jsaez@proimi.org.ar

C.S. Benimeli
Planta Piloto de Procesos Industriales y Microbiológicos (PROIMI-CONICET), Avenida Belgrano y Pasaje Caseros, 4000 Tucumán, Argentina

Universidad del Norte Santo Tomás de Aquino, Tucumán, Argentina
e-mail: cbenimeli@yahoo.com.ar

M.J. Amoroso (✉)
Planta Piloto de Procesos Industriales y Microbiológicos (PROIMI-CONICET), Avenida Belgrano y Pasaje Caseros, 4000 Tucumán, Argentina

Universidad Nacional de Tucumán, Tucumán, Argentina

Universidad del Norte Santo Tomás de Aquino, Tucumán, Argentina
e-mail: amoroso@proimi.org.ar

E.M. Goltapeh et al. (eds.), *Fungi as Bioremediators*, Soil Biology 32,
DOI 10.1007/978-3-642-33811-3_10, © Springer-Verlag Berlin Heidelberg 2013

economic reasons. Thus, new sites are continuously being contaminated by γ-HCH and its stereoisomers (Blais et al. 1998; Kidd et al. 2008; Fuentes et al. 2010). Although only lindane is insecticidal, HCH as a group is toxic and considered as potential carcinogens (Walker et al. 1999).

For the supply of γ isomer, the other stereoisomers are separated from γ-HCH and dumped as waste at different spots on the production sites causing serious soil pollution (Li 1999). HCH continues to pose a serious toxicological problem at industrial sites where post-production of lindane along with unsound disposal practices has led to serious contamination, and HCH contamination continues to be a global issue (Kidd et al. 2008). These compounds have moderate volatility and can be transported by air to remote locations (Galiulin et al. 2002). Therefore, a possible pathway for bioremediation of contaminated soils is the use of indigenous microorganisms. It is known that the microbial degradation of chlorinated pesticides such as HCH is usually carried out by using either pure or mixed culture systems. There have been some reports regarding aerobic degradation of γ-HCH by Gram-negative bacteria like *Sphingomonas* (Singh et al. 2000; Kidd et al. 2008) and by the white-rot fungi *Trametes hirsuta*, *Phanerochaete chrysosporium*, *Cyathus bulleri*, and *Phanerochaete sordida* (Mougin et al. 1999; Singh and Kuhad 2000). However, little information is available on the ability of organochlorine pesticide biotransformation by Gram-positive microorganisms and particularly by actinomycete species, the main group of bacteria present in soils and sediments (De Schrijver and De Mot 1999). These Gram-positive microorganisms have a great potential for biodegradation of organic and inorganic toxic compounds and also could remove different organochlorine pesticides when other carbon sources are present in the medium as energy source (Ravel et al. 1998; Amoroso et al. 1998; Benimeli et al. 2003). Therefore, the ability of actinomycetes to transform organochlorine pesticides has not been widely investigated, despite studies demonstrating that actinomycetes, specifically of the genus *Streptomyces*, have been able to oxidize, partially dechlorinate, and dealkylate aldrin, DDT, and herbicides like metolachlor or atrazine (Ferguson and Korte 1977; Radosevich et al. 1995). In addition to their potential metabolic diversity, strains of *Streptomyces* may be well suited for soil inoculation as a consequence of their mycelial growth habit, relatively rapid rates of growth, colonization of semi-selective substrates, and their ability to be genetically manipulated (Shelton et al. 1996). One additional advantage is that the vegetative hyphal mass of these microorganisms can differentiate into spores that assist in spread and persistence.

Recent studies demonstrate significantly enhanced dissipation and/or mineralization of persistent organic pollutants (such as organochlorine pesticides; Kidd et al. 2008) at the root–soil interface or rhizosphere (rhizodegradation) (Kuiper et al. 2004; Chaudhry et al. 2005; Krutz et al. 2005). This rhizosphere effect is generally attributed to an increase in microbial density, diversity and/or metabolic activity due to the release of plant root exudates, mucigel, and root lysates (Curl and Truelove 1986). A summary of potential root zone carbon sources is given in Table 10.1. Rhizodeposits not only provide a nutrient-rich habitat for microorganisms but can potentially enhance biodegradation in different ways: they may facilitate the

Table 10.1 Chemical compound observed in plant root exudates and extracts

Compound	Examples	References
Sugars	Glucose, xylose, mannitol, maltose, oligosaccharides	Pandya et al. (1999) and Curl and Truelove (1986)
Amino acids	Glutamate, isoleucine, methionine, tryptophan	Pandya et al. (1999)
Aromatics	Benzoate, phenols, *l*-carvone, limonene, *p*-cymene	Hegde and Fletcher (1996) and Tang and Young (1982)
Organic acids	Acetate, citrate, malate, propionate	Curl and Truelove (1986)
Enzymes	Nitroreductase, dehalogenase, laccase	Schnoor et al. (1995)

Adapted from Rentz et al. (2005)

co-metabolic transformation of pollutants with similar structures, induce genes encoding enzymes involved in the degradation process, increase contaminant bioavailability, and/or selectively increase the number and activity of pollutant degraders in the rhizosphere (Schnoor et al. 1995; Burken and Schnoor 1998; Miya and Firestone 2001; Shaw and Burns 2003).

The objective of this chapter is to study the bioremediation capacity of indigenous actinomycete strain and the effect of root exudates of *Zea mays* on this process.

10.2 Lindane Removal by *Streptomyces* sp. M7 in a Soil Extract Medium

Benimeli et al. (2007a) studied the growth of *Streptomyces* sp. M7 in a soil extract medium (SE) with and without lindane addition. Carbon and nitrogen composition of SE were 0.5 and 0.01 g L^{-1}, respectively. Nevertheless and despite the poor organic matter in SE, *Streptomyces* sp. M7 was able to grow in this medium for limited time.

When the effect of the temperature (25, 30, and 35 °C) on the growth of *Streptomyces* sp. M7 in SE was analyzed, it was observed that 25 °C was the optimal temperature of microbial growth. When *Streptomyces* sp. M7 was cultured in SE supplemented with lindane 100 μg L^{-1}, at different incubation temperatures (Fig. 10.1), a maximum growth of 0.11 mg mL^{-1} was observed at 25 °C. Significant differences in the biomass were not observed at 30 and 35 °C. These results would indicate that the optimal temperature for the growth of *Streptomyces* sp. M7 in SE, in presence as well as in absence of the pesticide, is 25 °C.

It is important to observe that the presence of lindane in culture medium did not inhibit the growth of *Streptomyces* sp. M7, since significant differences in bacterial growth in SE with and without the pesticide were not observed ($p > 0.05$). Similar results were obtained previously (Benimeli et al. 2007b), when *Streptomyces* sp. M7 was cultured in minimal medium supplemented with lindane 100 μg L^{-1},

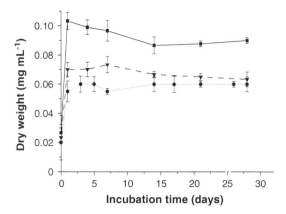

Fig. 10.1 Effect of temperature on the bacterial growth of *Streptomyces* sp. M7 in SE medium amended with lindane 100 μg L^{-1}. 25 °C (*filled squares*), 30 °C (*filled circles*), and 35 °C (*filled inverted triangles*). *Bars* represent standard deviations

suggesting that the pesticide could not be toxic for this microorganism and that would not either be accumulated toxic intermediary metabolites that had an inhibiting effect on the growth.

Streptomyces sp. M7 was able to grow in SE over a relatively wide range of initial pH. No significant differences were observed when the microorganism was cultured at initial pH of 5, 7, or 9 ($p > 0.05$). When *Streptomyces* sp. M7 was cultured in SE added with lindane 100 μg L^{-1}, at different initial pH, a maximum growth of 0.06 mg mL^{-1} was observed at pH 7; nevertheless, the microorganism was not able to grow at pH 5 and 9 (Fig. 10.2). The obtained results would indicate that the optimal initial pH for the growth of *Streptomyces* sp. M7 in SE with lindane is 7.

Figure 10.3 shows the impact of the incubation temperature on the lindane removal by *Streptomyces* sp. M7. The maximum lindane removal was 70.4 % when the microorganism was incubated in SE at 30 °C. Although the optimum temperature for *Streptomyces* sp. M7 growth, with and without lindane, was 25 °C, the optimal temperature for the pesticide removal was 30 °C.

Bachmann et al. (1998) reported that temperature of 30 °C was most favorable for the biodegradation of lindane in soil slurry by the mixed native microbial population of the soil. Arisoy and Kolankaya (1997) observed that the suitable incubation temperature for maximum growth and degradation activity of lindane by the fungus *Pleurotus sajor-caju* was 30 °C. Manonmani et al. (2000) also observed the degradation of the γ-HCH isomer by a microbial consortium under a wide range of temperatures (4–40 °C) in a liquid culture medium, and 30 °C was the optimum for γ-HCH degradation. Siddique et al. (2002) obtained similar results studying the effect of incubation temperature in the biodegradation of lindane by *Pandoraea* sp.; an incubation temperature of 30 °C was optimum for degradation of γ-HCH (57.7 %) in liquid culture and soil slurry (51.9 %).

The effect on lindane removal by *Streptomyces* sp. M7 in SE at initial pH of 5, 7, and 9 is presented in Fig. 10.4. Removal of pesticide (47.2 and 38.0 %) was observed

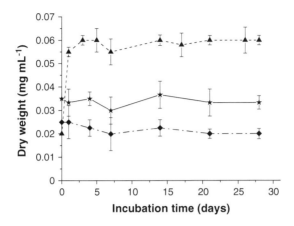

Fig. 10.2 Effect of pH on the bacterial growth of *Streptomyces* sp. M7 in SE amended with lindane 100 μg L^{-1}. pH 5 (*stars*), pH 7 (*filled triangles*), and pH 9 (*filled diamonds*). *Bars* represent standard deviations

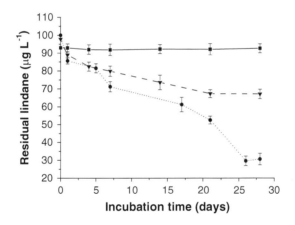

Fig. 10.3 Effect of temperature on the lindane removal by *Streptomyces* sp. M7 in SE amended with lindane 100 μg L^{-1}. 25 °C (*filled squares*), 30 °C (*filled circles*), and 35 °C (*filled inverted triangles*). *Bars* represent standard deviations

at initial pH of 5 and 9, respectively, at 28 days of incubation. The highest removal ability of *Streptomyces* sp. M7 (70.4 %) was noted at an initial pH = 7 at 28 days of incubation. The fate of organic pollutants in the environment is influenced by environmental factors, such as pH and temperature, affecting the activity of microorganisms. *Streptomyces* sp. M7 was able to remove lindane over a wide range of pH in SE.

Arisoy and Kolankaya (1997) reported that medium pH = 5 was the optimum for both growth and degradation activity of lindane by the fungus *Pleurotus sajor-caju*. Manonmani et al. (2000) examined the influence of pH on the degradation of

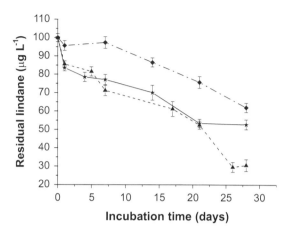

Fig. 10.4 Effect of pH on the lindane removal by *Streptomyces* sp. M7 in SE amended with lindane 100 µg L^{-1}. pH 5 (*stars*), pH 7 (*filled triangles*), and pH 9 (*filled diamonds*). *Bars* represent standard deviations

the γ-HCH isomer in a basal mineral medium by an acclimated consortium of microorganisms. They found that a pH range of 6–8 was most favorable for growth and degradation of the pesticide. Siddique et al. (2002) reported that *Pandoraea* species showed the highest degradation of α- and γ-HCH at an initial pH of 8 in broth culture.

10.3 Lindane Removal by *Streptomyces* sp. M7 in Sterile Soil Samples

Benimeli et al. (2008) studied the growth of *Streptomyces* sp. M7 in sterile soil samples by adding different lindane concentrations in 4 weeks. Simultaneously, lindane removal by *Streptomyces* sp. M7 was determined by gas chromatography; similar experiments were carried out without lindane as controls. As shown in Fig. 10.5, the cell concentration increased during incubation and significant differences in the growth were not observed at different lindane concentrations added, as in the control without lindane. These results would indicate that the growth of *Streptomyces* sp. M7 in soil was not affected by the lindane concentrations assayed, suggesting that the microorganism could tolerate or may degrade the pesticide by producing the necessary dehalogenase enzymes as was demonstrated by Nagata et al. (1999) for *Sphingomonas paucimobilis* UT26.

Influence of different initial concentrations of lindane on pesticide removal was evaluated by determining residual lindane in the soil samples. At initial lindane concentrations of 100, 150, 200, and 300 µg kg^{-1}, removal of pesticide after 4-week incubation was 29.1, 78.0, 38.8, and 14.4 %, respectively. Lindane concentrations in uninoculated control soils were unchanged after 4 weeks of incubation. This inhibition in the removal ability of *Streptomyces* sp. M7 could be due to the toxicity of metabolites, which might have former but not detected in this study by the

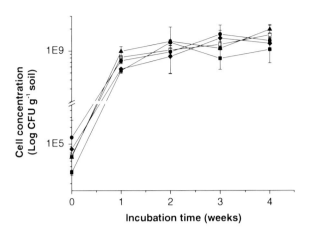

Fig. 10.5 *Streptomyces* sp. M7 growth in sterile soil samples, during 4 weeks of incubation, without (*open squares*) and with 100 (*open squares*), 150 (*filled circles*), 200 (*filled triangles*), and 300 (*filled diamonds*) μg g^{-1} of lindane. *Bars* represent standard deviations

methods employed. For demonstrating that, it could be necessary to measure the intermediary metabolites of lindane degradation as was determined by Nagata et al. (1999) for *Sphingomonas paucimobilis* UT26.

Awasthi et al. (2000) found that at low initial concentrations of endosulfan in soil (50 and 100 μg g^{-1} soil), the degradation was very rapid in inoculated soils; at higher concentrations, the degradation rates were slower, leading to a total inhibition of the degradation activity at the initial concentration of 10 mg g^{-1}. Similar findings were reported by Okeke et al. (2002), who studied the ability of a *Pandoraea* sp. strain to remove lindane in liquid and soil slurry cultures. Their results indicated that the rates and extent of lindane removal increased with increasing concentrations up to 150 mg L^{-1} but declined at 200 mg L^{-1}, after 4–6 weeks of incubation. However, in a similar study, Liu et al. (1991) found opposite results when they inoculated a *Streptomyces* sp. strain in sterile soil samples spiked with metolachlor and observed that the amount of residual metolachlor was higher at lower herbicide concentrations, after 1-week incubation.

This result indicates that *Streptomyces* sp. M7 could remove lindane from soil but at limited pesticide concentration and the viable bacterial count of the soil culture indicated growth and survival of *Streptomyces* sp. M7. This is not surprising because the soil contains available organic nutrients that the bacteria may prefer for growing. Then, lindane could be used as a secondary substrate source as it was demonstrated previously by Benimeli et al. (2007b) in culture medium where glucose at limited concentration was added.

To arrive at the optimal number of bacterial cells for effective removal of lindane, the influence of different inoculum sizes (0.5–4.0 g kg^{-1} ww soil) was studied in sterile soil samples with lindane 100 μg kg^{-1} ww soil (Fig. 10.6). 28 days after inoculation with different *Streptomyces* sp. M7 concentrations into lindane-amended autoclaved soil, the cell population increased rapidly in 2 weeks and was followed by a stationary phase from 2 to 4 weeks. This growth profile was observed for all assayed bacterial concentrations and was similar in contaminated as well as non-contaminated soil samples. These results reinforce the hypothesis according to

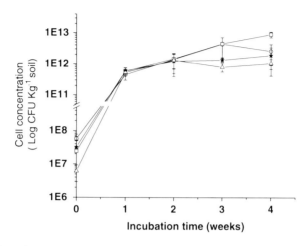

Fig. 10.6 Effect of inoculum size on the growth of *Streptomyces* sp. M7 in sterile soil samples, during 4 weeks of incubation. Symbols used: (*triangles*) 0.5 g kg^{-1} soil, (*stars*) 1.0 g kg^{-1} soil, (*open squares*) 2.0 g kg^{-1} soil, and (*open circles*) 4.0 g kg^{-1} soil

Table 10.2 Effect of inoculum size on the lindane removal by *Streptomyces* sp. M7 in sterile soil samples, after 4 weeks of incubation. Lindane initial concentration: 100 μg kg^{-1} ww soil

Inoculum size g kg^{-1} soil (ww)	Lindane removal (%)
0.5	24.4
1.0	30.8
2.0	56.0
4.0	53.6

which lindane concentrations present in soil were not toxic for *Streptomyces* sp. M7. The inoculum 2 g cells kg^{-1} soil (ww) significantly led the maximum bacterial count (9.0 × 10^{12} CFU kg^{-1} ww of soil).

A substantial decline of the residual lindane at different inoculum concentrations was observed within 0–2 weeks of incubation, whereas the compound did not disappear from the uninoculated sterile control. However, the percentage of lindane removal was not entirely proportional to the amount of *Streptomyces* sp. M7 initially added to the soils, indicating that the two parameters are not in direct proportion. Maximal pesticide depletion (56.0 %) was observed at 2 g cells kg^{-1} soil (ww) and thereafter decreased at 4 g cells kg^{-1} soil (Table 10.2).

In a similar work, Liu et al. (1991) found that approximately 80 % of the pesticide metolachlor was transformed in a sterile soil inoculated with a *Streptomyces* sp. strain after 1 week of incubation; however, the rate of metolachlor transformation was not proportional to the inoculum size. Kumar-Ajit et al. (1998) studied the inoculation of 3-chloro and 4-chlorobenzoate-treated sterile soil with a chlorobenzoate-degrading *Pseudomonas aeruginosa* 3mT; they probed three inoculum sizes (1, 2, and 4 mg cells kg^{-1} soil), and these were effective in degrading the

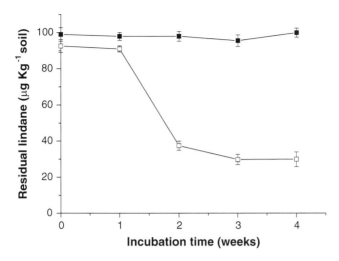

Fig. 10.7 Removal of 100 μg kg^{-1} of lindane by *Streptomyces* sp. M7 in non-sterile soil samples, during 4 weeks of incubation. 2.0 g cells (ww) kg^{-1} soil was added as inoculum. Symbols used: (*open squares*) residual lindane in uninoculated control soil, (*filled squares*) residual lindane in inoculated soil

chemical in the soil, but the degradation was faster with a larger inoculum. On the other hand, Johri et al. (2000) found that an increase in the inoculum size of *Sphingomonas paucimobilis* resulted in an increase in the degradation rate of the HCH isomers in the culture medium, indicating that the two parameters are in direct proportion.

10.4   Bioremediation of Lindane-Contaminated Soil Samples by *Streptomyces* sp. M7 and Effect of Pesticide on Maize Plants

Nonsterile soil spiked with lindane 100 μg kg^{-1} was inoculated with 2 g kg^{-1} *Streptomyces* sp. M7, and after 14 days incubation, *Zea mays* seeds were grown in this soil. Concentrations of residual lindane in soil were monitored periodically to assess the effectiveness of the inoculated strain in the bioremediation of the soil.

A strong decrease in the residual concentration of added lindane was observed in *Streptomyces* sp. M7-inoculated soil. After 14 days of incubation, 68 % of lindane was removed (Fig. 10.7). There were no evident changes in the concentration of the pesticide in the uninoculated soil, indicating that the native microorganisms were not involved into the pesticide removal.

In soil bioremediated with *Streptomyces* sp. M7, normal germination (100 %) and an increased in the seedling vigor were observed, compared to the control maize seeds (Table 10.3). It is possible that the pesticide was removed by the

Table 10.3 Growth, seed germination, and seedling vigor of maize seeds in lindane bioremediated soil by *Streptomyces* sp. M7

	Control	Treated[a]	Bioremediated[b]
Root (cm)	14.2 ± 4.7	15.4 ± 3.9	16.0 ± 2.8
Shoot (cm)	5.3 ± 1.1	5.1 ± 1.1	5.3 ± 1.0
Leaves (cm)	13.7 ± 2.1	14.7 ± 2.1	14.0 ± 1.5
Germination (%)	100.0	84.0	100.0
Vigor index	195.0 ± 6.0	172.0 ± 5.0	213.0 ± 4.0

[a]Soil treated with γ-HCH (100 μg kg^{-1} soil)

[b]Soil treated with γ-HCH (100 μg kg^{-1} soil) and bioremediated with *Streptomyces* sp. M7 (2 g kg^{-1} soil)

inoculated microorganism before seed germination began and the involvement of indigenous microflora capable of metabolizing lindane was not observed.

In a similar study by Krueger et al. (1991), soybean and pea seedlings susceptible to the herbicide dicamba were protected from its deleterious effect by inoculating soils with dicamba-degrading microorganisms. Kumar-Ajit et al. (1998) demonstrated that chlorobenzoates adversely affect the seed germination and seedling vigor of tomato. However, the bioremediation of the soil with *Pseudomonas aeruginosa* 3mT protected the tomato seeds, resulting in the normal germination and seedling vigor. Bidlan et al. (2004) observed that the effect of tech-HCH on germinating radish and green gram seeds was nullified by treatment of contaminated soil with a HCH-degrading microbial consortium.

Although no maize plant growth promotion was observed when *Streptomyces* sp. M7 was inoculated into the contaminated lindane soil samples, the fact that this pesticide can be removed and therefore its uptake by the plant into biomass could be lowered, avoiding the contamination through the food chain. The low water solubility and high hydrophobicity of lindane make their uptake and translocation within the plant unlikely (Kidd et al. 2008).

10.5 Lindane Removal by *Streptomyces* sp. M7 Cultured on Medium Amended with Root Exudates of Maize

Root exudates are products released in the root zone of plants with significant levels of carbon (and possibly nitrogen). The exudates are likely to favor fast-growing microbes, provided they have the corresponding metabolic abilities. In this context, Alvarez et al. (2010; Alvarez et al. 2012) studied the growth of *Streptomyces* sp. M7 in minimal medium supplemented with root exudates of maize and/or glucose (0.8 g L^{-1}) and spiked with lindane (1.66 mg L^{-1}). Cultures were incubated at 30 °C at 150 rpm for 7 days. Similar experiments were carried out without lindane as controls. Root exudates were collected from solution cultures of maize according to Luo et al. (2006). Exudates were lyophilised and stored at 4 °C until sample processing.

Regarding composition, carbohydrate and organic carbon of root exudates were 41.6 mg g^{-1} and 40 mg g^{-1}, respectively. The highest level of pesticide removal

Table 10.4 Biomass production and lindane removal by *Streptomyces* sp. M7 after 7 days of growth on minimal medium supplemented with different carbon sources

Carbon source	Biomass production[a] (g L^{-1} ± SD)	Lindane removal[b] (% ± SD)
Lindane + exudates	0.6 ± 0.1	30.7 ± 0.7
Lindane + glucose	1.0 ± 0.1	10.4 ± 1.2
Lindane	0.1 ± 0.03	18.5 ± 1.7
Exudates	0.3 ± 0.2	
Glucose	1.4 ± 0.2	

SD standard deviation

[a]Dry weight

[b]Lindane initial concentration: 1.66 mg L^{-1}

was obtained when *Streptomyces* sp. M7 was cultured in presence of root exudates (Table 10.4). In contrast, although biomass production was highest in the glucose treatment, the percentage of lindane removal under this condition was lower.

An increase in degradation of a wide range of organic pollutants was observed as result of modified microbial activity in the rhizosphere of several plants (Pandya et al. 1999; White 2001; Chekol et al. 2002; Corgié et al. 2003; Muratova et al. 2003; Shaw and Burns 2005; Phillips et al. 2005). Enhanced transformation of the explosive trinitrotoluene (TNT) by the forage grasses *Phalaris arundinacea* and *Panicum virgatum* was shown by Chekol et al. (2002). Concentrations of the organochlorine *p,p'* DDE (2,2-bis (*p*-chlorophenyl) 1,1-dichloroethylene), a metabolite of DDT, were significantly reduced in the rhizosphere of field-grown zucchini, pumpkin, and spinach compared to bulk soil (White 2001). In contrast, phenanthrene-degrading activity of *Pseudomonas putida* ATCC 17484 was repressed after incubation with plant root extracts (Rentz et al. 2004, 2005). However, these authors suggested that the enhanced microbial growth on rhizodeposits is likely to compensate for this partial repression since a larger microbial population leads to a faster degradation rate.

Little is known about the mechanisms involved in interaction of root exudates with bacteria. In this respect, Chen and Aitken (1999) and Hegde and Fletcher (1996) found that terpenes (such as cymene, α-pyrene, and α-terpinene) and phenolics (such as salicylate) released by roots of certain plants induced biphenyl dioxygenase in polychlorinated biphenyls-degrading bacteria. It could be possible that root exudates of maize enhanced removal of lindane by *Streptomyces* sp. M7 by means of a similar mechanism.

10.6   Lindane and Metabolite Determination in Cell-Free Extract by GC-Mass Spectrometry Analysis

The first signs of the aerobic lindane degradation were determined by Nagata et al. (1999). It was demonstrated that *Sphingobium japonicum* UT26 possesses a dechlorinase enzyme, LinA (γ-hexachlorocyclohexane dehydrochlorinase, EC 4.5.1), encoded by the linA gene that catalyzes two dehydrochlorination steps: γ-HCH to 1,3,4,6-tetrachloro-1,4-cyclohexadiene (1,4-TCDN) via γ-pentachlorocyclohexene

Table 10.5 Relative abundance analysis by GC-Mass of γ-HCH and intermediates γ-PCCH and 1,4-TCDN during aerobic degradation of γ-HCH by *Streptomyces* sp. M7

	Lindane and metabolites	Retention time (min)	Relative abundance[a]
Cell-free extract at 0 h	γ-HCH	7.78	732,750.96 ± 1,041.37
Cell-free extract at 48 h	γ-HCH	7.78	674,895.41 ± 1,961.84
	γ-PCCH	6.26	38,270.56 ± 650.79
	1,4-TCDN	5.29	14,795.47 ± 451.97
Cell-free extract at 96 h	γ-HCH	7.78	561,952.29 ± 2,310.10
	γ-PCCH	6.26	70,372.05 ± 3,268.05
	1,4-TCDN	5.29	25,988.30 ± 992.09

[a]The relative abundance is referred to the ions of different atomic or molecular mass (mass-to-charge ratio) within a sample. It frequently refers to the measured relative abundances of isotopes of a given element

(γ-PCCH). In addition to γ-HCH and γ-PCCH, α- and δ-isomers of HCH were also dehydrochlorinated by LinA, whereas β-HCH was not (Nagata et al. 1993). Furthermore, it was experimentally confirmed that dehydrochlorination of γ-HCH proceeds by a 1,2-ante dehydrochlorination reaction (Nagata et al. 2007).

Regarding the environmental problems caused by lindane and the current lack of information about the presence of dechlorinase activity in *Streptomyces*, the aim of this point was to demonstrate, for the first time, a specific dechlorinase activity in *Streptomyces* using lindane as substrate. In order to determine lindane and metabolites in cell-free extract of *Streptomyces* sp. M7, the strain was grown in flasks with 250 mL of MM containing γ-HCH 100 μg mL^{-1} and incubated at 30 °C at 100 rpm for 48 and 96 h. At the beginning of the experiment, the inoculum was 150 μl of concentrated spore suspension (10^9 CFU mL^{-1}). Supernatants were extracted by solid phase extraction (SPE) using C18 columns, evaporated to dryness under reduced pressure, and the residue was resuspended in hexane. Routine quantitative determinations of lindane (γ-HCH) γ-pentachlorocyclohexene (γ-PCCH) and 1,3,4,6-tetrachloro-1,4-cyclohexadiene (1,4-TCDN) were carried out with gas chromatograph-micro-electron capture detection (GC-μECD) (Cuozzo et al. 2009).

The gas chromatography results of the cell-free extracts obtained at 48 and 96 h of growth of *Streptomyces* sp. M7 revealed the appearance of γ-PCCH (Rt 6.26 min) and 1,4-TCDN,(Rt 5.29 min), the first and second product of the lindane catabolism by the specific dechlorinase in the catabolic way proposed by Nagata et al. (2007). The relative abundance of γ-PCCH and the 1,4-TCDN increased one and half times, at 96 h compared to 48 h of growth (Table 10.5).

However, these results indirectly demonstrated the presence of one specific enzyme in the lindane degradation way from *Streptomyces* sp. M7. It is the first report on dehalogenase activity in actinomycetes with lindane as specific substrate. This has only been reported in *Sphingomonas* (Nagata et al. 2007; Normand et al. 2007) detected a putative 2,5-dichloro-2,5cyclohexadiene-1,4-diol dehydrogenase (2,5-DDOL dehydrogenase) in *Frankia*. Genetic studies of this strain are necessary for a proper understanding of the principle of its ability to degrade different chlorinated hydrocarbon compounds.

10.7 Conclusions and Perspectives

The review clearly suggests that *Streptomyces* sp. M7 has the capacity to grow in a soil extract broth, a nutritionally poor medium, in the presence of lindane, and to remove the pesticide. Since streptomycetes are metabolically diverse and relatively resistant to adverse conditions that may occur in the soil environment, *Streptomyces* sp. M7 could be considered as attractive targets for lindane degradation in situ. In addition, the development of bioremediation processes using indigenous microorganisms is advantageous, as strains isolated are already adapted to the substrate, and to local soil and climatic conditions, and regulatory and legislation issues are simpler, compared to the introduction and release of exogenous microorganisms and genetically modified organisms into the environment. On the other hand, *Streptomyces* sp. M7 bioremediation activity is not inhibited by the natural soil microbial flora. Moreover, *Streptomyces* sp. M7 growth was not inhibited by 300 μg kg^{-1} of lindane.

Regarding the use of plants (or their products) for the treatment of contaminated sites, preliminary results suggest that *Streptomyces* sp. M7 was stimulated by the root exudates of maize amended to the culture. In this context, phytostimulation may be a promising strategy for the remediation of HCH-contaminated soils. It is important to continue these studies not only at the laboratory but also in field trails in order to evaluate the potential of *Streptomyces* sp. M7 in the bioremediation of natural habitats. Further studies evaluating the soil–plant–microbe system and its influence on HCH biodegradation are necessary so as to better explore and exploit an undoubtedly huge potential.

Acknowledgments This work was supported by Consejo de Investigaciones de la Universidad Nacional de Tucumán (CIUNT), Agencia Nacional de Promoción Científica y Tecnológica (ANPyCT), Consejo Nacional de Investigaciones Científicas y Técnicas (CONICET) and Fundación Bunge y Born.

References

Alvarez A, Benimeli CS, Sesto Cabral ME, Amoroso MJ (2010) Efecto de los exudados radiculares de plantas de maíz en la remoción de lindano por la cepa de Actinomycetes nativa *Streptomyces* sp. M7. In: Revista Argentina de Microbiología 2010: abstracts of the annual meeting of the Argentinean Society of Microbiology. Argentinean Society of Microbiology, Bs As, p 24

Alvarez A, Yañez LM, Benimeli CS, Amoroso MJ (2012). Maize plants (Zea mays) root exudates enhance lindane removal by native Streptomyces strains. Int Biodeterior Biodegrad 66:14–18.

Amoroso MJ, Castro G, Carlino F, Romero NC, Hill R, Oliver G (1998) Screening of actinomycetes isolated from Salí river tolerant to heavy metal. J Gen Appl Microbiol 44:29–32

Arisoy M, Kolankaya N (1997) Biodegradation of lindane by *Pleurotus sajor-caju* and toxic effects of lindane and its metabolites on mice. Environ Contam Toxicol 59:352–359

Awasthi N, Ahuja R, Kumar A (2000) Factors influencing the degradation of soil applied endosulfan isomers. Soil Biol Biochem 32:1697–1705

Bachmann A, Wijnen P, DeBruin W, Huntjens JL, Roelofsen W, Zehnder AJ (1998) Biodegradation of alpha- and beta-hexachlorocyclohexane in a soil slurry and under different redox conditions. Appl Environ Microbiol 54:143–149

Benimeli CS, Amoroso MJ, Chaile AP, Castro GR (2003) Isolation of four aquatic streptomycetes strains capable of growth on organochlorine pesticides. Bioresour Technol 89:133–138

Benimeli CS, Castro GR, Chaile AP, Amoroso MJ (2007a) Lindane uptake and degradation by aquatic *Streptomyces* sp. M7 Strain. Int Biodeter Biodegr 59:148–155

Benimeli CS, Gonzalez AJ, Chaile AP, Amoroso MJ (2007b) Temperature and pH effect on lindane removal by *Streptomyces* sp. M7 in soil extract. J Basic Microbiol 47:468–473

Benimeli CS, Fuentes MS, Abate CM, Amoroso MJ (2008) Bioremediation of lindane contaminated soil by *Streptomyces* sp M7 and its effects on *Zea mays* growth. Int Biodeter Biodegr 61:233–239

Bidlan R, Afsar M, Manonmani HK (2004) Bioremediation of HCH-contaminated soil: elimination of inhibitory effects of the insecticide on radish and green gram seed germination. Chemosphere 56:803–811

Blais JM, Schindler DW, Mair DC, Kimpe LE, Donald DB, Rosenberg B (1998) Acclimation of persistent organochlorine compounds in mountains of Western Canada. Nature 395:585–588

Burken JG, Schnoor JL (1998) Predictive relationships for uptake of organic contaminants by hybrid poplar trees. Environ Sci Technol 32:3379–3385

Chaudhry Q, Blom-Zandstra M, Gupta S, Joner EJ (2005) Utilising the synergy between plants and rhizosphere microorganisms to enhance breakdown of organic pollutants in the environment. Environ Sci Pollut Res 12:34–48

Chekol T, Vough LR, Chaney RL (2002) Plant-soil-contaminant specificity affects phytoremediation of organic contaminants. Int J Phytoremediation 4:17–26

Chen SH, Aitken MD (1999) Salicylate stimulates the degradation of high molecular weight polycyclic aromatic hydrocarbons by Pseudomonas saccharophila P15. Environ Sci Technol 33:435–439

Corgié SC, Joner EJ, Leyval C (2003) Rhizospheric degradation of phenanthrene is a function of proximity to roots. Plant Soil 257:143–150

Cuozzo SA, Rollán GC, Abate CM, Amoroso MJ (2009) Specific dechlorinase activity in lindane degradation by *Streptomyces* sp. M7. World J Microbiol Biotechnol 25:1539–1546

Curl EA, Truelove B (1986) The rhizosphere. Springer, Heidelberg

De Schrijver A, De Mot R (1999) Degradation of pesticides by actinomycetes. Crit Rev Microbiol 25:85–119

Ferguson JA, Korte F (1977) Epoxidation of aldrin to exo-dieldrin by soil bacteria. Appl Environ Microbiol 34:7–13

Fuentes MS, Benimeli CS, Cuozzo SA, Amoroso MJ (2010) Isolation of pesticide-degrading actinomycetes from a contaminated site: bacterial growth, removal and dechlorination of organochlorine pesticides. Int Biodeter Biodegr. doi:10.1016/j.biod.2010.05.001

Galiulin RV, Bashkin VN, Galiulina RA (2002) Behaviour of persistent organic pollutants in the air–plant–soil system. Water Air Soil Pollut 137:179–191

Hegde RS, Fletcher JS (1996) Influence of plant growth stage and season on the release of root phenolics by mulberry as related to development of phytoremediation technology. Chemosphere 32:2471–2479

Johri AK, Dua M, Saxena DM, Sethunathan N (2000) Enhanced degradation of hexachlorocyclohexane isomers by *Sphingomonas paucimobilis*. Curr Microbiol 41:309–311

Kidd PS, Prieto-Fernández A, Monterroso C, Acea MJ (2008) Rhizosphere microbial community and hexachlorocyclohexane degradative potential in contrasting plant species. Plant Soil 302:233–247

Krueger JP, Butz RG, Cork DJ (1991) Use of dicamba-degradating microorganisms to protect dicamba susceptible plant species. J Agric Food Chem 39:1000–1003

Krutz LJ, Beyrouty CA, Gentry TJ, Wolf DC, Reynolds CM (2005) Selective enrichment of a pyrene degrader population and enhanced pyrene degradation in Bermuda grass rhizosphere. Biol Fertil Soils 41:359–364

Kuiper I, Lagendijk EL, Bloemberg GV, Lugtenberg BJ (2004) Rhizoremediation: a beneficial plant-microbe interaction. Mol Plant Microbe Interact 17:6–15

Kumar-Ajit PV, Gangadhara KP, Manilal P, Kunhi AAM (1998) Soil inoculation with *Pseudomonas aeruginosa* 3mT eliminates the inhibitory effect of 3-chloro- and 4-chlorobenzoate on tomato seed germination. Soil Biol Biochem 30:1053–1059

Li YF (1999) Global technical hexachlorocyclohexane usage and its contamination consequences in the environment from: 1948–1997. Sci Total Environ 232:121–158

Liu SY, Freyer AJ, Bollag JM (1991) Microbial dechlorination of the herbicide metolachlor. J Agric Food Chem 39:631–636

Luo L, Zhang S, Shan X, Zhu Y (2006) Oxalate and root exudates enhance the desorption of p, p'-DDT from soils. Chemosphere 63:1273–1279

Manonmani HK, Chandrashekariah DH, Sreedhar Reddy N, Elecy CD, Kunhi AA (2000) Isolation and acclimation of a microbial consortium for improved aerobic degradation of α-hexachlorocyclohexane. J Agric Food Chem 48:4341–4351

Miya KR, Firestone MK (2001) Enhanced phenanthrene biodegradation in soil by slender oat root exudates and root debris. J Environ Qual 30:1911–1918

Mougin C, Pericaud C, Malosse C, Laugero C, Ashter M (1999) Biotransformation of the insecticide lindane by the white-rot basidiomycete Phanerochaete chrysosporium. Pept Sci 47:51–59

Muratova A, Thorsten H, Narula N, Wand H, Turkovskaya O, Kuschk P, Jahn R, Merbach W (2003) Rhizosphere microflora of plants used for the phytoremediation of bitumen-contaminated soil. Microbiol Res 158:151–161

Nagata Y, Nariya T, Ohtomo R, Fukuda M, Yano K, Takagi M (1993) Cloning and sequencing of a dehalogenase gene encoding an enzyme with hydrolase activity involved in the degradation of gamma-hexachlorocyclohexane (γ-HCH) in *Pseudomonas paucimobilis*. J Bacteriol 175:6403–6410

Nagata Y, Futamura A, Miyauchi K, Takagi M (1999) Two different types of dehalogenases, Lin A and Lin B, involved in γ-hexachlorocyclohexane degradation in *Sphingomonas paucimobilis* UT26 are localized in the periplasmic space without molecular processing. J Bacteriol 181:5409–5413

Nagata Y, Endo R, Ito M, Ohtsubo Y, Tsuda M (2007) Aerobic degradation of lindane (γ-hexachlorocyclohexane) in bacteria and its biochemical and molecular basis. Appl Microbiol Biotechnol 76:741–752

Normand P, Queiroux C, Tisa LS, Benson DR, Rouy Z, Cruvellier S, Médigue C (2007) Exploring the genomes of *Frankia*. Physiol Plant 130:331–343

Okeke BC, Siddique T, Arbestain MC, Frankenberger WT (2002) Biodegradation of γ-Hexachlorocyclohexane (lindane) and α-Hexachlorocyclohexane in water and a soil slurry by a *Pandoraea* species. J Agric Food Chem 50:2548–2555

Pandya S, Iyer P, Gaitonde V, Parekh T, Desai A (1999) Chemotaxis of Rhizobium sp. S2 towards Cajanus cajan root exudates and its major components. Curr Microbiol 38:205–209

Phillips TM, Seech AG, Lee H, Trevors JT (2005) Biodegradation of hexachlorocyclohexane (HCH) by microorganisms. Biodegradation 16:363–392

Radosevich M, Traina SJ, Hao YL, Tuovinen OH (1995) Degradation and mineralization of atrazine by a soil bacterial isolate. Appl Environ Microbiol 61:297–302

Ravel J, Amoroso MJ, Colwell RR, Hill RT (1998) Mercury-resistant actinomycetes from Chesapeake Bay. FEMS Microbiol Lett 162:177–184

Rentz JA, Alvarez PJ, Schnoor JL (2004) Repression of *Pseudomonas putida* phenanthrene-degrading activity by plant root extracts and exudates. Environ Microbiol 6:574–583

Rentz JA, Alvarez PJ, Schnoor JL (2005) Benzo[a]pyrene co-metabolism in the presence of plant root extracts and exudates: implications for phytoremediation. Environ Pollut 136:477–484

Schnoor JL, Licht LA, McCutcheon SA, Wolfe NL, Carreira LH (1995) Phytoremediation of organic and nutrient contaminants. Environ Sci Technol 29:318–323

Shaw LJ, Burns RG (2003) Biodegradation of organic pollutants in the rhizosphere. Adv Appl Microbiol 53:1–60

Shaw LJ, Burns RG (2005) Rhizodeposition and the enhanced mineralization of 2,4 dichlorophenoxyacetic acid in soil from the Trifolium pratense rhizosphere. Environ Microbiol 7:191–202

Shelton DR, Khader S, Karns JS, Pogell BM (1996) Metabolism of twelve herbicides by *Streptomyces*. Biodegradation 7:129–136

Siddique T, Okeke BC, Arshad M, Frankerberger WT Jr (2002) Temperature and pH effects on biodegradation of hexachlorocyclohexane isomers in water and soil slurry. J Agric Food Chem 50:5070–5076

Singh BK, Kuhad RC (2000) Degradation of the insecticide lindane (γ-HCH) by white-rot fungi *Cyathus bulleri* and *Phanerochaete sordida*. Pest Manag Sci 56:142–146

Singh BK, Kuhad RC, Singh A, Tripathi KK, Ghosh PK (2000) Microbial degradation of the pesticide lindane (gamma-hexachlorocyclohexane). Adv Appl Microbiol 47:269–298

Tang CS, Young C (1982) Collection and identification of allelopathic compounds from the undisturbed root system of Bigalta limpograss (Hemarthria altissima). Plant Physiol 69:155–160

Walker K, Vallero DA, Lewis RG (1999) Factors influencing the distribution of lindane and other hexachlorocyclohexanes in the environment. Environ Sci Technol 33:4373–4378

White JC (2001) Plant-facilitated mobilization and translocation of weathered 2,2-bis (p chlorophenyl)-1,1-dichloroethylene (p, p′-DDE) from an agricultural soil. Environ Toxicol Chem 20:2047–2052

Part III
Mycoremediation of Inorganic Pollutants

Chapter 11
Mycoremediation of Heavy Metals

Younes Rezaee Danesh, Mehdi Tajbakhsh, Ebrahim Mohammadi Goltapeh, and Ajit Varma

11.1   Introduction

Heavy metal pollution of soils is one of the most serious problems of present day agriculture which negatively affects both crop yields and quality. Heavy metal pollution results by the disposal of concentrated metal wastes. When the large adverse effects of emissions of heavy metals from smelters on surrounding ecosystems were observed in the 1960s–1970s, then it was realized that how severely soil microorganisms and soil microbial processes can become disrupted by elevated metal concentrations. Extreme metal contamination in the vicinity of smelters caused clearly visible effects such as accumulation of deep layers of organic matter on the soil surface through inhibition of the activity of soil microorganisms and soil fauna (Freedman and Hutchinson 1980). When measures to limit the metal loading rates of soils due to the use of sewage sludge in agriculture were first introduced in many European countries during the 1970s, these limits were focused on protecting against negative effects on crop plants, on animals grazing on land to which sewage sludge had been applied and to protect

Y.R. Danesh (✉)
Department of Plant Protection, Faculty of Agriculture, Urmia University, Urmia, Iran
e-mail: Y.rdanesh@urmia.ac.ir; Younes_rd@yahoo.com

M. Tajbakhsh
Department of Agronomy and Plant Breeding, Faculty of Agriculture, Urmia University,
Urmia, Iran
e-mail: mehditajbakhsh@gmail.com

E.M. Goltapeh
Department of Plant Pathology, College of Agriculture, Tarbiat Modares University, Tehran, Iran
e-mail: Emgoltapeh@modares.ac.ir; Emgoltapeh@yahoo.com

A. Varma
Amity Institute of Microbial Technology (AIMT), Amity University Uttar Pradesh, Noida,
Uttar Pradesh, India
e-mail: ajitvarma@amity.edu

E.M. Goltapeh et al. (eds.), *Fungi as Bioremediators*, Soil Biology 32,   245
DOI 10.1007/978-3-642-33811-3_11, © Springer-Verlag Berlin Heidelberg 2013

man from metal exposure through the food chain. It was not until 20 years later that the effects of elevated heavy metal concentrations on soil microorganisms were taken into consideration in the drafting of legislation to regulate the agricultural use of sewage sludge (Witter 1992). EU mandatory limits were also established to prevent the buildup of metal concentrations in agricultural soils. Several heavy metals are presently emitted in great quantities as a result of human activities. Heavy metals (HM) occur mainly in terrestrial aquatic ecosystem as well as they can be also emitted in to the atmosphere. Man-made soil contamination resulting from mining industry, agricultural, and military activities resulting in high local concentration varies in orders of magnitudes, but on average the concentrations were not marked, whereas in polluted soil dramatically higher concentrations were found. Since HM are not biodegradable and many enter the food chain, they are a long-term threat to both the environment and human health. Metalliferous soils with abnormally high concentrations of some of the elements that are normally present as minor constituents (200–2,000 mg/kg, e.g., Mn) or trace constituents (0.01–200 mg/kg, e.g., Zn, Cu, Ni, Cr, Pb, As, Co, Se, and Cd) vary widely in their effects on plants. These effects depend on plant species, the particular assemblage of enriched elements, and the physical and chemical characteristics of the soil (Reeves 2006). There are some extreme examples of soils being toxic to almost all species of higher plants, whereas in other cases a characteristic flora of metal-tolerant species may develop (Reeves 2006).

11.2 Toxicity of Heavy Metals

Heavy metals are toxic because of their ionic properties. They bind to many cellular ligands and displace native essential metals from their normal binding sites (Wittekind et al. 1996). For example, arsenate can replace phosphate in the cell. Metals also disrupt protein by binding to sulfhydryl groups and nucleic acids by binding to phosphate or hydroxyl groups. As a result, protein and DNA conformations are changed, and their function is disrupted (Bruce et al. 2003). For example, cadmium competes with cellular zinc and nonspecifically binds to DNA, inducing single-strand breaks (Alloway 1995). Metals may also affect oxidative phosphorylation and membrane permeability, as seen with vanadate and mercury (Muller et al. 2001). Microorganisms generally use specific transport pathways to bring essential metals across the cell membrane into the cytoplasm. Toxic metals can also cross membranes via diffusion or via pathways designed for other metals (Konopka et al. 1999). For instance, Cd^{2+} transport occurs via the Mn^{2+} active transport system in *Staphylococcus aureus*. These metal–microbe interactions result in decrease microbial growth, abnormal morphological changes, and inhibition of biochemical processes in individual (Akmal et al. 2005a, b). The toxic effects of metals can be seen on a community level as well. In response to metal toxicity, overall community numbers and diversity decrease. Soil is a living system where all biochemical activities proceed through enzymatic processes. Heavy metals have also adverse effects on enzyme activities.

11.3   Heavy Metal–Microbe Interactions

Microbial communities in soil are extremely diverse, with estimates of as on ecological systems, experimental evidence to demonstrate such a link is scarce. Microbial biomass which represents the living component of the organic matter of soil usually makes up less than 5 % of soil organic matter (Dalal 1998), but it carries out many critical functions in the soil ecosystem. Microbial biomass is both a source and sink for nutrients in the soil. It participates in the C, N, P, and S transformations and plays an active role in the degradation of xenobiotic organic compounds. It also helps in the mobilization and immobilization of heavy metals and participates in the formation of soil structure, etc. (Nannipieri et al. 2002). As soil microorganisms play a vital role in maintaining soil productivity, thus, anything that disrupts these microorganisms and their functions in soil could be expected to affect the long-term soil productivity and sustainability and even the ecosystem stability. The associated heavy metal can affect the bioavailability of the metal in question by additive, synergistic, or antagonistic effects. These interactions can be positive, negative, or nonexistent. However, the gross microbial biomass and activity measurements seldom indicated whether the observed effects were due to changes in species composition or to reduced physiological capacities of the microbial community (Frostegard et al. 1996; Knight et al. 1997). Studies, using the plate count techniques, have demonstrated a shift in the composition of fungal species toward a more metal-tolerant community in the metal-contaminated soils (Yamamoto et al. 1985; Ueda et al. 1988). Usually a decrease in the commonly isolated genera as *Penicillium*, *Oidiodendron*, and *Mortierella* spp. was observed in the metal-polluted soils by Nordgren et al. (1983). Others, such as *Geomyces* and *Paecilomyces*, increased in abundance toward the metal source. *Penicillium* spp. were mostly dominant in soils polluted by copper mine drainage (Yamamoto et al. 1985). Like fungi, soil bacteria also vary in their sensitivity to the metal pollution. There have been reports of effects on the bacterial community composition, generally showing an increase in Gram-negative bacteria in metal-contaminated soils (Zelles et al. 1994). An exception to this has been reported by Ross et al. (1981), who observed that Gram-negative bacteria were slightly more sensitive to Cr than Gram-positive ones. Since bacteria were seldom identified up to species level, conclusions on the effects of metals on bacterial species are hard to be drawn (Frostegard et al. 1996). Generally, the degree of tolerance of microorganisms to metal pollutants determines the dominance of particularly competitive species, whereas moderate stresses may decrease the likelihood of competitive exclusion. Fewer studies have attempted to examine more subtle effects of heavy metal pollution on the structure of microbial communities or on the genetic diversity of particular groups of organisms. Most of the studies used a physiological approach in which the ability of the bacterial microbial community to utilize a variety of substrates was tested to compare the relative activities of different groups of microorganisms, and this ability has been related to metal tolerance (Reber 1992; Doelman et al. 1994). These studies have served to highlight the subtle effects of heavy metals on the soil microbial community. Evidence from the field experiments suggests that under long-term metal stress, a change in the genetic structure of the soil microbial community is produced (Amann et al. 1996). A decrease in the total soil microbial biomass under chronic metal stress has been observed in many field experiments but is likely to be preceded by changes in community structure (Kozdroj and van Elsas 2000). A decreased size of the microbial biomass could at least partially be explained by physiological causes such as a decrease in the microbial substrate utilization efficiency, and an increased maintenance energy require varies in the order: fungi > bacteria > actinomycetes (Frostegard et al. 1993). A decrease in bacterial number within 24 h of incubation in a Zn-spiked soil was observed by Ohya et al. (1985). In contrast, Frostegard et al. (1996) reported an increase in the overall fungal populations in the Cr- or Zn-contaminated soil. Therefore, metal pollution of soils often results in an increase in the fungal to bacterial ratio in soils (Hattori 1992). Genetic diversity is always present within species and may be crucial in determining the response of a population to changing conditions (Young 1994). A decrease in the number of substrates which can be utilized and thus a reduction in the efficient exploitation of all ecological niches may also explain the decrease in the size of the biomass.

11.3.1 Biosorption of Heavy Metals

Microorganisms have the ability to bind metals from aqueous solution. This phenomenon is known as biosorption, and the microorganisms responsible for the process are considered biosorbents. A wide variety of living and dead biomass of bacteria, algae, fungi, and plants is capable of sequestering toxic metals. This is the foundation of biosorption technology, which offers a promising and economical alternative for the treatment of discharges of a wide variety of metal-containing industrial effluents. Conservative estimates of new biosorbents in the North America environmental market amount to $27 million per year (Volesky 2001). Yeasts and fungi are unique in metal biosorption, and this process is known as mycosorption. The fungal biomass used in mycosorption is termed mycosorbent. Mycosorption is a topic of great interest for researchers all over the world (Paknikar et al. 1998; Tobin 2001; Malik 2004). The process of biosorption and bioaccumulation of metals by microorganisms is not new. The accumulation of metals by fungi has received more attention in recent years because of its applications in environmental protection and recovery of metals. The biological removal of metals from solutions can be divided into three categories: (1) biosorption of metal ions on the surface of fungi, (2) intracellular uptake of metal ions, and (3) chemical transformation of metal ions by fungi. Living fungal biomass is required in the last two categories. Nonliving fungal biomass does not depend on requirements for growth, metabolic energy, and transport. In addition, nonliving biomass shows a strong affinity for metal ions due to the lack of protons produced during metabolism. The problem of toxicity of metals does not affect this type of biomass, which is seen as one of the major advantages of biosorption. Fungal biomass can be generated as a waste by-product of large-scale industrial fermentation and is pretreated by washing with acids and/or bases before final drying and granulation. All these factors contribute to reducing the final cost of the process. Biosorption is a pseudo-ion-exchange process in which metal ion is exchanged for a counterion in the biomass or resin. In general, the filamentous fungi possess higher adsorption capacities for heavy metal removal. Aquatic fungi are also known to accumulate heavy metals. Uptake of metals was described by Michelot et al. (1998), and a tentative approach related to mechanisms of bioaccumulation in mushrooms was projected. The marine fungi *Corollospora lacera* and *Monodictys pelagica* have been found to accumulate lead and cadmium extracellularly in mycelia (Taboski et al. 2005). Biosorption involves a number of external factors (e.g., type of metal, ionic form in solution, and the functional site) and tends to be exothermic. Other factors, such as pH, temperature, biomass concentration, type of biomass preparation, initial metal ion concentration and metal characteristics, and concentration of other interfering ions, are also important in evaluating the extent of biosorption. Biosorption and recovery can be intensified in the presence of stirring induced by magnetic field (Gorobets et al. 2004).

11.3.2   Fungal Heavy Metals Biosorption

Biosorption consists of several mechanisms that differ according to the fungal species used, the origin of the biomass, and its processing. These mechanisms include ion exchange, chelation, adsorption, crystallization, and precipitation, followed by ion entrapment in inter- and intrafibrillar capillaries, spaces of the polysaccharide material, and diffusion through the cell wall and membranes of fungi. Cell walls of fungi are composed of chitins, chitosans, and glucans and also contain proteins, lipids, and other polysaccharides. Yeast cell walls consist mainly of glucans and an outer layer of mannoprotein. Precipitation can occur in the cell wall components. The biomass usually contains a larger number and variety of functional groups or sites than those in monofunctional group ion-exchange resins. These sites include carboxyl, sulfate, phosphate, hydroxyl, amino, imino, sulfonate, imidazole, sulfhydryl, carbonyl, thioether, and other moieties. Certain fungal species are more effective and selective than others in removing particular metal ions from solution. The mechanisms of fungal biosorption can be divided into two categories: metabolism independent and metabolism dependent. The first category employs live or dead biomass, and the second category transforms the metal internally coupled with the production of extracellular metabolites. The mechanisms of fungal biosorption are a topic of great interest to many authors; however, significant differences are recognized in the biosorbent mechanisms of fungal biomass. Electron microscopy, X-ray energy diffraction analysis, and infrared (IR) spectroscopy are also used to study binding mechanisms. The amino group of the cell walls of *Rhizopus nigricans* is involved in Cr(VI) binding from solution and from wastewater (Bai and Abraham 2002). Chemical modification increases the number of active binding sites on the surface area that enhances the chromium adsorption capacity. The amine functional groups of *Mucor* cell walls also contribute to the removal of chromium from tanning effluent (Tobin and Roux 1998). Fourier transform infrared (FTIR) spectroscopic analysis reveals the involvement of −COOH groups of acetone-washed yeast biomass in lead biosorption (Ashkenazy et al. 1997). The −COOH groups also contribute to the binding sites of metals in the cell walls of *Mucor rouxii* (Gardea-Torresdey et al. 1996). Electron microscopy, X-ray energy diffraction analysis, and IR spectroscopy reveal three hypotheses of the mechanism of lead uptake (Zhang et al. 1998). NaOH-treated and NaOH-untreated biomass establish that biosorption occurs in the chitin structure of cell walls. Electron microscopy reveals the localization of nickel on the cell surface of *Rhizopus* sp. (Mogollon et al. 1998). Transmission electron microscopy (TEM) shows that Pb(II) is associated in the cell wall and membrane after 3 min and cytoplasm after 2 h in *Saccharomyces cerevisiae* (Suh et al. 1998). A three-step mechanism of Pb(II) accumulation is advocated. The first step is metabolism independent, the second step is metabolism dependent, and the third step is metabolism dependent or independent after 24 h. Two mechanisms govern the removal of Cr(VI) from the aqueous solution by dead biomass of *Aspergillus niger* (Park et al. 2005). During mechanism I, Cr(VI) is reduced directly to Cr(III) by contact with the biomass. Mechanism II consists of three steps: the binding of Cr(VI) to positively charged groups in the cell wall, reduction of Cr(VI) to Cr(III) by adjacent functional groups, and release of Cr(III) by electron repulsion. Fungi can remove both soluble and insoluble metals from solution and can leach metals from solid wastes. Fungi produce protons, organic acids, phosphatases, and other metabolites for solubilization. Many heterotrophic fungi produce organic acids that assist in solubility and complexing of metal cations. Several fungi are known to produce large amounts of different kinds of acids that assist in metal leaching purposes. Oxalic acid is a leaching agent for a variety of metals, such as Al, Fe, and Li, forming soluble metal oxalate complexes. Metal oxalates are also produced by a wide range of fungi, including mycorrhizas and lichenicolous fungi. The white-rot fungi *Bjerkandera fumosa*, *Phlebia radiata*, and *Trametes versicolor* and the brown-rot fungus *Fomitopsis pinicola* produce oxalate crystals in high levels on ZnO, $Co_3(PO_4)_2$, and $CaCO_3$ (Jarosz-Wilkolazka and Gadd 2003). In brown-rot fungi, induction of oxalic acid is related to copper tolerance (Green and Clausen 2003). Brown-rot fungi can maintain oxalic acid concentrations as high as 600 μM/g. Oxalic acid is also produced by brown-rot fungi during leaching of metals from the treated wood (Humar et al. 2004). One-third of the isolates of soil fungi are able to solubilize at least one toxic metal compound, ZnO, $Co_3(PO_4)_2$, and $Zn_3(PO_4)_2$, and 10 % solubilize all three (Sayer et al. 1995). In *Penicillium simplicissimum*, adsorption of zinc is accompanied by the production of citric acid (Franz et al. 1991). The cultural filtrate of *Aspergillus niger* can render the solubility of 18 % Cu, 7 % Ni, and 4 % Co, and these amounts are enhanced by the addition of HCl (Sukla et al. 1992). Fe(III) can be solubilized by a low-molecular-weight chelating compound known as the ferrichrome (Crichton 1991). The structure of the hyphae in the form of hard compact pellets is altered due to a manganese deficiency (Schreferl et al. 1986). *Penicillium janthinellum* F-13 on different media reduces Al toxicity, but tolerance of the high external concentration of Al appears to be due to a different mechanism (Zhang et al. 2002). The mechanisms of fungal transformation are reduction, methylation, and dealkylation of metals; certain species of *Penicillium* are also known to remove iron from alloys (Siegel et al. 1990). *Alternaria alternata* causes volatilization of substantial amounts of selenium to the dimethylselenide form (Thompson-Eagle et al. 1991). The volatilization process is optimized and used in the fungal bioremediation of contaminated water and land (Thompson-Eagle and Frankenberger 1992). The mechanism of fungal selenium transformations has been described (Thompson-Eagle and Frankenberger 1992; Gadd 1993). Several fungal species (i.e., *Candida humicola*, *Gliocladium roseum*, and *Penicillium* sp.) have been established to methylate arsenic compounds such as arsenate [As(V), AsO_4^{3-}], arsenite [As(III), AsO^{2-}], and methylarsonic acid [$CH_3H_2AsO_3$] to volatile dimethyl [$(CH_3)_2HAs$] or trimethylarsine [$(CH_3)_3As$] (Tamaki and Frankenberger 1992). *Fusarium oxysporum* reduces silver ions in solution, thus forming stable silver hydrosol (Ahmad et al. 2003). Silver nanoparticles of 5–15 nm are stabilized by proteins of the fungus. It seems that the reduction of silver ions occurs due to an enzymatic process. A $FeCl_3$-pretreated waste tea fungal mat is an effective biosorbent for As(III) and As(V), and an autoclaved fungal mat is effective for Fe(II) removal from a groundwater sample (Murugesan et al. 2006). Modification of mobility and toxicity of metalloids by these processes can lead to their biotechnological potential in bioremediation. For more discussion of the mechanisms of mycotransformation of metals, the reader is referred to a review by Gadd (2001).

11.4 Mycorrhizal Fungi and Rhizosphere Remediation

Mycorrhizas are symbiotic associations between certain soil fungi and plant roots and are ubiquitous in the natural environments. Their role in nutrient transport in ecosystems and protection of plants against environmental and cultural stress has long been known. The majority of the mycorrhizas are obligate symbionts because they have little or no ability for independent growth. Autobionts are not much involved in associations. By combining the structure and functional aspects of symbiosis, Trappe (1996) defined mycorrhizas as "dual organs of absorption formed when symbiotic fungi inhabit healthy organs of most terrestrial plants." Smith and Read (1997) defined a mycorrhiza as "a symbiosis in which an external mycelium of a fungus supplies soil-derived nutrients to a plant root." The arbuscular mycorrhiza (AM) is the most ancient type of association and colonizes the plants by scavenging for phosphate. About two-thirds of plants are known to have an AM type of association. The AM association belongs to nearly 150 taxa of the order Glomales. The second common type in the environmental systems is the ectomycorrhizal (ECM) fungi. The important differences between the ECM and AM fungi have been summarized (Colpaert and Van Tichelen 1996). The ECM fungi are more specialized in nutrient capture, and both fungi differ in the quantity of production of external biomass. These fungi are also called extremophiles, due to their occurrence in extreme habitats, including high or low temperature, pH, salt and metal concentration, drought, and so on. The growth of mycorrhizas is also known in the presence of low availability of nutrients that could increase plant growth. The symbiotic relationship allows the fungus to tide over the harsh conditions of toxic contamination. The rhizosphere zone around the roots of plants is the active seat of metabolic substrates responsible for the growth and survival of fungi in soil ecosystems. The fungi and plant can extend nutrients to each other. However, the supply of nutrients in nutrient-deficient ecosystems by mycorrhizal fungi to the host plant is quite interesting. The role of mycorrhizas in rhizosphere remediation and of ECM fungi to degrade a wide variety of persistent organic compounds has finally been established (Meharg and Cairney 2000a). Several different types of mycorrhizal associations have been classified from time to time by different mycologists. Certain species of fungi exhibit a narrow host range of plants, whereas others have a broad range. Worldwide distribution and a broad host range are known for the ECM fungus *Pisolithus tinctorius*. Some associations in forests are specific, whereas others are nonspecific. Some mycorrhizal associations influence the pattern of plant communities. The diverse communities of ECM, AM, and ericoid mycorrhizal (ERM) fungi are known at many host plant root systems. The specificity of ECM fungi is well expressed. In a soil environment, levels of colonization depend on seasonal variations and activities related to soil disturbance. These variations lead to reduction in propagule density and mycelial systems. Seven different types of associations have been classified: arbuscular, ectomycorrhiza, ectendomycorrhiza, ericoid, arbutoid, orchid, and monotropoid (Smith and Read 1997). These categories are based on the criteria of the type of the fungus involved and the resulting structures produced by the root–fungus association. The host cells are not penetrated by ectotrophic mycorrhiza and form an intercellular hyphal network in a sheath around the plant roots. The host cells are penetrated by endomycorrhiza with a superficial hyphal network on the root surface.

11.4.1   Toxic Metals Uptake and Resistance in Mycorrhizal Symbiosis

Plants that grow on heavily polluted soils are known as mycotrophic. ECM, ERM, and AM fungi can increase plant tolerance to heavy metals at toxic concentrations. This is due to the accumulation of metals in extramatrical hyphae and extrahyphal slime. This leads to the immobilization of metals in or near roots and decreases uptake to shoots. Increased levels of metal tolerance to plants are not known in all mycorrhizal associations, due to differences in influence on plant metal tolerance. ECM fungi may be useful as bioindicators of pollution (Haselwandter et al. 1988; Aruguete et al. 1998). Mine spoils are good target sites from which to study the application of mycorrhizal fungi. The influence of soil acidification is generally difficult to separate from the influence of metals. Many ECM fungi are associated with acidic, nutrient-poor soils in coniferous forests, and these acidic soils increase the solubility of Al and Mn ions (Dighton and Jansen 1991). Species of ECM fungi exhibit constitutive tolerance to acidic conditions and high metal concentrations. Inoculation with *Pisolithus tinctorius* at the root of *Pinus massoniana* increases its ability to resist toxicity due to Al stress (Kong et al. 2000). ECM fungi are metal sensitive in pure solid and liquid cultures. Based on soil concentration, *Pinus* and *Pices* seedlings can be protected from heavy metal toxicity by *Suillus luteus* (Dixon and Buschena 1988). AM fungi decrease the concentrations of Cd, Mn, and Zn in leaves of host plants in highly metal-polluted soils (Heggo et al. 1990; Hetrick et al. 1994). Several studies are known for in vitro tolerance of ECM fungi against heavy metals. Perhaps no correlation exists between in vitro tolerance of fungi and their expression of increased metal tolerance of plants. A combination of in vitro, in vivo, and field studies is important for the evaluation of metal tolerance of ECM fungi. The percentage of mycorrhizal root tips can be related to the measurement of metal tolerance in vitro and in vivo. Heavy metal tolerance under in vivo conditions indicates a potential for future inoculations of *Eucalyptus* on ultramafic soils (Aggangan et al. 1998). Mycorrhizas of *Cortinarius semisanguinea* exhibit the highest mean levels of Cu and Zn, while those of *Russula* spp. and *Suillus* spp. display the highest mean levels of Pb and Cd, respectively (Berthelsen et al. 1995). These authors also discuss the various morphological types. Levels of Zn (0–38 %), Cd (0–33 %), and Pb (0–2 %), respectively, were accumulated in the fungal biomass. Increasing the Pb concentration produced different ECM morphotypes in soil (Chappelka et al. 1991). In glasshouse investigations, five ECM morphotypes were identified in association with seedlings of *Pinus sylvestris* (Hartley 1997). However, the addition of Cd > 20 mg/kg soil or Zn > 200 mg/kg soil revealed only two morphotypes. Smith and Read (1997) propounded several different mechanisms for the interactions between colonization of mycorrhizas and accumulation of heavy metals, including interactions with phosphorus nutrition leading to tissue dilution of the toxic element, metal sequestration, and tolerance development by the fungus. However, the results presented by Zhu et al. (2001) cannot be explained according to their mechanisms. Brooks (1998) reported increased transport of metal to the shoots by hyperaccumulator plant species that developed tolerance to high metal concentrations. Additional studies are needed on mycorrhiza hyperaccumulators as a mechanism of protection against heavy metals in the soil. The protective effect of mycorrhiza against plant Zn uptake has been observed (Li and Christie 2001). This correlates with changes in Zn solubility due to changes in soil solution pH or by immobilization of Zn in extraradical mycelium. Cairney and Meharg (1999) reported that the diversity of ECM fungi and infections could be maintained in multiple contaminated sites. In this way, fungus provides protection to a plant against contaminants and assists the plant to grow in hazardous areas. During sand culture, *Glomus caledonium* shows the highest infection rates and the poorest sporulating ability in heavy metal (Cu and Cd) treatments (Liao et al. 2003).

11.4.2   Metal Resistance Response Mechanisms

The mechanisms of the mycorrhizal fungi against metal toxicity and host plants and to them are not well understood. However, several mechanisms have been postulated. Many of the principles are similar to the biosorption of other classes of fungi. The mechanisms can be divided into two categories: avoidance and sequestration. Avoidance can reduce the concentration of metal by precipitation, biosorption, and uptake or efflux. Sequestration involves the formation of compounds for intracellular chelation. The binding of heavy metals on extraradical mycelium is suggested as the mechanism for the tolerance of heavy metals in ericaceous mycorrhizal and ectomycorrhizal plants. Abundant production of extramatrical mycelium can provide the best protection to the host, as it provides increased capacity for metal retention by reducing exposure to the individual hyphae. X-ray microanalysis has located high Zn concentrations in extramatrical hyphae in cell walls and extrahyphal polysaccharide slime (Denny and Wilkins 1987). The same is also true in ericoid mycorrhizal endophytes (Denny and Ridge 1995). Dense mycelium can also provide an increased supply of nutrients to the host and will infect new plant roots due to the larger biomass. In certain mycorrhizal associations, the fungal sheath can prevent metals from reaching the root surface. In general, these fungi produce loosely packed sheaths with large interhyphal spaces. The permeability of the apoplastic pathway to solute movement of the fungal sheath can play an important role in determining fungus-mediated host metal tolerance. The high metal storage capacity of *Xerocomus badius–Picea abies* in acidic soil is related to both the activity of the hyphal sheath and the frequent occurrence of vacuoles (Kottke et al. 1998). In ECM fungi, polyphosphates are produced in the vacuole that exists in the form of insoluble granules complexed with a variety of cations. Polyphosphate granules bind the metal cations, and this appears to be a method of intracellular metal detoxification in these fungi. Electron spectroscopic imaging and electron energy loss spectroscopy reveal two types of electron-opaque granules detected in vacuoles of Hartig net hyphae in *Paxillus involutus–Pinus sylvestris* mycorrhizas of heavily polluted sites (Turnau et al. 1993a, b, 1996a, b). One type of granule detects high concentrations of P with S, Ca, and Al, and the other type, low concentration of P with more N, S, and Cd. Polyphosphate granules are localized inside the cytoplasm of the fungal mantle and Hartig net and contain P and Ca (Grellier et al. 1989; Moore et al. 1989). The ability of *Paxillus involutus* to accumulate Cd in different compartments has been recognized (Blaudez et al. 2000). Two mechanisms of metal detoxification can be considered: binding of Cd onto cell walls and accumulation of Cd in the vacuolar compartment. *Oidiodendron maius* Cd 8-*Vaccinium myrtillus* mycorrhizas of polluted soils display the forma-tion of insoluble Zn crystals (Martino et al. 2003). Isolates of *O. maius* produce citric, malic, and fumaric acids that solubilize the insoluble Zn compounds. Strains isolated from unpolluted soils are more efficient in solubilization from both ZnO and $Zn_3(PO_4)_2$ than strains from polluted soils. Increased tolerance to arsenate has been found for *Hymenoscyphus ericae* collected from polluted soils

(Sharples et al. 2001). Arsenic enters the cell via the phosphate transporter. A higher tolerance of Zn and Cd of *Suillus luteus* isolates from a polluted habitat is expressed in vitro than of isolates from an unpolluted habitat (Colpaert et al. 2000). Nearly 70–90 % of ERM fungi solubilize Cd and Cu phosphates and cuprite (Fomina et al. 2005). Also, toxic metal minerals are better solubilized by metal-tolerant isolates. Other metal chelators are found in the vacuoles of fungi: inorganic ions, phytochelatins, metallothioneins, and organic acids. Chelators may vary from ECM species to species and nutrient status of the fungus. Different metals also have varying affinities for chelators. Zn and Al show higher affinity for organic acids and Cu and Cd for phytochelatins. Metallothionein-like peptides are induced in *Pisolithus tinctorius* in the presence of Cu, Cd, and Zn (Morselt et al. 1986). Cd exposure leads to increased activity of adenosine 3'-phosphate 5'-phosphosulfate sulfotransferase, sulfate reduction, and acid-soluble thiols in *Laccaria laccata* in symbiosis with *Picea abies* (Galli et al. 1993). Forming complexes with glutathione and γ-glutamylcysteine detoxifies Cd. Polyamine formation in response to metals is also known. Exposure of Cu to ECM fungi propels an increase in activity of extra- and intracellular tyrosinase (Gruhn and Miller 1991). The activity of tyrosinase enhances melanin, which limits the entry of Cu and other ions into the cells. On the contrary, little polyamine formation is noticed in *Paxillus involutus* due to Zn exposure (Zarb and Walters 1995). Metallothioneins and phytochelatins are involved in constitutive tolerance (Meharg 1994). The plasmalemma of ECM fungi also appears to protect the host plant from metal toxicity. Metallothioneins are isolated and characterized from the ECM fungi *L. laccata* and *Paxillus involutus* (Howe et al. 1997). Some open reading frames encoding for putative metallothioneins have been identified on the chromosomes of the AM fungus *Gigaspora margarita* (Lanfranco et al. 2002) and the ECM fungus *Tuber borchii* (Pierleoni et al. 2004). AM fungi have also been recognized on hosts colonizing metal-contaminated soils (Meharg and Cairney 2000b). It appears that AM fungi have evolved resistance to act as the nutrient supply or provide increased resistance to host plants. AM fungi from arsenic mine spoils also contribute to plant tolerance (Gonzalez-Chavez 2000). This study indicated an increase in the arsenate resistance of the arsenate-resistant grass *Holcus lanatus*.

11.5 Arbuscular Mycorrhiza in Metal-Contaminated Environments

11.5.1 Mycorrhizal Status of Plants on Metal-Contaminated Soils

In general, initial colonizers of heavily disturbed and metal-contaminated soils are metal-tolerant plant species, which tend to be nonmycorrhizal or develop low arbuscular mycorrhizal (AM) colonization levels, with important impacts on the increase of soil organic matter content and improvement of the soil microclimate.

This tends to be conducive to the establishment of plant species favoring higher AM colonization levels and/or favoring other mycorrhizal types, particularly ericoid and ectomycorrhizal, and the mycorrhizal succession can therefore be seen as a gradual replacement of nonmycorrhizal by mycorrhizal plant species (Allen 1991; Leyval and Joner 2001; Regvar et al. 2006). The reduced arbuscular mycorrhizal fungal (AMF) spore diversity, spore density, and infectivity commonly observed in metal-polluted soils were frequently seen as the main factor influencing the observed low mycorrhizal colonization levels (Pawlowska et al. 1996; Leyval and Joner 2001; Regvar et al. 2001). Despite that, several *Glomus* species (e.g., *G. mosseae*, *G. fasciculatum*, *G. intraradices*, *G. aggregatum*, and *G. constrictum*) along with *Scutellospora dipurpurescens*, *Gigaspora* sp., and *Entrophospora* sp. were frequently identified on the basis of spore morphology in metal-polluted habitats (Griffioen 1994; Pawlowska et al. 1996, 2000; Regvar et al. 2001). Studies of the occurrence of mycorrhizal symbiosis in the metal-rich lateritic soils of New Caledonia, where almost 75 % of vegetation is endemic, showed that all examined plants were consistently colonized by AMF. The presence of spores of AMF found in their rhizosphere soil indicated the existence of specific mechanisms to control high Fe, Ni, Cr, Co, Mn and Al and to cope with low nutrient levels in soil, resulting from the coevolution of plants and their fungal symbionts over millions of years. Regardless of the high fungal metal tolerance, however, the lowest AMF spore densities observed in soils with the highest Ni concentrations (Perrier et al. 2006) seem to be a feature frequently found on metal-polluted sites, indicating strong selection pressure(s) on the indigenous AMF populations in metal-enriched soils. Some plants are clearly more mycorrhizal than others, and despite frequently observed seasonal variations, a consistent hierarchy is maintained (Peat and Fitter 1993; Fitter and Merryweather 1992; Regvar et al. 2006). Recent studies show that mycorrhizal grasses (e.g., *Agrostis capillaris*, *Sesleria caerulea*, *Calamagrostis varia*) that usually colonize Zn-, Cd-, and Pb-polluted mining sites (Leyval and Joner 2001; Regvar et al. 2006) are able to maintain relatively constant, though moderate (50–80 %), levels of mycorrhizal colonization (Iestwaart et al. 1992; Leyval et al. 1997; Regvar et al. 2006), whereas highly mycorrhizal plant species are mostly found dominating less polluted sites (Pawlowska et al. 1996; Regvar et al. 2006). The functional significance of plant colonization levels is still a matter of debate, highlighting the lack of understanding of the formation of particular AM structures in plant roots (Fitter and Merryweather 1992; Allen 2001; Regvar et al. 2006). Vesicles, frequently hard to distinguish from intraradical spores, are more frequently formed on the most polluted locations and are seen as a part of the mycorrhizal survival strategy on metal-polluted sites (Pawlowska et al. 1996; Turnau et al. 1996a, b; Regvar et al. 2006). The increased proportion of colonized root length at low soil metal concentrations suggests that plants invest increasingly more in AM symbiosis at low soil metal levels (Audet and Charest 2007). However, plant and fungal metal tolerance mechanisms as well as specific edaphic conditions (e.g., soil metal concentrations, metal speciation, soil pH, and organic matter contents) should be taken in account when interpreting such results (Leyval et al. 1997; Leyval and Joner 2001; Audet and Charest 2007). In conclusion, despite the commonly observed low AMF spore diversities and densities in metal-enriched soils, the existing fungal colonizers are presumably the best suited to cope with the existing microclimatic/microedaphic conditions and should therefore be investigated in contemporary conservation practices. The reduced mycorrhizal colonization levels frequently seen within plant communities from metal-contaminated sites seem to arise from both the composition of the plant communities, usually comprised of plant species with low to moderate mycorrhizal colonization levels, and low levels of mycorrhizal colonization of the specimens of particular plant species on the more polluted locations, which are, however, still within the limits for that plant species, thus maintaining the consistent hierarchy. In addition, intense formation of intraradical fungal spores is frequently found at the most polluted locations. All these characteristics of mycorrhizal colonization may be seen as a plant "mycorrhizal strategy" on highly metal-contaminated environments, contributing significantly to increased plant fitness.

11.5.2 Functional Significance of Arbuscular Mycorrhizas in Metal-Tolerant Plants

Arbuscular mycorrhizal fungi (Glomeromycota) (Schüssler et al. 2001) are ubiquitous soil microbes considered essential for the survival and growth of plants in nutrient-deficient soils. They expand the interface between plants and the soil environment and contribute to plant uptake of macronutrients and micronutrients, as well as significantly to stress alleviation (Allen 1991; Smith and Read 1997). In addition, they were shown to protect plants from harmful effects of excess metals, connected mainly to fungal metal tolerance mechanisms (Gildon and Tinker 1983; Dehn and Schüepp 1989; Hetrick et al. 1994; Hildebrandt et al. 1999; Chen et al. 2003; Gaur and Adholeya 2004). They are therefore frequently seen as a tolerance mechanism of plants in highly metal-polluted soils (Turnau et al. 1996a, b; Hall 2002; Regvar et al. 2006). Studies aiming to explain the physiological mechanism (s) involved in metal tolerance of AMF to elevated metals face difficulties in demonstrating their possible role(s) in metal uptake or contribution to the metal tolerance of the host due to their obligate symbiotic character. Numerous effects of AMF on plant metal accumulation may be seen as an array of responses from decreased toward neutral or even increased metal uptake, depending on different plant–AMF combinations and the metals involved (Leyval et al. 1997; Leyval and Joner 2001; Malcova et al. 2003), thus making it difficult to draw any firm conclusions. Despite that, numerous studies have demonstrated that the selection of indigenous metal-tolerant isolates seems to serve the aim of reducing the endogenous concentrations of metals in plants better than non-tolerant ones. For example, AMF inoculum from metal-tolerant *Viola calaminaria* was efficient in sequestering metals (e.g., Cd and Zn) in the roots of subterranean clover (Tonin et al. 2001). *Glomus intraradices* Br isolated from the roots of *V. calaminaria* improved maize growth in polluted soil and reduced root and shoot metal concentrations in comparison to a common *G. intraradices* isolate or non-colonized controls (Hildebrandt et al. 1999; Kaldorf et al. 1999), and AMF isolates from As-contaminated soil were proved to confer enhanced arsenate resistance on the grass species *Holcus lanatus* by suppressing As uptake (Gonzalez-Chavez et al. 2002). Greater accumulation of Cd, Ti, and Ba was observed in fungal structures than in the host plant cells (Turnau et al. 1993a, b), and measurements of the metal-binding capacity of mycorrhizal mycelium showed that AMF hyphae have a high metal adsorption capacity (e.g., for Cd), potentially representing a barrier for metal translocation to plant tissues (Joner et al. 2000). Surface adsorption mechanisms involve ion exchange, complexation, precipitation, and crystallization of metals on the extra- and intraradical hyphal cell wall components (e.g., chitin, cellulose derivatives, and melanin) or extracellular slime, which may reduce the intracellular accumulation of metals and their effects on cytoplasmic processes (Turnau et al. 1993a, b; Galli et al. 1994; Denny and Ridge 1995). Cd-induced transcript levels of a putative Zn transporter gene (GintZnT1) of the cation diffusion facilitator family (CDF) were observed in symbiotic mycelia of *Gigaspora margarita* and of an ABC putative transporter gene (GintABC1) of the external mycelia of *Glomus intraradices*, thought to be involved in Cd detoxification in metal-contaminated soils (González-Guerrero et al. 2005, 2006). Mechanisms within fungal cells, however, involve chelation of metal ions by ligands like polyphosphates and metallothioneins, compartmentation within vacuoles, and coping with oxidative stress (Turnau et al. 1993a, b; Kaldorf et al. 1999; Joner et al. 2000; Leyval and Joner 2001; Ouziad et al. 2005; Hildebrandt et al. 2007), and vesicles might potentially serve as storage compartments for heavy metals (Turnau 1998; Weiersbye et al. 1999). A study on the significance of AM as a function of low/high nutrient availability demonstrated higher biomass production and nitrogen content of mycorrhizal plants under low nutrient treatment (Cruz et al. 2004). Similar trends seem to apply for uptake of the nonessential elements. The greater volume of the mycorrhizosphere, compared to the rhizosphere alone, provides an increased access to soil resources, including macro-, micro-, but also nonessential elements at low soil concentrations. On the other hand, at higher soil metal levels, AMF are expected to reduce soil metal bioavailability since metals are sequestered in extraradical hyphae (Joner et al. 2000), therefore resulting in lower metal uptake in AM than non-AM plants. In a literature survey, Audet and Charest (2007) correlated metal uptake and relative plant growth parameters of AM plants in heavy metal-polluted soils at low and high soil metal concentration intervals. They were able to determine (1) an increased mycorrhizospheric metal uptake at low metal concentrations and (2) a reduced metal uptake via AMF "metal-binding" processes at high soil metal levels, hence resulting in enhanced plant tolerance through metal stress avoidance, whereas in the area of transition between the two zones, (3) there is kinetic equilibrium that shows no detectable differences between AM and non-AM plants. Interestingly, far smaller differences are seen on the basis of biomass, with the higher plant biomass of AM plants in the zone of high metal levels being the most prominent (Audet and Charest 2007). Taken together, the tolerance mechanisms AMF develop at metal-contaminated sites may play a crucial role(s) in mediating metal uptake and translocation to plants (Leyval et al. 1997), with immobilization of metals in the fungal biomass as the main mechanism involved (Li and Christie 2001; Zhu et al. 2001; Gaur and Adholeya 2004). Because of the lower sensitivity of fungal hyphae to metals compared to plant roots (Joner and Leyval 2001), functional symbiosis with metal-tolerant AMF strains may confer improved metal tolerance on plants, while maintaining an adequate supply of nutrients like P and N through active hyphal uptake (Gaur and Adholeya 2004), thus contributing to improved plant fitness in metal-contaminated soils. However, a range of environmental factors including soil metal concentrations and their bio-availability, soil absorption/desorption characteristics, as well as endogenous factors (e.g., the fungal properties and inherent heavy metal uptake capacity of plants) may influence the uptake of metals by mycorrhizal plants (Leyval et al. 1997; Pawlowska and Charvat 2004). Considering the compromise between plant growth and metal tolerance, AM plants most likely invest more in a stress-avoidance strategy via metal binding by AMF rather than in metabolically more costly stress-resistance alternatives, such as metal chelation and sequestration (Audet and Charest 2007). This was already indicated by a study on tomato gene expression in metal-polluted soils with induction of plant Lemt2 (metallothionein) and LeNramp1 (metal transporter) genes, whose products are putatively involved in metal stress alleviation and that were not induced in the otherwise better growing AM plants. It was suggested that colonization lowered the concentration of heavy metals in plant cells to a level insufficient to induce the expression of these genes (Ouziad et al. 2005; Hildebrandt et al. 2007), which may also apply to more tolerant plant species.

11.6 Conclusion and Future Perspective

Mycorrhizas constitute a bridge for nutrient transport from soils to plant roots. Their role in the formation of soil aggregates and the protection of host plants against drought and root pathogens is well established. A slow process of remediation by these fungi is predicted, which depends on the development of the host root system and associated fungal biomass. The selection of compatible host–fungus–substrate combinations can be exploited by inoculating with ECM or ERM fungi for sustaining the infections. A survival rate of years is well known when inoculated experimentally with ECM fungi (Selosse et al. 1999). Multitrophic interactions are useful in rapid remediations. Fast remediation of persistent organic compounds may be possible with interaction between ECM fungi or bacteria. High variability and lack of homogeneity in experimental protocols, bioassays, and field data are quite familiar in mycorrhizal research. Variability in concentration, duration, and exposure of constituent, coupled with soil and edaphic features, multiple constituents, short-lived/long-lived infections, and colonization and greenhouse/natural conditions, will greatly affect the generation of data. Such inconsistencies make interpretations and conclusions more difficult. A fungus growing continuously in the field indicates an ample supply of carbon coming from the host plant. This can lead to the development of certain enzyme activities and a fungal biomass that aids in the degradation of toxic compounds. Experiments should be designed so as to decipher the intricate fungal degradations in the field in symbiotic association with the host plant as to how the influence of environmental factors on degradation and the degradation rate are affected. Due to prevailing mycorrhizal fungal diversity in ecosystems, it is difficult to predict the effect of soil acidification, metal or hydrocarbon pollution, and other factors on mycorrhizal communities. Because of this effect, the loss of fungal species may occur in soils, resulting in increased predisposition of the host plant to environmental stress. Current knowledge on the functional diversity of mycorrhizal communities is fragmentary. Therefore, the ultimate interest shifts back to the effects of pollutants on the extramatrical mycelial systems of these fungi in soil. This can result in reduction or gain of fungal biomass in soil. Study of the effects of pollutants on the growth of extramatrical mycelium of these fungi is the first step in understanding the stability or sustainability in the ecosystems. This knowledge on the behavior of such fungi in association with the host plant can be deciphered collectively. Ultrastructural details related to soil acidification and contamination, nutrient transfer, and structural alterations can answer questions of general ecological importance. Trees can be planted on soils low in contamination where there is no risk of human health. Trees can be inoculated with the suitable fungi. This may result in gradual decontamination after the establishment of mycorrhizal fungi in the soil. Important gaps exist in understanding multitrophic interactions, and such complexity is far greater than described. Little work has been performed on screening of these fungi related to persistent organic compounds, and comprehensive screening is required. Mineralization of certain compounds has been demonstrated in certain cases, and the intermediate products are known in only a few cases. Production of extracellular enzymes by mycorrhizal fungi is now known and has a great potential in soil mycoremediation. Unfortunately, no example is known that elucidates the mechanism of action of mycotransformation of any pollutant. It is necessary to identify matching host plant species, and efforts need to be intensified to perform large-scale greenhouse field trials to update the process. Soil characteristics related to contaminants, fungal mycelia, host plants, and the efficiency of the process of mycoremediation need to be optimized. Tree growth and its management, including fertilization and irrigation, will also be important in the future of rhizosphere mycoremediation (Meharg and Cairney 2000a). AM fungi can express anastomosis formation and nuclear and protoplasmic exchange. Genetic exchange during anastomosis formation can open the door to vegetative compatibility in natural populations of AM fungi. Somatic incompatibility (Dahlberg 1999) and genetic transformation in ECM fungi (Lemke et al. 1999) have been described. Success can be obtained by cloning transport-system genes in plants by complementation of yeast mutants (D'Enfert et al. 1995). This will assist in our understanding of the molecular aspects of transport systems in ectomycorrhizal fungi and ectomycorrhizas. Transport-system-specific antibody development against protein sequences derived from cDNAs or against purified transport proteins can be helpful in this regard (Chalot and Brun 1998). Genetic transformation system is highly complex, due to the symbiotic nature of these fungi. However, it is now possible to design molecular techniques to characterize signals, genes, and proteins necessary for symbiosis. Only then can the involvement of these molecules and signals in the ectomycorrhizal process be demonstrated functionally. Genetically engineered mycorrhizal fungi can be developed and applied for the manipulation of host plants in many physiological and ecological ways. Such novel fungi can express metal-chelating factors for the reduced uptake of soluble metals by roots. In addition, an adverse environment around root systems due to acidity or toxic pollutants can be changed.

References

Aggangan NS, Dell B, Malajczuk N (1998) Effects of chromium and nickel on growth of the ectomycorrhizal fungus *Pisolithus* and formation of ectomycorrhizas on *Eucalyptus urophylla* S.T. Blake. Geoderma 84:15–27

Ahmad A, Mukherjee P, Senapati S, Mandal D (2003) Extracellular biosynthesis of silver nanoparticles using the fungus *Fusarium oxysporum*. Colloids Surf B 28:313–318

Akmal M, Xu JM, Li ZJ, Wang HZ, Yao HY (2005a) Effects of lead and cadmium nitrate on biomass and substrate utilization pattern of soil microbial communities. Chemosphere 60:508–514

Akmal M, Wang HZ, Wu JJ, Xu JM, Xu DF (2005b) Changes in enzymes activity, substrate utilization pattern and diversity of soil microbial communities under cadmium pollution. J Environ Sci 17:802–807

Allen MF (1991) The ecology of mycorrhizae. Cambridge University Press, Cambridge

Allen MF (2001) Modelling arbuscular mycorrhizal infection: is percent infection an appropriate variable? Mycorrhiza 10:255–258

Alloway BJ (1995) Cadmium. In: Alloway BJ (ed) Heavy metals in soils. Blackie, Glasgow, pp 122–151

Amann R, Anaidr J, Wagner M, Ludwig W, Schleifer KH (1996) In situ visualization of high genetic diversity in a natural microbial community. J Bacteriol 178:3496–3500

Aruguete DH, Aldstadt JH, Mueller GM (1998) Accumulation of several heavy metals and lanthanides in mushrooms (Agaricales) from the Chicago region. Sci Total Environ 224:43–56

Ashkenazy R, Gottlieb L, Yannai S (1997) Characterization of acetone washed yeast biomass functional groups involved in lead biosorption. Biotechnol Bioeng 55:1–10

Audet P, Charest C (2007) Dynamics of arbuscular mycorrhizal symbiosis in heavy metal phytoremediation: meta analytical and conceptual perspectives. Environ Pollut 147:609–614

Bai RS, Abraham TE (2002) Studies on enhancement of Cr(VI) biosorption by chemically modified biomass of *Rhizopus nigricans*. Water Res 36:1224–1236

Berthelsen BO, Olsen RA, Steinnes E (1995) Ectomycorrhizal heavy metal accumulation as a contributing factor to heavy metal levels in organic surface soils. Sci Total Environ 170:141–149

Blaudez D, Botton B, Chalot M (2000) Cadmium uptake and subcellular compartmention in the ectomycorrhizal fungus *Paxillus involutus*. Microbiology 146:1109–1117

Brooks RR (1998) General introduction. In: Brooks RR (ed) Plants that hyperaccumulate heavy metals. CAB, Wallingford, pp 1–14

Bruce F, Moffett FA, Nicholson NC, Uwakwe BJ, Chambers JAH, Tom CJH (2003) Zinc contamination decreases the bacterial diversity of agricultural soil. FEMS Microbiol Ecol 43:13–19

Cairney JWG, Meharg AA (1999) Influences of anthropogenic pollution on mycorrhizal fungal communities. Environ Pollut 106:169–182

Chalot M, Brun A (1998) Physiology of organic nitrogen acquisition by ectomycorrhizal fungi and ectomycorrhizas. FEMS Microbiol Rev 22:21–44

Chappelka AH, Kush JS, Runion GB, Meier S, Kelly WD (1991) Effects of soil-applied lead on seedling growth and ectomycorrhizal colonisation of Loblolly pine. Environ Pollut 72:307–316

Chen BD, Tao HQ, Christie P, Wong MH (2003) The role of arbuscular mycorrhiza in zinc uptake by red clover growing in a calcareous soil spiked with various quantities of zinc. Chemosphere 50:839–846

Colpaert JV, Van Tichelen KK (1996) Mycorrhizas and environmental stress. In: Frankland J (ed) Fungi and environmental change. Cambridge University Press, Cambridge, pp 109–128

Colpaert JV, Vandenkoornhuyse P, Adriaensen K, Vangronsveld J (2000) Genetic variation and heavy metal tolerance in the ectomycorrhizal basidiomycete *Suillus luteus*. New Phytol 147:367–379

Crichton RR (1991) Inorganic biochemistry of iron metabolism. Ellis Horwood, Chichester

Cruz C, Green JJ, Watson CA, Wilson F, Martins-Loucao MA (2004) Functional aspects of root architecture and mycorrhizal inoculation with respect to nutrient uptake capacity. Mycorrhiza 14:177–184

D'Enfert C, Minet M, Lacroute F (1995) Cloning plant genes by complementation of yeast mutants. Methods Cell Biol 49:417–430

Dahlberg A (1999) Somatic incompatibility in ectomycorrhizas. In: Varma A, Hock B (eds) Mycorrhiza structure, function, molecular biology and biotechnology. Springer, Berlin, pp 111–132

Dalal PC (1998) Soil microbial biomass: what do the numbers really mean? Aust J Exp Agric 38:649–665

Dehn B, Schüepp H (1989) Influence of VA mycorrhizae on the uptake and distribution of heavy metals in plants. Agric Ecosyst Environ 29:79–83

Denny HJ, Ridge I (1995) Fungal slime and its role in the mycorrhizal amelioration of zinc toxicity to higher plants. New Phytol 130:251–257

Denny HJ, Wilkins DA (1987) Zinc tolerance in *Betula* spp. IV. The mechanism of ectomycorrhizal amelioration of zinc toxicity. New Phytol 106:545–554

Dighton J, Jansen AE (1991) Atmospheric pollutants and ectomycorrhizae: more questions than answers. Environ Pollut 73:179–204

Dixon RK, Buschena CA (1988) Response of ectomycorrhizal *Pinus banksiana* and *Picea glauca* to heavy metals in soil. Plant Soil 105:265–271

Doelman P, Jansen E, Michels M, Van-Til M (1994) Effects of heavy metals in soil on microbial diversity and activity as shown by the sensitivity-resistance index, an ecological relevant parameter. Biol Fertil Soils 17:177–184

Fitter AH, Merryweather JW (1992) Why are some plants more mycorrhizal than others? An ecological enquiry. In: Read DJ, Lewis DH, Fitter AH, Alexander IJ (eds) Mycorrhizas in ecosystems. CABI, Wallington, pp 26–36

Fomina MA, Alexander IJ, Colpaert JV, Gadd GM (2005) Solubilization of toxic metal minerals and metal tolerance of mycorrhizal fungi. Soil Biol Biochem 37:851–866

Franz A, Burgstaller W, Schinner F (1991) Leaching with *Penicillium simplicissimum*: influence of metals and buffers on proton extrusion and citric acid production. Appl Environ Microbiol 57:769–774

Freedman B, Hutchinson TC (1980) Pollutant inputs from the atmosphere and accumulations in soils and vegetation near a nickel-copper smelter at Sudbury, Ontario, Canada. Can J Bot 58:108–132

Frostegard A, Tunlid A, Baath E (1993) Phospholipids fatty acid composition, biomass, and activity of microbial communities from two soil types experimentally exposed to different heavy metals. Appl Environ Microbiol 59:3605–3617

Frostegard A, Tunlid A, Baath E (1996) Changes in microbial community structure during long term incubation in two soils experimentally contaminated with metals. Soil Biol Biochem 28:55–63

Gadd GM (1993) Microbial formation and transformation of organometallic and organometalloid compounds. FEMS Microbiol Rev 11:297–316

Gadd GM (2001) Metal transformations. In: Gadd GM (ed) Fungi in bioremediation. Cambridge University Press, Cambridge, pp 359–382

Galli U, Meier M, Brunold C (1993) Effects of cadmium on nonmycorrhizal and mycorrhizal Norway spruce seedlings and its ectomycorrhizal fungus *Laccaria laccata* Bk and Br: sulfate reduction, thiols and distribution of the heavy metal. New Phytol 125:837–843

Galli U, Schuepp H, Brunold C (1994) Heavy metal binding by mycorrhizal fungi. Physiol Plant 92:364–368

Gardea-Torresdey J, Cano-Aguilera I, Webb R, Tiemann KJ, Gutierrez-Corona F (1996) Copper adsorption by inactivated cells of *Mucor rouxii*: effect of esterification of carboxyl groups. J Hazard Mater 48:171–180

Gaur G, Adholeya A (2004) Prospects of arbuscular mycorrhizal fungi in phytoremediation of heavy metal contaminated soils. Curr Sci 86:528–534

Gildon A, Tinker PB (1983) Interactions of vesicular-arbuscular mycorrhizal infection and heavy metals in plants. I. The effects of heavy metals on the development of vesicular-arbuscular mycorrhizas. New Phytol 95:247–261

Gonzalez-Chavez C (2000) Arbuscular mycorrhizal fungi from As/Cu polluted soils, contribution to plant tolerance and importance of external mycelium. Ph.D dissertation, University of Reading, UK

Gonzalez-Chavez C, Harris PJ, Dodd J, Meharg AA (2002) Arbuscular mycorrhizal fungi confer enhanced arsenate resistance on *Holcus lanatus*. New Phytol 155:163–171

González-Guerrero M, Azcon-Aguilar C, Mooney M, Valderas A, MacDiarmid CW, Eide DJ, Ferrol N (2005) Characterization of a *Glomus intraradices* gene encoding a putative Zn transporter of the cation diffusion facilitator family. Fungal Genet Biol 42:130–140

González-Guerrero M, Azcon-Aguilar C, Ferrol N (2006) GintABC1 and GintMT1 are involved in Cu and Cd homeostasis in *Glomus intraradices*. In: Abstracts of the 5th international conference of mycorrhiza, Granada

Gorobets S, Gorobets O, Ukrainetz A, Kasatkina T, Goyko I (2004) Intensification of the process of copper ions by yeast of *Saccharomyces cerevisiae* 1968 by means of a permanent magnetic field. J Magn Magn Mater 272–276:2413–2414

Green F, Clausen CA (2003) Copper tolerance of brown-rot fungi: time course of oxalic acid production. Int Biodeterior Biodegrad 51:145–149

Grellier B, Strullu DG, Martin F, Renaudin S (1989) Synthesis in-vitro microanalysis and phosphorus-31 NMR study of metachromatic granules in birch mycorrhizas. New Phytol 112:49–54

Griffioen WAJ (1994) Characterization of a heavy metal-tolerant endomycorrhizal fungus from the surroundings of a zinc refinery. Mycorrhiza 4:197–200

Gruhn CM, Miller JR (1991) Effect of Cu on tyrosinase activity and polyamine content of some ectomycorrhizal fungi. Mycol Res 95:268–272

Hall JL (2002) Cellular mechanisms for heavy metal detoxification and tolerance. J Exp Bot 366:1–11

Hartley J (1997) Effects of heavy metal pollution on Scots pine (*Pinus sylvestris* L.) and its ectomycorrhizal symbionts. Ph.D dissertation, University of Leeds, Leeds

Haselwandter K, Berreck M, Brunner P (1988) Fungi as bioindicators of radiocaesium contamination: pre-and post-Chernobyl activities. Trans Br Mycol Soc 90:171–174

Hattori H (1992) Influence of heavy metals on soil microbial activities. Soil Sci Plant Nutr 38:93–100

Heggo A, Angle JS, Chaney RL (1990) Effects of vesicular–arbuscular mycorrhizal fungi on heavy metals uptake by soybeans. Soil Biol Biochem 22:865–869

Hetrick BAD, Wilson GWT, Figge DAH (1994) The influence of mycorrhizal symbiosis and fertilizer amendments on establishment of vegetation in heavy metal mine spoils. Environ Pollut 86:171–179

Hildebrandt U, Kaldorf M, Bothe H (1999) The zinc violet and its colonization by arbuscular mycorrhizal fungi. J Plant Physiol 154:709–717

Hildebrandt U, Regvar M, Bothe H (2007) Arbuscular mycorrhiza and heavy metal tolerance. Phytochemistry 68:139–146

Howe R, Evans RL, Ketteridge SW (1997) Copper-binding proteins in ectomycorrhizal fungi. New Phytol 135:123–131

Humar M, Bokan M, Amartey SA, Sentjurc M, Kalan P, Pohleven F (2004) Fungal bioremediation of copper, chromium and boron treated wood as studied by electron paramagnetic resonance. Int Biodeterior Biodegrad 53:25–42

Iestwaart JH, Griffioen WAJ, Ernst WHO (1992) Seasonality of VAM infection in three populations of *Agrostis capillaris* (Gramineae) on soil with or without heavy metal enrichment. Plant Soil 139:67–73

Jarosz-Wilkolazka A, Gadd GM (2003) Oxalate production by wood-rotting fungi growing in toxic metal-amended medium. Chemosphere 52:541–547

Joner EJ, Leyval C (2001) Bioavailability of heavy metals in the mycorrhizosphere. In: Gobran GR, Wenzel WW, Lombi E (eds) Trace elements in the rhizosphere. CRC, Boca Raton, FL, pp 165–185

Joner E, Briones R, Leyval C (2000) Metal-binding capacity of arbuscular mycorrhizal mycelium. Plant Soil 226:227–234

Kaldorf M, Kuhn AJ, Schroder WH, Hildebrandt U, Bothe H (1999) Selective element deposits in maize colonized by a heavy metal tolerance conferring arbuscular mycorrhizal fungus. J Plant Physiol 154:718–728

Knight BP, McGrath SP, Chaudri AM (1997) Biomass carbon measurements and substrate utilization patterns of microbial populations from soils amended with cadmium, copper, or zinc. Appl Environ Microbiol 63:39–43

Kong FX, Liu Y, Hu W, Shen PP, Zhou CL, Wang LS (2000) Biochemical responses of the mycorrhizae in *Pinus massoniana* to combined effects of Al, Ca and low pH. Chemosphere 40:311–318

Konopka A, Bercot T, Nakatsu C (1999) Bacterioplankton community diversity in a serious of thermally stratified lakes. Microb Ecol 38:126–135

Kottke I, Qian XM, Pritsch K, Haug I, Oberwinkler F (1998) *Xerocomus badius–Picea abies*, an ectomycorrhiza of high activity and element storage capacity in acidic soil. Mycorrhiza 7:267–275

Kozdroj J, van Elsas JD (2000) Response of the bacterial community to root exudates in soil polluted with heavy metals assessed by molecular and cultural approaches. Soil Biol Biochem 32:1405–1417

Lanfranco L, Bolchi A, Ros EC, Ottonello S, Bonfante P (2002) Differential expression of a metallothionein gene during the presymbiotic versus the symbiotic phase of an arbuscular mycorrhizal fungus. Plant Physiol 130:58–67

Lemke PA, Singh NK, Temann UA (1999) Genetic transformation in ectomycorrhizal fungi. In: Varma A, Hock B (eds) Mycorrhiza structure, function, molecular biology and biotechnology. Springer, Berlin, pp 133–152

Leyval C, Joner EJ (2001) Bioavailability of heavy metals in the mycorrhizosphere. In: Gobran GR, Wenzel WW, Lombi E (eds) Trace elements in the rhizosphere. CRC, Boca Raton, FL, pp 165–185

Leyval C, Turnau K, Haselwandter K (1997) Effect of heavy metal pollution on mycorrhizal colonization and function, physiological, ecological and applied aspects. Mycorrhiza 7:139–153

Li XL, Christie P (2001) Changes in soil solution Zn and pH and uptake of Zn by arbuscular mycorrhizal red clover in Zn-contaminated soil. Chemosphere 42:201–207

Liao JP, Lin XG, Cao ZH, Shi YQ, Wong MH (2003) Interactions between arbuscular mycorrhizae and heavy metals under sand culture experiment. Chemosphere 50:847–853

Malcova R, Vosatka M, Gryndler M (2003) Effects of inoculation with *Glomus intraradices* in lead uptake by *Zea mays* L. and *Agrostis capillaris* L. Appl Soil Ecol 23:55–67

Malik A (2004) Metal bioremediation through growing cells. Environ Int 30:261–278

Martino E, Perotto S, Parsons R, Gadd GM (2003) Solubilization of insoluble inorganic zinc compounds by ericoid mycorrhizal fungi derived from heavy metal polluted sites. Soil Biol Biochem 35:133–141

Meharg AA (1994) Integrated tolerance mechanisms: constitutive and adaptive plant responses to elevated metal concentrations in the environment. Plant Cell Environ 17:989–993

Meharg AA, Cairney JWG (2000a) Ectomycorrhizas: extending the capabilities of rhizosphere remediation? Soil Biol Biochem 32:1475–1484

Meharg AA, Cairney JWG (2000b) Co-evolution of mycorrhizal symbionts and their hosts to metal contaminated environments. Adv Ecol Res 30:69–112

Michelot D, Siobud E, Dore JC, Viel C, Poirier F (1998) Update on metal content profiles in mushrooms: toxicological implications and tentative approach to the mechanisms of bioaccumulation. Toxicon 36:1997–2012

Mogollon L, Rodriguez R, Larrota W, Ramirez N, Torres R (1998) Biosorption of nickel using filamentous fungi. Appl Biochem Biotechnol 70–72:593–601

Moore AEP, Massicotte HB, Peterson RL (1989) Ectomycorrhiza formation between *Eucalyptus pilularis* Sm. and *Hydnangium carneum* Wallr. in Dietr. New Phytol 112:193–204

Morselt AFW, Smits WTM, Limonard T (1986) Histochemical demonstration of heavy metal tolerance in ectomycorrhizal fungi. Plant Soil 96:417–420

Muller AK, Westergaard K, Christensen S, Sorensen SJ (2001) The effect of long term mercury pollution on the soil microbial community. FEMS Microbiol Ecol 36:11–19

Murugesan GS, Sathishkumar M, Swaminathan K (2006) Arsenic removal from groundwater by pretreated waste tea fungal biomass. Bioresour Technol 97:483–487

Nannipieri P, Kandeler E, Ruggiero P (2002) Enzyme activities and microbiological and biochemical processes in soil. In: Burns RG, Dick RP (eds) Enzymes in the environment. Marcel Dekker, New York, pp 1–34

Nordgren A, Baath E, Soderstrom B (1983) Microfungi and microbial activity along a heavy metal gradient. Appl Environ Microbiol 45:1829–1837

Ohya H, Komai Y, Yamaguchi M (1985) Zinc effects on soil microflora and glucose metabolites in soil amended with 14C-glucose. Biol Fertil Soils 1:117–122

Ouziad F, Hildebrandt U, Schmelzer E, Bothe H (2005) Differential gene expressions in arbuscular mycorrhizal-colonized tomato grown under heavy metal stress. J Plant Physiol 162:634–649

Paknikar KM, Puranik PR, Agate AD, Naik SR (1998) Metal biosorbents from waste fungal biomass: a new bioremedial material for control of heavy metal pollution. In: Sikdar SK, Irvine RT (eds) Bioremediation: principles and practices, bioremediation technologies, vol III. Technomic, Lancaster, PA, pp 557–576

Park D, Yun YS, Jo JH, Park JM (2005) Mechanism of hexavalent chromium removal by dead fungal biomass of *Aspergillus niger*. Water Res 39:533–540

Pawlowska TE, Charvat I (2004) Heavy metal stress and developmental patterns in arbuscular mycorrhizal fungi. Appl Environ Microbiol 70:6643–6649

Pawlowska TE, Blaszkowski J, Ruhling A (1996) The mycorrhizal status of plants colonising a calamine spoil mound in southern Poland. Mycorrhiza 6:499–505

Pawlowska TE, Chaney RL, Chin M, Charavat I (2000) Effects of metal phytoextraction practices on the indigenous community of arbuscular mycorrhizal fungi at a metal-contaminated landfill. Appl Environ Microbiol 66:2526–2530

Peat HJ, Fitter AH (1993) The distribution of arbuscular mycorrhizas in the British Flora. New Phytol 125:843–854

Perrier N, Amir H, Colin F (2006) Occurrence of mycorrhizal symbioses in the metal-rich lateritic soils of the Koniambo Massif, New Caledonia. Mycorrhiza 16:449–458

Pierleoni R, Buffalini M, Vallorani L (2004) *Tuber borchii* fruit body: 2-dimensional profile and protein identification. Phytochemistry 65:813–820

Reber HH (1992) Simultaneous estimates of the diversity and the degradative capability of heavy-metal-affected soil bacterial communities. Biol Fertil Soils 13:181–186

Reeves RD (2006) Hyperaccumulation of trace elements by plants. In: Morel JL, Echevarria G, Goncharova N (eds) Phytoremediation of metal contaminates soils, Nato science series: IV: Earth and environmental sciences. Springer, Heidelberg, pp 25–52

Regvar M, Groznik N, Goljevšček N, Gogala N (2001) Diversity of arbuscular mycorrhizal fungi at various differentially managed ecosystems in Slovenia. Acta Biol Sloven 44:27–34

Regvar M, Vogel-Mikuš K, Kugonič N, Turk B, Batič F (2006) Vegetational and mycorrhizal successions at a metal polluted site: indications for the direction of phytostabilisation? Environ Pollut 144:976–984

Ross DS, Sjogren RE, Bartlett RJ (1981) Behavior of chromium in soils: IV. Toxicity to microorganisms. J Environ Qual 10:145–148

Sayer JA, Raggett SL, Gadd GM (1995) Solubilization of insoluble metal compounds by soil fungi: development of a screening method for solubilizing ability and metal tolerance. Mycol Res 99:987–993

Schreferl G, Kubicek CP, Rohr M (1986) Inhibition of citric acid accumulation by manganese ions in *Aspergillus niger* mutants with reduced citrate control of phosphofructokinase. J Bacteriol 165:1019–1022

Schüssler A, Schwarzott D, Walker C (2001) A new phylum, the Glomeromycota: phylogeny and evolution. Mycol Res 105:1413–1421

Selosse MA, Martin F, Bouchard D, le Tacon F (1999) Structure and dynamics of experimentally introduced and naturally occurring *Laccaria* sp. discrete genotypes in a Douglas fir plantation. Appl Environ Microbiol 65:2006–2016

Sharples JM, Meharg AA, Chambers SM, Cairney JWG (2001) Arsenate resistance in the ericoid mycorrhizal fungus *Hymenoscyphus ericae*. New Phytol 151:265–270

Siegel SM, Galun M, Siegel BZ (1990) Filamentous fungi as metal biosorbents: a review. Water Air Soil Pollut 53:335–344

Smith SE, Read DJ (1997) Mycorrhizal symbiosis, 2nd edn. Academic, London

Suh JH, Kim DS, Yun JW, Song SK (1998) Process of Pb^{2+} accumulation in *Saccharomyces cerevisiae*. Biotechnol Lett 20:153–156

Sukla LB, Kar RN, Panchanadikar V (1992) Leaching of copper converter slag with *Aspergillus niger* culture filtrate. Biometals 5:169–172

Taboski MAS, Rand TG, Piorko A (2005) Lead and cadmium uptake in the marine fungi *Corollospora lacera* and *Monodictys pelagica*. FEMS Microbiol Ecol 53:445–453

Tamaki S, Frankenberger WT (1992) Environmental biochemistry of arsenic. Rev Environ Contam Toxicol 124:79–110

Thompson-Eagle ET, Frankenberger WT (1992) Bioremediation of soils contaminated with selenium. In: Lal R, Stewart BA (eds) Advances in soil science, vol 17. Springer, New York, pp 261–309

Thompson-Eagle ET, Frankenberger WT, Longley KE (1991) Removal of selenium from agricultural drainage water through soil microbial transformations. In: Dinar A, Zilberman D (eds) The economics and management of water and drainage in agriculture. Kluwer, New York, pp 169–186

Tobin JM (2001) Fungal metal biosorption. In: Gadd GM (ed) Fungi in bioremediation. Cambridge University Press, Cambridge, pp 424–444

Tobin JM, Roux JC (1998) *Mucor* biosorbent for chromium removal from tanning effluent. Water Res 32:1407–1416

Tonin C, Vandenkoornhuyse P, Joner EJ, Straczek J, Leyval C (2001) Assessment of arbuscular mycorrhizal fungi diversity in the rhizosphere of *Viola calaminaria* and effect of these fungi on heavy metal uptake by clover. Mycorrhiza 10:161–168

Trappe JM (1996) What is a mycorrhiza? In: Proceedings of the fourth European symposium on mycorrhiza, EC Report EUR 16728, Granada, pp 3–9

Turnau K (1998) Heavy metal content and localisation in mycorrhizal *Euphorbia cyparissias* from zinc wastes in southern Poland. Acta Soc Bot Pol 67:105–113

Turnau K, Kottke I, Oberwinkler F (1993a) Element localization in mycorrhizal roots of *Pteridium aquilinum* (L.) Kuhn collected from experimental plots with cadmium dust. New Phytol 123:313–324

Turnau K, Kottke I, Oberwinkler F (1993b) *Paxillus involutus–Pinus sylvestris* mycorrhizae from heavily polluted forest. I. Element localization using electron energy loss spectroscopy and imaging. Bot Acta 106:213–219

Turnau K, Miszalski Z, Trouvelot A, Bonfante P, Gianinazzi S (1996a) *Oxalis acetosella* as a monitoring plant on highly polluted soils. In: Azcon-Aguilar C, Barea JM (eds) Mycorrhizas in integrated systems, from genes to plant development. Proceedings of the fourth European symposium on mycorrhizas (COST edition). European Commission, Brussels, Luxemburg, pp 483–486

Turnau K, Kottke I, Dexheimer J (1996b) Toxic element filtering in *Rhizopogon roseolus/Pinus sylvestris* mycorrhizas collected from calamine dumps. Mycol Res 100:16–22

Ueda K, Kobayashi M, Takahashi E (1988) Effect of chromate and organic amendments on the composition and activity of the microorganisms flora in soil. Soil Sci Plant Nutr 34:233–240

Volesky B (2001) Detoxification of metal-bearing effluents: biosorption for the next century. Hydrometallurgy 59:203–216

Weiersbye IM, Straker CJ, Przybylowicz WJ (1999) Micro-PIXE mapping of elemental distribution in arbuscular mycorrhizal roots of the grass, *Cynodon dactylon*, from gold and uranium mine tailings. Nucl Instrum Method B 158:335–343

Wittekind E, Werner M, Reinicke A, Herbert A, Hansen P (1996) A microtiter plate urease inhibition assay-sensitive rapid and cost-effective screening for heavy metals in water. Environ Technol 17:597–603

Witter E (1992) Heavy metal concentrations in agricultural soils critical to microorganisms. Swedish Environmental Protection Agency Report 4079, Stockholm

Yamamoto H, Tatsuyama K, Uchiwa T (1985) Fungal flora of soil polluted with copper. Soil Biol Biochem 17:785–790

Young JPW (1994) Sex and the single cell: the population ecology and genetics of microbes. In: Ritz K, Dighton J, Giller KE (eds) Beyond the biomass: compositional and functional analyses of soil microbial communities. Wiley, Chichester, pp 101–107

Zarb J, Walters DR (1995) Polyamine biosynthesis in the ectomycorrhizal fungus *Paxillus involutus* exposed to zinc. Lett Appl Microbiol 21:93–95

Zelles L, Bai QY, Ma RX, Rackwitz R, Winter K, Beese F (1994) Microbial biomass, metabolic activity and nutritional status determined from fatty acid patterns and poly-hydroxybutyrate in agriculturally-managed soils. Soil Biol Biochem 26:439–446

Zhang L, Zhao L, Yu Y, Chen C (1998) Removal of lead from aqueous solution by non-living *Rhizopus nigricans*. Water Res 32:1437–1444

Zhang D, Duine JA, Kawai F (2002) The extremely high Al resistance of *Penicillium janthinellum* F-13 is not caused by internal or external sequestration of Al. Biometals 15:167–174

Zhu YG, Christie P, Laidlaw AS (2001) Uptake of Zn by arbuscular mycorrhizal white clover from Zn-contaminated soil. Chemosphere 42:193–199

Chapter 12
Mycorrhizae Adsorb and Bioaccumulate Heavy and Radioactive Metals

Hassan Zare-Maivan

12.1 Introduction

Mycorrhizae are mutual symbiosis (associations) between plant roots and a wide group of soil-inhabiting, filamentous fungi. Both partners exchange essential nutrients required for their growth and survival (Sapp 2004). The fungal partner acquires nitrogen, phosphorus, and other nutrients from the soil environment and exchanges them with the plant partner for photosynthetically derived carbon compounds that are essential for its metabolism. Primary function of mycorrhizal fungi is nutrient exchange, and this provides the basis for broad classification into seven groups: ectomycorrhizae, ericoid mycorrhizae, ectendomycorrhizae, arbuscular mycorrhizae, arbutoid mycorrhizae, monotropoid mycorrhizae, and orchid mycorrhizae.

In temperate and boreal forest ecosystems, trees typically form ectomycorrhizal (ECM) symbioses, whereas the major constituents of the understory and rangeland vegetation form arbuscular (AM) and sometimes ericoid (ERM) or arbutoid (ARM) mycorrhizae. The ECM and AM mycorrhizae will be considered in more detail. These two groups of mycorrhizae occur in a wide variety of plants, particularly those plant species capable of propagating in reclaimed areas, such as abandoned mines and contaminated fields. Ectomycorrhizal (ECM) and arbuscular mycorrhiza (AM) are the mutual symbiosis between some fungi and roots of terrestrial plants. Mycorrhizal fungi have been reported to be direct physical links between soil and plant roots increasing soil nutrient exploitation and transfer of minerals to the roots. As such, mycorrhizae have provided an efficient system to advance the stabilization of heavy metals by plants (phytostabilization). This is achieved through secretion of compounds (e.g., enzymes) by the fungal partner that precipitate into polyphosphate granules in the soil and subsequent adsorption to fungal cell walls and chelating of

H. Zare-Maivan (✉)
Department of Plant Biology, Tarbiat Modares University, Tehran, Iran
e-mail: zare897@yahoo.com; zaremaih@modares.ac.ir

E.M. Goltapeh et al. (eds.), *Fungi as Bioremediators*, Soil Biology 32,
DOI 10.1007/978-3-642-33811-3_12, © Springer-Verlag Berlin Heidelberg 2013

heavy metals inside the fungal cells. Therefore, the use of mycorrhizal plants for land remediation and reclamation has been proposed with results including mycorrhizae influencing metal transfer in plants by increasing plant biomass and plant phosphorus nutrition and reducing metal toxicity to plants by decreasing root to shoot heavy metal translocation and or inside cell stabilization. The effect of mycorrhizae on metal uptake in plants is still controversial, and the issue of heavy metal uptake and bioaccumulation in plants needs further research.

12.2 Soil Habitat

Conditions within soil habitats vary by orders of magnitude over micrometer distances, in response to physical (structure and texture), chemical (pH, O_2, pollutants, soluble substances, and plant residues), and biological (soil biota, community fauna, plant roots, and interactions) variables. Such variability provides ample opportunity for establishing infinite variety of microscale habitats and means of interactions among consumer groups in the mycorrhizosphere. It is perhaps because of this, any soil sample contain representative species from major genera of the known microbiota of terrestrial ecosystems. Among these, mycorrhizal fungi form a distinctive and widespread group of organisms that establish mutual symbiotic relationships with fine roots of plant species. In light of growing urbanization, industrialization, and greater environmental disturbances, much effort is being directed toward friendly interaction of man and the environment. As such, considering the new technological and biotechnological advances in the field of ecology and environmental studies, from a management perspective, the genetic potential to mediate virtually any biogeochemical reaction and the habitat needed to support it exist in most soils which with developing specialized capabilities could be investigated further.

12.3 Ectomycorrhizae

Associations between ECM fungi and the roots of woody plants are characterized by three structural components: the mantle, the Hartig net, and the extraradical mycelium (Smith and Read 1997). The mantle is a sheath of fungal tissue that covers the highly active tips of the lateral roots of the plant and forms the boundary between the root and the soil environment. Its compact, but also variable, morphological nature provides a buffering capacity that helps to prevent root cell dehydration or penetration by pathogenic organisms (Brundrett 1991). Fungal hyphae extend from the outer mantle form a web of extraradical mycelia which grow into the rhizosphere. These mycelia extend into micropore areas and absorb nutrients that may otherwise be inaccessible, usually biochemically to roots (Perez-Moreno and Read 2000). Some ECM fungi also form rhizomorphs, which are thick linear aggregates of hyphae that are specialized for long-distance translocation of nutrients and water (Agerer 2001). Lipids, phenolic compounds, proteins, and polyphosphates may accumulate in the hyphae of the outer mantle, which may also bind heavy metals and thereby prevent their uptake into roots (Peterson et al. 2004). The inner mantle consists of repeatedly branched hyphae, suggesting a role in nutrient exchange such as enabling absorption of simple sugars, such as glucose and fructose from the root, and conversion to fungal sugars such as trehalose, mannitol, or glycogen (Peterson et al. 2004). Hartig net is a highly branched hyphal structure growing between epidermal and cortical cells of the root and is the probable site for exchange of nutrients between symbionts (Peterson et al. 2004). Subtle variations in morphological attributes viewed using light microscopy are often used to distinguish between ECM fungal taxa; development and differentiation of extraradical mycelia has been used to define features relevant to the ecological classification of ECMs (Agerer 1987–2002, 2001). ECM fungi are usually classified among Ascomycetes and Basidiomycetes, unlike AM fungi which are established primarily by Glomeromycetes. ECM plant species represent only about 8,000 species (mostly in the families Pinaceae, Betulaceae, Fagaceae, Dipterocarpaceae, Salicaceae, and Myrtaceae which form extended forest ecosystems and woody habitats). These species are of global importance because of their disproportionate occupancy and domination of terrestrial ecosystems in boreal, temperate, and subtropical forests (Smith and Read 1997). It has been estimated that 5,000–6,000 species of fungi (of the classes Basidiomycetes, Ascomycetes, and few Zygomycetes) form ECM symbioses (Molina et al. 1992; Horton and Bruns 2001), but these numbers are expected to rise as more regions are progressively explored in detail (Cairney 2000). ECM fungal communities exhibit high species richness and diversity, even within small areas with little heterogeneity in plant communities, soil properties, climate, and disturbance patterns (Bruns 1995; Robertson et al. 2006).

12.4 Arbuscular Mycorrhizae Fungi

Arbuscular mycorrhizae (AM) fungi, unlike ECMs fungi, do not induce distinctive changes in the root morphology of partner plant. AMs are generally established in fine lateral roots of plants that are composed of a vascular cylinder, few rows of cortical cells, and an epidermal layer. AM fungi do not form mantles or Hartig nets, but rather penetrate through the epidermal layer of thin roots or from root hairs and develop intracellular hyphal branch-like coils, called arbuscules, that are specialized for nutrient exchange and function as the site of bidirectional nutrient exchange. The intracellular fungal symbiont is separated from the plant cytoplasm by a plant-derived membrane, which invigilates to follow fungal growth and coil formation. Usually, within 2 weeks of arbuscule formation and subsequent growth and development of fungal hyphae in the cortex of the root, thick-walled inter- or intracellular vesicles are formed that act as endophytic storage compartments for fungal partner. Arbuscules are often formed progressively within a short distance behind the penetrating hyphal tip. AM presence in the roots of plants requires dissecting and sectioning of roots, proper tissue staining, and microscopic examination. However, AM fungi are primarily identified on the basis of morphological features of their spores and sporocarps and chlamydospores. These propagules are produced outside the root on the extended fungal mycelium and act as dormant propagules for mycorrhizal establishment under proper environmental conditions. Besides, protein profiling, isozyme polymorphism, and DNA analysis have also been applied to identify AM fungi.

12.5 Evolution and Diversity

Mycorrhizal symbioses have been an important force in evolution (Pirozynski and Malloch 1975; Sapp 2004). Based on reconstructions of evolutionary lineages phylogenies from fungal DNA and the fossil record, it is currently accepted that the first mycorrhizal associations were pivotal in allowing plants to colonize the terrestrial environment about 600 million years ago, and they form the evolutionary basis of present plant communities (Pirozynski and Malloch 1975). ECM fungal diversity appears to have arisen about 200 million years ago, corresponding to changes in climate that allowed for colonization of the land with trees and increased organic matter content of some ancient soils (Cairney 2000). Although phylogenetic analyses reveal that ECM fungi have originated from several independent lineages and that symbiosis with plants has been convergent derived (and perhaps lost) many times over millions of years (Hibbett et al. 2000). Greater diversity and high species richness and abundance in many ecosystems may represent ecological adaptation of mycorrhizae to local environmental heterogeneity and are thought to provide forests with a range of strategies to maintain efficient functioning under an array of environmental conditions (Cairney 1999). In general, soil microbial communities appear to comprise groups of organisms that fulfill broadly similar ecosystem functions, and their diversity represents the potential value and spectrum of capabilities that are possessed by organisms present in a given ecosystem and play a functional role in ecosystem processes (Allen et al. 2003). Knowledge of the individual roles of mycorrhizal fungal species, or of their distribution either in relation to each other or to the physical and chemical environments of the soil, is limited (Rosling et al. 2003) and insufficient for determination of community needs and responses by building up from the species level. Further research will reveal the functional heterogeneity and role as well as effect of fungal community diversity on the functioning of the ecosystems as a whole. These findings will allow environmental managers to adopt a more clearly defined plan of action to remedy a disturbed or polluted ecosystem, to restore and rehabilitate abandoned habitats and to sustain pollution stressed plant communities.

12.6 Soil and Rhizosphere

Soils are living, open, dynamic systems in which constant integrating and disintegrating (i.e., maintaining and releasing) of inorganic and organic substrates and plant residues occur. Soils contain structured and heterogeneous matrices, generally store nutrients and energy, and support high microbial diversity and biomass (Nannipieri et al. 2003). To thrive, soil microorganisms must mobilize energy and nutrients stored in soil. Soil structure provides a complex and variable set of microbial microhabitats ranging from energy-rich to barren, or aerobic to anaerobic, over minuscule distances. Soils are composed of sand, silt, and clay particles that are held together by organic matter (i.e., humus), precipitated inorganic materials, microorganisms, and the products of chemical reactions and interactions by plant roots and other substances. Soil particles adsorb important biological molecules (e.g., DNA, enzymes, etc.), and many soil reactions are catalyzed at the surfaces of soil minerals such as clays, Mn oxides, and Fe oxides (Nannipieri et al. 2003). Water occupies the aggregate pore spaces and forms a meniscus around a central pocket of air, which provides an aerobic and aqueous habitat suitable for supporting microbial communities. Water-logged soils usually create anaerobic microsites which retard gas exchange and limit the distribution of aerobic organisms such as mycorrhizal fungi. When mycorrhizae are present in the soil, even under suboptimal conditions, the rhizosphere is more metabolically active. Fungal metabolic activities produce organic acids that percolate with rainwater down through the soil profile and contribute to accelerated weathering of mineral of the soil, and this provides an opportunity for mycorrhizal propagules to adsorb and/or to interact with metal ions under proper matrix potential of the soil. Although less than 5 % of the soil volume is occupied by microorganisms, including mycorrhizal fungi, increased biological activity takes place in these sites, and this is where the majority of soil reactions are mediated (Diaz 2004). The availability and nutrient content of organic matter are key factors influencing microbial biomass and community composition. Other major factors controlling the distribution and abundance of soil microbial communities include (1) chemical properties of the soil environment (e.g., pH, O_2 supply and capacity for gas exchange and availability of water and nutrients such as N, P, and Fe), (2) physical properties of the soil affecting dispersal (e.g., soil structure and texture), and (3) biotic properties and recycling (i.e., turnover) capacity of soil. If heavy metals and other contaminants that alter soil chemical and physical properties limit resources and/or disturb the ongoing biological processes and are introduced into soils in established ecosystems, the functional capacities of mycorrhizal fungi have proven to be effective in ecological risk management of such ecosystems. These capacities are as follows: (1) altering of the morphology and physiology of plant roots as for the permeability of root membranes, (2) changing root exudation patterns as well as the types of C substrates exuded (Linderman 1988), (3) generating extra volume on colonized roots via mantle in ECMs and extraradical mycelia for increased absorption surface and microbial colonization platform, and (4) generating competitive advantage between interacting microbial species in the immediate microscale distance.

12.7   Phytoremediation

Phytoremediation is the use of plants to extract, sequester, and detoxify pollutants from contaminated ecosystems. It was once regarded as an effective, nonintrusive, inexpensive, and socially accepted technology to remediate polluted soils. Plants that are growing in mine sites are usually tolerant to heavy metals. Plants that survive on metalliferous soils can be grouped into one of three categories: (1) excluders, where metal concentrations in the shoots are maintained up to a critical value at a low level across a wide range of metal concentrations in soil (Baker and Brooks 1989); (2) accumulators, where metals are concentrated in aboveground plant parts from low to high soil concentrations (Baker and Brooks 1989); and (3) indicators, where internal concentrations in the plant reflect external levels such as soil concentrations (McGrath and Zhao 2003). However, heavy metal tolerance in all these plants is heavily dependent on various biological, chemical, and physiological adaptations in contaminated sites. Mycorrhizal formation in plants may contribute by providing a metal excluder barrier and improving nutritional status (Turnau et al. 1993; Wissenhorn et al. 1995). Severe metal pollution does have an evolutionary impact on the plant mycorrhizal interaction. Evolutionary adaptation of these AM-colonized plants to the polluted conditions can be accepted based on numerous lines of evidence, including genetic variation in tolerance, habitability of tolerance, higher fitness of tolerant individuals on polluted sites, and higher fitness of non-tolerant individuals on unpolluted sites. Therefore, the AM-colonized plant grown on mine sites has its own survival strategy such as chelating heavy metals by forming organic complexes by creating a symbiotic relationship with mycorrhizal fungi. It is a fact that mycorrhizal fungi are associated with a majority of the plants in the industrially polluted sites and support plant survival in acidic soils polluted with industrial effluent containing heavy metals. However, the interrelationship between indigenous fungi and heavy metal accumulation in plants and also in mycorrhizosphere soil is not known.

12.8   AM Fungal Spores as Bioindicators and Biomonitors of Contaminating Heavy Metals

Recent studies have shown that ecosystems are constantly contaminated through wet and dry deposition of atmospheric pollutants such as gases, heavy metals, and radioactive isotopes. Studies with pollutants applied singly in controlled fumigation or mixed with soil provide the basis for much of our present understanding of the effects of gases, toxic metals, and contaminants on plants. Additional studies have shown that effects of any particular pollutant on plants may not necessarily be similar to those when it is applied simultaneously or in combination with other pollutants (Runeckles 1984). Increased mining activities as well as disposal of increasing quantities of waste containing heavy metals and use of fossil fuels introduce considerable amounts of contaminating pollutants to ecosystems (Ritchie and Thingvold 1985; Baes and McLaughlin 1987). Although adverse effects of toxic metals on pollen germination, seedling growth, and revegetation schemes are known nowadays, many plants are capable of ameliorating such contaminations to a certain extent (Marx and Artman 1979; Seaward and Richardson 1990; Wissenhorn et al. 1994, 1995). Further research has shown that symbiotic relationships of plants, such as mycorrhizae, play a significant part in stabilization and biodegradation of toxic metal compounds and amelioration of contaminated ecosystems (Tam and Griffith 1993; Tam 1995; Baghvardani 1997; Baghvardani and Zare-maivan 1998). Tam (1995) investigated heavy metal tolerance by ectomycorrhizal fungi and metal amelioration by *Pisolithus tinctorius* in vitro. Baghvardani and Zare-maivan (1998) showed that as many as 18 heavy and radioactive metals were adsorbed by mycorrhizal fungi in two separated field investigations on the Hyrcanian broadleaf forests of Iran. Natural occurrence of some radioactive and heavy metals in parts of Hyrcanian forests, south of Caspian Sea, prompted the idea of this research with the following objectives: (1) to determine differences in capability of adsorbing radioactive and heavy metals by AM fungal species isolated from naturally contaminated forest soil and (2) to determine effects fungal community composition might have on metal adsorbing capability.

In recent decades, interest to having healthy environment has increased. Phytoremediation of contaminated sites, abandoned mining areas, and waste disposal landfills is taking place widely in many parts of the world. Similarly, application of biomonitoring techniques is expanding in many aquatic and terrestrial ecosystems. Access to new, accurate, fast, and reliable techniques, such as that of Mc Kenny and Donald (1987), has always been appealing to scientists and environmental managers. For example, indicator species have been one of approaches taken by many researchers and environmental decision makers for environment impact assessments of contaminated or disturbed ecosystems. Mycorrhizal fungi have been used extensively in mine reclamation and landfill reforestation. These fungi have been indicated as biotracers, bioaccumulators, and biodegraders of toxic compounds. For example, adsorption of radioactive isotopes and heavy metals has been demonstrated in Caspian (Hyrcanian) deciduous forests of Iran (Baghvardani 1997; Baghvardani and Zare-maivan 1998). These researchers investigated capability of AM spores adsorption in areas of high background radiation near Ramsar, a city located on the southern coast of the Caspian Sea and on northern slopes of the Alborz mountain ranges, Iran, and well known for its higher background radiation which is usually five times higher than the 20 mSv year^{-1} that is permitted for radiation workers. Tam (1995) indicated the capability of ectomycorrhizal fungus *Pisolithus tinctorius* for amelioration of heavy metal contaminated sites (Vare 1990; Tam 1995) and involves extrahyphal mucilaginous substances (Denny and Wilkins 1987a, b; Tam 1995) or chelators (Bradley et al. 1982). Brown and Wilkins (1985) investigated the bioavailability and uptake of heavy metals by AM fungi in areas polluted *via* atmospheric depositions from a smelter or from a waste sludge. In this study, however, adsorption capability of various species of AM fungi exposed to naturally occurring chronic heavy metal contamination and background radiation was investigated. Many researchers have indicated that heavy metal uptake in mycorrhizal fungal species is achieved by polyphosphate linkage of copper and zinc (Vare 1990; Tam 1995) and involves extrahyphal mucilaginous substances (Denny and Wilkins 1987a, b; Tam 1995) or chelators (Bradley et al. 1982; Brown and Wilkins 1985). Findings of this research, however, indicated that spore surface and ornamentation contributed to metal adsorption capability of AM fungal spores. *G. multicaule*, for instance, with its larger and highly ornamented spores, demonstrated the highest adsorption readings of many cations. Therefore, adsorbing range of heavy and radioactive metals may be species specific and within a certain range of tolerance. There are many ways that one can measure and track pollutants of greater concentration in the environment; however, AM spores can be used as bioindicators of traces of contaminating metals and as biomonitors for fluctuations in the concentration of metals throughout the year, particularly in environments with definite arid and wet or warm and cold seasons. Further research regarding the mechanisms of amelioration of contaminating metals and adaptation to chronic background radioactive radiation is necessary.

In recent decades, a great attention has been paid to the problem of radioactive and heavy waste disposals. Trace and heavy metal depositions in sediments and metal accumulation in forest soils suggest that forests have been exposed increasingly to greater concentrations of atmospheric pollutants (Baes and McLaughlin 1987). It has been shown that severe contamination by pollutants such as heavy metals can result in pollen malfunctioning and seedling mortality and thus in several decades cause delays in revegetation schemes (Ritchie and Thingvold 1985). Similarly, it has been known for long that exposure to radioactive waste or radiation, such as that from Chernobyl and Fukushima, Japan, power plant accident, contaminates environment and affects genetic material of impacted species. This, in turn, influences plant species composition and process of succession in impacted ecosystems. In contrast, many fungal species exhibit tolerance to radioactive and heavy metals in concentrations that normally can be toxic to higher plants (Tam 1995). Similarly, mycorrhizal fungi have also been shown to accumulate a greater concentration of heavy metals. For example, Bargagli and Baldi (1984) demonstrated that mycorrhizal fungi could accumulate up to 63 times the concentration of mercury in soil from a mercury mining area. Chemical analysis of shortleaf pine tree cores from east Tennessee, USA (Baes and McLaughlin 1987), and of *Ziziphus* trees from Hormozgan Province, Iran (Korury et al. 1999), have shown trace metal accumulation in tree rings with local and regional increase in combustion of fossil fuels. Crowded roadsides, reclaimed areas, and landfills exhibit greater concentration of heavy metals, particularly lead, copper, zinc, and nickel (Seaward and Richardson 1990). Revegetation of such areas usually requires planting of mycorrhizal plant species. Such a phytoremediation practice has become common in many parts of the world; however, there are dense deciduous forest ecosystems, for example, Caspian forest near Ramsar in Northern Iran, whereby radioactive and heavy metals occur naturally (Baghvardani and Zaremaivan 1998). Previous studies by Akbarloo (1994), Nourbakhsh (1994), and

Nourbakhsh and Zare-maivan (1995) demonstrated the widespread occurrence of mycorrhizae in plant communities south of Caspian Sea. These authors indicated that there was a succession pattern of vegetation from seaside toward upper mountains.

Salt-tolerant aquatic weeds dominated sandy coastal areas followed by grasses, shrubs, and trees as distance from the coast and altitude from sea level increased. Accordingly, there was a mycorrhizal succession pattern, with no mycorrhizae on aquatic weeds, endomycorrhizae on grasses and few shrub species and ectomycorrhizae on tree species. Occurrence of mycorrhizal propagules has been reported in contaminated sites by many researchers (Marx and Artman 1979; Wissenhorn et al. 1994, 1995). Previous research indicated that ectomycorrhizal fungi were capable of establishing symbiotic relationship with various Eucalypt, pine, beech, and oak tree species (Marx and Artman 1979; Zare-maivan 1983; Chan and Griffith 1991; Tam and Griffith 1993). Also, research has shown that polyphosphate granules may be responsible for detoxification of radioactive and heavy metals at high concentrations; for example, Vare (1990) demonstrated that aluminum polyphosphate granules were located in the ectomycorrhizal fungus, *Suillus variegatus*. Tam (1995) also demonstrated through polyphosphate linkage of copper and zinc and by energy dispersive X-ray spectroscopy that heavy metal amelioration mechanism in the metal-tolerant fungal species, *Pisolithus tinctorius*, involved extrahyphal slime. Amelioration of heavy metals by mycorrhizal fungi improves plant species ability to establish in new environments. This might explain, to some extent, why so many plant species are mycorrhizal. In this context, while the possibility that forest trees may be important indicators of trends in atmospheric deposition of pollutants (Baes and McLaughlin 1987; Korury et al. 1999) offers a potentially useful tool for regressing historical trends and characterizing of metal depositions in an area, there is far less knowledge regarding the role of mycorrhizal fungi for metal accumulation in natural ecosystems. Natural occurrence of unexpected concentrations of radioactive and heavy metals and dense and diverse vegetation in parts of Hyrcanian (Caspian) forest prompted the idea of present investigation. The purpose of this study was first to determine the accumulation (adsorption) spectrum of radioactive and heavy metals by mycorrhizal fungal structures in nature and second to examine the trend in succession of metal adsorption along the succession gradient of plant communities.

Vegetation and mycorrhizae succession have been investigated in Kheiroud Kenar region (Assadi 1984; Nourbakhsh and Zare-maivan 1995) which is in general agreement with succession pattern of vegetation in South Caspian Sea forest realm. Succession pattern of mycorrhizae in this ecosystem follows general trend observed elsewhere with similar climatic and edaphic conditions (Barbour et al. 1999). Presence of barium and chloride in soil and mycorrhizal propagules, respectively, indicated the intrusion of marine water into sandy coastal areas; however, presence of eight radioactive and heavy metals on mycorrhizal plants in coastal area demonstrated the capability of mycorrhizal propagules as potential bioaccumulators and, therefore, as bioindicators of contaminating pollutants. This finding is supported further by the fact that similar trends are seen in other plant communities along the succession gradient.

Except for the coastal sand in station 1 which exhibited only barium, soil in all other stations contained three to five elements. This might be attributed to presence of clay particles in soil; because, it is widely accepted that cations have a greater tendency to be adsorbed to clay particles. Since mycorrhizal propagules throughout the succession gradient have greater number of elements on them than that in soil, presumably, it seems that presence of mycorrhizae might have contributed to the ability of plant species to tolerate radioactive and heavy metals more extensively. Availability of sufficient nutrients and chronic exposure to radioactive and heavy metals may have been responsible for this adaptation. Adaptive response in humans can be induced by chronic exposure to natural background radiation as opposed to acute exposure to higher levels of radiation in the laboratory. Our knowledge of prevalence of adaptive response phenomenon in the Hyrcanian (Caspian) forest ecosystems is minimal and needs further research. There are many radioactive and heavy metals in the mycorrhizosphere on all sites. Many of these elements are known to be toxic and have adverse effects on plant growth and regeneration; for example, presence of aluminum in forest soil and on mycorrhizal propagules indicates its potential adverse effects on forest trees. Aluminum reduces root growth and limits availability of essential plant nutrients such as phosphorus and calcium. Favoring role of mycorrhizae for facilitated absorption of nutrients in phosphorus poor soils and in reclaimed mining areas has been indicated frequently in the literature (Marx and Artman 1979; Zare-maivan 1983; Marshner and Dell 1994). Therefore, planting mycorrhizal plant species for phytoremediation of aluminum and other metal contaminated sites becomes a formidable possibility. Presence of radioactive metals on ectomycorrhizal roots of Hyrcanian forest trees has been indicated in the past (Baghvardani and Zare-maivan 1998). Previous research has demonstrated that ectomycorrhizal fungi improve metal tolerance of their host plant by primarily accumulating metals in the extra-matrical hyphae and extrahyphal slime (Brown and Wilkins 1985; Denny and Wilkins 1987a, b; Tam 1995). Prevalence of mycorrhizae has been indicated in Hyrcanian forest plant communities recently (Nourbakhsh and Zare-maivan 1995; Baghvardani and Zare-maivan 1998). Findings of this research showed that many herbaceous and shrubby plants were AM and all other tree species were ectomycorrhizal. Considering the fact that occurrence of high levels of natural background radiation has been reported by Baghvardani and Zare-maivan (2000) for Ramsar, a city located about 60 km west of Kherood Kenar Forest Research Station (KKFR) station, this research implicates KKFR as well. However, scale of this study does not permit one to draw a comprehensive conclusion before further research is undertaken. Widespread occurrence of contaminating metals on mycorrhizal plants and in greater frequencies than that in soil throughout the succession gradient suggest the possibility of application of mycorrhizal propagules for identifying (and quantifying) radioactive and heavy metals. Therefore, a reliable and accurate technique is at hand to biomonitor traces of contaminating elements and to show trends in contaminant depositions. However, there are substantial gaps in our understanding of the linkages between metal inputs to ecosystems and their uptake by trees or by their mycorrhizal symbionts. Similarly, the possible role of mycorrhizal fungi in the metal tolerance of higher plants and mechanism of detoxification and biodegradation of toxic elements and tolerance limits by mycorrhizal fungi is poorly understood. But natural occurrence of such elements and high background radiation in parts of Caspian (Hyrcanian) forest in Iran provides a very opportunistic situation for further and more detailed studies.

12.9 Future Perspective

Evolutionary biology and ecology of plant species has broaden our knowledge of origin and adaptive capability of plants and symbiosis in plant ecosystems in the past. Using mycorrhizal plants in phytoremediation practices in contaminated sites has been an effective restoration tool in many parts of the world. As such, taking advantage of mycorrhizal fungal spores or mycelial structures for tracing residual and chronic distribution of contaminating elements would provide an efficient and less costly way of decision making for environmental managers. For those interested in the world geological and ecosystem development, perhaps mycorrhizal propagules and their functioning under different magnetic fields of the earth throughout the geologic time could prove a useful research tool.

References

Agerer R (ed) (1987–2002) Colour atlas of ectomycorrhizae. Einhorn-Verlag Eduard Dietenberger, Schwäbisch Gmünd

Agerer R (2001) Exploration types of ectomycorrhizae: a proposal to classify ectomycorrhizal mycelial systems according to their patterns of differentiation and putative ecological importance. Mycorrhiza 11:107–114

Akbarloo S (1994) Mycorrhizal distribution in Kheiroud Kenar Forest Research Station. MSc thesis, Tarbiat Modarres University, Tehran (In Persian)

Allen MF, Swenson W, Querejeta JI, Egerton-Warburton LM, Treseder KK (2003) Ecology of mycorrhizae: a conceptual framework for complex interactions among plants and fungi. Annu Rev Phytopathol 41:271–303

Assadi M (1984) Plant sociology of Kheiroud Kenar. MSc thesis, Tehran University, Iran (In Persian)

Baes CF III, McLaughlin SB (1987) Trace metal uptake and accumulation in trees as affected by environmental pollution. In: Hutchinson TC, Meems KM (eds) Effects of atmospheric pollutants on forests, wetlands and agricultural ecosystems, vol G16, NATO ASI series. Springer, Berlin, pp 307–319

Baghvardani M (1997) Determining adsorption of radioactive and heavy metals by mycorrhizal fungi in Ramsar Forests. MSc thesis, Tarbiat Modares University, Tehran (in Farsi with English Abstract)

Baghvardani M, Zare-maivan H (1998) Determining adsorption of radioactive and heavy metals by mycorrhizal fungi. In: 13th Iranian plant pathology congress (abstract)

Baghvardani M, Zare-maivan H (2000) Determining adsorption capacity of radioactive and heavy metals by mycorrhizal fungi. Iran J Plant Pathol 36:1–5

Baker AJM, Brooks RR (1989) Terrestrial higher plants which hyperaccumulate metallic elements: a review of their distribution, ecology and phytochemistry. Biorecovery 1:81–126

Barbour M, Burk JH, Pitts DW, Gilliam FS, Schwartz MW (1999) Terrestrial plant ecology. Addison-Wesley Longman, New York, 678 p

Bargagli R, Baldi F (1984) Mercury and methyl mercury in higher fungi and their relation with sulfur in a Cinnabar mining area. Chemosphere 13:1059

Bradley R, Burt AJ, Read DJ (1982) The biology of mycorrhiza in the Ericaceae. VIII. The role of mycorrhizal infection in heavy metal resistance. New Phytol 91:197–209

Brown MT, Wilkins DA (1985) Zinc tolerance of mycorrhizal Betula. New Phytol 99:101–106

Brundrett M (1991) Mycorrhizas in natural ecosystems. In: Begon M, Fitter AH, MacFayden A (eds) Advances in ecological research, vol 21. Academic, Harcourt Brace Jovanovich, New York, pp 171–313

Bruns TD (1995) Thoughts on the processes that maintain local species diversity of ectomycorrhizal fungi. Plant Soil 170:63–73

Cairney JWG (1999) Intraspecific physiological variation: implications for understanding functional diversity in ectomycorrhizal fungi. Mycorrhiza 9:125–135

Cairney JWG (2000) Evolution of mycorrhiza systems. Naturwissenschaften 87:467–475

Chan WK, Griffith DA (1991) The induction of mycorrhizas in Eucalyptus microcorys and E. torrelliana grown in Hong Kong. Forest Ecol Manag 43:15–24

Denny HJ, Wilkins DA (1987a) Zinc tolerance in Betula spp. II. Microanalytical studies of zinc uptake into root tissues. New Phytol 106:525–534

Denny HJ, Wilkins DA (1987b) Zinc tolerance in Betula spp. IV. The mechanism of ectomycorrhizal amelioration of zinc toxicity. New Phytol 106:546–553

Diaz E (2004) Bacterial degradation of aromatic pollutants: a paradigm of metabolic versatility. Int Microbiol 7:173–180

Hibbett DS, Gilbert L-B, Donoghue MJ (2000) Evolutionary instability of ectomycorrhizal symbioses in basidiomycetes. Nature 407:506–508

Horton TR, Bruns TD (2001) The molecular revolution in ectomycorrhizal ecology: peeking into the black box. Mol Ecol 10:1855–1871

Korury SAA, Teimuri M, Khoshnevis M, Salehi P, Salahi A, Matinzadeh M (1999) Determining heavy metal accumulation in *Ziziphus spina-christi using Peroxidase analysis.* Iranian Forest and Rangeland Institute Report. Ministry of Jihad-Agriculture, Tehran (In Farsi)

Linderman RG (1988) Mycorrhizal interactions with the rhizosphere microflora: the mycorrhizosphere effect. Phytopathology 78:366–371

Marshner H, Dell B (1994) Nutrient uptake in mycorrhizal symbiosis. Plant Soil 159:89–102

Marx DH, Artman JD (1979) Pisolithus tinctorius ectomycorrhizae improve survival and growth of pine seedlings on acid coal soils in Kentucky and Virginia. Reclam Rev 2:23–31

Mc Kenny MC, Donald DL (1987) Improved method for quantifying endomycorrhizal fungi: spores from soil. Mycologia 79:179–187

McGrath SP, Zhao FJ (2003) Phytoremediation of metals and metalloids from soils. Curr Opin Biotechnol 14:277–282

Molina R, Massiocotte H, Trappe JM (1992) Specificity phenomena in mycorrhizal symbioses: community-ecological consequences and practical implications. In: Allen MF (ed) Mycorrhizal functioning: an integrative plant-fungal process. Chapman and Hall, New York, pp 357–423

Nannipieri P, Ascher J, Ceccherni MT, Landi L, Pietramellara G, Renella G (2003) Microbial diversity and soil functions. Eur J Soil Sci 54:655–670

Nourbakhsh A (1994) Succession of vegetation and mycorrhizae in Kheiroud Kenar Forest. MSc thesis, Tarbiat Modares University, Tehran (In Farsi with English abstract)

Nourbakhsh A, Zare-maivan H (1995) Succession pattern of vegetation and mycorrhizae in Kheiroud Kenar Forest. In: 11th Iranian plant pathology congress (abstract)

Perez-Moreno J, Read DJ (2000) Mobilization and transfer of nutrients from litter to tree seedlings via the vegetative mycelium of ectomycorrhizal plants. New Phytol 145:301–330

Peterson RL, Massiocotte HB, Melville LH (2004) Mycorrhizas: anatomy and cell biology. NRC Research Press, Ottawa

Pirozynski KA, Malloch DW (1975) The origin of land plants: a matter of mycotrophism. Biosystems 6:153–164

Ritchie IM, Thingvold DA (1985) Assessment of atmospheric impacts of large-scale copper-nickel development in northern Minnesota. Water Air Soil Pollut 25:145–160

Robertson SJ, Tackabery LE, Egger KN, Massiocotte HB (2006) Ectomycorrhizal fungal communities of black spruce differ between wetland and upland forests. Can J Forest Res 36:972–985

Rosling A, Landeveert R, Lindhall BD, Larsson KH, Kuyper TW, Taylor AFS, Finlay RD (2003) Vertical distribution of ectomycorrhizal fungal taxa in a podzol soil profile. New Phytol 159:775–783

Runeckles VC (1984) Impact of air pollution combinations on plants. In: Treshaw M (ed) Air pollution and plant life. Wiley, Chichester, p 239

Sapp J (2004) The dynamics of symbiosis: an historical overview. Can J Bot 82:1046–1056

Seaward MRD, Richardson DHS (1990) Atmospheric source of metal pollution and effects on vegetation. In: Shaw AJ (ed) Heavy metal tolerance in plants: evolutionary aspects. CRC, Boca Raton, FL, pp 75–92

Smith SE, Read DJ (1997) Mycorrhizal symbiosis, 2nd edn. Academic, London

Tam PCF (1995) Heavy metal tolerance by ectomycorrhizal fungi and metal amelioration by Pisolithus tinctorius. Mycorrhiza 5:181–187

Tam PCF, Griffith DA (1993) Mycorrhizal associations in Hong Kong Fagaceae. IV. The mobilization of organic and poorly soluble phosphates by ectomycorrhizal fungus Pisolithus tinctorius. Mycorrhiza 2:133–139

Turnau K, Kotke I, Oberwinkler F (1993) Element localization in mycorrhizal roots of *Pteridium aquilinum* (L) Kuhn collected from experimental plots treated with cadmium dust. New Phytol 123:313–324

Vare H (1990) Aluminum polyphosphate in the ectomycorrhizal fungus Suillus variegatus (Fr) O. Kunze as revealed by energy dispersive spectroscopy. New Phytol 116:663–668

Wissenhorn I, Leyval C, Berthelin J (1994) Bioavailability of heavy metals and abundance of arbuscular mycorrhizae in soil polluted by atmospheric deposition from a smelter. Biol Fertil Soils 19:22–28

Wissenhorn L, Leyval C, Belgy G, Berthelin J (1995) Arbuscular mycorrhizal contribution to heavy metal uptake by maize(Zea mays L.) in pot culture with contaminated soil. Mycorrhiza 5:245–251

Zare-maivan H (1983) Root and mycorrhizal distribution of healthy and declining English oaks (Quercus robur L.). MSc thesis, Western Illinois University, Macomb, IL

Part IV
Mycoremediation: Agricultural and Forest Ecosystem Sustainability

Chapter 13
Upscaling the Biogeochemical Role of Arbuscular Mycorrhizal Fungi in Metal Mobility

A. Neagoe, Virgil Iordache, and Erika Kothe

13.1 Introduction

The target of environmental management is to optimize the production of natural resources and services at scales relevant for humans (usually environmental management units like ecosystems—10^4–10^6 m^2, simple landscapes (10^6–10^8 m^2), and larger landscapes—river basins and ecoregions). The operational measures of such management require knowledge about the role of each species. In order to describe this role, one can use the term service production unit (SPU, Luck et al. 2003) referring to an aggregate of populations of a certain species or to a part of such a population directly involved in the production of a certain natural service or resource. The scale of an SPU is usually different from that of the environmental management units (either smaller or larger), and when the scale is of the same order, the SPU does not exactly overlap in space with the institutionally delineated environmental management unit. In the case of small-scale organisms like fungi, one is additionally confronted with the problems inherent to upscaling the scientific knowledge on the role of fungi from the scale of the SPU to the scale of the management units.

Usually, natural services are related to the production of natural resources as structural elements of the ecosystems (e.g., renewable resources provided by a forest)

The contribution of the first two authors to the writing of this chapter was equal.

A. Neagoe (✉) • V. Iordache
Research Centre for Ecological Services (CESEC), Faculty of Biology, University of Bucharest, Bucharest, Romania
e-mail: aurora.neagoe@unibuc.ro; virgil.iordache@g.unibuc.ro

E. Kothe
Institute of Microbiology – Microbial Phytopathology, Friedrich Schiller University, Neugasse 25, 07743 Jena, Germany
e-mail: erika.kothe@uni-jena.de

E.M. Goltapeh et al. (eds.), *Fungi as Bioremediators*, Soil Biology 32,
DOI 10.1007/978-3-642-33811-3_13, © Springer-Verlag Berlin Heidelberg 2013

and about services related to other functions of the ecosystems (e.g., hydrological or biogeochemical services). In this chapter, we focused on some of the biogeochemical services supported by AMF and on the methodology of their evaluation. However, the mentioned analytical and managerial distinction between types of services is not supported by a more basic insight in what happens in an ecosystem, because the production of biomass (supporting the exploitation of renewable resources) is inseparable from the circulation of substances, as well as energy between abiotic and biotic compartments (supporting the biogeochemical services).

Scientific disciplines dealing with element cycling and its effects on ecosystems can be seen to have developed in two historical phases. During the first, biogeochemistry has been conceived as a discipline dealing with ecosystem element cycling (pool and flux approach), where the cycling of each element was analyzed separately. In the particular case of toxic elements, like heavy metals or organic pollutants, one would have to firstly describe the cycling of these substances in ecosystems and then (in the frame of a separate scientific discipline, ecotoxicology) to describe their effects on the functioning of ecosystems. The basic problem with this approach is that the circulation by organisms (and their role as reservoirs of elements in the cycling of elements) is not separable from the effects of the elements taken up on the production of biomass, in particular from the ecotoxicological effects of toxic elements or substances. As a reaction to the limitations of this approach, there are holistic proposals (in systems ecology line of thought) that biogeochemistry is a part of ecotoxicology (or vice versa, depending on the institutional location of the researchers developing the ideas) and that the methodology should approach at the ecosystems scale both the circulation and the effects of elements. From a basic science point of view, the problem with this second approach is that ecosystems are concepts imposing a scale of analyses relevant for human management interests and not giving the attention due to the scale of each group of organisms (the role of organisms is evaluated directly at the scale of management units without an explicit upscaling or downscaling from the specific scale of the SPU, Iordache et al. 2011). Also, the partial decoupling between processes occurring at different scales in an environmental management unit and the multi-scale character of the ecosystems are ignored in the traditional (holistic systems analyses) research methodology of this approach (Iordache et al. 2012). From a practical point of view, this holistic variant of biogeochemistry (or ecotoxicology) is not workable because it tends to overconcentrate the existing resources in single large-scale projects (and for this reason faces institutional resistance) and, in the frequent case of resource scarcity, may also lead to a superficial (from a good science standards perspective) approach of the SPU scale mechanisms involved in the functioning of ecosystems.

In this context, we introduced elsewhere (Iordache et al. 2012) a concept for the role of an organism in the mobility of elements giving attention to its specific scale as follows: "By *role* of a variable in the mobility of elements, we understand the causal influence of a variable in a coupling mechanism in producing the outcome of a process involving it. The *role* of a subsystem of the coupled entities (characterized by a variable) is specific to the coupling scale. The fluxes resulted from this role at the coupling scale propagate to larger or smaller scales (these effects at distance could eventually be labeled as indirect roles)." Characterizing the direct role is a matter of understanding the functioning of the SPU, and characterizing the indirect role at ecosystem scale is a matter, for small-scale organisms like fungi, of upscaling from the SPU scale to the ecosystem scale. In this framework, one can recover a part of the holistic idea. Characterizing the direct role supposes the construction of a minimal homomorphic model of the SPU (e.g., Iordache et al. 2011 for ectomycorrhizal fungi), much less complex than a model of an environmental management unit (ecosystem), which makes the approach workable from a practical point of view. It also supposes the integrated research of the circulation of elements as resources (or toxicants) by the target organisms and the other environmental objects directly interacting with the target organisms. The study of the circulation of elements and substances is put in the frame of their effects on the biomass production of the target organisms. In the particular case of heavy metals (HM) from contaminated areas, implications are that the ecotoxicological effects of metals on mycorrhizal plants are not studied separately from their circulation and, for instance, from the effects of macronutrients like nitrogen and phosphorus (at least as a research program) or any other significant resources needed for the functioning of the fungi–plant systems. To give a name to this approach, we call it *objective scale integrated biogeochemistry* (OSIB). We mean here by "objective" that the scale of analysis is imposed by the environmental objects involved in the cycling, and not by an a priori delineation of environmental management units (ecosystems) based on human (institutional) interests. By integrated, we refer to the integrated investigation of the fluxes of nutrients and toxic elements through the target organisms and through the biotic and abiotic compartments directly connected with them. OSIB is contrasted to institutional scale integrated biogeochemistry (the holistic approach described above) and with classic biogeochemistry (by element).

The question tackled is to what extent the existing knowledge on AMF allows an OSIB approach of the role of AMF in the mobility of metals in contaminated areas. We start by surveying the existing knowledge about AMF, describe the structure of an ideal methodology for investigating the biogeochemical role of AMF, show how the limits of the existing knowledge impact on the implementation of the ideal methodology, and propose an operational methodology and illustrate it with some of our research results.

13.2 Survey of the Literature on AMF

The recent secondary literature (review type and meta-analyses) directly and indirectly relevant for an assessment of the biogeochemical role of AMF is large (Table 13.1). From taxonomic to structural and functional aspects, comprehensive reviews are available. Especially striking is the large body of recent reviews concerning the influence of AMF on plants growing in areas contaminated with metals, in the context of phytoremediation studies, the focus of our interest in this chapter.

Table 13.1 Reviews and chapters relevant for the biogeochemical role of AMF (use of AMF in agriculture and organic farming not included)

Topic	Review or chapter
Basic issues	
Species, individuals, populations and communities, hyphal networks of AMF	Taylor et al. (2000), Glass et al. (2004), Rosendahl (2008), and Young (2009)
Evolutionary ecology of AMF selection	Cairney (2000), Mehard and Cairney (2000), Helgason and Fitter (2009), and Hartmann et al. (2009)
Succession of AMF	Hart et al. (2001) and Piotrowski and Rillig (2008)
Interaction with other environmental objects	
Interactions with bacteria	Boer et al. (2005) and Bonfante and Anca (2009)
Interactions with rocks and minerals	Hoffland et al. (2004), Gadd (2007, 2010), Rosling et al. (2009), and Martino and Perotto (2010)
Influence on aboveground consumers	Moore et al. (2003), Gehring and Bennett (2009), and Koricheva et al. (2009)
Interaction with genetically modified plants	Liu (2010)
Use of AMF for induced phytoextraction of metals	Lebeau et al. (2008, 2011)
Effects on ecological processes	
Processes and functions of AMF in ecosystems	Allen et al. (2003), Rillig (2004), Simard and Durall (2004), Goltapeh et al. (2008), and Garg and Chandel (2011)
Role of AMF in water circulation	Auge (2001) and Allen (2007, 2009, 2010)
Role of AMF in phosphorous cycling	Jansa et al. (2011)
Effect of AMF on plant root systems	Berta et al. (2002) and Hodge et al. (2009)
Effect of AMF on soil structure, glomalin	Rillig and Mammey (2006) and Treseder and Turner (2007)
Scale and modeling	
Scale of AMF, heterogeneity in space, and modeling	Miller and Kling (2000), Johnson et al. (2006), Chaudhary et al. (2008), Wolfe et al. (2009), and Johnson (2010)
Biogeochemistry of trace elements, ecotoxicology, and *management of contaminated sites*	
Effects of metals on AMF	Pawlowska and Charvat (2004) and Gadd (2005)
Effect of AMF on translocation of metals from soil to plants	Audet and Charest (2007a, b, 2008)
Effect of AMF on improving trace elements deficiency	Cavagnaro (2008)
Effect of AMF on heavy metal tolerance of plants, use in phytoremediation	McGrath et al. (2001), Khan (2005, 2006), Hullebusch et al. (2005), Göhre and Paszkowski (2006), Hildebrandt et al. (2007), Mathur et al. (2007), Giasson et al. (2008), Gamalero et al. (2009), Khan et al. (2009), Wenzel (2009), Smith et al. (2010), Vamerali et al. (2010), and Turnau et al. (2010)
Effect of AMF on organic pollutants, use in phytoremediation	Leyval et al. (2002), Brar et al. 2006, and Joner and Leyval (2009)

On the other hand, if one looks only at biogeochemical studies done in an ecosystem (holistic) paradigm, for instance, by searching for the keyword "arbuscular" in the online versions of the journals Biogeochemistry and Ecosystems, with the purpose to identify articles about the role of AMF in metal biogeochemistry, the picture is different. At the date of our search, there were 24 primary literature articles dealing with AMF in the whole collection of Biogeochemistry, of which 12 related to nitrogen; 3 to organic matter and carbon; 4 to carbon and nitrogen; one to organic matter and minerals; 1 to carbon and phosphorous; 2 to carbon, nitrogen, and phosphorus; and one to nitrogen, phosphorous, and sulfur. In the journal Ecosystems, we found 15 articles, most of them dealing with nitrogen and carbon and several with disturbances not related to the contamination with metals. Complementary, we searched for the same key term in two ecotoxicological journals: Ecotoxicology, and Environmental Toxicology and Chemistry. We found only 8 articles in the first journal and 13 in the second one, of which about half were dealing with metals, usually in an experimental setting. One can conclude that the information about the role of AMF in holistic (large-scale) type approaches of metal biogeochemistry was lacking at the level of January 2011. Despite the large apparent interest on their use in phytoremediation of contaminated areas, most of the information is at small scale. The roles of AMF in the mobility of metals based on this literature can be synthesized in Table 13.2 (adapted from Neagoe et al. 2012; details in Sect. 13.2.1 of this chapter).

The problematic situation of basic issues like identifying individuals, populations, and communities (reviewed by Rosendahl 2008) should also be underlined, because it has profound effects on the practicability of a systems ecology type approach with respect to the role of AMF and on the methodology of upscaling the information about functions at small scale to functions at large scale. Currently, we do not have methodological access to the delineation of AMF physiological individuals in space in real field ecosystems, and implicitly, we cannot have true estimations of the number of individuals in populations and the numeric and biomass abundances in communities needed for characterizing succession changes, as we do for aboveground communities. The situation seems to be worse than that for ectomycorrhizal fungi (Iordache et al. 2011). In this context, the discussions about the succession of AMF, and the influence of AMF on plant succession, should be interpreted keeping in mind the basic problems outlined above. Although soil pH, nutrients and organic matter, as well as plant community structural changes are considered to be the most important drivers of AMF succession (Piotrowski and Rillig 2008), and the role of AMF especially in primary succession has been proven (Gange et al. 1993), it is not yet clear at what scale these statements hold up in real systems.

Before exploring the consequences of these limitations in knowledge on the possibility to characterize the role of AMF in the frame of an objective scale integrated biogeochemistry, we provide some extra information on the available knowledge concerning the influence of AMF on the transfer of metals to plants and to lower soil layers.

Table 13.2 Examples of direct roles of AMF and organic matter observable at microscales and of indirect roles observable at scales ranging from pot to soil column (lysimeter) and to field plot (adapted from Neagoe et al. 2012)

Soil layer	Roles	Direct role by immobilization metals	Direct role by mobilization of metals	Direct role by supporting the mobilization or immobilization of metals	Indirect roles
Soil layer relevant for plants	AMF	Biosorption, intracellular accumulation	Chemoorganotrophic leaching, bioweathering	Organic matter decomposition, organic acid and glomalin production	Transfer of metals to plants and to lower soil layers
	Organic matter	Immobilization in litter, immobilization in soil aggregates, chelates in fine pores	Organochemical weathering, soluble chelates, organocolloids, free enzymatic degradation of immobile organic carbon	Energy source for microorganisms, buffering of soil solution	

13.2.1 Influence of AM Fungi on the Bioaccumulation of Heavy Metals in Plants

In the last decade, a vast number of publications on the effect of AMF on the uptake of HM by plants became available. Interestingly, this effect is not uniform; both stimulation and inhibition of uptake have been reported. The outcome depends on the selected plants, the species, and strain of the fungus used; on the degree of pollution; and on the pollutant's binding capacity to soil constituents. In particular, heavy metals (HM) have been shown to have positive, negative, or neutral effects on mycorrhizal colonization in soil or culture solution, depending on their host plants. Metal accumulation by plant shoots is reported to be lower under elevated soil metal concentrations or higher under normal metal conditions (Toler et al. 2005); when this is applied to phytoremediation technologies using AMF, HM from polluted areas, in the majority of cases, were retained in the rhizosphere. The retention capacity differs as a function of the degree of pollution and of the nature of the element. In such cases, HM are not accumulated in big quantities in the aboveground part of plants, although the root absorption area can increase up to 47-fold using mycorrhizal fungi (Turnau et al. 2006 after Smith and Read 1997; Soares and Siqueira 2008). Metal acquisition by plants is reduced, and plant growth is enhanced; thus, metal dilution occurs in plant tissue as a consequence of increased root or shoot growth, but also due to the HM uptake, reduction, exclusion by precipitation, or chelation in the rhizosphere system, and also due to the phosphorus (P)-mediated effects on the host plants (Soares and Siqueira 2008).

The influence of AMF on metal plant uptake depends on many factors such as "fungal genotype, uptake of metal by plant via AM symbiosis, root length density, competition between AMF communities, seasonal variation in AM, association with soil microorganisms, chemical properties of the soil outside the rhizosphere (pH, CEC, etc.), the metal itself, concentrations of available metals, soil contamination conditions (contaminated or artificially contaminated vs. non-contaminated soil, interactions between P and metals (addition of P fertilizers), experimental conditions (light intensity, plant growth stage, available N and P), litter inputs, plant species and plant size" (Giasson et al. 2008). The toxic effect could be diminished by covalent binding, compartmentalization, or extrusion; moreover, a physico-chemical barrier can be formed to protect crucial organs from toxicity (avoidance), or prevention of HM uptake by excluder plant species, HM precipitation on the surface of extraradical mycelium, production of metallothioneins (metal-binding proteins), and metabolism alterations like increasing production of scavengers for reactive oxygen such as proline through symbiotic fungi are known. None of these mechanisms so far have been proven with molecular techniques (Giasson et al. 2008; Turnau et al. 2010).

According to Gadd (2005), both live and dead components of the fungal cell wall can be involved in HM binding with help of free amino, hydroxyl, carboxyl, and other groups. AMF form extraradical mycelium and intraradical hyphae that penetrate the intercellular spaces and enter cortical root cells. In the case of reduction of

HM uptake, an important role in retention, binding, and immobilization seems to be associated to *fungal vacuoles*. They are involved in the regulation of cytosolic metal ion concentrations and the detoxification of potentially toxic metal ions (Gadd 2005). The *fungal cell walls*, respectively, chitin and glomalin from the fungal wall (Christie et al. 2004), are also important due to the presence of free amino, hydroxyl, carboxyl, and other functional groups (Gadd 1993, 2005; Kapoor and Viraraghavan 1995). While chitin is an important component of fungal cell walls acting as an effective biosorbent for radionuclides (Tobin et al. 1994), glomalin is a glycoprotein produced by the hyphae of AMF. It plays an important role in fungal physiology and in the soil environment with a negative effect on soil aggregate stability. It seems to be efficient in sorbing potentially toxic elements by sequestering As, Cu, Cd, Zn, Pb (Alexander 1994; Gonzalez-Chavez et al. 2004; Carrasco et al. 2009), Cr (Alexander 1994; Estaún et al. 2010), Hg (Alexander 1994; Yu et al. 2010), etc., and reducing the level of toxicity in soil by converting HM into their organic form thus making them more bioavailable. As already noted, the toxic effect of metals entering into the cells can be counteracted by synthesizing complex organic molecules such as metallothioneins and phytochelatins or HM transporters of various families (Turnau et al. 2010 citing Hall 2002). Van Keulen et al. (2008) described some advanced proteomic studies in which an array of proteins was found to be expressed under stress conditions including chitinases.

HM enter the roots, where are deposited mainly in the inner root parenchyma cells. There also are most of the fungal structures (intraradical hyphae, arbuscules, and vesicles) in which metals can be stored (Turnau 1998). Due to their potential in sequestering HM, AM fungi play a significant ecological role in phytostabilization of toxic HM in polluted soils and, at the same time, help mycorrhizal plants to survive in these hostile soils. The migration of HM is prevented also by microbial production of organic acids, acid phosphatases, or pigments which precipitate HM outside of the mycelium (Turnau and Dexheimer 1995). To that end, chelation of metals is possible with the help of different compounds excreted from the extraradical mycelium of mycorrhizal fungi (Turnau et al. 2006) or by metal retention in the root system, a result of surface complexation of metals with cysteine-containing ligands of fungal proteins, a phenomenon which seems to play an important role in resistance of plants to heavy metals (Christie et al. 2004).

Water deficiency is a parameter which may affect the intra- and extraradical development of AMF through a decrease in plant photosynthesis and stomatal closure, leading to a decreased supply of the fungal symbiont with carbohydrates. A low soil moisture regime leads also to a decrease in phosphorus and other major nutritional elements in plants lowering in turn plant biomass production and increasing concentrations of HM *in planta*. In this case, HM can be bound within the AMF cells to metallothionein or be stored in vacuoles (Neumann et al. 2009).

In conclusion, as Vamerali et al. (2010) underlined, the beneficial effect of AMF on HM uptake into plants consists of a higher biomass production which could lead to a metal "dilution effect," an increase in HM plant tolerance, as well as a greater HM concentrations in plant tissues. However, information about the variability in space of these effects in large-scale contaminated ecosystems is lacking.

13.2.2   Particular Effects of AM Fungi on Different HM

AMF inoculation with or without fertilizer application leads to a decrease of toxicity symptoms of plants. It is well documented that AMF and P nutrition can produce larger plant biomass with a resulting "growth dilution" effect. Thus, shoots **As** concentrations are slightly reduced and soil inorganic As is transformed into the less toxic organic form by AMF (Adriano 2001; Ultra et al. 2007). Inorganic arsenic (e.g., arsenite, arsenate) is very toxic for membranes and can also inhibit seed germination or lead to death of plant cells (Barua et al. 2010 after Carbonell et al. 1998). Giasson et al. (2008) noted that arsenate may be accumulated in the cytoplasm in the same way as polyphosphates. There are more studies in which clear indications are given for mycorrhizal inoculation reducing harmful effects of arsenic on the initial growth of different plant species improving plant tolerance to toxicity (Sanon et al. 2006; Chern et al. 2007; Ultra et al. 2007). The tolerance to arsenic toxicity is variable from one plant species to another. Moreover, a low level of As has been found to stimulate plant growth in spite of the fact that As is not a nutritional element for plants. Barua et al. (2010) underlined that plant roots contain higher proportion of arsenic than any other plant parts. Under As exposure, the length of roots is reduced and thus the spreading of plants is disturbed. High arsenic contamination in soil, however, leads to a strong growth suppression, stopping the normal functioning of roots (e.g., water transport, gas diffusion, and nutrient uptake) in addition to the shorter roots and disassembled root cap (Ahmed et al. 2006). In the presence of AMF, the biomass production was increased up to 2.4 times due to a better root development. Chemical changes of root exudates under AMF action provided conditions for more mycelium to develop, while they act as nutrient source for rhizosphere microorganisms (Leung et al. 2006; Ultra et al. 2007). Barua et al. (2010) underlined the increase of roots as a result of AMF association with plants, after the transformation of As from inorganic to organic forms by biomethylation. Dimethylarsine was detected in colonized *Pteris vittata* but also in other plant species, and the mycorrhization was not inhibited by As (Liu et al. 2005; Gadd 2005; Ultra et al. 2007; Giasson et al. 2008).

The alleviation of **Zn** toxicity toward plants by using AMF was reported in Christie et al. (2004) and Chen et al. (2004), and this phenomenon was shown to be dependent on direct and indirect mechanisms. As an example for a direct mechanism, Zn was bound in mycorrhizal structures and immobilized in mycorrhizosphere, while for an indirect effect, an influence of mycorrhiza on the plant's mineral nutrition, especially for P, leads to increased plant growth and enhanced metal tolerance. The mobility of Zn is greatly affected by the changes in soil pH. The Zn immobilization through the fungal activity might be an effect of these changes, contributing to the inhibition of Zn uptake into the mycorrhizal plant by storage in the arbuscules but also in hyphae (Christie et al. 2004). In highly contaminated soil, Zn was found in higher concentration in roots, while a decrease in the shoots was seen as an effect of AMF. When Zn amounts in soil increased, a critical threshold exists, below which Zn uptake is enhanced, while above this level

Zn translocation to the aboveground parts of host plants is inhibited. In some plant species, higher translocation rates may occur, but at the cost of poor plant biomass development and probable early death of the individuals (Chen et al. 2005b). Giasson et al. (2008) found that $ZnCO_3$ can be solubilized by hyphae and Zn is then translocated into the plant roots. Chen et al. (2004) provided information about the effects of chelating agents such as EDTA on mycorrhizal development and on Zn uptake in plants. It is known that EDTA application on a polluted soil leads to the release of absorbed HM increasing the mobility of metals in the soil. In the case of AMF inoculation, the uptake differed in function of the soil Zn. In general, mycorrhizal colonization decreased both deficiency and the effects of pollution with Zn. However, neither EDTA nor AM colonization stimulated Zn translocation from roots to shoots. In conclusion, the combined effects of both did not promote metal removal from the soil. Zn depositions were found in spores of *Glomus intraradices* in the periplasmic space between the inner layer of the wall and the plasmalemma. The fungal alkaline phosphatase activity as an indicator of fungal viability was strongly reduced in the presence of EDTA, also affecting Zn uptake (Turnau et al. 2006). Göhre and Paszkowski (2006) and Cavagnaro (2008) explored the effects of Zn on the colonization of roots by AMF and the molecular Zn physiology for both plant and fungal gene products. These authors concluded that the changes in the Zn status of plants due to AMF colonization were influenced by the expression of a specific gene of *Glomus intraradices*. They recommended the identification of further genes involved in interactions between AMF and Zn, as well of the factors which control their expression.

Yu et al. (2010) reported that in the case of **Hg**, the uptake is lower by mycorrhizal than by non-mycorrhizal roots of maize, and AMF inoculation significantly decreased the total and extractable Hg concentrations in soil as well as the ratio of extractable to total Hg. Calculating mass balances for Hg in soil indicated a loss of Hg which can be attributed to Hg volatilization as a result of AMF influence. No significant difference of Hg concentrations was found between mycorrhizal and non-mycorrhizal shoots of maize which suggest that contribution of root uptake to shoot accumulation of Hg is very limited. The release of Hg into soil gases or into the atmosphere is a result of methylation (CH_3Hg^+), which leads to phytovolatilization, seen also with As and other metalloids (Giasson et al. 2008).

Various studies show that AM symbiosis may alleviate also **Cd** stress and decrease Cd uptake by plants growing in contaminated soils. A negative effect of Cs on root colonization and the development of extraradical mycelium was seen (Rivera-Becerril et al. 2002; Janoušková et al. 2006; Biró and Takács 2007; de Andrade and da Silveira 2008; Rashid et al. 2009; Janoušková and Pavlíková 2010; Regvar et al. 2010). Redon et al. (2008) as well as Janoušková and Pavlíková (2010) reported that in the mycorrhizosphere the concentration of Cd was lower as compared to non-mycorrhizal plants with extraradical hyphae being responsible for the decrease of Cd toxicity. Janoušková et al. demonstrated in 2006 that extraradical hyphae have a high Cd-binding capacity which is specific for HM (Hildebrandt et al. 2007). The toxic and nonessential metal Cd is bound more strongly to ligands compared with essential metals. This phenomenon leads to the displacement of essential metals from their normal sites, and the Cd toxic effect is exercised by binding to other sites. The small plant peptides called phytochelatins (PC) which have a protective role in plant cells help Cd to be sequestered via cysteinyl residues. Similarly, the sequestration of Cd in fungal structures could be responsible for the retention of Cd in the roots. The molecular responses of AMF plants to Cd were a research subject for Aloui et al. (2009) who worked at the proteomic level and found downregulation of several proteins upon Cd stress. Cd can be also adsorbed on the spores. After saturation of fungi with Cd, increased translocation of this metal to shoots occurred (Giasson et al. 2008).

It is known that **Pb** has low mobility (even less than P and Zn) and is strongly complexed with the organic matter in soil, forming organic complexes which are unavailable for plants. Similarly, plants possess mechanisms to precipitate Pb in the rhizosphere, in highly insoluble forms such as $PbSO_4$. However, Pb that has entered the roots can be sequestrated in fungal vesicles, which can provide an additional detoxification mechanism. However, sequestration of Pb in roots was not found to be correlated with an increase in the number of vesicles (Göhre and Paszkowski 2006). Similar results were found by Chen et al. (2005c) where, after growth of plants in the presence of AMF in an artificially polluted soil, Pb had a positive effect on vesicular abundance which was not correlated with a lower concentration of Pb as it was expected. At the same time, enhanced Pb concentrations both in roots and shoots and a higher root/shoot ratio were found for highly mycorrhizal plant species (*K. striata, I. denticulate*, and *E. crus-galli* var. mitis). Marschner (1995) reported lower Pb concentration in the cortical cell walls of Norway spruce (*Picea abies* L. var. Karst), while Chen et al. (2005c) reported the contrary, an elevated Pb concentration, and suggested that under this higher concentration of Pb, mycorrhiza could promote plant growth by increasing P uptake and mitigate Pb toxicity by sequestrating more Pb in roots. Also, a higher amount of Pb absorbed and accumulated also in *Eucalyptus* mycorrhizal variants (Bafeel 2008). The shoot biomass was higher, while root biomass was significantly lower than those of non-mycorrhizal trees. Other experimental variants included the growth of *Eucalyptus* together with legumes (*F. vulgaris*) and AMF. This association with heavy-metal-resistant legume varieties and AMF further increased the positive effects on shoot biomass and helped to improve the resistance of *Eucalyptus* to HM. This effect was correlated to an increase of Pb absorption and accumulation. An explanation was given with the statement that AMF lead to increased plant biomass, and thus a larger potential absorption area for Pb, of accessible soil, and a higher efficiency in hyphal translocation. Similar studies were described in Chen et al. (2009) and Regvar et al. (2010) who found, however, that the indirect ordination analysis of restriction fragment length polymorphisms data showed no statistically significant correlations with environmental gradients.

Estaún et al. (2010) investigated the influence of the AMF *Glomus intraradices* (BEG 72) on the transfer of **Cr** in *Plantago lanceolata* grown in a high Cr soil. Although *Plantago lanceolata* is known to be tolerant toward many environmental stress factors while showing a high degree of mycorrhization, under Cr pollution, the plants' survival decreased quickly. When Cr was supplied as $CrCl_3$, this mycorrhizal model plant accumulated Cr, both in the shoots and roots. Therefore, AMF seem to act as a barrier, decreasing Cr uptake, in comparison to non-mycorrhizal plants. The effect consisting in a decrease of HM stress is due to selective immobilization in the root system colonized by the fungus or to a very high Cr sorption capacity of the extraradical mycelium of the AMF. This mechanism may explain the decrease of HM concentrations in root cells and through a different allocation of the metal within the root cells may have an effect on fungal colonization. Similar research was performed also on *Helianthus annuus* in the presence of Cr(III) and Cr(VI) with similar results: the uptake of Cr decreased both in shoots and roots upon mycorrhization (Davies et al. 2001). On the other part, the non-hyperaccumulator plants tend to avoid Cr uptake and translocation, and in this case, the higher concentrations are distributed in roots, rather than leave or stem (Estaún et al. 2010 after Glosh and Sigh 2005). Similar investigations were performed by Carrasco et al. (2009) who remarked that the soluble content of Cr in soil was negatively correlated with the glomalin concentration. Testing the synergistic effect of mycorrhizal and saprobic fungi (*Trichoderma pseudokoningii*) on HM uptake into *Cynodon dactylon*, it was found that the amount of Cr uptake was highest in dual inocula. Nevertheless, the amount of Cr uptake in single saprobic fungi was more significant as compared to the mycorrhizal treatment alone (Bareen and Nazir 2010).

Studies related by Rufyikiri et al. (2004) demonstrated that the mobility of U in soil depends on the organic compound content, the bioavailability being highly dependent on soil pH. The same author found that the most mobile U forms are U (VI) salts, predominantly as $UO2^{2+}$ and carbonate complexes, while other forms are less bioavailable and remain bound to soil particles. The role of AM fungi in translocating U as uranyl cations to roots through fungal tissues is related to fungal mycelium HM-binding capacity (Chen et al. 2005a). Chen et al. (2005b, cited by Babula et al. 2008) performed and confirmed such studies using *Medicago truncatula* as a model plant, inoculated with *Glomus intraradices*. They found higher concentrations of U in roots than in shoots of mycorrhizal plant, suggesting that the AM fungus has a potential to reduce the translocation of U from roots to shoots.

Some research has been carried out on Cs, with, e.g., Leyval et al. (2002) reporting that [134]Cs radioactivity increased twofold in leaf tissue of *Paspalum notatum* in symbiosis with AMF while in the case of mycorrhizal *Melilotus officinalis*, 1.7–2 times increased ^{137}Cs was found. *Sorghum sudanense* revealed only insignificantly increase. A significant decrease of ^{137}Cs in mycorrhizal *Festuca ovina* and *Agrostis tenuis* was found; this finding underlines that soil fungi represent a potential for Cs immobilization. On the other hand, Rosén et al. (2005) working with mycorrhizal ryegrass and leek found an enhanced ^{137}Cs uptake by leek, but no effect on the uptake by ryegrass. Similar studies were performed on mycorrhizal *Festuca ovina* in which shoots showed higher ^{137}Cs concentration than roots, as well as on *Trifolium repens*, and AM plants took up less Cs with no increase in translocation of ^{137}Cs to the shoots being found. In conclusion, AMF seem to play a role, with regard to both immobilization and phytoextraction being represented depending on plant species. Specifically grasses seem to respond with decreased uptake into shoot biomass.

Turnau and Mesjasz-Przybylowicz (2003) have found an increased **Ni** content in shoots of *Berkheya coddii*, a hyperaccumulator from the Asteraceae family, with well-developed mycorrhization. There is only very scare literature relating details on other elements, excepting the nutrients P, Ca, or Mn.

13.2.3   The Effects of Poly-metal Pollution on HM Uptake in Plants Inoculated with AMF and Combination to Other Microbial Communities

When research focusses on a single pollutant, the response of AMF on HM uptake into plants is much more clear as compared to soil containing more than two pollutants (poly-metal pollution). In such cases, it is very useful to include mechanisms of HM toxicity like the producing of ROS, which can damage biomolecules such as membrane lipids, proteins, chloroplast pigments, enzymes, and nucleic acids. Normally, plants have mechanisms for protecting against oxidative stress. As an effect of AMF inoculation, decreased oxidative stress could be shown either by assaying enzymatic activity (of, e.g., SOD, superoxide dismutase; POD, peroxidases; CAT, catalase) or nonenzymatic systems acting as free radical scavengers (e.g., soluble protein or pigments such as chlorophyll and carotenoids). Zhang et al. (2006) found that inoculation with *Glomus mosseae* decreased POD activity and DNA damage in AM *Vicia faba* plants. The high level of such antioxidative systems as well as nonenzymatic antioxidants in mycorrhizal plants could be an effect of arbuscule senescence (Fester and Hause 2005).

It is well documented that AMF coexist with other soil microorganisms which might have positive or negative effects on HM uptake (Giasson et al. 2008). When the inoculation with AMF is combined with saprobic fungi, there is a synergistic effect. Bareen and Nazir (2010) used a dual inoculation with mycorrhizal fungi and a saprobic fungus (*Trichoderma pseudokoningii*) applied in a poly-metal soil using the host plant *Cynodon dactylon*. The results showed an increased plant biomass and phytoextraction ability for metals like Cd, Cr, Cu, as well as Na, with a greater volume of HM being stored in the mycorrhizal structures and spores. These studies were performed using autochthonous fungi, because it is known that the results are much better than in the case of using allochthonous fungi. Similar results using saprobic fungi and *Glomus deserticola* in symbiosis with tomato and alfalfa were related by Sampedro et al. (2008).

Wang et al. (2005) used only one fungal strain (*Glomus caledonium* 90036) and compared that to a consortium of AMF *Gigaspora margarita* ZJ37, *Gigaspora decipens* ZJ38, *Scutellospora gilmori* ZJ39, *Acaulospora* spp. and *Glomus* spp. The results indicated that the inoculation with the consortium led to a significant increase of shoot biomass and shoot P, Cu, Zn, and Pb concentrations, but did not alter shoot Cd concentrations, resulting in higher Cu, Zn, Pb, and Cd extracted by *E. splendens*. In poly-metal pollution, it has been documented that Cu mainly accumulated in the spore vacuoles, while Cd was contained in vacuoles of the mycelium (Turnau et al. 2010). Barea et al. (2005) described the tripartite symbiosis among legume–mycorrhiza–rhizobia underlining that they established positive interactions reflected in the increase of Zn, Cu, Mo, Ca, and P uptake in plants, making it possible to use the combination in revegetation programs to restore HM-polluted sites.

13.2.4 Influence of Succession on AMF Communities and Bioaccumulation of HM in Plants

It is known that at HM-polluted sites, either individual plants or the entire communities and populations are affected. It is assumed that the response includes changes in species composition and diversity in terms of *succession* and higher stress tolerance in terms of natural *selection*, both of them being processes at the long time scale (Turnau et al. 2010). According to these authors, at polluted sites first appeared the non-mycorrhizal plant species, which are not able to establish a vegetative cover, while second stages are dominated by facultative mycorrhizal species. A very pure vegetation cover could be developed only after several decades with spontaneous succession. The succession is thought to influence parameters like relative mycorrhizal colonization, relative arbuscule formation, or arbuscule richness, which could affect HM bioaccumulation in plants allowing a differentiation between restored and non-restored sites. Rowe et al. (2007) suggested using native inoculum rather than commercial ones since they seem to be more effective in establishing the late-successional associations. However, all discussions on succession should be considered with care, because of the methodological limitations underlined in the end of Sect. 13.2.

13.2.5 Influence of AMF on Leaching in the Soil Column

There are few articles looking for the effects of plants on the leaching of elements with the seepage water (Dudley et al. 2008), and even fewer about the effect of AMF on leaching in mesocosms extracted from the field (Heijden 2010). Several studies on the effect of AMF on metal leaching were performed in the context of investigations dealing with contaminated sites (Banks et al. 1994; Iordache et al. 2006; Neagoe et al. 2009).

The mechanisms underlying the effects of plants on leaching are the creation of preferential flow paths (by roots) and the direct and indirect influence of soil pore size distribution (Halabuck 2006). Simulation of metal transport in the soil profile as a result of plant influences is a matter of current research (Dusek et al. 2010), but in principle, AMF might exert an influence which can be modeled by their effects on preferential flow paths and pore size distribution. Such effects have not been studied explicitly to our knowledge. A synthesis of the knowledge on the influence of organic matter on soil profile leaching, which can be eventually related to the role of AMF on the distribution of organic matter in soil, can be found in Neagoe et al. (2012).

The main problem of the existing studies in terms of upscaling is that the scale of these experiments is not explicitly related to the size of AMF populations in natural ecosystems. Thus, it cannot be used to extrapolate the function of control variables in the field distribution. In the existing studies, the development of AMF was restricted in space by the experimental design, and the results cannot meaningfully be used to predict the effects in the field under variable soil and vegetation conditions. A need for an integrated study of the role of AMF in metal mobility in situ by both leaching to lower soil horizons and bioaccumulation in plants is in need.

13.3 Methodology for Investigating the Biogeochemical Role of AMF

Upscaling the role of AMF depends, in general terms, on the development of community–environment relationships (where the environment is "stratified by multiple gradients"; Lilleskov and Parrent 2007). In particular, exactly how this stratification should be set up (which is the scale of the elementary unit of extrapolation and implicitly the sampling scale for characterizing the gradients) is to be described. Graham (2008) states that "experimental design should either integrate multiple mechanisms of the landscape scale and include such measures as mycorrhizal influences on net primary production, evapotranspiration and nutrient cycling, or integrate measures of [. . .] fungal diversity into assessment of ecosystem function." We believe that a functional dynamic approach (already applied to ectomycorrhizal fungi—Iordache et al. 2011) would be useful also for AMF.

Pool-flux classical ecosystem type research can be associated with this method by the construction of a minimally complex homomorphic model, and it is also compatible with the stoichiometric approach proposed by Johnson (2010), who considers that a "stoichiometric perspective of C, N and P fluxes through mycorrhizas may provide a "common currency" to facilitate cross-scale communication among a diversity of scientists interested in understanding AM symbioses from genes to ecosystems." Instead of C, N, and P fluxes studied separately, and metals studied separately (Audet and Charest 2007a), the integrated research of macronutrients, micronutrients, and toxic metals influence on soil–plant–AMF system would be of greater relevance.

An ideal methodology for investigating the biogeochemical roles of AMF is based on a conceptual model of productive systems involving both biotic and abiotic objects that would involve the following steps:

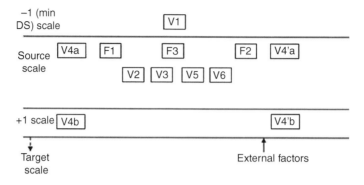

Fig. 13.1 Homomorphic model of a community of arbuscular mycorrhizal fungi with the entities with value for them—biotic and abiotic objects influencing them (relationships not represented for reason of visibility; a connectivity matrix can be easily constructed using the information presented in the text). The scale refers to scale in space, not in time. The physical part is not represented. The part of the model within the source scale (within a stratum) is the homomorphic model for upscaling under the constraints from organisms of different scales and external factors external to the AM functional dynamic modules. F1, fungal parts on plant roots 1; F2, fungal parts at plant roots 2; F3, hyphae in the extraradical mycelium; V1, bacteria; V2, mineral P and N; V3, organic P and N; V4, plants (a, belowground; c, aboveground parts); V5, micronutrients and toxic substances; V6, soil fungi or micro-invertebrates

1. Identification of the system of productive objects at the source scale. Description of the space-time scales of the biotic and abiotic objects involved in the productivity of AMF and of the main processes coupling these objects. The result of this step is a homomorphic model of minimal multi-scale complexity needed to understand the processes occurring at the time and space scale of AMF.
2. Characterizing the functioning of the system identified in step 1 in terms of relationships between the productivity of AMF–plant systems and macronutrients, micronutrients, and toxic elements and substances. The result of this step is a model of the circulation of elements and substances due to biological processes.
3. Characterizing the transport of elements and substances by larger scale abiotic objects (e.g., hydro-systems) coupled to the productive objects investigated in step 2. The result of this step is a model of the effects at distance in space-time (at the target scale) of the relatively smaller scale processes occurring at the source scale. For more details on effects at large distance, see Iordache et al. (2012).

Within step 1 by applying the methodology described in detail in Iordache et al. (2011), one would obtain the structural model presented in Fig. 13.1.

Step 2 involves performing the following activities (adapted from Iordache et al. 2011): (1) describing the biomass production functions of the plant–AMF systems at source scale; (2) characterizing the mathematical functions describing the influence of larger scale biological objects on the productivity of the original objects (e.g., plant consumers); (3) characterizing the mathematical relationships underlying the influence of large-scale abiotic parameters on productive system at the original scale, the space distribution of the abiotic parameters in the system of target scale with a resolution of original scale, and the mathematical functions predicting their space-time dynamic and the abiotic processes involved at the target scales); (4) estimating the large-scale abiotic factor's influence on large-scale DSs connected to the source scale (the same as at point 1 done for the large-scale biological objects mentioned at point 2); (5) stratifying the system of target scale system into strata having the scale of the system identified as in Fig. 13.1; (6) characterizing the production function by strata as controlled by large-scale biotic and abiotic factors; and (7) model upscaling (extrapolation of production function to the target area of interest by types of strata).

The problems with putting this ideal methodology into operation start from the very beginning, with the system identification. As seen from the previous chapter, we do not know the exact size in space of the AMF physiological organisms, which means that we will not be able to quantify the surface occupied by the extraradical parts of the individuals (F3 compartment in Fig. 13.1). Based on the current information in literature, F3 compartment in Fig. 13.1 can be limited in space to several meters around the plants or based on experimental work (Mikkelsen et al. 2008), or field observations like those summarized by Jakobsen 2004) could link plants of the same or different species as far as 10 m apart, although usually the distance would be smaller. It seems reasonable at the current state of knowledge to assume that the source for upscaling has to be a plot of a maximum of 100 m^2, preferably smaller, down to 10 m^2. As we go to smaller discretization units, the quantity of data needed will increase accordingly. Another problem is related to the estimation of species diversity of AMF in the extraradical (hyphal) compartment (Taylor et al. 2000; Rosendahl 2008).

As for the practical delineation in space of the types of strata to be studied separately for the final purpose of upscaling the productivity to target scale, they can be obtained by a mapping of the main entities with values for AMF, at the needed resolution. The strata will result from an intersections of the (10–100 m^2 scale) maps for C, N, P, micronutrients and toxic substances, pH and other soil parameters, as well as the map of plant associations, if plant cover is present (in the case of soils impacted by pollution it may lack). The intersection of layers leads to types of areas to be studied for effects of AM on mobility of metals, with an associated homomorphic model for each one, or to potential effects in case of myco-/phytoremediation.

According to our knowledge, there is no article yet dealing with such a locally integrated approach, useful for farther extrapolation at ecosystem scale of the processes occurring in contaminated sites still covered by plants. Characterizing the AMF source scale system structurally, systems of two smaller and larger scales (Fig. 13.1), and mapping the target scale system in discretization units relevant for system identification is simpler than an institutional scale (ecosystem, site) integrating biogeochemistry for a truly holistic approach, but still sufficiently complex. It is an important research direction in objective scale integrated biogeochemistry and provides a framework for testing research hypotheses resulting from meta-analyses of existing results on nutrient-dependent plant response to AMF (Hoeksema et al. 2010) and from field studies of the relative role of AMF and ectomycorrhiza in highly dynamic systems (Piotrowski et al. 2008).

Step 3 is not a matter of ecological research, but rather of modeling the larger scale abiotic processes of transport to distance (e.g., by vertical flow to groundwater or by runoff) with the right resolution using the results from step 1, by type of strata as input variables in the discretization units needed for modeling the abiotic process.

What is possible at present based on existing knowledge about AMF biology and ecology? A research opportunity is provided by the challenge of myco-/phytoremediation of contaminated areas using selected plants and commercial inoculums. In this case, the problem of characterizing the species diversity and scale of AMF is eliminated (although the functional diversity within the same species will still be potentially existing—Munkvold et al. 2004), at least in the first part of the experiments. Using this opportunity, we devised an operational, three-scale experimental approach as described in Table 13.2. A research program started in 2003 allowed us to investigate patterns and processes mentioned in this table, including the relative role of AMF in the export of metals by plants and leaching from soil columns. A preliminary synthesis of the results of this approach was performed by Kothe et al. (2005). We summarize here only the main findings of the program (methods and extended results can be found in the sources mentioned in Table 13.3): (1) the plant response to inoculation is species specific, but there is a general pattern of positive response in terms of biomass increase and decrease of oxidative stress (illustrated in Fig. 13.1); (2) the plant-specific response depends on environmental conditions and plant community structure; plants classified as nonhost may benefit from inoculation in highly contaminated soils; (3) large-scale (field plot) response to artificial inoculation and monospecific cultivation is modulated by seed bank and natural inoculation from clean soil amendment; and (4) the inoculation may increase the relative importance of plants compared to leachate in the export of metals, but this depends on the rhizosphere microbial community structure (bacteria co-inoculated with AMF influenced the dynamics of soil redox potential and metal toxicity) and on soil variable heterogeneities at meter scale.

13.4 Conclusions

We introduced a concept of the biogeochemical role explicitly related to cross-scale effects in the frame of an approach labeled as *objective scale integrated biogeo-chemistry*. By objective, we mean that the scale of analyses is that of the environmental objects involved in the cycling and is not assumed a priori based on human (institutional) interests. By integrated, we refer to the consideration of multielement

Table 13.3 Experimental approach at three scales for the study of AMF roles in metal mobility (adapted from Iordache et al. 2012)

Name of the system and usual scales	Environmental complex system studied at these scales	Patterns studied/control variables	Reference for detailed methodology
Pot 10^{-2} m^2	Soil + plants + AMF	Exploration by root, bioaccumulation/ organic carbon, other microorganisms, level and spatial structure of amendments	Neagoe et al. (2004), Iordache et al. (2004), and Stancu et al. (2010)
Lysimeter 10^{-1}–10^0 m^2	Soil + plants + AMF + small-scale hydro-system	Same as in pots + leaching, internal redistribution, net outputs/same as in pots + soil structure, hydraulic conductivity, humidity, redox potential in soil profile	Neagoe et al. (2006, 2009), Iordache et al. (2006), Visan et al. (2008), and Nicoara et al. (2010)
Plot 4×10^0–10^2 m^2	Soil + plants + AMF + other organisms	Same as in pots + heterogeneity in space, margin effects, other effects due to external entities (natural seed bank, consumers)/ same as in pots + variables for external entities	Neagoe et al. (2005, 2010)

fluxes (of nutrients and toxic elements and substances) through the target organisms and through the biotic and abiotic compartments directly connected with them, as control factors of the productivity of the system. After a critical analysis of the knowledge on AMF relevant for estimating their biogeochemical role in the mobility of metals, we proposed a methodology for characterizing this role and underlined the practical limitations linked to current state of knowledge concerning AMF. These limitations are linked to basic issues such as estimating the dimension of a physiological individual, the size in space of a population of AMF, and the species diversity of AMF in the hyphal compartment. We underlined then a research opportunity provided by the challenge of myco/phytoremediation of contaminated areas using selected plants and commercial inoculums. Operationally, we developed a multi-scale (pot, lysimeter, field plot) experimental approach of the biogeochemical role of AMF in metal mobilities tackling scale-specific issues. The main findings of the operational research program are summarized.

Acknowledgments We warmly thank Prof. Dr. Younes Rezaee Danesh who invited us to write the present chapter for this book and the reviewers for constructive criticism. The theoretical research presented here was done with financing from National University Research Council (CNCSIS) by project 291/2007 MECOTER, from National Center for the Management of Projects (CNMP) by projects 52175/2008 METAGRO, and in the international consortium of the FP7 project UMBRELLA, grant agreement 226870.

References

Adriano DC (2001) Trace elements in the terrestrial environments. Biogeochemistry, bioavailability, and risks of heavy metals, 2nd edn. Springer, New York

Ahmed FRS, Killham K, Alexander I (2006) Influences of arbuscular mycorrhizal fungus *Glomus mosseae* on growth and nutrition of lentil irrigated with arsenic contaminated water. Plant Soil 258:33–41

Alexander M (1994) Biodegradation and bioremediation. Academic, San Diego, CA

Allen MF (2007) Mycorrhizal fungi: highways for water and nutrients in arid soils. Vadose Zone J 6:291–297

Allen MF (2009) Bidirectional water flows through the soil–fungal–plant mycorrhizal continuum. New Phytol 182:290–293

Allen MF (2010) Dynamics of arbuscular mycorrhizae through drought cycles: restoring functional arid land ecosystems. In: COST870 meeting in Jyvaskyla, 13–15 December, Book of abstracts, p 13

Allen MF, Swenson W, Querejeta JI, Egerton-Warburton LM, Treseder KK (2003) Ecology of mycorrhizae: a conceptual framework for complex interactions among plants and fungi. Annu Rev Phytopathol 41:271–303

Aloui A, Recorbet G, Gollotte A, Robert F, Valot B, Gianinazzi-Pearson V, Aschi-Smiti S, Dumas-Gaudot E (2009) On the mechanisms of cadmium stress alleviation in *Medicago truncatula* by arbuscular mycorrhizal symbiosis: a root proteomic study. Proteomics 9:420–433

Audet P, Charest C (2007a) Dynamics of arbuscular mycorrhizal symbiosis in heavy metal phytoremediation: meta-analytical and conceptual perspectives. Environ Pollut 147:609–614. doi:10.1016/j.envpol.2006.10.006

Audet P, Charest C (2007b) Heavy metal phytoremediation from a meta-analytical perspective. Environ Pollut 147:231–237. doi:10.1016/j.envpol.2006.08.011

Audet P, Charest C (2008) Allocation plasticity and plant-metal partitioning: meta-analytical perspectives in phytoremediation. Environ Pollut 156:290–296. doi:10.1016/j.envpol.2008.02.010

Auge RM (2001) Water relations, drought and vesicular-arbuscular mycorrhizal symbiosis. Mycorrhiza 11:3–42

Babula P, Adam V, Apatrilova R, Zehnalek J, Havel L, Kizek R (2008) Uncommon heavy metals, metalloids and their plant toxicity: a review. Environ Chem Lett 6:189–213

Bafeel SO (2008) Contribution of mycorrhizae in phytoremediation of lead contaminated soils by *Eucalyptus rostrata* plants. World Appl Sci J 5:490–498

Banks MK, Schwab AP, Fleming GR, Hetrick BA (1994) Effects of plants and soil microflora on leaching of zinc from mine tailings. Chemosphere 29:1691–1699

Barea JM, Werner D, Azcón-Guilar C, Azcón R (2005) Interactions of arbuscular mycorrhiza and nitrogen-fixing symbiosis in sustainable agriculture. In: Werner D and Newton WE (eds.), Nitrogen Fixation in Agriculture, Forestry Ecology and the Environment, Volume 4, Springer. pp. 199–222

Bareen F, Nazir A (2010) Metal decontamination of tannery solid waste using *Tagetes patula* in association with saprobic and mycorrhizal fungi. Environmentalist 30:45–53

Barua A, Gupta SD, Mridha MAU, Bhuiyan MK (2010) Effect of arbuscular mycorrhizal fungi on growth of *Gmelina arborea* in arsenic-contaminated soil. J Forest Res 21:423–432

Berta G, Fusconi A, Hooker JE (2002) Arbuscular mycorrhizal modifications to plant root systems: scale, mechanisms and consequences. In: Gianinazzi S, Schuepp H, Barea JM, Haselwandter K (eds) Mycorrhiza technology in agriculture: from genes to bioproducts. Birkhauser, Basel, pp 71–85

Biró I, Takács T (2007) Effects of *Glomus mosseae* strains of different origin on plant macro- and micronutrients uptake in Cd-polluted and unpolluted soils. Acta Agronomica Hungarica 55(2):183–192. doi:10.1556/AAgr.55.2007

Boer W, Folman LB, Summerbell RC, Boddy L (2005) Living in a fungal world: impact of fungi on soil bacterial niche development. FEMS Microbiol Rev 29:795–811

Bonfante P, Anca IA (2009) Plants, mycorrhizal fungi, and bacteria: a network of interactions. Annu Rev Microbiol 63:363–383

Brar SK, Verma M, Surampalli RY, Misra K, Tyagi RD, Meunier N, Blais JF (2006) Bioremediation of hazardous wastes – a review. Pract Period Hazard Toxic Radioact Waste Manag 10:59–73

Cairney JWG (2000) Evolution of mycorrhiza systems. Naturwissenschaften 87:467–475

Carbonell AA, Aarabi MA, DeLaune RD, Gambrell RP, Patrick WH Jr (1998) Arsenic in wetland vegetation: availability, phytotoxicity, uptake and effects on plant growth and nutrition. Sci Total Environ 217:189–199

Carrasco L, Gattinger A, Fließbach A, Roldán A, Schloter M, Caravaca F (2009) Estimation by PLFA of microbial community structure associated with the rhizosphere of *Lygeum spartum* and *Piptatherum miliaceum* growing in semiarid mine Tailings. Microb Ecol 60:265–271

Cavagnaro TR (2008) The role of arbuscular mycorrhizas in improving plant zinc nutrition under low soil zinc concentrations. Rev Plant Soil 304:315–325

Chaudhary VB, Lau MK, Johnson NC (2008) Macroecology of microbes – biogeography of the glomeromycota. In: Varma A (ed) Mycorrhiza: genetics and molecular biology, eco-function, biotechnology, eco-physiology, structure and systematics, 3rd edn. Springer, Heidelberg, pp 529–563

Chen B, Shen H, Li X, Feng G, Christie P (2004) Effects of EDTA application and arbuscular mycorrhizal colonization on growth and zinc uptake by maize (*Zea mays* L.) in soil experimentally contaminated with zinc. Plant Soil 261:219–229

Chen B, Jakobsen I, Roos P, Borggaard OK, Zhu YG (2005a) Mycorrhiza and root hairs enhance acquisition of phosphorus and uranium from phosphate rock but mycorrhiza decreases root to shoot uranium transfer. New Phytol 165:591–598

Chen B, Thang X, Zhu Y, Christie P (2005b) Metal concentrations and mycorrhizal status of plants colonizing copper mine tailings: potential for revegetation. Sci China C Life Sci 48 (Suppl I):156–164

Chen X, Wu CH, Tang JJ, Hu SJ (2005c) Arbuscular mycorrhizae enhance metal uptake and growth of host plants under a sand culture experiment. Chemosphere 60:665–671

Chen S, Sun L, Chao L, Zhou Q, Sun T (2009) Estimation of lead bioavailability in smelter contaminated soils by single and sequential extraction procedure. Bull Environ contam Toxicol 82:43–47

Chern ECW, Tsai AI, Gunseitan OA (2007) Deposition of glomalin related soil protein and sequestered toxic metals into watersheds. Environ Sci Technol 41:3566–3572

Christie P, Li X, Chen B (2004) Arbuscular mycorrhiza can depress translocation of zinc to shoots of host plants in soils moderately polluted with zinc. Plant Soil 261:209–217, Kluwer Academic Publishers. Printed in the Netherlands

Davies FT Jr, Puryear JD, Newton RJ, Egilla JN, Grossi JAS (2001) Mycorrhizal fungi enhance accumulation and tolerance of chromium in sunflower (Helianthus annuus). Plant Physiol 158:777–786

de Andrade SAL, da Silveira APD (2008) Mycorrhiza influence on maize development under Cd stress and P supply. Braz J Plant Physiol 20(1):39–50

Dudley LM, Ben-Gal A, Shani U (2008) Influence of plant, soil, and water on the leaching fraction. Vadose Zone J 7:420–425

Dusek J, Vogel T, Lubomir L, Cipakova A (2010) Short-term transport for cadmium during a heavy-rain event simulated by a dual-continuum approach. J Plant Nutr Soil Sci 173:536–547

Estaún V, Cortés A, Velianos K, Camprubí A, Calvet C (2010) Effect of chromium contaminated soil on arbuscular mycorrhizal colonisation of roots and metal uptake by *Plantago lanceolata*. Span J Agric Res 8(S1):S109–S115, ISSN: 1695-971-X

Fester T, Hause G (2005) Accumulation of reactive oxygen species in arbuscular mycorrhizal roots. Mycorrhiza 15:373–379

Gadd GM (1993) Interactions of fungi with toxic metals. New Phytol 124:25–60. doi:10.1111/j.1469-8137.1993.tb03796.x

Gadd MG (2005) Microorganisms in toxic metal-polluted soils. In: Buscot F, Varma A (eds) Microorganisms in soils: roles in genesis and functions, vol 3, Soil biology. Springer, Berlin, pp 325–356

Gadd GM (2007) Geomycology: biogeochemical transformations of rocks, minerals, metals and radionuclides by fungi, bioweathering and bioremediation. Mycol Res 3:3–49

Gadd GM (2010) Metals, minerals and microbes: geomicrobiology and bioremediation. Microbiology 156:609–643

Gamalero E, Lingua G, Berta G, Glick BR (2009) Beneficial role of plant growth promoting bacteria and arbuscular mycorrhizal fungi on plant responses to heavy metal stress. Can J Microbiol 55:501–514

Gange AC, Brown VK, Sinclair GS (1993) Vesicular-arbuscular mycorrhizal fungi: a determinant of plant community structure in early succession. Funct Ecol 7:616–622

Garg N, Chandel S (2011) Arbuscular mycorrhizal networks: process and functions. In: Lichtfouse E et al (eds) Sustainable agriculture, Vol 2, Springer, Dordrecht, pp 907–930

Gehring C, Bennett A (2009) Mycorrhizal fungal-plant-insect interactions: the importance of a community approach. Environ Entomol 38:93–102

Giasson P, Karam A, Jaouich A (2008) Arbuscular mycorrhizae and alleviation of soil stresses. In: Siddiqui ZA, Akhtar MS, Futai K (eds) Mycorrhizae: sustainable agriculture and forestry. Springer, Dordrecht, pp 99–134, ISBN: 978-1-4020-8769-1, e-ISBN: 978-1-4020-8770-7

Glass NL, Rasmussen C, Roca G, Read ND (2004) Hyphal homing, fusion and mycelia interconnectedness. Trends Microbiol 12:135–141

Glosh M, Sigh SP (2005) Comparative uptake and phytoextraction study of soil induced chromium by accumulator and high biomass weed species. Appl Ecol Environ Res 3:67–79

Göhre V, Paszkowski U (2006) Contribution of the arbuscular mycorrhizal symbiosis to heavy metal phytoremediation. Planta 223:1115–1122

Goltapeh EM, Danesh YR, Prasad R, Varma A (2008) Mycorrhizal fungi: what we know and what should we know? In: Varma A (ed) Mycorrhiza. Springer, Berlin, pp 3–28

Gonzalez-Chavez MC, Carrillo-Gonzalez R, Wright SF, Nichols K (2004) The role of glomalin, a protein produced by arbuscular mycorrhizal fungi, in sequestering potentially toxic elements. Environ Pollut 130:317–323. doi:10.1016/j.envpol.2004.01.004

Graham JH (2008) Scaling up evaluation of field functioning of arbuscular mycorrhizal fungi. New Phytol 180:1–2

Halabuck A (2006) Influence of different vegetation types on saturated hydraulic conductivity in alluvial topsoils. Biologia 61(S19):S266–S269

Hall JL (2002) Cellular mechanisms for heavy metal detoxification. J Exp Bot 53:1–11

Hart MM, Reader RJ, Klironomos JN (2001) Life-history strategies of arbuscular mycorrhizal fungi in relation to their successional dynamics. Mycologia 93:1186–1194

Hartmann A, Schmid M, Tuinen D, Berg G (2009) Plant-driven selection of microbes. Plant Soil 321:235–257

Heijden van der MG (2010) Mycorrhizal fungi reduce nutrient loss from model grassland ecosystems. Ecology 91:1163–1171

Helgason T, Fitter AH (2009) Natural selection and the evolutionary ecology of the arbuscular mycorrhizal fungi (Phylum Glomeromycota). J Exp Bot 60:2465–2480

Hildebrandt U, Regvar M, Bothe H (2007) Arbuscular mycorrhiza and heavy metal tolerance. Phytochemistry 68:139–146

Hodge A, Berta G, Doussan C, Merchan F, Crespi M (2009) Plant root growth, architecture and function. Plant Soil 321:153–187

Hoeksema JD, Chaudhary VB, Gehring CA, Johnson NC, Karst J, Koide RT, Pringle A, Zabinski C, Bever JD, Moore JC, Wilson GWT, Klironomos JN, Umbanhowar J (2010) A meta-analysis of context-dependency in plant response to inoculation with mycorrhizal fungi. Ecol Lett 13:394–407

Hoffland E, Kuyper TW, Wallander H, Plassard C, Gorbushina AA, Haselwandter K, Holmström S, Landeweert R, Lundström US, Rosling A, Sen R, Smits MM, van Hees PA, van Breemen N (2004) The role of fungi in weathering. A review. Front Ecol Environ 2:258–264

Hullebusch van ED, Lens PNL, Tabak HH (2005) Developments in bioremediation of soils and sediments polluted with metals and radionuclides. 3. Influence of chemical speciation and bioavailability on contaminants immobilization/mobilization bio-processes. Rev Environ Sci Biotechnol 4:185–212

Iordache V, Neagoe A, Bergman H (2004) Effects of mycorrhization of *Phacelia tanacetifolia* on metals accumulation and oxidative stress. Proceedings of the 5th international symposium on "metal elements in environment, medicine and biology", Timisoara, pp 105–113

Iordache V, Neagoe A, Bergman H, Kothe E, Buechel G (2006) Factors influencing the export of metals by leaching in bioremediation experiments. 23. Arbeitstagung in Jena, Agricultural, biological, environmental, nutritional and medical importance of macro, trace and ultra trace elements, Friedrich Schiller Universität, pp 288–295

Iordache V, Kothe E, Neagoe A, Gherghel F (2011) A conceptual framework for up-scaling ecological processes and application to ectomycorrhizal fungi. In: Rai M, Varma A (eds) Diversity and biotechnology of ectomycorrhizae. Springer, Berlin, pp 255–299

Iordache V, Lăcătusu R, Scrădeanu D, Onete M, Jianu D, Bodescu F, Neagoe A, Purice D, Cobzaru I (2012) Contributions to the theoretical foundations of integrated modeling in biogeochemistry and their application in contaminated areas. In: Kothe E, Varma A (eds) Bio-geo interactions in metal-contaminated soils, vol 31, Soil biology. Springer, Berlin, pp 385–416

Jakobsen I (2004) Hyphal fusion to plant species connections – giant mycelia and community nutrient flow. New Phytol 164:4–7

Janoušková M, Pavlíková D (2010) Cadmium immobilization in the rhizosphere of arbuscular mycorrhizal plants by the fungal extraradical mycelium. Plant Soil 332:511–520. doi:10.1007/s11104-010-0317-2

Janoušková M, Pavlíková D, Vosátka M (2006) Potential contribution of arbuscular mycorrhiza to cadmium immobilisation in soil. Chemosphere 65:1959–1965

Jansa J, Finlay R, Wallander H, Smith FA, Smith SE (2011) Role of mycorrhizal symbioses in phosphorus cycling. In: Bunemann EK et al (eds) Phosphorus in action, vol 26, Soil biology. Springer, Berlin, pp 137–168

Johnson NC (2010) Resource stoichiometry elucidates the structure and function of arbuscular mycorrhizas across scales. New Phytol 185:631–647

Johnson CN, Hoeksema JD, Bever JD, Chaudhary VB, Gehring C, Klironomos J, Koide R, Miller RM, Moore J, Moutoglis P, Schwartz M, Simard S, Swenson W, Umbanhowar J, Wilson G, Zabinski C (2006) From lilliput to brobdingnag: extending models of mycorrhizal function across scales. Bioscience 56:889–900

Joner EJ, Leyval C (2009) Phytoremediation of organic pollutants using mycorrhizal plants: a new aspect of rhyzosphere interactions. In: Lichtfouse E et al (eds) Sustainable agriculture, vol Part 7. Springer, New York, pp 885–894

Kapoor A, Viraraghavan T (1995) Fungal biosorption – an alternative treatment option for heavy metal bearing wastewater. A review. Bioresour Technol 53:195–206

Khan AG (2005) Role of soil microbes in the rhizospheres of plants growing on trace metal contaminated soils in phytoremediation. J Trace Elem Med Biol 18:355–364

Khan AG (2006) Mycorrhizoremediation – an enhanced from of phytoremediation. Review. J Zhejiang Univ Sci B 7:503–514

Khan MS, Zaidi A, Wani PA, Oves M (2009) Role of plant growth promoting rhizobacteria in the remediation of metal contaminated soils. Environ Chem Lett 7:1–19

Koricheva J, Gange AC, Jones T (2009) Effects of mycorrhizal fungi on insect herbivores: a meta-analysis. Ecology 90:2088–2097

Kothe E, Bergmann H, Büchel G (2005) Molecular mechanism in bio-geo-interactions: from a case study to general mechanisms. Chem Erde 65(S1):7–27

Lebeau T, Braud A, Jezequel K (2008) Performance of bioaugmentation-assisted phytoextraction applied to metal contaminated soils. A review. Environ Pollut 153:497–522

Lebeau T, Jezequel K, Braud A (2011) Bioaugmentation-assisted phytoextraction applied to metal-contaminated soils: state of the art and future prospects. In: Ahmad I et al (eds) Microbes and microbial technology: agricultural and environmental applications. Springer, New York. doi:10.1007/978-1-4419-7931-5_10

Leung HM, Ye ZH, Wong MH (2006) Interactions of mycorrhizal fungi with Pteris vittata (as hyperaccumulator) in as contaminated soils. Environ Pollut 139:1–8

Leyval C, Joner EJ, del Val C, Haselwandter K (2002) Potential of arbuscular mycorrhizal fungi for bioremediation. In: Ginanazzi S et al (eds) Mycorrhizal technology in agriculture. Birkhauser, Basel, pp 175–186

Lilleskov EA, Parrent JL (2007) Can we develop general predictive models of mycorrhizal fungal community-environment relationships? New Phytol 174:250–256

Liu W (2010) Do genetically modified plants impact arbuscular mycorrhizal fungi? Ecotoxicology 19:229–338

Liu Y, Zhu YG, Chen BD, Christie P, Li XL (2005) Yield and arsenate uptake of arbuscular mycorrhizal tomato colonized by *Glomus mosseae* BEG167 in a spiked soil under greenhouse conditions. Environ Int 31:867–873

Luck GW, Daily GC, Ehrlich PR (2003) Population diversity and ecosystem services. Trends Ecol Evol 18:331–336

Marschner H (1995) Mineral nutrition of higher plants, 2nd edn. Academic, London, p 889 pp

Martino E, Perotto S (2010) Mineral transformations by mycorrhizal fungi. Geomicrobiol J 27:609–623

Mathur N, Bohra JSS, Quaizi A, Vyas A (2007) Arbuscular mycorrhizal fungi: a potential tool for phytoremediation. J Plant Sci 2:127–140

McGrath SP, Zhao FJ, Lombi E (2001) Plant and rhizosphere processes involved in phytoremediation of metal-contaminated soils. Plant Soil 232:207–214

Mehard AA, Cairney JWG (2000) Co-evolution of mycorrhizal symbionts and their hosts to metal-contaminated environments. Adv Ecol Res 30:70–102

Mikkelsen BL, Rosendahl S, Jakobsen I (2008) Underground resource allocation between individual networks of mycorrhizal fungi. New Phytol 180:890–898

Miller RM, Kling M (2000) The importance of integration and scale in the arbuscular mycorrhizal symbiosis. Plant Soil 226:295–309

Moore JC, McCann K, Setala H, De Ruiter PC (2003) Top-down is bottom-up: does predation in the rhizosphere regulate aboveground dynamics? Ecology 84:846–857

Munkvold L, Kjoller R, Vestberg M, Rosendahl S, Jakobsen I (2004) High functional diversity within species of arbuscular mycorrhizal fungi. New Phytol 164:357–364

Neagoe A, Mascher R, Iordache V, Voigt K, Knoch B, Bergmann H (2004) The influence of vesicular arbuscular mycorrhiza *Glomus intraradiceae* on mustard (*Sinapis alba* L.) grown on a soil contaminated with heavy metals. 22. Arbeitstagung in Jena, Lebensnotwendigkeit und Toxizität der Mengen-, Spuren- und Ultraspurenelemente, Friedrich Schiller Universität, pp 597–606

Neagoe A, Ebenå G, Carlsson E (2005) The effect of soil amendments on plant performance in an area affected by acid mine drainage. Chem Erde 65:115–129

Neagoe A, Iordache V, Mascher R, Knoch B, Kothe E, Bergmann H (2006) Lysimeters experiment using soil from a heavy metals contaminated area. 23. Arbeitstagung in Jena, Agricultural, biological, environmental, nutritional and medical importance of macro, trace and ultra trace elements, Friedrich Schiller Universität, pp 568–575

Neagoe A, Merten D, Iordache V, Buechel G (2009) The effect of bioremediation methods involving different degrees of soil disturbance on the export of metals by leaching and by plant uptake. Chem Erde 69:57–73

Neagoe A, Iordache V, Kothe E (2010) Effects of the inoculation with AM fungi on plant development and oxidative stress in areas contaminated with heavy metals. Presentation at COST870 meeting in Jyvaskyla, 13–15 December, Book of abstracts, p 22

Neagoe A, Iordache V, Farcasanu IC (2012) The role of organic matter in the mobility of metals in contaminated catchments. In: Kothe E, Varma A (eds) Bio-geo interactions in metal-contaminated soils, vol 31, Soil biology. Springer, Berlin, pp 297–326

Neumann E, Schmid B, Römheld V, George E (2009) Extraradical development and contribution to plan performance of an arbuscular mycorrhizal symbiosis exposed to complete or partial rootzone drying. Mycorrhiza 20:13–23

Nicoara A, Neagoe A, Donciu R, Iordache V (2010) The effects of mycorrhizal fungi, streptomycetes and plants on heavy metal mobility and bioaccumulation in an industrially enriched soil: preliminary results of a lysimeter experiment. In: Proceedings of the 10th international symposium on "metal elements in environment, medicine and biology", Timisoara, November 2010

Pawlowska TE, Charvat I (2004) Heavy-metal stress and developmental patterns of arbuscular mycorrhizal fungi. Appl Environ Microbiol 70:6643–6649

Piotrowski JS, Rillig MC (2008) Succession of arbuscular mycorrhizal fungi: patterns, causes and considerations for organic agriculture. Adv Agron 97:111–130

Piotrowski JS, Lekberg Y, Harner MJ, Ramsey PW, Rillig MC (2008) Dynamics of mycorrhizae during development of riparian forests along an unregulated river. Ecography 31:245–253

Rashid A, Ayub N, Ahmad T, Gul J, Khan AG (2009) Phytoaccumulation prospects of cadmium and zinc by mycorrhizal plant species growing in industrially polluted soils. Environ Geochem Health 31:91–98. doi:10.1007/s10653-008-9159-8

Redon PO, Béguiristain T, Leyval C (2008) Influence of Glomus intraradices on Cd partitioning in a pot experiment with Medicago truncatula in four contaminated soils. Soil Biol Biochem 40:2710–2712. doi:10.1016/j.soilbio. 2008.07.018

Regvar M, Likar M, Piltaver A, Kugonič N, Smith JE (2010) Fungal community structure under goat willows (Salix caprea L.) growing at metal polluted site: the potential of screening in a model phytostabilisation study. Plant Soil 330:345–356

Rillig MC (2004) Arbuscular mycorrhizae and terrestrial ecosystem processes. Ecol Lett 7:740–754

Rillig MC, Mammey DL (2006) Mycorrhizas and soil structure. New Phytol 171:41–53

Rivera-Becerril F, Calantzis C, Turnau K, Caussane JP, Belimov AA, Gianinazzi S, Strasser RJ, Gianinazzi-Pearson V (2002) Cadmium accumulation and buffering of cadmium-induced stress by arbuscular mycorrhiza in three Pisum sativum L. genotypes. J Exp Bot 53:1177–1185

Rosén K, Weiliang Z, Mårtensson A (2005) Arbuscular mycorrhizal fungi mediated uptake of Cs in leek and ryegrass. Sci Total Environ 338:283–290

Rosendahl S (2008) Communities, populations and individuals of arbuscular mycorrhizal fungi. New Phytol 178:253–266

Rosling A, Roose T, Herrmann AM, Davidson FA, Finlay RD, Gadd GM (2009) Approaches to modeling mineral weathering by fungi. Fungal Biol Rev 23:138–144

Rowe HI, Brown CS, Claassen VP (2007) Comparisons of mycorrhizal responsiveness with field soil and commercial inoculum for six native Montane species and Bromus tectorum. Restor Ecol 15:44–52

Rufyikiri G, Declerck S, Thiry Y (2004) Comparison of 233U and 33P uptake and translocation by the arbuscular mycorrhizal fungus *Glomus intraradices* in root organ culture conditions. Mycorrhiza 14:203–207

Sampedro I, Aranda E, Díaz R, García-Sanchez M, Ocampo JA, García-Romera I (2008) Saprobe fungi decreased the sensitivity to the toxic effect of dry olive mill residue on arbuscular mycorrhizal plants. Chemosphere 70:1383–1389

Sanon A, Martin P, Thioulouse J, Plenchette C, Spichigeer R, Lepage M, Duponnois R (2006) Displacement of an herbaceous plant species community by mycorrhizal and non-mycorrhizal *Gmelina arborea*, an exotic tree, grown in a microcosm experiment. Mycorrhiza 16:125–132

Simard SW, Durall DM (2004) Mycorrhizal networks: a review of their extent, function, and importance. Can J Bot 82:1140–1165

Smith SE, Read DJ (1997) Mycorrhizal symbiosis, 2nd edn. Academic, London, 605 pp

Smith SE, Facelli E, Pope S, Smith FA (2010) Plant performance in stressful environments: interpreting new and established knowledge of the roles of arbuscular mycorrhizas. Plant Soil 326:3–20

Soares CRFS, Siqueira JO (2008) Mycorrhiza and phosphate protection of tropical grass species against heavy metal toxicity in multi-contaminated soil. Biol Fertil Soils 44:833–841

Stancu P, Neagoe A, Jianu D, Iordache V, Nicoară A, Donciu R (2010) Testing phytoremediation methods for the zlatna tailing dams, Romania. In: Proceedings fot the 10th international symposium on "metal elements in environment, medicine and biology", Timisoara, November 2010

Taylor JW, Jacobson DJ, Kroken S, Kasuga T, Geiser DM, Hibbett DS, Fisher MC (2000) Phylogenetic species recognition and species concepts in fungi. Fungal Genet Biol 31:21–32

Tobin JM, White C, Gadd GM (1994) Metal accumulation by fungi – applications in environmental biotechnology. J Ind Microbiol 13:126–130

Toler HD, Morton JB, Cumming JR (2005) Growth and metal accumulation of mycorrhizal sorghum exposed to elevated copper and zinc. Water Air Soil Pollut 164:155–172

Treseder KK, Turner KM (2007) Glomalin in ecosystems. Soil Sci Soc Am 71:1257–1266

Turnau K (1998) Heavy metal uptake and arbuscular mycorrhiza development of *Euphorbia cyparissias* on zinc wastes in South Poland. Acta Soc Bot Pol 67:105–113

Turnau K, Dexheimer J (1995) Acid phosphatase activity in *Pisolithus arrhizus* mycelium treated with cadmium dust. Mycorrhiza 5:205–211

Turnau K, Mesjasz-Przybylowicz J (2003) Arbuscular mycorrhiza of *Berkheya coddii* and other Ni-hyperaccumulating members of Asteraceae from ultramafic soils in South Africa. Mycorrhiza 13:185–190

Turnau K, Orlowska E, Ryszka P, Zubek S, AnielskaT GS, Jurkiewicz A (2006) Role of mycorrhizal fungi in phytoremediation and toxicity monitoring of heavy metal rich industrial wastes in southern Poland. In: Twardowska I et al (eds) Soil and water pollution monitoring, protection and remediation. Springer, Dordrecht, pp 3–23

Turnau K, Ryszka P, Wojtczak G (2010) Metal tolerant mycorrhizal plants: a review from the perspective on industrial waste in temperate region. In: Koltai H, Kapulnik Y (eds) Arbuscular mycorrhizal: physiology and function, vol Part 4. Springer, Heidelberg, pp 257–276

Ultra VUY Jr, Tanaka S, Sakurai K, Iwasaki K (2007) Arbuscular mycorrhizal fungus (*Glomus aggregatum*) influences biotransformation of arsenic in the rhizosphere of sunflower (*Helianthus annuus* L.). Soil Sci Plant Nutr 53:499–508

Vamerali T, Bandiera M, Mosca G (2010) Field crops for phytoremediation of metal-contaminated land. A review. Environ Chem Lett 8:1–17

Van Keulen H, Cutright T, Wei R (2008) Arsenate-induced expression of a class III chitinase in the dwarf sunflower *Helianthus annuus*. Environ Exp Bot 63:281–288

Visan L, Sandu R, Iordache V, Neagoe A (2008) Influence of microorganisms community structure on the rate of metals percolation in soil. Analele stiintifice ale UAIC 53:79–88

Wang F, Lin X, Yin R (2005) Heavy metal uptake by arbuscular mycorrhizas of *Elsholtzia splendens* and potential for phytoremediation of contaminated soil. Plant Soil 269:225–232

Wenzel WW (2009) Rhizosphere processes and management in plant-assisted bioremediation (phytoremediation) of soils. Plant Soil 321:385–408

Wolfe BE, Parrent JL, Koch AM, Sikes BA, Gardes M, Klironomos JN (2009) Spatial heterogeneity of mycorrhizal populations and communities: scales and mechanisms. In: Azcón-Aguilar C et al (eds) Mycorrhizas – functional processes and ecological impact. Springer, Berlin, pp 167–186

Young JPW (2009) Kissing cousins. New Phytol 181:751–753

Yu Y, Zhanh S, Huang H (2010) Behavior of mercury in a soil-plant system as affected by inoculation with the arbuscular mycorrhizal fungus *Glomus mosseae*. Mycorrhiza 20:407–414

Zhang X, Lin A, Chen B, Wang Y, Smith FA (2006) Effect of *Glomus mosseae* on the toxicity of heavy metals to Vicia faba. J Environ Sci 18(4):721–726

Chapter 14
Fungi and Their Role in Phytoremediation of Heavy Metal-Contaminated Soils

Mozhgan Sepehri, Habib Khodaverdiloo, and Mehdi Zarei

14.1 Introduction

Contamination of soil and water with heavy metals (HM) and metalloids is an increasing environmental problem worldwide that has accelerated dramatically since the beginning of industrial revolution and represents an important environmental problem due to their toxicity, and accumulation throughout the food chain leads to serious ecological and health problems. The primary source of this pollution includes the industrial operations such as mining, smelting, metal forging, combustion of fossil fuels, and sewage sludge application in agronomic practices. The metals released from these sources accumulate in soil and, in turn, adversely affect the microbial composition and their metabolic activities. In addition, the elevated concentration of metals in soils and their uptake by plants adversely affect the growth, symbiosis, and consequently the yields of crops (Moftah 2000; Wani et al. 2007a) by disintegrating cell organelles and disrupting the membranes (Stresty and Madhava Rao 1999), acting as genotoxic substance (Sharma and Talukdar 1987) disrupting the physiological process such as photosynthesis (Van Assche and Clijsters 1990; Wani et al. 2007b), or inactivating the respiration, protein synthesis, and carbohydrate metabolism (Shakolnik 1984). The remediation of metal-contaminated soils thus becomes important as these soils usually cover large areas that are rendered unsuitable for sustainable agriculture. Therefore,

M. Sepehri (✉)
Department of Soil Science, College of Agriculture, Isfahan University of Technology, Isfahan 84156-83111, Iran
e-mail: msepehri@cc.iut.ac.ir

H. Khodaverdiloo
Department of Soil Science, College of Agriculture, Urmia University, Urmia 57135-165, Iran
e-mail: h.khodaverdiloo@urmia.ac.ir

M. Zarei
Department of Soil Science, College of Agriculture, Shiraz University, Shiraz 71441-45186, Iran
e-mail: mehdizarei@shirazu.ac.ir

E.M. Goltapeh et al. (eds.), *Fungi as Bioremediators*, Soil Biology 32,
DOI 10.1007/978-3-642-33811-3_14, © Springer-Verlag Berlin Heidelberg 2013

increasing attention has been paid in recent years to the remediation of polluted soils, among which the use of plants and microbes to remove hazardous metal ions is particularly emphasized (Winge et al. 1985; Mehra and Winge 1991).

The HMs in general cannot be biologically degraded to more or less toxic products and, hence, persist in the environment indefinitely. Conventional methods through common physicochemical techniques that include excavation and land fill, thermal treatment, acid leaching, and electro-reclamation are ineffective for metal detoxification because of the high cost, low efficiency, large destruction of soil structure and fertility, and also production of large quantities of toxic products. The advent of bioremediation technology which is the use of microbial metabolic potential has provided a safe and economic alternative to conventional methods for remediating the metal-poisoned soils. The other effective and promising approach is phytoremediation, which is the use of plants to extract, sequester, and detoxify pollutants to clean up the contaminated soils (Brooks 1998).

Phytoremediation involves the use of metal-accumulating plants to remove, transfer, or stabilize the contaminants from soils, but this technique is time consuming (Wenzel et al. 1999). The success of phytoremediation depends on the extent of soil contamination, bioavailability of the metal, and the ability of the plant to absorb and accumulate metals in shoots. However, plants with exceptionally high metal-accumulating capacity often have a slow growth rate and produce limited amounts of biomass when the concentration of metal in the contaminated soil is very high and toxic. To maximize the chance of success of phytoremediation, plant growth-promoting rhizobacteria (PGPR) and arbuscular mycorrhiza fungi (AMF), soil microbes that inhabit the rhizosphere, are utilized in the nutrient poor agricultural soils. They increase HM sequestration capacity of plants by recycling nutrients, maintaining soil structure, detoxifying chemicals, and controlling pests while decreasing toxicity of metals by changing their bioavailability. Meanwhile, plants provide the microorganisms with root exudates such as free amino acids, proteins, carbohydrates, alcohols, vitamins, or hormones which are important sources of nutrient (Winge et al. 1985).

The aim of this chapter is phytoremediation of HM-contaminated soils by using the fungi with the emphasis of arbuscular mycorrhizal fungi.

14.2 HM Pollutants

HM pollution is a global concern. The levels of metals in all environments, including air, water, and soil, are increasing in some cases to toxic levels with contributions from a wide variety of industrial and domestic sources. Metal pollution results when human activity disrupts normal biogeochemical activities or results in disposal of concentrated metal wastes. Mining, ore refinement, nuclear processing, the industrial manufacture of batteries, metal alloys, paints, preservatives, and insecticides are examples of processes that produce metal by-products. Thus, while metals are ubiquitous in nature, human activities have caused metals to accumulate in soil. Such contaminated soils provide a metal sink from which surface waters and groundwaters can become contaminated. Contaminated soil contributes to high metal concentrations in the air through metal volatilization. In addition, industrial emissions and smelting activities cause release of substantial amounts of metals to the atmosphere. Naturally, high metal concentrations can also occur as a result of weathering of parent materials containing high levels of metals.

Although some HMs are essential plant micronutrients since they are required for plant growth and development (Zn, Cu, Fe, Mn, Ni, Mo, Co), high contents of HMs, as well as the long-term presence of potentially toxic metals (Cd, Pb) and metalloids (As) in surface horizon of agricultural soils, are generally considered a matter of concern, as they may adversely affect the quality of soils and surface water and compromise sustainable food production (Pandolfini et al. 1997; Kabata-Pendias 2001; Keller et al. 2002; Voegelin et al. 2003; Kabata-Pendias and Mukherjee 2007). HMs exert their toxicity in a number of ways including the displacement of essential metals from their normal binding sites on biological molecules (e.g., arsenic and cadmium compete with phosphate and zinc, respectively), inhibition of enzymatic functioning, and disruption of nucleic acid structure. It is important to note that the toxicity of a metal depends to a large extent on its speciation which in turn influences metal bioavailability. The chemical nature and, thus, bioavailability of a metal can be changed through oxidation or reduction; however, the elemental nature remains the same because metals are neither thermally decomposable nor microbiologically degradable. Consequently, metals are difficult to remove from the environment. In addition, total metal concentrations in the environment do not necessarily reflect the degree of biological metal toxicity or bioavailability, making it difficult to assess accurately the extent of risk posed by metals.

14.2.1   Detrimental Effects of HMs on Soil Biota

Microbial communities play important roles in soil because of the many functions they perform in nutrient cycling, plant symbioses, decomposition, and other eco-system processes (Nannipieri et al. 2003). Large HM contents in soil are of concern because of their toxicity to soil microorganisms and impairment of ecosystem functions (Giller et al. 1997). First observations of the effects of HMs on soil microbial processes date back to the beginning of this century (Lipman and Burgess 1914; Brown and Minges 1916). But only when the large adverse effects of HMs emissions from smelters on surrounding ecosystems were observed in the 1960s–1970s was it realized how severely soil microorganisms and soil microbial processes can become disrupted by elevated metal concentrations, sometimes resulting in severe ecosystem disturbance.

Short-term responses of microbial communities to HM contamination are well known (Shi et al. 2002; Ranjard et al. 2000; Gremion et al. 2004; Rajapaksha et al. 2004), but medium- and long-term effects of HM in the field have been less frequently investigated (Pennanen et al. 1996; Kandeler et al. 2000; Sandaa et al. 1999; Renella et al. 2004). However, a considerable body of information has now been accumulated on the effects of HMs on soil microorganisms and microbially mediated soil processes from both laboratory studies and field experiments (Bååth 1989). HMs exert toxic effects on soil microorganism (Pawlowska and Charvat 2004), hence results in the change of the diversity, population size, and overall activity of the soil microbial communities (Smejkalova et al. 2003; Gupta 1992; Hattori 1996; Kelly et al. 2003).

Gasper et al. (2005) reported that the aftereffect of the observed HMs (Cr, Zn, and Cd) pollution influenced the metabolism of soil microbes in all cases. In general, an increase of metal concentration adversely affects soil microbial activities, for example, soil microbial biomass (Fritze et al. 1996), weak enzyme activity (Kandeler et al. 1996), and increasing microbial respiration rate (Bogomolov and Chen 1996), which appears to be very useful indicators of soil pollutions (Brookes 1995; Szili-Kovács et al. 1999). Given a sufficiently high rate of addition, HMs added to soil in laboratory ecotoxicological studies result in a decrease in the amount of microbial biomass and a change in community structure (Maliszewska et al. 1985; Ohya et al. 1985; Naidu and Reddy 1988; Aoyama et al. 1993; Leita et al. 1995; Speir et al. 1995; Kandeler et al. 1996; Knight et al. 1997). This is not surprising; microorganisms differ in their sensitivity to metal toxicity, and sufficient metal exposure will result in immediate death of cells due to disruption of essential functions, and to more gradual changes in population sizes due to changes in viability or competitive ability. What is perhaps more surprising is that soil microorganisms subject to long-term metal stress, even at modest levels of exposure, are not able to maintain the same overall biomass as in unpolluted soils.

Development of tolerance and shifts in community structure could be expected to compensate for less of more sensitive populations. Instead, results from laboratory ecotoxicological studies suggest that changes in community structure go hand in hand with a decrease in the soil microbial biomass (Frostegård et al. 1993, 1996). There is now a considerable amount of evidence documenting a decrease in the soil microbial biomass as a result of long-term exposure to HM contamination from past application of sewage sludge (McGrath 1994; McGrath et al. 1995). Analysis of soil contaminated with HMs from other sources such as Cu and Zn in animal manures (Christie and Beattie 1989), runoff from timber treatment plants (Bardgett et al. 1994; Yeats et al. 1994), past application of Cu-containing fungicides (Zelles et al. 1994; Filser et al. 1995), and analysis of soils in the vicinity of metal-contaminated army disposal sites (Kuperman and Carreiro 1997) confirms that a decrease in the microbial biomass occurs at a relatively modest and sometimes even at a surprisingly low (Dahlin et al. 1997) metal loading. The widespread occurrence of this effect of metal toxicity suggests that there may be a common physiological explanation.

Enzyme activity is a soil property that is chemical in nature but has a direct biological origin. This activity arises from the presence of many types of enzymes that are present in the soil and within soil microorganisms. From an assortment of enzymes present and active in soil, phosphatases are interesting groups of enzymes that catalyze the hydrolysis of phosphate from organic monoester linkages (Dmitri and Begonia 2008). Phosphates released from such phosphatase action are very important to the plants and microorganisms that depend on soil for their phosphorus requirements. Indications of specific inhibitory action of HMs have been produced in microbes as well (Fulladosa et al. 2005a, b). Such selective targeting of specific enzymatic systems and pathways suggests that certain members of the microbial community would be more sensitive to HM exposure than others, depending on the sensitivity of their critical metabolic pathways. Thus, while toxicity of HMs to microbes is a well-established phenomenon, the effects of those metals upon specific enzymatic systems at lower ("subacute") concentrations are not well known. Denitrification is a natural microbial process converting nitrate to dinitrogen gas during anaerobic respiration. Such reduction occurs sequentially, with nitrate converted to nitrite, nitric oxide, nitrous oxide, and, finally, nitrogen gas. A number of enzyme classes, mostly located in the periplasmic space, are involved in denitrification (Dmitri and Begonia 2008), with a number of corresponding genes that can be used as genetic markers for presence and expression of such enzymes in the soil metagenome. As denitrification-related enzymes are generally located within the cell membrane or periplasmic space, expelling HM ions out of the cell would place them in the immediate contact with denitrification-related enzymes, thus limiting utility of such a resistance strategy. The fact that denitrification enzymes are located on or near the outer cell surfaces further increases the vulnerability of the entire denitrification pathway to chemical disruption. Recent work has suggested a direct effect of HMs upon extracellular enzyme activities (Begonia et al. 2004; Hinojosa et al. 2004). Combined with the fact that scavenging/pumping systems are unlikely to protect the denitrification pathway from HM effects (and may, in fact, exacerbate the situation), it is expected that denitrification pathway would be uniquely sensitive to HMs. The notion of selective inhibition of denitrification steps by HMs has been supported by work of Holtan-Hartwig et al. (2002), suggesting the potential for production of undesirable by-products, such as nitrous oxide.

The second mechanism of microbial resistance to metals is evolution of enzyme forms resistant to metals. This resistance pathway is expected to be the predominant in the denitrifying bacteria, due to inability to use metal pumps for the reasons described above. The metal-resistant forms of enzymes present in metal-stressed denitrifying community are expected to be readily identifiable by their gene sequence and therefore their genetic signature. Disruption of denitrification by HMs could lead to a number of undesirable consequences, influencing the human health at both global and local levels. Suppressed denitrification in the soil could lead to enhanced nitrogen retention and flushing, resulting in nonpoint nutrient pollution in waterways receiving overland or subsurface flow from impacted locations. Nutrient pollution, in turn, leads to eutrophication and massive algal blooms, including those of toxic algae and cyanobacteria (e.g., *Microcystis*), affecting human populations relying on surface waters for municipal, recreational, or agricultural needs. Specific inhibition of nitrous oxide reductase by metal has been observed recently (Holtan-Hartwig et al. 2002), resulting in incomplete denitrification leading to emission of nitrous (and possibly nitric) oxides. As nitrous oxide is a potent greenhouse gas that also damages ozone layer (Crutzen 1970; Dickinson and Cicerone 1986), denitrification disruption via metal contamination could act as a link between local metal contamination and global climate change phenomena.

Another features of HM polutes soils are impeded litter decomposition and soil respiration (Marschner and Kalbitz 2003; Illmer and Schinner 1991). The degree of impedance, however, is determined by the rate of carbon and nitrogen mineraliza-tion. Thus, under HM pollution, the rates of such activities are impaired and carbon and nitrogen accumulate in the soil. Assay of soil respiration also helps to quantify the effects of metals on the total biological activity of soils. Addition of HM salts to soils usually causes an immediate decrease in respiration rates, but responses are determined by the properties of both the metal and the soil. The response of base respiration to metals is dependent on the nature of the substrates mineralized at the time of measurement. The response of base respiration to increasing doses of Cu, Cr, Ni, and Zn can be inconsistent, with increases in base respiration sometimes occurring even though both higher and lower doses of the same metal resulted in a decrease in base respiration (Doelman and Haanstra 1984). These bizarre and inexplicable responses probably result from strong interactive effects between both abiotic and biotic factors. A potential difficulty is that it is not possible to distinguish a metal toxicity effect from an effect of metal addition on substrate availability. Some metals such as Pb may decrease the amount of substrate avail-able for respiration through the formation of complexes and thus decrease respira-tion, whereas death of microbial cells as a result of metal addition may explain the increase of the base respiration in response to metal addition (Leita et al. 1995). The initial response in soil respiration due to metal addition may therefore bear little relation to long-term effects, and possibly even less relation to the typical field situation where there often is an increasing amount of metal contamination over a period of many years.

When metal toxicity data to soil microbial processes and populations from the literature is summarized, an enormous variability in the data becomes apparent. In principle there are only two factors which may contribute to the discrepancies between studies: (1) factors which modify the toxicity of the metals and (2) differences in sensitivity of the microorganism(s) or microbial process(es). It is extremely difficult to separate these factors when metal toxicity is studied in soils, both because of the difficulties in determining the "bioavailability" of metals in soils and because of the complexity of soil microbial communities.

In microbial investigations, the term "bioavailability" is usually ill defined and is rarely quantified. Bioavailability is dependent on soil characteristics such as min-eralogy, pH, texture, organic matter, iron oxide, and HM content as well as plants and microorganisms and can be assessed by the growth of the organism of interest and an evaluation of the uptake or toxicity of a metal after the fact (Wolt 1994). Plant root exudates both directly (e.g., Fe^{3+}) or through the effects exudates have on microbial activity and resulting rhizosphere chemistry. As bacteria are present within colonies in soil (Harris 1994) or protected by clays (Van veen et al. 1985; Ladd et al. 1995), they may often not be exposed to the equilibrium solution activity of HMs. Metals may become bound to bacterial or fungal cell walls or on extracellular polysaccharides of bacteria, and the ingestion of such bacteria by protozoa or nematodes will result in vastly different exposures to metals in the predators than would result simply from exposure to the metals present in the soil solution. Microorganisms may also alter metal availability in their vicinity due to localized acidification on the environment or production of compounds which complex metals. Species of microorganisms (e.g., Berdicevsky et al. 1993), strains of the same species (e.g., Romandini et al. 1992), and also activities of the same microbial species (e.g., Balsalobre et al. 1993; Torslov 1993) can all show considerable differences in their sensitivity to metal toxicity.

14.3 Remediation Techniques of HM-Contaminated Soils

Since the industrial revolution, anthropogenic impacts have caused more and more hazardous HMs releasing to environment. Soils, being the basic and most essential part of the ecological system, are heavily contaminated, too. Compared to organic pollutants, the remediation of toxic metals in porous matrices (soil and sediment) requires a specific approach since hazardous HMs are indestructible, as they cannot be chemically or biologically degraded, hence require appropriate methods for their removal. Treatments make necessary metal extraction (e.g., by solubilization or complexation) to avoid their dissemination in the environment and/or the food chain contamination. Therefore, increasing attention has been paid in recent years to the remediation of polluted soils. To date, main four methods, chemical or physical remediation, animal remediation such as earthworm, phytoremediation, and microremediation, were proposed by researchers. Because of the obvious disadvantages and deficiency in feasibility, wide application of the former two methods is restricted. The latter two, namely, the use of plants and microbes, are preferred because of their cost-effectiveness, environmental friendliness, and fewer side effects. Using transgenic technology is a tendency in the future to create an ideal species purposely. In the future crop hyperaccumulators will be a better choice due to its feasibility, in the field of which current emphasis is scarce. Microbes, in many cases, are more efficient in accumulating and absorbing HMs because of their astronomical amount and specific surface area. Furthermore, technique of genetic engineering in microbes is easier and more mature than in plant cells. Therefore, using transgenic technology to create an optimum plant + soil + microbes combination would be a promising way in the future development (Gang et al. 2010).

14.3.1 Conventional Methods

Chemical or physical method which is named "conventional method" is early used and even endemically commercialized in America. The in situ or ex situ remediation of these methods is more often based on (1) improvement of the solubility and bioavailability of HMs by synthetic chelators such as ethylenediaminetetraacetic acid (EDTA); (2) solidification/stabilization by either physical inclusion or chemical interactions between the stabilizing agent and the pollutant; (3) vitrification using thermal energy for soil fusion, allowing physical or chemical stabilization; (4) electrokinetical treatment which ionic species of the pollutant migrate to electrodes inserted into the soil; (5) chemical oxidation or reduction of the pollutant to attain chemical species with lower toxicity that are more stable and less mobile; and (6) excavation and off-site treatment or storage at a more appropriate site (Saxena et al. 1999). Most of these conventional remediation technologies are expensive and labor intensive, are technically limited to relatively small areas, and cause further disturbance to the already damaged environment (Alloway and Jackson 1991; Mench et al. 1994). These techniques for soil remediation may render the land useless for plant growth as they remove all biological activities, including useful microbes such as nitrogen-fixing bacteria, mycorrhiza, fungi, as well as fauna in the process of decontamination. Furthermore, natural soil, structure, texture, and fertility can be impaired by the method itself and by the regent added. Additionally, excessive use of chelators like EDTA which is both toxic and nonbiodegradable would poison both plants and microbes (Gang et al. 2010). Therefore, due to improved knowledge of the mechanisms of uptake, transport, tolerance, and exclusion of contaminants in microorganisms and plants, development of alternative technologies, named bioremediation and phytoremediation which respectively refer to the use of microbes and plants, has been promoted.

14.3.2   Biological Methods

Bioremediation is based on the potential of living organisms, mainly microorganisms and plants, to detoxify the environment (Anderson and Coats 1994). Bioremediation technologies could be classified under two main categories, namely, "microbial-based" and "plant-based" remediation methods. For organic pollutants, the goal of phytoremediation is to completely mineralize them into relatively nontoxic constituents, such as carbon dioxide, nitrate, chlorine, and ammonia (Cunningham et al. 1997). However, HMs are essentially immutable by any biological or physical process short of nuclear fission and fusion, and thus their remediation presents special scientific and technical problems. Furthermore, in the case of uptake of HMs by microbes, there is no cost-effective method to collect the microbes from soil body. Plant-based bioremediation technologies have been collectively termed as "phytoremediation" that refers to the use of green plants and their associated microbiota for the in situ treatment of contaminated soil and groundwater. While the use of plants for remediation of contaminated soils has been developed much more recently, it was not until the 1990s that the concept of phytoremediation emerged as a promising technology that uses plants for decontamination of polluted sites (Barceló and Poschenrieder 2003). With a few notable exceptions, the best scenarios for the phytoremediation of HMs involve plants extracting and translocating a toxic cation or oxyanion to aboveground tissues for later harvest, converting the element to a less toxic chemical species (i.e., transformation), or at the very least sequestering the element in roots to prevent leaching from the site (Meagher 2000).

Although phytoremediation offers cost advantages and is comparable to in situ bioremediation and natural attenuation (Cunningham et al. 1997), it has its own limitations, for example, the difficulty with treating wastes greater than three meters deep, possible uptake of contaminants into leaves and release during litter fall, inability to assure cleanup below action levels in a short period of time, difficulty in establishing the vegetation due to toxicity at the site, and possible migration of contaminants off-site by preferential flow or by binding with soluble plant exudates (Schnoor 1997). Therefore, and most likely due low bioavailability of PTEs and/or low biomass of hyperaccumulators, phytoremediation method usually remains as a time-consuming process (Cunningham et al. 1997; Khodaverdiloo and Homaee 2008). However, numerous studies have indicated that soil microbial community such as arbuscular mycorrhizal (AM) fungi could help to overcome these limitations, for example, by enhancing uptake of nutrient elements as well as water by host plants through their extraradical mycelial networks (Marschner and Dell 1994) and protecting the host plants against HM toxicity (Leyval et al. 1997). Therefore, inoculation of plants with AM fungi can be a potential biotechnological tool for successful restoration of degraded ecosystems (Dodd and Thompson 1994; Mathur et al. 2007).

14.3.2.1 Phytoremediation

Phytoremediation is a solar-driven remediation technology with greatly reduced costs and minimum adverse side effects (Cunningham and Ow 1996; Cunningham et al. 1997; Garbisu et al. 2002; Glick 2003). Within the field of phytoremediation, different categories have been defined such as phytofiltration, phytostabilization, phytovolatilization, phytodegradation, phytostimulation, and phytoextraction; among them phytoextraction and phytostabilization are of great concern for remediation of HM-contaminated soils. Phytoextraction is the use of hyperaccumulating/high-biomass plants to uptake the contaminants in their aboveground tissues with subsequent harvest, recovery, and disposal or recycling of the metals (Geiger et al. 1993; Kayser et al. 2000; Hammer et al. 2003). Hyperaccumulators are wild species that can accumulate large amounts of specific metals in their shoots, but they are often with low biomass. The fast-growing, high-biomass plants are usually not metal specific and have low to average HM concentrations (Hammer et al. 2003). Phytoextraction has been proposed as a suitable alternative to destructive techniques used so far to clean up soils contaminated with HMs. Indeed, the use of plants to remove metals from soils is environmental friendly, and its cost is much lower compared to engineering-based techniques (Cunningham and Ow 1996; Cunningham et al. 1997; Garbisu et al. 2002; Glick 2003). Although phytoextraction is a promising option to remediate contaminated soils, so far, no suitable method is yet available to remove metals in a reasonably short time. Indeed, the potential for phytoextraction depends not only on bioaccumulation factor but also on plant biomass. However, the hyperaccumulator and high-biomass plant species fulfill only one of these conditions.

In phytostabilization, plants are used for immobilizing contaminant metals in soils or sediments by root uptake, adsorption onto roots, or precipitation in the rhizosphere. By decreasing metal mobility, these processes prevent leaching and groundwater pollution, and bioavailability is reduced and fewer metals enter the food chain (Barceló and Poschenrieder 2003).

Metal Hyperaccumulators

Selection of plants for phytoremediation of metals depends on the type of application (Schnoor 1997). Plants show several response patterns to the presence of potentially toxic concentrations of HMs. Most are sensitive even to very low concentrations, others have developed resistance, and a reduced number behave as hyperaccumulators of HMs (e.g., Brooks 1998; Salt et al. 1998). Hyperaccumulators have opened up the possibility to use phytoextraction for remediation of HM-contaminated environments (Barceló and Poschenrieder 2003) and provide valuable tools for reclamation of polluted soils, enhancement of soil quality, and recovery and reestablishment of biotic. Plants with metal resistance mechanisms based on exclusion can be efficient for phytostabilization technologies. Hyperaccumulator plants, in contrast, may become useful for extracting toxic elements from the soil and thus decontaminate and restore fertility in polluted areas. In recent years, improved knowledge of the mechanisms of uptake, transport, and tolerance of high metal concentrations in these plants (e.g., Assunçao et al. 2001; Hall 2002) has opened up new avenues for remediation by phytoextraction.

At least 400 species distributed in 45 botanical families are considered metal hyperaccumulators (Brooks 1998). By definition, hyperaccumulators are herbaceous or woody metallophytes, belong to the natural vegetation of metal-enriched soils, and accumulate and tolerate without visible symptoms a hundred times or greater metal concentrations in shoots than those usually found in non-accumulators. Baker and Brooks established 0.1 % as the minimum threshold tissue concentrations for plants considered Co, Cu, Cr, Pb, or Ni hyperaccumulators, while for Zn or Mn the threshold is 1 % (Baker and Brooks 1989). As discussed in the next section, these species have evolved internal mechanisms that allow them to take up and tolerate large metal concentrations that would be exceedingly toxic to other organisms.

An ideal plant species for remediation purposes should grow easily and produce high biomass quickly on HM-contaminated soils, have high root-to-shoot translocation and high bioconcentration factors, and tolerate high shoot metal concentrations (Barceló and Poschenrieder 2003). Unfortunately, most metal hyperaccumulator plants grow quite slowly and have a low biomass, while plants that produce a high biomass quickly are usually sensitive to high metal concentrations.

HM complexes in hyperaccumulators plants are mainly associated with carboxylic acids like citric, malic, and malonic acids. These organic acids are implicated in the storage of HMs in leaf vacuoles. Amino acids like cysteine, histidine glutamic acids, and glycine also form HM complexes in hyperaccumulators (Homer et al. 1997). These complexes are more stable than those with carboxylic acids. They are mostly involved in HM transport through xylem. Moreover, hyperaccumulator plants can increase availability of metals like Fe and also Zn, Cu, and Mn by releasing chelating phytosiderophores. Hyperaccumulation mechanisms may then be related to rhizosphere processes such as to the release of chelating agents (phytosiderophores and organic acids) and/or to differences in the number or affinity of metal root transporters (Lombi et al. 2001).

Although hyperaccumulator plants are widely used in phytoextraction, they are generally of low biomass, inconvenient for phytoremediation. However, arbuscular mycorrhizae fungi (AMF), especially *Glomus intraradices*, and colonized *Festuca* and *Agropyron* species have shown higher HM (Zn, Cd, As, and Se) content than non-colonized controls (Giasson et al. 2006). As for hyperaccumulators, fungi can synthesize cysteine-rich metal-binding proteins called metallothioneins (Gadd and White 1989). AMF might therefore be directly implicated in HM hyperaccumulation in plants.

Cellular Mechanisms of Plant Metal Detoxification and Tolerance

Although many metals are essential, all metals are toxic at higher concentrations, because they cause oxidative stress by formation of free radicals and/or they can replace essential metals in pigments or enzymes disrupting their function. Thus, metals render the land unsuitable for plant growth and destroy the biodiversity. However, as discussed earlier, some specific plant species preferentially grow on metalliferous soils and are capable to accumulate very high levels of specific metals.

These plants are perfectly adapted to the particular environmental conditions of their habitat, and high metal accumulation may contribute to their defense against herbivores and fungal infections (Barceló and Poschenrieder 2003). However, usually, the metabolic and energetic costs of their adaptation mechanisms do not allow them to compete efficiently on uncontaminated soil with non-metallophytes. Metal hyperaccumulators are highly specialized models of plant mineral nutrition. As it has been discussed by Barceló and Poschenrieder (2003), several hypotheses have been proposed to explain the mechanisms of metal hyperaccumulation including (1) complex formation and compartmentation, (2) deposition hypothesis, (3) inadverted uptake, and (4) hyperaccumulation as a defense mechanism against abiotic or biotic stress conditions (Barceló and Poschenrieder 2003).

Plants may use several potential cellular/molecular mechanisms for detoxification of and tolerance to excess concentrations of specific HMs in the environment (Hall 2002). Generally, the strategy adopted by plants aims to avoid the buildup of excess metal levels in the cytosol and thus to prevent the onset of toxicity symptoms. This is achieved by the use of various mechanisms that are present and likely to be employed in general metal homeostasis in all plants. It appears likely that specific mechanisms are employed for specific metals in particular species. Potential cellular mechanisms for metal detoxification and tolerance in higher plants include (but not limited to) (a) restriction of metal movement to roots by mycorrhizas, (b) binding to cell wall and root exudates, (c) reduced influx across plasma membrane, (d) active efflux into apoplast, (e) chelation in cytosol by various ligands, (f) repair and protection of plasma membrane under stress conditions, (g) transport of PC–Cd complex into the vacuole, and (h) transport and accumulation of metals in vacuole (Hall 2002). It is also possible that more than one mechanism may be involved in reducing the toxicity of a particular metal (Hartley-Whitaker et al. 2001; Hall 2002). These processes involved in reducing toxicity are of considerable current interest because an understanding of the means of manipulating metal tolerance could be important in the development of crops for phytoremediation of, for example, HM-contaminated soils (Salt et al. 1998). However, as discussed by others (e.g., Hall 2002), there is no single mechanism that can account for tolerance to a wide range of metals (Macnair et al. 2000).

Although not always considered in general reviews of plant metal tolerance mechanisms, mycorrhizas, and particularly ectomycorrhizas that are characteristic of trees and shrubs, can be effective in ameliorating the effects of metal toxicity on the host plant (e.g., Hüttermann et al. 1999; Jentschke and Godbold 2000). However, the mechanisms involved in conferring this increase in tolerance have proved difficult to resolve; they may be quite diverse and show considerable species and metal specificity since large differences in response to metals have been observed, both between fungal species and to different metals within a species (e.g., Hüttermann et al. 1999; Rahmanian et al. 2011).

The mechanisms employed by the fungi at the cellular level to tolerate HMs are probably similar to some of the strategies employed by higher plants, namely, binding to extracellular materials or sequestration in the vacuolar compartment. Regarding the role of ectomycorrhizas in metal tolerance by the host plant, most mechanisms that have been proposed involve various exclusion processes that restrict metal movement to the host roots. These have been extensively reviewed and assessed (Jentschke and Godbold 2000) and include absorption of metals by the hyphal sheath, reduced access to the apoplast due to the hydrophobicity of the fungal sheath, chelation by fungal exudates, and adsorption onto the external mycelium.

14.3.2.2 Bioremediation

Gadd (2001) defined bioremediation as an area of environmental biotechnology and as the application of biological processes to the treatment of pollution. Applications of fungi in environmental protection and recovery of metals have received more attention in recent years. Biosorbent fungi are engaged microorganisms for the process of biosorption of metal ions on their surface. Biosorption to *Rhizopus*,

Mucor, *Penicillium*, and *Aspergillus* genera is well documented. Biosorption is the non-metabolic sorption process. Many potential binding sites are present in fungal cell walls, including chitin, chitosan, amino, carboxyl, phosphate, sulfhydryl, and other functional groups (Volesky and Holan 1995; Gadd 2001). Fungal solubilization of insoluble metal compounds occurs by several mechanisms such as protonation of the anion of the metal compound, the production of organic acids, siderophores (it is also as extracellular metal-binding molecules), and chelating agents (Morley et al. 1996; Singh 2006). Metal sequestration in the cytosol by induced metal-binding molecules such as metallothioneins and phytochelatins is an intracellular detoxification in fungi (Cobbett 2000). Metal(loid)s may be transformed by fungal reduction, methylation, and dealkylation, so through this mobility and toxicity of metals modified (Gadd 2001). Mechanisms of fungal biosorption, solubilization, transformation, and immobilization of metal(loid)s are of potential for bioremediation.

14.3.2.3 Mycoremediation: Fungal Bioremediation

Mycoremediation is a form of bioremediation, which more broadly refers to degrading or removing organic and inorganic toxicants in the environment using biological processes. Mycoremediation went from the theoretical to the practical just over a decade ago. The term "mycoremediation" was coined by the American mycologist Paul Stamets, who has studied many potential uses of mushrooms.

Mycoremediation is the process of using fungi to return an environment contaminated by pollutants to a less or without contaminated state. It can apply to contaminated soil, oil spills, industrial chemicals, contaminated surface water, and farm waste. It is not widely used at present, but the below-noted applications suggest its broader potential. Some examples of used fungi included the following: *Lentinus edodes* can degrade pentachlorophenol (PCP), *Pleurotus pulmonarius* can degrade atrazine, *Phanerochaete chrysosporium* can degrade biphenyl and triphenylmethane, and some fungi have also proven useful in remediation of HMs that are not degraded further but fungi can extract them from soil or water and accumulate them in their or host tissues (Singh 2006). Some of them are hyperaccumulators, capable of absorbing and concentrating HMs in the mushroom fruit bodies. The mushrooms can be used to remediate the metal-polluted soil. Many studies carried out to evaluate the possible danger to human health from the ingestion of mushrooms containing HMs (Gast et al. 1998; Ouzouni et al. 2007; Elekes et al. 2010). Numerous data on metal contents in fungal fruiting bodies were published previously (Alonso et al. 2003; Soylak et al. 2005; Svoboda et al. 2006; Elekes et al. 2010), and the reported metal concentrations in the fruiting body of mushrooms vary from one species to another, because of many factors affecting the accumulation rate (Elekes et al. 2010). Elekes et al. (2010) indicated that HM concentrations in the fruiting body of mushrooms were mean values of 11.94 mg kg^{-1} for Ti, 1.07 mg kg^{-1} for Sr, 1,163.86 mg kg^{-1} for Bi, and 17.49 mg kg^{-1} for Mn. The bioconversion factor of HMs represented the level of metals concentration in the mushrooms body correlated with the metallic element in the soil on which the fungus grow and had the highest values in *Marasmius oreades* species for bismuth and titanium. Totally, fungi perform a wide variety of ecosystem functions such as the important role in mycoremediation and may be a simple and relatively cheap method of environmental remediation, especially if indigenous species of each site are isolated, identified, and used.

Mycorrhizoremediation with the Emphasis on Arbuscular Mycorrhizal Fungi

Mycorrhizas are mutualistic associations of plant roots and fungi. The symbiotic fungi are provided with carbon by the photobionts, while the fungi may protect the symbiosis from harsh environmental conditions, increase the absorptive area, and provide increased access to inorganic nutrients and water (Gadd 2010). The mycelium of mycorrhizal fungi is more resistant to abiotic agents than the root itself, and this may compensate for reduced root growth. They increase tolerance to extreme conditions. They are crucial in the ecology and physiology of terrestrial plants and are the rule in nature, not the exception (Khan 2006). Mycorrhizal associations vary widely in structure and function and included arbuscular mycorrhiza, ectomycorrhiza, ectendomycorrhiza, arbutoid mycorrhiza, monotropoid mycorrhiza, ericoid, and orchid mycorrhiza. Mycorrhizal fungi act on ecosystems in widely different ways. Amongst them the arbuscular mycorrhizal fungi (AMF) are of ecological and economical importance. AMF are universal and ubiquitous rhizosphere microflora forming symbiosis with plant roots of Bryophyta, Pteridophyta, Gymnospermae, and Angiospermae in nature (Smith and Read 2008). The AMF are as biofertilizers and bioprotectants. They cannot be cultured in the absence of their host, and the extracellular hyphal network is not as extensive as ectomycorrhiza associations. These fungi belong to Glomeromycota (Schubler et al. 2001).

Occurrence of AMF has been reported in relation to plants growing on HM-polluted soils (Leyval et al. 1995; Göhre and Paszkowski 2006; Khade and Adholeya 2007; Zarei et al. 2008a, b). Many of plants are highly dependent on arbuscular mycorrhiza. Use of arbuscular mycorrhizal symbiosis has multidirectional effects such as excretion of chelating agents, producing of plant growth-promoting factors and increasing of plant biomass, extending of soil rhizosphere (mycorrhizosphere), and increasing of uptake per unit surface area. AMF can help in ecosystem remediation (Gaur and Adholeya 2004). Rhizoremediation by mycorrhiza symbiosis, that is, mycorrhizoremediation, is an enhanced form of phytoremediation (Khan 2006). In some cases, AMF have generally such a strong influence on plant biomass and can increase HMs uptake and root-to-shoot transport (phytoextraction), while in other cases AMF contribute to HM stabilization within the soil/root and reduce their uptake (phytostabilization) (Zarei and Sheikhi 2010).

It was proved that the AMF are effective in immobilization of metals in the plant rhizosphere and help in HM stabilization by their accumulation in a nontoxic form in plant roots and extracellular mycelia (Zarei and Sheikhi 2010). There are similar strategies in decreasing of the toxic effects of HMs for fungi and host plants that include immobilization of these elements by the fungal exudations, their deposit in polyphosphate granular, adsorption of elements on the cell wall, and chelation in the fungal organs (Göhre and Paszkowski 2006). Glomalin is a glycoprotein produced abundantly on hyphae and spores of AMF in soil and in roots and is able to link with HMs and extract them from the soil. Therefore, it can be said the fungal strains that secrete more glomalin are more suitable for biological stabilization (Göhre and Paszkowski 2006). Binding HMs with chitin in the cell wall of fungal organs reduces their concentration in the soil solution, and broad absorption surface of extraradical mycelia is considered an important source of discharged HMs from the soil solution. The vesicles of fungi also have a role in accumulation of toxic compounds and in this way can help in the detoxification of metals. High concentration of HMs in mycorrhizal roots than non-mycorrhizal ones showed that the fungus could maintain HMs in surface and/or within mycelia, for example, zinc concentration in fungal mycelia in comparison with plant tissues was reported more than ten times (Chen et al. 2001). Kaldorf et al. (1999) showed that most of zinc was accumulated within the fungal tissue, such as vesicles inside the cells of root cortex. It seems that the immobilization of HMs in fungal tissue is one of the mechanisms of reducing HMs toxicity in mycorrhizal plants. The results of Rufyikiri et al. (2004) indicated the accumulation of uranium in the plant root of mycorrhizal plant that was exposed to the high levels of uranium and the supportive effect of this fungus for the host plant. High uptake of HMs by mycorrhizal roots and the possible role of AM fungus in phytostabilization were demonstrated by Wang et al. (2007b) and Wang et al. (2007c).

Plant colonized by AMF can also increase the uptake and accumulate of HMs in plant shoots or phytoextraction (Leung et al. 2006; Wang et al. 2007a). AMF increased the uptake and accumulation of arsenic in hyperaccumulator plant of *Pteris vittata* (Leung et al. 2006). It was shown dynamic and mobilization of zinc and transferring to shoots of corn and clover colonized by AMF (Chen et al. 2003). In a pot experiment, Zarei and Sheikhi (2010) illustrated that for corn and *Festuca* plants and under the high soil pollution (500 mg Zn kg^{-1}), *Glomus mosseae* (a noticeable indigenous fungus in HM-contaminated soil) was the most effective fungal species in Zn extraction and translocation.

Overall, it is possible to enhance and improve the capabilities of plants in different types of phytoremediation processes by inoculating with appropriate arbuscular AMF (i.e., mycorrhizoremediation).

The potential role of mycorrhizoremediation in HM-contaminated soils is becoming an interest, and it needs to completely understand the ecological complexities of the plant–microbe–soil interactions, mechanisms for how AMF are involved in HM absorption and transportation in plants and the tolerance to HM. Multidisciplinary investigations using molecular, biochemical, and physiological techniques and employment of appropriate combination of plant–fungus in remediation strategies for HM-contaminated soils may be helpful.

Diversity of AMF in Contaminated Soils

The presence of AMF propagules in the HM-polluted soils was abundantly reported (Bohn and Liberta 1982; Diaz and Honrubia 1994; Pawlowska et al. 1996; Gaur and Adholeya 2004). Weissenhorn et al. (1993) measured root colonization in the polluted soils with 1,220 and 895 mg cadmium and lead per kg, respectively, up to 40 %. Root colonization rate in the dominant native plants of an HM-contaminated site was measured 35–85 % and the spore numbers 80–1306 per 200 g dry soil along the transect (Zarei et al. 2008a). Many plant species, such as *Fragaria vesca*, *Viola calaminaria*, *Veronica rechingeri*, *Solidago gigantea*, *Thymus polytrichus*, *Holcus lanatus*, and *Thlaspi praecox*, growing well at natural HM-polluted areas were colonized by diverse AMF. *Acaulospora*, *Entrophospora*, *Gigaspora*, and *Glomus* genera in symbiosis with plant species grown in the HM-contaminated soils were identified (Tonin et al. 2001; Turnau et al. 2001; Gonzalez-Chavez et al. 2002; Whitfield et al. 2004; Vallino et al. 2006; Vogel-Mikus et al. 2006; Zarei et al. 2008a, b; Long et al. 2010; Zarei et al. 2010). Some AMF species and sequences types may be exclusively found in the high HM pollution levels (Zarei et al. 2010). Long et al. (2010) studied the diversity of AMF communities associated with five selected plant species (*Phytolacca americana*, *Rehmannia glutinosa*, *Perilla frutescens*, *Litsea cubeba*, and *Dysphania ambrosioides*) from severely HM-polluted soils in Dabaoshan Mine region, China, using molecular methods. DGGE and sequence analysis revealed that *Glomus* dominated all of the samples except for the roots of *D. ambrosioides*, while *Kuklospora* and *Ambispora* dominated the roots of *D. ambrosioides* and the rhizosphere of *P. americana*. The studies indicated that diverse AMF are associated with plants grown in HM-polluted soils.

Arbuscular Mycorrhiza and HMs

High amounts of HMs can delay, reduce, or even completely eliminate AMF spore germination and AM colonization (Gildon and Tinker 1981; Del Val et al. 1999). Similarly, Boyle and Paul (1988) reported a negative correlation between Zn concentrations in a soil treated with urban-industrial sludge and AM colonization in barley. In other studies, however, the addition of metal containing sludge did not affect AM development under field conditions (Arnold and Kaputska 1987). These contrasting results may be explained due to the fact that different AMF ecotypes can exhibit varying degrees of tolerance to metals (Haselwandter et al. 1994). A higher tolerance to Cu, Zn, Cd, and Pb of indigenous fungi from sludge-polluted sites, in comparison to reference isolates from unpolluted soils, has been reported (del Val et al. 1999). AMF species isolated from the HM-polluted soil could be more adapted and tolerated to HM pollution (Gildon and Tinker 1983; Weissenhorn et al. 1993; Diaz et al. 1996). Gildon and Tinker (1981) isolated a strain of *Glomus mosseae* from HM-contaminated soils that could tolerate 100 mg Zn kg^{-1}. Dueck et al. (1986) reported the presence of some strains of *Glomus fasciculatum* as tolerant strains to HMs in the several HM-contaminated areas in the Netherlands. Weissenhorn et al. (1993) isolated the spores belonging to *Glomus mosseae* group from contaminated soils with HMs in France. They showed that the two *Glomus mosseae* strains isolated from the polluted soils with cadmium had the ability to tolerate the cadmium concentrations from 50 to 70 and 200 to 500 mg l^{-1}, respectively. Sensitivity of different AM species or isolates and even propagules to different HMs may be varied. This can be dependent on phenotypes or genotypes characteristics of fungal species and HM type. Zarei (2008) demonstrated spore numbers were more affected by Zn and Pb concentrations than root colonization. The variations of AM fungi propagules were more related to available than total concentration of both metals. Metal-adapted AMF have a more efficient protecting effect on metal tolerance of host.

Mechanisms for HM Tolerance in AMF

Because of being compulsive of AMF symbiosis with plants and lack of its growth in conventional culture media, less information is available about the tolerance mechanisms of these fungi to HMs. The more information was based on plant response to HMs and observation of fungal structures within colonized roots, which were difficult to separate of fungal and plant responses. Different mechanisms were proposed for explanation of plant responses to the high concentrations of HMs. Primary effects can be diagnosed in the molecular, biochemical, and cellular levels, and next effects are visible in the physiological and organelle levels. Generally, suggested tolerance mechanisms of AMF to HMs are reviewed by Göhre and Paszkowski (2006), Gonzalez-Chavez et al. (2006), and Hildebrandt et al. (2007). In these fungi, the tolerance does not have a general pattern and may be different among species in the response to a particular metal. There are also large changes in the rate of tolerance to HMs among different populations of a species or ecotype. The mechanisms included extracellular chelation, binding of HM to the cell wall components of fungi, control of HMs transferring to the cell by metals' specific and nonspecific carriers in their plasma membrane, chelation in the cytoplasm as an intracellular buffer system, HM export via specific or nonspecific active or passive transport from cells and metal sequestration in the vacuoles, transport of HMs in the hyphae of the fungus, and active and passive transport of metals from the arbuscules to plant cells. Effects of AMF on the plant nutrition, root exudations, rhizosphere microbial communities, soil structure, and protection against environmental stresses can be considered indirect mechanisms in the increasing tolerance to HMs (Chen et al. 2003; Turnau et al. 2006; Vivas et al. 2006).

AMF colonization of the roots has a significant impact on the expression of several plant genes coding for proteins presumably involved in HM tolerance/ detoxification. A novel metallothioneins (MT)-like polypeptide designated *GmarMT1* that is modulated in a metal and life cycle stage-dependent manner and may afford protection against HMs (and other types of stress) to both partners of the AM symbiosis (Lanfranco et al. 2002). Hildebrandt et al. (2007) described genes expression in extraradical mycelia (ERM) of in vitro cultured *Glomus intraradices* Sy167 supplemented with different HMs (Cd, Cu, or Zn). The expression of several genes encoding proteins potentially involved in HM tolerance varied in their response to different HMs. Such proteins included a Zn transporter, a metallothionein, a 90-kDa heat shock protein, and a glutathione S-transferase (all assignments of protein function are putative). Studies on the expression of the selected genes were also performed with roots of *Medicago truncatula* grown in either a natural, Zn-rich HM soil or in a non-polluted soil supplemented with 100 µM $ZnSO_4$. The transcript levels of the genes analyzed were enhanced up to eightfold in roots grown in the HM-containing soils. The data obtained demonstrate the HM-dependent expression of different AMF genes in the intra- and extraradical mycelium. The HM-dependent induction of genes encoding a heat shock protein and a glutathione S-transferase in the mycelium of the AMF *G. intraradices* Sy167 suggests that alleviating the HM-induced oxidative stress might be of primary concern for AMF exposed to elevated HM. Other strategies possibly contributing to HM tolerance appear to be involved as well, which is indicated by the significantly enhanced expression of the metallothioneins and the Zn transporter gene, particularly under Cu stress. Molecular bases of HM tolerance in AM symbiotic system may also help the selection of the most effective AMF isolates (Tonin et al. 2001; Turnau et al. 2001) and plant–fungus combinations for bioremediation and soil protection purposes.

14.4 Biotechnological Approaches to Improve Phyto- and Bioremediation Efficiencies

With advances in biotechnology, biological remediation techniques, including phyto- and bioremediation, have become one of the most rapidly developing fields of environmental restoration and have been commercially applied for the treatment of hazardous wastes and contaminated sites. HM-hyperaccumulating plant species which possess a unique ability to accumulate metals to extremely high concentrations without suffering any toxic effects are unsuitable for phytoextraction purposes, due to their slow growth rate and low biomass. Genetic modification of fast-growing plants might be a viable alternative and provides a powerful method of improving the capacity of these plants to remediate various contaminants including HMs. Additionally, hyperaccumulators can provide an important resource of genes which are responsible for trace element hyperaccumulation and detoxification through unique biochemical and genetic mechanisms (Glazer and Nikaido 2007).

One of the important approaches using genetic engineering to enhance phytoremediation potential is to transform fast-growing host plants with unique genes from natural hyperaccumulators. One such gene encodes the enzyme selenocysteine methyltransferase (SMT), which has been cloned from the Se hyperaccumulator, *Astragalus bisulcatus* (Arshad et al. 2007). SMT converts the amino acid, selenocysteine, to the nonprotein amino acid, methylselenocysteine (MetSeCys). By doing so, it diverts the flow of Se from the Se-amino acids that may otherwise be incorporated into protein, leading to alterations in enzyme structure and function and possible toxicity. Additionally, Se-Cys may also cause oxidative damage. Transgenic plants overexpressing SMT show enhanced tolerance to Se, particularly selenite, and produced three- to sevenfold more biomass than the wild type plants.

Metallothioneins (MTs) and phytochelatins (PCs) are well-known HM-chelating proteins and peptides that play important roles in the detoxification of toxic HMs and the regulation of intracellular concentrations of essential metals in various organisms. Therefore, the expression of MTs and PCs in higher plants in order to enhance tolerance to HMs and their accumulation has great potential for phytoremediation of toxic HMs from contaminated soil and water (Meagher 2000; Mejare and Bülow 2001). Researchers expected that increasing the concentrations of metal-binding proteins or peptides in plant cells would increase metal-binding capacity and tolerance. In higher plants, PCs mainly function for detoxification of toxic HMs rather than MTs. Moreover, PCs have a higher metal-binding capacity rather than do MTs (Mehra and Mulchandani 1995). Therefore, modification or overexpression of PC synthase for accumulation of high levels of PCs seems to be a more practical approach to enhance HM accumulation in plants (Kazumasa et al. 2005). Overexpression of genes involved in PC synthesis, such as GSH1, GSH2, and PCS, encoding gamma-glutamylcysteine synthetase, glutathione synthetase, and PC synthase, respectively, has been shown to increase Cd tolerance in various heterologous expression systems. Likewise, heterologous MT overexpression often leads to increased tolerance to Cu and, occasionally, Cd and Zn. In general, overexpression of metal sequestration traits is associated with marginally to moderately increased accumulation of the metals concerned, presumably due to a delayed downregulation of the transporters involved in their uptake.

The approach of overexpressing genes that catalyze rate-limiting steps can also be used for the phytoremediation of HMs. GSH (Glu-Cys-Gly) plays an essential role in HM detoxification by plants. GSH is the direct precursor of PCs, which are metal-binding peptides involved in HM tolerance and sequestration (Steffens 1990). Additionally, GSH is a major component of the active oxygen scavenging system of the cell (Thomas 2008) and can protect the plant cell from Cd-induced oxidative stress (Gallego et al. 1996; Wecks and Clisjsters 1997). It is also possible that GSH detoxifies Cd by directly forming a GSH–Cd complex such as that reported for yeast (Litz and Lavi 1997). The role of GSH and PCs in HM tolerance is illustrated by the Cd hypersensitivity of Arabidopsis mutants defective in GSH and PC biosynthesis (Howden et al. 1995). γ-glutamylcysteine synthetase (γ-ECS) catalyzes the first step in the ATP-dependent synthesis of GSH. This is considered to be the rate-limiting step in the biosynthesis of GSH since the activity of this enzyme is subject to feedback regulation by GSH and is dependent upon the availability of cysteine (Steffens 1990). Zhu et al. (1999) studied the effect of overexpression of *E. coli*-γ-ECS, targeted to the chloroplasts of Indian mustard. The transgenic plants had two- to threefold higher levels of γ-EC as well as GSH and PC when subjected to Cd. Their increased Cd tolerance was almost certainly due to their higher production of PCs or GSH. In addition to conferring tolerance to Cd, overexpression of γ-ECS led to an increase in total shoot S suggesting an added advantage of enhanced S assimilation (Zhu et al. 1999). Similar results were also obtained in the case of poplar plants overexpressing γ-ECS (Arisi et al. 2003; Noctor et al. 1998). Overexpression of glutathione synthetase in Indian mustard also led to enhanced levels of GSH and PC2 in the presence of HMs (Zhu et al. 1999).

In the field of bioremediation, advances in genetic and protein engineering techniques have opened up new avenues to move towards the goal of genetically engineered microorganisms (GEMs) to function as "designer biocatalysts," in which certain desirable biodegradation pathways or enzymes from different organisms are brought together in a single host with the aim of performing specific reactions. A number of opportunities for improving degradation performance using GEMs have been described (Timmis and Piper 1999). Genetic engineering also permits the combination of several degradative activities within a single host organism. If a single strain is constructed to perform several related or unrelated metabolic activities, the efficiency and predictability of the process may be significantly enhanced. Such recombinant strains may be useful for the bioremediation of recalcitrant compounds (Brenner et al. 1994). Requirements for the design of bacteria with multiple pathways for use in bioremediation have been described (Lau and Lorenzo 1999; Gibson and Parales 2000). Timmis and Piper (1999) suggested a strategy for designing organisms with novel pathways and the creation of a bank of genetic modules encoding broad specificity enzymes or pathway segments that can be combined at will to generate new or improved activities. The use of appropriate regulatory circuits can enhance substrate flux through these designed pathways, and rationally engineering the pathway branch points can avoid or reduce substrate misrouting (Timmis and Piper 1999).

The diversity and adaptability of microorganisms allows them to thrive in harsh, toxic environments that prevent the growth of higher plants. For example, solar evaporation ponds, which are used to collect Se-contaminated agricultural drainage water, have extremely high concentrations of salt, Se, and other toxic trace elements. The specific composition of the microbial communities present in these ponds may themselves be useful for the bioremediation of Se since bacteria are able to produce volatile Se (Danika et al. 2005). Additionally, they may serve as reservoirs of unique genes involved in tolerance and volatilization of Se. Identification of the genes involved in these processes could pave the way for generating highly efficient plants by transferring these genes to the plants (McIntyre 2003).

Nowadays, developing methods to accelerate natural processes used in bioremediation of contaminated environments and also scientific understanding needed to harness these processes is necessary. Except few limiting factors, this technology has the ability to rejuvenate the contaminated environments effectively. However, rapid advances in the last few years have helped us in the understanding of process of bioremediation. The use of culture-independent molecular techniques has definitely helped us to understand the microbial community dynamics and structure and assisted in providing the insight in to details of bioremediation which has surely facilitated to make the technology safer and reliable. Bioremediation in relation to process optimization, validation, and its impact on the ecosystem can be performed, and by judicious use of the models that can predict the activity of microorganisms that are involved in bioremediation with existing geochemical and hydrological models, transformation of bioremediation from a mere practice into a science is now a reality. With the exciting new development in this field and focus on interdisciplinary research and using it on gaining the fundamental knowledge necessary to overcome the obstacles facing current technologies and also with respect to ethical, legal, and social issues involved, this technology will go a long way in cleaning the environment in near future (Keshav et al. 2010).

14.4.1 Assessment of Remediation Efficiency by Microbial Indicators of Soil Health

Despite the current great interest in improving the HM extraction capacity of hyperaccumulating plants, their influence on soil microorganisms has been rarely investigated (Delorme et al. 2001; Gremion et al. 2004). In fact, up to date, when evaluating the success of a phytoextraction process, emphasis has mostly been placed on metal removal. But it is most important to emphasize that the ultimate goal of any soil remediation process (physicochemical or biological processes) must be not only to remove the contaminant(s) from the polluted site or to render their harmless but also, most importantly, to restore the capacity of the soil to perform or function according to its potential as well (i.e., its health) (Hernandez-Allica et al. 2006). After all, some traditional methods of soil "remediation" irreversibly alter the functionality of the soil ecosystem while removing the contaminants, which is clearly not desirable and must be avoided at all costs. Soil quality or soil health (both terms are often used interchangeably) can be defined as the capacity of a given soil to successfully and sustainably perform its functions and ecosystem services from both an anthropocentric and ecocentric point of view and, most importantly, to properly recover its functionality after a disturbance. Regarding the recovery of soil health/functioning derived from the phytoextraction process, an ideal target should be to return to the conditions of a valid control soil (i.e., a vegetated, unpolluted soil of similar physicochemical properties and subjected to the same edaphoclimatic conditions). In this respect, indicators of soil health are needed to properly assess the efficiency of a phytoextraction process (Alkorta et al. 2003). Although to date, much more emphasis has been placed on physicochemical indicators of soil health, particularly, when evaluating the impact of agricultural practices on soil fertility and quality. Nonetheless, in the last years biological indicators such as enzyme activities, microbial biomass, basal- and substrate-induced respiration, potentially mineralizable N, and structural and functional biodiversity are most promising due to their being more sensitive to changes in the soil as well as to their capacity to provide information that integrates many environmental factors (Alkorta et al. 2003). Moreover, biological monitoring has special relevance to human health because it evaluates the effects of environmental changes on key elements of the food chain (Pandolfini et al. 1997). From all of the above, it is concluded that the success of metal phytoremediation procedures (phytoextraction and phytostabilization) must be evaluated not only in relation to the reduction of the concentration of total and bioavailable HMs but, most importantly, through the careful monitoring of the recovery of soil health using, among others, soil microbial properties as bioindicators, as microorganisms have a vital role in the functioning of the soil ecosystem.

Dehydrogenase activity, an intracellular process that occurs in every viable microbial cell, is used to determine overall microbiological activity of soil (Nannipieri et al. 2002). Soil microbial activity can also be measured through the determination of soil basal respiration (ISO 16072 Norm). In turn, soil microbial functional diversity can be determined through the utilization of community level physiological profiles (CLPPs) which reflect the potential of the cultivable portion of the heterotrophic microbial community to respond to carbon substrates (Bending et al. 2004). Substrate (glucose)-induced respiration (SIR) is a suitable indicator of potentially active microbial biomass (ISO 17155 Norm). The addition of carbon sources, other than glucose, commonly reported as constituents of root exudates might convert SIR into an ecologically more relevant parameter for testing rhizospheric microbial communities (Dedourge et al. 2004). The microbial respiration quotient (QR ¼ basal soil respiration to SIR ratio) has been used to assess the effects of various perturbations in soil ecosystems (Insam and Domsch 1988).

Evaluating soil microbial biomass through numerous methods is another bioindicator to access the success of remediation processes. Traditional enrichment culture-based techniques, such as heterotrophic plate counts, are frequently used; however, biases may be introduced by media type and richness, presence or absence of oxygen, and numerous other factors. Such techniques are thought to reveal as little as 10 % of the total microbial diversity in soil. For this reason, innovative methods have been developed to more completely describe soil microbial diversity. Recent scientific advances have made it possible to use molecular biological techniques for assessment of microbial communities in complex environmental systems. Several molecular biological techniques, such as PCR amplification, cloning, and sequencing of ribosomal RNA genes, denaturing gradient gel electrophoresis (DGGE), phospholipid fatty acid (PLFA) analysis, and thermal-GGE (TGGE), have recently been embraced by the environmental science community as important tools for predicting soil and water remediation success. Molecular methods, such as denaturing gradient gel electrophoresis (DGGE), rely on genetic differences to draw distinctions between microbes and microbial populations. Chemical extraction of phospholipid-fatty acids from soil can provide both a description of the diversity in that soil and an estimate of the microbial biomass present. Finally, most probable number (MPN), a specialized enrichment technique utilizing substrates of interest, gives an estimate of the number of organisms in an environment capable of degrading specific contaminants. Taken individually,

DGGE, phospholipid-fatty acid analysis, and MPN are all useful tools for understanding microbial communities. In combination, however, they are likely to yield extensive information on microbial biomass and community diversity. Furthermore, they provide the capability to pinpoint dominant groups of organisms and to assess the microbial community's ability to degrade contaminants. Integration of these diverse methods represents a potentially powerful tool for characterization—and, ultimately, optimization—of bioremediation systems.

Well-characterized techniques such as DGGE and TGGE separate amplification products by sequence-dependent helix denaturation and the accompanying change in electrophoretic mobility. Another approach, single-strand conformation polymorphism (SSCP), takes advantage of sequence-dependent conformational differences between reannealed single-stranded products, which also results in electrophoretic mobility changes. The T-RFLP was developed most recently and has three clear advantages. First, direct reference to the sequence database is possible. Second, the nucleic acid sequencing technology has considerably greater resolution than the electrophoretic systems of either DGGE or SSCP. Third, the T-RFLP gel analysis is instantaneous and the output is digital. So, T-RFLP is a molecular approach that can assess subtle genetic differences between strains as well as provide insight into the structure and function of microbial communities. The technique has both high sensitivity and throughput; it is an ideal tool for comparative analyses.

Amplified ribosomal DNA restriction analysis (ARDRA) is another community analysis technique that provides a representation of the microbial community through restriction analysis of clones in an rDNA library. Although ARDRA was effective for identifying phylogenetic groups in a highly diverse community, it is prohibitively expensive and time consuming for this research project, because of the construction of clones and the identification of environmental clones by sequence analysis. In conclusion, T-RFLP provides a sensitive and rapid technique for assessing amplification-produced diversity within a community as well as comparative distribution across communities.

14.5 Conclusions and Perspective

The soil will become more and more valuable as a commodity. The preservation of soil quality, including its restoration in case the latter has been lost, certainly is a business opportunity, which will keep growing in all developed and industrialized countries. Phytoremediation is the processes of cleaning contaminated soils by making use of the metabolic properties of plants. During the last two decades, this field of soil remediation holds great potential as an environmental cleanup technology which has been extensively researched and developed. The search for hyperaccumulating plants has recently involved the identification of metal-excluding plants to learn more about plants that ensure that at least the edible part of the plant will be free of toxic metals. Scientists expect that the mechanisms responsible for storing or excluding metals in plants are interrelated and that knowledge of one will lead to an understanding of the other. Genetic engineering approaches are currently being used to optimize the metabolic and physiological processes that enable hyperaccumulating plants to phytoremediate sites contaminated with HMs and metalloids. Someday, genetically modified plants will be developed to extract all types of contamination from soil and water and, at the same time, eliminate the largest drawback to current phytoremediation technology—the time required for plants to remediate contaminated sites. Microbes isolated from highly contaminated environments represent another potentially huge reservoir of new genes and unique metabolic capabilities that could be transferred to plants to enhance their phytoremediation potential. Microbes, in many cases, are more efficient in accumulating and absorbing HMs because of their astronomical amount and specific surface area. Furthermore, technique of genetic engineering in microbes is easier and more mature than in plant cells. Just as in pristine sites, there is always a close interaction between the microorganisms in the rhizosphere like plant growth-promoting rhizobacteria (PGPR) and arbuscular mycorrhizal fungi (AMF) and host plants which can lead to an increased activity related to soil remediation (Compant et al. 2010). Thus, understanding and controlling the combination of soil, a beneficial rhizo- and/or endospheric microbial community and plants systems, and, even more important, their interactions provides a great opportunity for various innovative approaches to improve soil cleaning and production processes. Recent research on PGPR and AMF combined with genetic engineering illustrates a promising vision for future research. While advances in remediation have increased the effectiveness of HM degradation, still little is known about the interactions between PGPR, AMF, plant roots, and other microorganisms. Also, the mechanism of mobilization and transfer of metals is not fully understood. Additionally, most phytoremediation studies with PGPR and AMF have been conducted in the lab or greenhouse, overlooking the more complicated natural ecosystem. A more comprehensive understanding of these microbes in their natural environment is needed for this technology to reach its full potential.

References

Alkorta I, Aizpurua A, Riga P, Albizu I, Amezaga I, Garbisu C (2003) Soil enzyme activities as biological indicators of soil health. Rev Environ Health 18:65–73

Alloway BJ, Jackson AP (1991) The behavior of heavy metals in sewage sludge amended soils. Sci Total Environ 100:151–176

Alonso J, Garcia AM, Pérez-López M, Melgar MJ (2003) The concentrations and bioconcentration factors of copper and zinc in edible mushrooms. Arch Environ Contam Toxicol 44:180–188

Anderson TA, Coats JR (1994) Bioremediation through rhizosphere technology, vol 563, ACS symposium series. American Chemical Society, Washington, DC, p 249

Aoyama M, Itaya S, Otowa M (1993) Effects of copper on the decomposition of plant residues, microbial biomass and beta-glucosidase activity in soils. Soil Sci Plant Nutr 39:557–566

Arisi ACM, Noctor G, Foyer CH, Jouanin L (2003) Modification of thiol contents in poplars (Populus tremula × P. alba) overexpressing enzymes involved in glutathione synthesis. Planta 203:362–372

Arnold PT, Kaputska LA (1987) VA mycorrhizal colonization and spore populations in abandoned agricultural field after five years of sludge additions. Ohio J Sci 87:112–114

Arshad M, Saleem M, Hussain S (2007) Perspectives of bacterial ACC deaminase in phytoremediation. Trends Biotechnol 25:356–362

Assunçao AGL, Da Costa MP, De Folter S, Vooijs R, Schat H, Aarts MGM (2001) Elevated expression of metal transporter genes in three accessions of the metal hyperaccumulator *Thlaspi caerulescens*. Plant Cell Environ 24:217–226

Bååth E (1989) Effects of heavy metals in soil on microbial processes and populations: a review. Water Air Soil Pollut 47:335–379

Baker AJM, Brooks RR (1989) Terrestrial higher plants which hyperaccumulate metallic elements – a review of their distribution, ecology and phytochemistry. Biorecovery 1:81–126

Balsalobre C, Calonge J, Jiménez E, Lafuente R, Mouriño M, Muño MT, Riquelme M, Mas-Castella J (1993) Using the metabolic capacity of Rhodobacter sphaeroides to assess heavy metal toxicity. Environ Toxicol Water Qual 8:437–450

Barceló J, Poschenrieder C (2003) Phytoremediation: principles and perspectives. Contrib Sci 2: 333–344

Bardgett RD, Speir TW, Ross DJ, Yeats GW, Kettles HA (1994) Impact of pasture contamination by copper, chromium, and arsenic timber preservative on soil microbial properties and nematodes. Biol Fertil Soils 18:71–79

Begonia MT, Begonia GB, Miller G, Gilliard D, Young C (2004) Phosphatase activity and populations of microorganisms from cadmium and lead contaminated soils. Bull Environ Contam Toxicol 73:1025–1032

Bending GD, Turner MK, Rayns F, Marx MC, Wood M (2004) Microbial and biochemical soil quality indicators and their potential for differentiating areas under contrasting agricultural management regimes. Soil Biol Biochem 36:1785–1792

Berdicevsky I, Duek L, Merzbach D, Yannai S (1993) Susceptibility of different yeast species to environmental toxic metals. Environ Pollut 80:41–44

Bogomolov DM, Chen SK (1996) An ecosystem approach to soil toxicity testing—A study of copper contamination in laboratory soil microcosms. Appl Soil Ecol 4:95–105

Bohn KS and Liberta AE (1982) In: Graves DH (ed) Symposium on surface mining hydrology, sedimentology and reclamation. University of Kentucky, Lexington

Boyle M, Paul EA (1988) Vesicular-arbuscular mycorrhizal associations with barley on sewage-amended plots. Soil Biol Biochem 20:945–948

Brenner V, Arensdorf JJ, Foght DD (1994) Genetic construction of PCB degraders. Biodegradation 5:359–377

Brookes PC (1995) The use of microbial parameters in monitoring soil pollution by heavy metals. Biol Fertil Soils 19:269–279

Brooks RR (1998) Geobotany and hyperaccumulators. In: Brook RR (ed) Plants that hyperaccumulate heavy metals. CAB, Walingford, pp 55–94

Brown PE, Minges GA (1916) The effect of some manganese salts on ammonification and nitrification. Soil Sci 1:67–85

Chen B, Christie P, Li X (2001) A modified glass bead compartment cultivation system for studies on nutrient and trace metal uptake by arbuscular mycorrhiza. Chemosphere 42:185–192

Chen BD, Li XL, Tao HQ, Christie P, Wong MH (2003) The role of arbuscular mycorrhiza in zinc uptake by red clover growing in a calcareous soil spiked with various quantities of zinc. Chemosphere 50:839–846

Christie P, Beattie JAM (1989) Grassland soil microbial biomass and accumulation of potentially toxic metals from long term slurry application. J Appl Ecol 26:597–612

Cobbett CS (2000) Phytochelatin biosynthesis and function in heavy-metal detoxification. Curr Opin Plant Biol 3:211–216

Compant S, Clément B, Sessitsch A (2010) Plant growth-promoting bacteria in the rhizo- and endosphere of plants: their role, colonization, mechanisms involved and prospects for utilization. Soil Biol Biochem 42:669–678

Crutzen PJ (1970) The influence of nitrogen oxides on the atmospheric ozone content. Q J R Meteorol Soc 96:320–325

Cunningham SD, Ow DW (1996) Promises and prospects of phytoremediation. Plant Physiol 110:715–719

Cunningham SD, Shann JR, Crowley DE, Anderson TA (1997) Phytoremediation of contaminated water and soil. In: Kruger EL, Anderson TA, Coats JR (eds.) Phytoremediation of soil and water contaminants. ACS symposium series 664. American Chemical Society, Washington, DC, pp. 2–19

Dahlin S, Witter E, Mårtensson AM, Turner A, Bååth E (1997) Where's the limit? Changes in the microbiological properties of agricultural soils at low levels of metal contamination. Soil Biol Biochem 29:1405–1415

Danika L, Duc L, Terry N (2005) Phytoremediation of toxic trace elements in soil and water. J Ind Microbiol Biotechnol 32:514–520

Dedourge O, Vong PC, Lasserre-Joulin F, Benizri E, Guckert A (2004) Effect of glucose and rhizodeposits (with or without cysteine-S) on immobilized-^{35}S, microbial biomass-^{35}S and arylsulphatase activity in a calcareous and an acid brown soil. Eur J Soil Sci 55:649–656

Del Val C, Barea JM, Azcon-Aguilar C (1999) Assessing the tolerance to heavy metals of arbuscular mycorrhizal fungi isolated from sewage sludge-contaminated soils. Appl Soil Ecol 11:261–269

Delorme TA, Gagliardi JV, Angle JS, Chaney RL (2001) Influence of the zinc hyperaccumulator *Thlaspi caerulescens* J. & C. Presl. and the nonmetal accumulator *Trifolium pratense* L. on soil microbial populations. Can J Microbiol 47:773–776

Diaz G, Honrubia M (1994) A mycorrhizal survey of plants growing on mine wastes in southeast Spain. Arid Soil Res Rehabil 8:59–68

Diaz G, Azcon-Aguilar C, Honrubia M (1996) Influence of arbuscular mycorrhizae on heavy metal (Zn and Pb) uptake and growth of Lygeum spartum and Anthillis cystisoides. Plant Soil 180: 1201–1205

Dickinson RE, Cicerone RJ (1986) Future global warming from atmospheric trace gases. Nature 319:109–115

Dmitri S, Begonia FT (2008) Effects of heavy metal contamination upon soil microbes: lead-induced changes in general and denitrifying microbial communities as evidenced by molecular markers. Int J Environ Res Public Health 5:450–456

Dodd JC, Thompson BD (1994) The screening and selection of inoculant arbuscular mycorrhizal and ectomycorrhizal fungi. Plant Soil 159:149–158

Doelman P, Haanstra L (1984) Short-term and long-term effects of cadmium, chromium, copper, nickel, lead and zinc on soil microbial respiration in relation to abiotic soil factors. Plant Soil 79:317–327

Dueck TA, Visser P, Ernst WHO, Schat H (1986) Vesicular-arbuscular mycorrhizae decrease zinc toxicity to grasses growing in zinc polluted soils. Soil Biol Biochem 18:331–333

Elekes CC, Busuoic G, Ionita G (2010) The mycoremediation of metals polluted soils using wild growing species of mushrooms. Latest trends on engineering education. Not Bot Horti Agrobot Cluj Napoca 38:147–151

Filser J, Fromm H, Nagel RF, Winter K (1995) Effects of previous intensive agricultural management on microorganisms and the biodiversity of soil fauna. Plant Soil 170:123–129

Fritze H, Vanhala P, Pietikäinen J, Mälkönen E (1996) Vitality fertilization of Scots pine stands growing along a gradient of heavy metal pollution: short-term effects on microbial biomass and respiration rate of the humus layer. Fresenius J Anal Chem 354:750–755

Frostegård Å, Tunlid A, Bååth E (1993) Phospholipid fatty acid composition, biomass and activity of microbial communities from two soil types experimentally exposed to different metals. Appl Environ Microbiol 59:3605–3617

Frostegård Å, Tunlid A, Bååth E (1996) Changes in microbial community structure during long-term incubation in two soils experimentally contaminated with metals. Soil Biol Biochem 28: 55–63

Fulladosa E, Murat JC, Martínez M, Villaescusa I (2005a) Patterns of metals and arsenic poisoning in *Vibrio fischeri*. Chemosphere 60:43–48

Fulladosa E, Murat JC, Villaescusa I (2005b) Study on the toxicity of binary equitoxic mixtures of metals using the luminescent bacteria Vibrio fischeri as a biological target. Chemosphere 58(5):551–557

Gadd GM (2001) Fungi in bioremediation. Cambridge University Press, Cambridge

Gadd GM (2010) Metals, minerals and microbes: geomicrobiology and bioremediation. Microbiology 156:609–643

Gadd GM, White C (1989) Heavy metal and radionuclide accumulation and toxicity in fungi and yeasts. In: Poole RK, Gadd GM (eds) Metal-microbe interactions. Special publication of the Society for General Microbiology, vol 26. IRL/Oxford University Press, New York, pp 19–38

Gallego SM, Benavides MP, Tomaro ML (1996) Effect of heavy metal ion excess on sun-flower leaves: evidence for involvement of oxidative stress. Plant Sci 121:151–159

Gang WU, Kang H, Xiaoyang Z, Hongbo S, Liye C, Chengjiang R (2010) A critical review on the bio-removal of hazardous heavy metals from contaminated soils: issues, progress, eco-environmental concerns and opportunities. J Hazard Mater 174:1–8

Garbisu C, Hernandez-Allica J, Barrutia O, Alkorta I, Becerril JM (2002) Phytoremediation: a technology using green plants to remove contaminants from polluted areas. Rev Environ Health 17:75–90

Gasper GM, Mathe P, Szabo L, Orgovanyl B, Uzinger N, Anton A (2005) After-effect of heavy metal pollution in brown forest soils. Proceedings of the 8th Hungarian Congress on Plant Physiology and the 6th Hungarian Conference on Photosynthesis. Acta Biologica Szegediensis 49:71–72

Gast CH, Jansen E, Bierling J, Haanstra L (1998) Heavy metals in mushrooms and their relationship with soil characteristics. Chemosphere 17:789–799

Gaur A, Adholeya A (2004) Prospects of arbuscular mycorrhizal fungi in phytoremediation of heavy metal contaminated soils. Curr Sci 86:528–534

Geiger G, Federer P, Sticher H (1993) Reclamation of heavy metal-contaminated soils: field studies and germination experiments. J Environ Qual 22:201–207

Giasson P, Jaouich A, Gagné S, Massicotte L, Cayer P, Moutoglis P (2006) Enhanced phytoremediation: a study of mycorrhizoremediation of heavy metal contaminated soil. Remediation 17:97–110

Gibson DT, Parales RE (2000) Aromatic hydrocarbons dioxygenases in environmental biotechnology. Curr Opin Biotechnol 11:236–243

Gildon A, Tinker PB (1981) A heavy metal-tolerant strain of mycorrhizal fungus. Trans Br Mycol Soc 77:648–649

Gildon A, Tinker PB (1983) Interactions of vesicular-arbuscular mycorrhizal infection and heavy metals in plants. The effects of heavy metals on the development of vesicular-arbuscular mycorrhizas. New Phytol 95:247–261

Giller KE, Beare MH, Lavelle P, Izac MN, Swift MJ (1997) Agricultural intensification, soil biodiversity and ecosystem function. Appl Soil Ecol 6:3–16

Glazer AN, Nikaido H (2007) Microbial biotechnology: fundamentals of applied microbiology, 2nd edn. Cambridge University press, Cambridge, pp 510–528

Glick BR (2003) Phytoremediation: synergistic use of plants and bacteria to clean up the environment. Biotechnol Adv 21:383–393

Göhre V, Paszkowski U (2006) Contribution of the arbuscular mycorrhizal symbiosis to heavy metal phytoremediation. Planta 223:1115–1122

Gonzalez-Chavez C, Harris PJ, Dodd J, Meharg AA (2002) Arbuscular mycorrhizal fungi confer enhanced arsenate resistance on Holcus lanatus. New Phytol 155:163–171

Gonzalez-Chavez MC, Vangronsveld J, Colpaert J, Leyval C (2006) Arbuscular mycorrhizal fungi and heavy metals: tolerance mechanisms and potential use in bioremediation. In: Prasad MNV,

Sajwan KS, Naidu R (eds) Trace elements in the environment. Biogeochemistry, biotechnology, and bioremediation. CRC, Boca Raton, FL, pp 211–234

Gremion F, Chatzinotas A, Kaufmann K, Von Sigler W, Harms H (2004) Impacts of heavy metal contamination and phytoremediation on a microbial community during a twelve-month microcosm experiment. FEMS Microbiol Ecol 48:273–283

Gupta SK (1992) Mobilizable metal in anthropogenic contaminated soils and its ecological significance. In: Vernet JP (ed) Impact of heavy metals on the environment. Elsevier, Amsterdam, pp 299–310

Hall JL (2002) Cellular mechanisms for heavy metal detoxification and tolerance. J Exp Bot 53: 1–11

Hammer D, Kayser A, Keller C (2003) Phytoextraction of Cd and Zn with Salix viminalis in field trials. Soil Use Manag 19:187–192

Harris PJ (1994) Consequences of the spatial distribution of microbial communities in soil. In: Ritz K, Dighton J, Giller KE (eds) Beyond the biomass. Compositional and functional analysis of soil microbial communities. Wiley, Chichester, pp 239–246

Hartley-Whitaker J, Ainsworth G, Vooijs R, Ten Bookum W, Schat H, Meharg AA (2001) Phytochelatins are involved in differential arsenate tolerance in Holcus lanatus. Plant Physiol 126:299–306

Haselwandter K, Leyval C, Sanders FE (1994) Impact of arbuscular mycorrhizal fungi on plant uptake of heavy metals and radionuclides from soil. In: Gianinazzi S, SchuÈepp H (eds) Impact of arbuscular mycorrhizas on sustainable agriculture and natural ecosystems. BirkhaÈuser, Basel, pp 179–189

Hattori H (1996) Decomposition of organic matter with previous cadmium adsorption in soils. Soil Sci Plant Nutr 42:745–752

Hernandez-Allica J, Becerril JM, Zarate O, Garbisu C (2006) Assessment of the efficiency of a metal phytoextraction process with biological indicators of soil health. Plant Soil 281:147–158

Hildebrandt U, Regvar M, Bothe H (2007) Arbuscular mycorrhiza and heavy metal tolerance. Phytochemisrty 68:139–146

Hinojosa MB, Carreira J, García-Ruíza R, Dick RP (2004) Soil moisture pre-treatment effects on enzyme activities as indicators of heavy metal-contaminated and reclaimed soils. Soil Biol Biochem 36:1559–1568

Holtan-Hartwig L, Bechmann M, Høyås TR, Linjordet R, Bakken LR (2002) Heavy metals tolerance of soil denitrifying communities: N_2O dynamics. Soil Biol Biochem 34:1181–1190

Homer FA, Reeves RD, Brooks RR (1997) The possible involvement of aminoacids in nickel chelation in some nickel-accumulating plants. Curr Top Phytochem 14:31–33

Howden R, Anderson CR, Goldsbrough PB, Cobbett CS (1995) A cadmium-sensitive, glutathione-deficient mutant of Arabidopsis thaliana. Plant Physiol 107:1067–1073

Hüttermann A, Arduini I, Godbold DL (1999) Metal pollution and forest decline. In: Prasad NMV, Hagemeyer J (eds) Heavy metal stress in plants: from molecules to ecosystems. Springer, Berlin, pp 253–272

Illmer P, Schinner F (1991) Effects of lime and nutrient salts on the microbiological activities of forest soils. Biol Fertil Soils 11:261–266

Insam H, Domsch KH (1988) Relationship between soil organic carbon and microbial biomass on chronosequences of reclamation sites. Microb Ecol 15:177–188

Jentschke G, Godbold DL (2000) Metal toxicity and ectomycorrhizas. Physiol Plant 109:107–116

Kabata-Pendias A (2001) Trace elements in soils and plants, 3rd edn. CRC, Boca Raton, FL

Kabata-Pendias A, Mukherjee AB (2007) Trace elements from soil to human. Springer, Heidelberg, 550 pp

Kaldorf M, Kuhn A, Schroder WH, Hildebrandt U, Bothe H (1999) Selective element deposits in maize colonized by a heavy metal tolerance conferring arbuscular mycorrhizal fungus. J Plant Physiol 154:718–728

Kandeler E, Kampichler C, Horak O (1996) Influence of heavy metals on the functional diversity of soil microbial communities. Biol Fertil Soils 23:299–306

Kandeler E, Tscherko D, Bruce KD, Stemmer M, Hobbs PJ, Bardgett RD, Amelung W (2000) Structure and function of the soil microbial community in microhabitats of a heavy metal polluted soil. Biol Fertil Soils 32:390–400

Kayser A, Wenger K, Keller A, Attinger W, Felix HR, Gupta SK, Schulin R (2000) Enhancement of phytoextraction of Zn, Cd, and Cu from calcareous soil: the use of NTA and sulfur amendments. Environ Sci Technol 34:1778–1783

Kazumasa H, Naoki T, Kazuhisa M (2005) Biosynthetic regulation of phytochelatins, heavy metal-binding peptides. J Biosci Bioeng 100:593–599

Keller C, McGrath SP, Dunham SJ (2002) Trace metal leaching through a soil–grassland system after sewage sludge application. J Environ Qual 31:1550–1560

Kelly JJ, Haggblom MM, Tate RL (2003) Effects of heavy metal contamination and remediation on soil microbial communities in the vicinity of a zinc smelter as indicated by analysis of microbial community phospholipids fatty acid profiles. Biol Fertil Soils 38:65–71

Keshav PS, Nand KS, Shivesh S (2010) Bioremediation: developments, current practices and perspectives. Genet Eng Biotechnol J 2010:1–20

Khade SW, Adholeya A (2007) Feasible bioremediation through arbuscular mycorrhizal fungi imparting heavy metal tolerance: a retrospective. Bioremediat J 11:33–43

Khan AG (2006) Mycorrhizoremediation—an enhanced form of phytoremediation. J Zhejiang Univ Sci B 7:503–514

Khodaverdiloo H, Homaee M (2008) Modeling of cadmium and lead phytoextraction from contaminated soil. Pol J Soil Sci 41:149–162

Knight B, McGrath SP, Chaudri AM (1997) Biomass carbon measurements and substrate utilization patterns of microbial populations from soils amended with cadmium, copper or zinc. Appl Environ Microbiol 63:39–43

Kuperman RG, Carreiro MM (1997) Soil heavy metal concentrations, microbial biomass and enzyme activities in a contaminated grassland ecosystem. Soil Biol Biochem 29:179–190

Ladd JNM, Amato M, Grace PR, van Veen JA (1995) Simulation of ^{14}C turnover through the microbial biomass in soils incubated with ^{14}C-labelled plant residues. Soil Biol Biochem 27:777–783

Lanfranco L, Bolchi A, Ros EC, Ottonello S, Bonfante P (2002) Differential expression of a metallothionein gene during the presymbiotic versus the symbiotic phase of an arbuscular mycorrhizal fungus. Plant Physiol 130:58–67

Lau PCK, Lorenzo VDE (1999) Genetic engineering: the frontier of bioremediation. Environ Sci Technol 4:124A–128A

Leita L, Denobili M, Muhlbachova G, Mondini C, Marchiol L, Zerbi G (1995) Bioavailability and effects of heavy metals on soil microbial biomass survival during laboratory incubation. Biol Fertil Soils 19:103–108

Leung HM, Ye ZH, Wong MH (2006) Interactions of mycorrhizal fungi with Pteris vittata (as hyperaccumulator) in As-contaminated soils. Environ Pollut 139:1–8

Leyval C, Singh BR, Joner EJ (1995) Occurrence and infectivity of arbuscular mycorrhiza fungi in some Norwegian soils influenced by heavy metals and soil properties. Water Air Soil Pollut 84: 203–216

Leyval C, Turnau A, Haselwandter K (1997) Effect of heavy metal pollution on mycorrhizal colonization and function: physiological, ecological and applied aspects. Mycorrhiza 7: 139–153

Lipman CB, Burgess PS (1914) The effects of copper, zinc, iron and lead salts on ammonification and nitrification in soils. Univ Calif Publ Agric Sci 1:127–139

Litz R, Lavi U (1997) Mango biotechnology (chapter 12). In: Litz R (ed) The mango. CRC, Boca Raton, FL, pp 401–424

Lombi E, Wenzel WW, Gobran GR, Adriano DC (2001) Dependency of phyto-availability of metals on indigenous and induced rhizosphere processes: a review. In: Gobran GR, Wenzel WW, Lombi E (eds) Trace elements in the rhizosphere. CRC, New York, pp 3–24

Long LK, Yao Q, Guo J, Yang RH, Huang YH, Zhu HH (2010) Molecular community analysis of arbuscular mycorrhizal fungi associated with five selected plant species from heavy metal polluted soils. Eur J Soil Biol 46:288–294

Macnair MR, Tilstone GH, Smith SE (2000) The genetics of metal tolerance and accumulation in higher plants. In: Terry N, Banuelos G (eds) Phytoremediation of contaminated soil and water. CRC, Boca Raton, FL, pp 235–250

Maliszewska W, Dec S, Wierzbicka H, Wozniakowska A (1985) The influence of various heavy metal compounds on the development and activity of soil micro-organisms. Environ Pollut A 37:195–215

Marschner H, Dell B (1994) Nutrient uptake in mycorrhizal symbiosis. Plant Soil 159:89–102

Marschner B, Kalbitz K (2003) Control of bioavailability and biodegradation of dissolved organic matter in soils. Geoderma 113:211–235

Mathur NJ, Singh S, Bohra A, Vyas A (2007) Arbuscular mycorrhizal fungi: a potential tool for phytoremediation. J Plant Sci 2:127–140

McGrath SP (1994) Effects of heavy metals from sewage sludge on soil microbes in agricultural ecosystems. In: Ross SM (ed) Toxic metals in soil-plant systems. Wiley, Chichester, pp 242–274

McGrath SP, Chaudri AM, Giller KE (1995) Long term effects of metals in sewage sludge on soils, microorganisms and plants. J Ind Microbiol 14:94–104

McIntyre T (2003) Phytoremediation of heavy metals from soils. Adv Biochem Eng Biotechnol 78:97–123

Meagher RB (2000) Phytoremediation of toxic elemental and organic pollutants. Curr Opin Plant Biol 3:153–162

Mehra RK, Mulchandani P (1995) Glutathione-mediated transfer of Cu(I) into phytochelatins. Biochem J 307:687–705

Mehra RK, Winge DR (1991) Metal ion resistance in fungi: molecular mechanisms and their regulated expression. J Cell Biochem 45:30–40

Mejare M, Bülow L (2001) Metal-binding proteins and peptides in bioremediation and phytoremediation of heavy metals. Trends Biotechnol 19:67–73

Mench MJ, Didier VL, Loffler M, Gomez A, Masson P (1994) A mimicked insitu remediation study of metalcontaminated soils with emphasis on cadmium and lead. J Environ Qual 23: 785–792

Moftah AE (2000) Physiological response of lead polluted tomato and eggplant to the antioxidant ethylene diurea. Menofia Agric Res 25:933–955

Morley GF, Sayer JA, Wilkinson SC, Gharieb MM, Gadd GM (1996) Fungal sequestration, solubilization and transformation of toxic metals. In: Frankland JC, Magan N, Gadd GM (eds) Fungi and environmental change. Cambridge University Press, Cambridge, pp 235–256

Naidu CK, Reddy TKR (1988) Effect of cadmium on microorganisms and microbe-mediated mineralization process in soil. Bull Environ Contam Toxicol 41:657–663

Nannipieri P, Kandeler E, Ruggiero P (2002) Enzyme activities and microbiological and biochemical processes in soil. In: Burns RG, Dick R (eds) Enzymes in the environment. Marcel Dekker, New York, pp 1–33

Nannipieri P, Ascher J, Ceccherini MT, Landi L, Pietramellara G, Renella G (2003) Microbial diversity and soil functions. Eur J Soil Sci 54:655–670

Noctor G, Arisi ACM, Jouanin L, Kuner KJ, Rennenberg H, Foyer C (1998) Glu-tathione biosynthesis metabolism and relationship to stress tolerance explored in transformed plants. J Exp Bot 49:623–647

Ohya H, Komai Y, Amaguchi MY (1985) Zinc effects on soil microflora and glucose metabolites in soil amended with 14C-glucose. Biol Fertil Soils 1:117–122

Ouzouni PK, Veltsistas PG, Paleologos EK, Riganakos KA (2007) Determination of metal content in wild edible mushrooms species from region of Greece. J Food Compos Anal 20:480–486

Pandolfini T, Gremigni P, Gabbrielli R (1997) Biomonitoring of soil health by plants. In: Pankhurst CE, Doube BM, Gupta VVSR (eds) Biological indicators of soil health. CAB, New York, pp 325–347

Pawlowska TE, Charvat I (2004) Heavy metal stress and developmental patterns in arbuscular mycorrhizal fungi. Appl Environ Microbiol 70:6643–6649

Pawlowska TE, Blaszkowski J, Rühling A (1996) The mycorrhizal status of plants colonizing a calamine spoil mound in southern Poland. Mycorrhiza 6:499–505

Pennanen T, Frostegård A, Fritze H, Bååth E (1996) Phospholipid fatty acid composition and heavy metal tolerance of soil microbial communities along two heavy metal-polluted gradients in coniferous forests. Appl Environ Microbiol 62:420–428

Rahmanian M, Khodaverdiloo H, Rezaee Danesh Y, Rasouli Sadaghiani MH (2011) Effects of heavy metal resistant soil microbes inoculation and soil Cd concentration on growth and metal uptake of millet, couch grass and alfalfa. Afr J Microbiol Res 5:403–410

Rajapaksha R, Tobor-Kaplon MA, Bååth E (2004) Metal toxicity affects fungal and bacterial activities in soil differently. Appl Environ Microbiol 70:2966–2973

Ranjard L, Nazaret S, Gourbiere F, Thioulouse J, Linet P, Richaume A (2000) A soil microscale study to reveal the heterogeneity of Hg (II) impact on indigenous bacteria by quantification of adapted phenotypes and analysis of community DNA fingerprints. FEMS Microbiol Ecol 31: 107–115

Renella G, Mench M, van der Lelie D, Pietramellara G, Ascher J, Ceccherini MT, Landi L, Nannipieri P (2004) Hydrolase activity, microbial biomass and community structure in long-term Cd-contaminated soils. Soil Biol Biochem 36:443–451

Romandini P, Tallandini L, Beltramini M, Salvato B, Manzano M (1992) Effects of copper and cadmium on growth, superoxide dismutase and catalase activities in different yeast strains. Comp Biochem Physiol 103C:255–262

Rufyikiri G, Huysmans L, Wannijin J, Hees MV, Leyval C, Jakobsen I (2004) Arbuscular mycorrhizal fungi can decrease the uptake of uranium by subterranean clover grown at high levels of uranium in soil. Environ Pollut 130:427–436

Salt DE, Smith RD, Raskin I (1998) Phytoremediation. Annu Rev Plant Physiol Plant Mol Biol 49: 643–668

Sandaa RA, Torsvik V, Enger O, Daae FL, Castberg T, Hahn D (1999) Analysis of bacterial communities in heavy metal-contaminated soils at different levels of resolution. FEMS Microbiol Ecol 30:237–251

Saxena PK, Raj SK, Dan T, Perras MR, Vettakkorumakankav NN (1999) Phytoremediation of heavy metal contaminated and polluted soils. In: Prasad MNV, Hagemayr J (eds) Heavy metal stress in plants. From molecules to ecosystems. Springer, Berlin, pp 305–329

Schnoor JL (1997) Phytoremediation. Technology evaluation report. Ground-Water Remediation Technologies Analysis Center. E Series TE-98-101

Schubler A, Schwarzott D, Walker C (2001) A new fungal phylum, the Glomeromycota: phylogeny and evolution. Mycol Res 105:1413–1421

Shakolnik MY (1984) Trace elements in plants. Elsevier, NewYork, pp 140–171

Sharma A, Talukdar G (1987) Effects of metals on chromosomes of higher organisms. Environ Mutagen 9:191–226

Shi W, Becker J, Bischoff M, Turco RF, Konopka AE (2002) Association of microbial community composition and activity with lead, chromium, and hydrocarbon contamination. Appl Environ Microbiol 68:3859–3866

Singh H (2006) Mycoremediation: fungal bioremediation. Wiley-Interscience, Hoboken, NJ

Smejkalova M, Mikanova O, Boruvka L (2003) Effect of heavy metal concentration on biological activity of soil microorganisms. Plant Soil Environ 49:321–326

Smith SE, Read DJ (2008) Mycorrhizal symbiosis, 3rd edn. Academic and Elsevier, London

Soylak M, Saraçoğlu S, Tüzen M, Mendil D (2005) Determination of trace metals in mushroom sample from Kayseri, Turkey. Food Chem 92:649–652

Speir TW, Kettles HA, Parshotam A, Searle PL, Vlaar LNC (1995) A simple kinetic approach to derive the ecological dose value, ED(50), for the assessment of Cr(VI) toxicity to soil biological properties. Soil Biol Biochem 27:801–810

Steffens JC (1990) The heavy metal-binding peptides of plants. Annu Rev Plant Physiol 41: 553–575

Stresty EV, Madhava Rao KV (1999) Ultrastructural alterations in response to zinc and nickel stress in the root cells of pigeonpea. Environ Exp Bot 41:3–13

Svoboda L, Havličková B, Kalač P (2006) Contents of cadmium, mercury and lead in edible mushrooms growing in a historical silver-mining area. Food Chem 96:580–585

Szili-Kovács T, Anton A, Gulyás F (1999) Effect of Cd, Ni and Cu on some microbial properties of a calcareous chernozem soil. In: Kubát J (ed) Proceedings of 2nd symposium on the pathways and consequences of the dissemination of pollutants in the biosphere, Prague, pp 88–102

Thomas KW (2008) Molecular approaches in bioremediation. Curr Opin Biotechnol 19:572–578

Timmis KN, Piper DH (1999) Bacteria designed for bioremediation. Trends Biotechnol 17: 201–204

Tonin C, Vandenkoornhuyse P, Joner EJ, Straczek J, Leyval C (2001) Assessment of arbuscular mycorrhizal fungi diversity in the rhizosphere of Viola calaminaria and effect of these fungi on heavy metal uptake by clover. Mycorrhiza 10:161–168

Torslov J (1993) Comparison of bacterial toxicity tests based on growth, dehydrogenase activity and esterase activity of Pseudomonas fluorescens. Ecotoxicol Environ Saf 25:33–40

Turnau K, Ryszka P, Gianinazzi PV, van Tuinen D (2001) Identification of arbuscular mycorrhizal fungi in soils and roots of plant colonizing zinc wastes in southern Poland. Mycorrhiza 10: 169–174

Turnau K, Jurkiewicz A, Lingua G, Barea JM, Gianinazzi-Pearson V (2006) Role of arbuscular mycorrhiza and associated microorganisms in phytoremediation of heavy metal-polluted sites (chapter 13). In: Prasad MNV, Sajwan KS, Naidu R (eds) Trace elements in the environment. Biogeochemistry, biotechnology, and bioremediation. CRC, Boca Raton, FL

Vallino M, Massa N, Lumini E, Bianciotto V, Berta G, Bonfante P (2006) Assessments of arbuscular mycorrhizal fungal diversity in roots of Solidago gigantea growing in a polluted soil in Northern Italy. Environ Microbiol 8:971–983

Van Assche F, Clijsters H (1990) Effect of metals on enzyme activity in plants. Plant Cell Environ 13:195–206

Van Veen JA, Ladd JN, Amato M (1985) Turnover of carbon and nitrogen through the microbial biomass in a sandy loam and a clay soil incubated with [^{14}C(U)]glucose and [^{15}N](NH$_4$)SO$_4$ under different moisture regimes. Soil Biol Biochem 17:747–756

Vivas A, Barea JM, Biro B, Azcon R (2006) Effectiveness of autochthonous bacterium and mycorrhizal fungus on Trifolium growth, symbiotic development and soil enzymatic activities in Zn contaminated soil. J Appl Microbiol 100:587–598

Voegelin A, Barmettler K, Kretzschmar R (2003) Heavy metal release from contaminated soils: comparison of column leaching and batch extraction results. J Environ Qual 32:865–875

Vogel-Mikus K, Pongrac P, Kump P, Necemer M, Regvar M (2006) Colonisation of a Zn, Cd and Pb hyperaccumulator Thlaspi praecox Wulfen with indigenous arbuscular mycorrhizal fungal mixture induces changes in heavy metal and nutrient uptake. Environ Pollut 139:362–371

Volesky B, Holan ZR (1995) Biosorption of heavy metals. Biotechnol Prog 11:235–250

Wang FY, Lin XG, Yin R (2007a) Effect of arbuscular mycorrhizal fungal inoculation on heavy metal accumulation of maize grown in a naturally contaminated soil. Int J Phytoremediation 9: 345–353

Wang FY, Lin XG, Yin R (2007b) Inoculation with arbuscular mycorrhizal fungus Acaulospora mellea decrease Cu phytoextraction by maize from Cu-contaminated soil. Pedobiologia 51: 99–109

Wang FY, Lin XG, Yin R (2007c) Role of microbial inoculation and chitosan in phytoextraction of Cu, Zn, Pb and Cd by Elsholtzia splendens – a field case. Environ Pollut 147:248–255

Wani PA, Khan MS, Zaidi A (2007a) Cadmium, chromium and copper in greengram plants. Agron Sustain Dev 27:145–153

Wani PA, Khan MS, Zaidi A (2007b) Impact of heavy metal toxicity on plant growth, symbiosis, seed yield and nitrogen and metal uptake in chickpea. Aust J Exp Agric 47:712–720

Wecks JEJ, Clisjsters HMM (1997) Zn toxicity induces oxidative stress in primary leaves of *Phaseolus vulgaris*. Plant Physiol Biochem 35:405–410

Weissenhorn I, Leyval C, Berthelin J (1993) Cd-tolerant arbuscular mycorrhizal (AM) fungi from heavy-metal polluted soils. Plant Soil 157:247–256

Wenzel WW, Adriano DC, Sal D, Smith R (1999) Phytoremediation: a plant-microbe-based remediation system. In: Adriano DC, Bollag JM, Frankenburger WT Jr, Sims RC (eds) Bioremediation of contaminated soils, vol 37, Agronomy monographs. ASA, CSSA, and SSSA, Madison, WI, pp 457–508

Whitfield L, Richards AJ, Rimmer DL (2004) Relationships between soil heavy metal concentration and mycorrhizal colonization in Thymus polytrichus in Northern England. Mycorrhiza 14:55–62

Winge DR et al (1985) Yeast metallothionein: sequence and metal-binding properties. J Biol Chem 260:14464–14470

Wolt J (1994) Soil solution chemistry. Wiley, New York

Yeats GW, Orchard VA, Speir TW, Hunt JL, Hermans MCC (1994) Impact of pasture contamination by copper, chromium, arsenic and timber preservative on soil biological activity. Biol Fertil Soils 18:200–208

Zarei M (2008) Diversity of arbuscular mycorrhizal fungi in heavy metal pollution soils and their roles in phytoremediation. Ph.D. dissertation in Soil Science (Soil Biology and Biotechnology), Agricultural Faculty, University of Tehran, Tehran, Iran, p 219 (In Persian with an English abstract)

Zarei M, Sheikhi J (2010) The role of arbuscular mycorrhizal fungi in phytostabilization and phytoextraction of heavy metal contaminated soils. In: Golubev IA (ed) Handbook of phytoremediation. Nova, New York

Zarei M, König S, Hempel S, Khayam Nekouei M, Savaghebi G, Buscot F (2008a) Community structure of arbuscular mycorrhizal fungi associated to Veronica rechingeri at the Anguran zinc and lead mining region. Environ Pollut 156:1277–1283

Zarei M, Saleh-Rastin N, Salehi Jouzani G, Savaghebi G, Buscot F (2008b) Arbuscular mycorrhizal abundance in contaminated soils around a zinc and lead deposit. Eur J Soil Biol 44:381–391

Zarei M, Hempel S, Wubet T, Schäfer T, Savaghebi G, Jouzani GS, Nekouei MK, Buscot F (2010) Molecular diversity of arbuscular mycorrhizal fungi in relation to soil chemical properties and heavy metal contamination. Environ Pollut 158:2757–2765

Zelles L, Bai QY, Ma RX, Rackwitz R, Winter K, Beese F (1994) Microbial biomass, metabolic activity and nutritional status determined from fatty acid patterns and poly-hydroxybutyrate in agriculturally-managed soils. Soil Biol Biochem 26:439–446

Zhu YL, Pilon-Smits EAH, Tarun AS, Weber SU, Jouanin L, Terry N (1999) Cadmium tolerance and accumulation in Indian mustard is enhanced by overexpressing γ-glutamyl-cysteine synthetase. Plant Physiol 121:1169–1176

Chapter 15
Sustainable Agriculture in Saline-Arid and Semiarid by Use Potential of AM Fungi on Mitigates NaCl Effects

Mohammad Javad Zarea, Ebrahim Mohammadi Goltapeh, Nasrin Karimi, and Ajit Varma

15.1 Introduction

The semiarid region encompasses a wide variety of agricultural systems where water is probably one of the main keys to productivity. Water frequently limits rainfed crop production in this area because of low annual precipitation (<450 mm) and an uneven interannual distribution (Zarea 2010). Addition, field salinization is a growing problem in these areas. For better nutrient management in semiarid areas, an increased use of the biological potential is important. Many keys to agricultural success in semiarid areas are to use the soil biology potential to maintain soil fertility and to guard against erosion and water limiting (Zarea 2010).Use of AM is important for producing several benefits of plant symbiosis under drought stress and limited water (Zarea 2010).

Field salinization is a growing problem worldwide. It was estimated that 10 % of the world's cropland and as much as 27 % of the irrigated land may be already affected by salinity (Shannon 1997). One-third of the world's arable land resources are affected by salinity (Qadir et al. 2000). Saline–alkaline soils occupy most arid and semiarid areas of the world land and represent a major limiting factor in crop production. Most of this salinity is natural, but the extent of saline soils is increasing in a significant decrease in irrigation practices because of climate changing, saline irrigation, and high evaporation which induce salt accumulation in the soil. Saline

M.J. Zarea (✉) • N. Karimi
Faculty of Agriculture, Ilam University, PO Box 69315516, Ilam, Iran
e-mail: mj.zarea@ilam.ac.ir; mjzarea@ymail.com; karimi.nasrin64@yahoo.com

E.M. Goltapeh
Department of Plant Pathology, Faculty of Agriculture, Tarbiat Modarres University, Tehran, Iran
e-mail: emgoltapeh@yahoo.com; emgoltapeh@modares.ac.ir

A. Varma
Amity Institute of Microbial Technology, Amity University Uttar Pradesh, Sector 125 New Super Highway, Noida 201303, Uttar Pradesh, India
e-mail: ajitvarma@amity.edu

E.M. Goltapeh et al. (eds.), *Fungi as Bioremediators*, Soil Biology 32,
DOI 10.1007/978-3-642-33811-3_15, © Springer-Verlag Berlin Heidelberg 2013

soils and saline irrigation constitute a serious production problem for vegetable crops as saline conditions are known to suppress plant growth (Kohler et al. 2009), particularly in arid and semiarid areas (Parida and Das 2005). Over 800 million ha of land throughout the world are salt affected, either by salinity (397 million ha) or the associated condition of sodicity (434 million ha) (Manchanda and Garg 2008; FAO 2005). This is over 6 % of the world's total land area. Most of this salinity and all of the sodicity are natural. However, a significant proportion of recently cultivated agricultural land has become saline because of land clearing or irrigation (Manchanda and Garg 2008). Of the 1,500 million ha of land farmed by dryland agriculture, 32 million (2 %) are affected by secondary salinity to varying degrees (Manchanda and Garg 2008). Of the current 230 million ha of irrigated land, 45 million ha are salt affected (20 %) (FAO 2005). Irrigated land is only 15 % of total cultivated land, but as irrigated land has at least twice the productivity of rainfed land, it produces one-third of the world's food (Munns 2005). Scientists have searched for new salt-tolerant crop plants (Aronson 1985; Glenn and O'Leary 1985), developed salt-tolerant crops through breeding (Shannon 1984), and continue to investigate the physiology of genetic alterations involved in salt tolerance (Apse et al. 1999). Other attempts to deal with saline soils have involved leaching excessive salts (Hamdy 1990a, b) or desalinizing seawater for use in irrigation (Muralev et al. 1997). Although these approaches have been successful, most are beyond the economic means of developing nations (Cantrell and Linderman 2001). Plant breeding may be available to those areas for some plant species, but they would not be available for all the crops being grown (Cantrell and Linderman 2001).

Arbuscular mycorrhizal fungi (AM fungi) can reduce the impact of environmental stresses such as salinity (Ruíz-Lozano et al. 1996). Endomycorrhizal associations often result in greater yields of crop plants even under saline conditions (Table 15.1). Arbuscular mycorrhizal fungi (AM fungi) widely exist in salt-affected soils (Juniper and Abbott 1993). To some extent, these fungi have been considered as bio-ameliorators of saline soils (Azcón-Aguilar and Barea 1997; Singh et al. 1997; Rao 1998).The use of plant symbiotic microorganisms, especially AM fungi, proves useful in developing strategies to facilitate plant growth in saline soils (Kohler et al. 2009). The use of mycorrhizal fungi to promote plant growth in saline soils is a developing technology (Bacilio et al. 2004). AM fungi can improve the performance of plants under salinity stress (Porras-Soriano et al. 2009; Ruíz-Lozano et al. 1996; Al-Karaki 2006).

Low concentrations of salts may be beneficial to plant growth, even though they reduce osmotic potential (Ben-Gal et al. 2009) and excessive concentrations that have harmful effects. Low levels of Na are beneficial, and Cl is essential to plant health (Marschner 1995). However, concentrations of Na and Cl in soil solution are often above these beneficial amounts, and, therefore, both reduction of (leading to reduced water uptake) and excess ion levels (leading to toxicity) are likely to occur (Bernstein 1975).While many studies have demonstrated that inoculation with AM fungi improves growth of plants under salt stress (Cho et al. 2006), relatively few mechanisms have been demonstrated that explain the increased resistance to salt stress of plants treated with AM fungi.

Table 15.1 Effect of AM fungi on enhancement tolerance of several plants growing under salt stress

AM fungi sp.	Plant species	References
Glomus mosseae	Cotton	Tian et al. (2004)
Glomus sp.	*Lactuca sativa*	Ruíz-Lozano et al. (1996)
Glomus sp.	Lettuce	Ruíz-Lozano and Azcón (2000)
Glomus deserticola	Lettuce	Ruíz-Lozano and Azcón (2000)
Glomus etunicatum	Soybean	Sharifi et al. (2007)
Glomus intraradices	*Atriplex nummularia* L.	Plenchette and Duponnois (2005)
Glomus spp.	Cucumber (*Cucumis sativus* L.)	Rosendahl and Rosendahl (1991)
Glomus clarum	Pepper (*Capsicum annum* cv. 11B 14)	Kaya et al. (2009)
Glomus etunicatum and *Glomus clarum*	Banana (*Musa* sp. cv. Pacovan)	Yano-Melo et al. (2003)
Glomus mosseae, Glomus intraradices	Olive tree	Porras-Soriano et al. (2009)
VA mycorrhizal fungi	Lettuce (*Lactuca sativa* L.) and onion (*Allium cepa* L.)	Cantrell and Linderman (2001)
Glomus fasciculatum	*Acacia nilotica*	Giri et al. (2007)
Glomus intraradices, Gigaspora margarita	Sorghum	Cho et al. (2006)
Glomus mosseae	Maize	Sheng et al. (2008)
Glomus intraradices	*Lotus glaber*	Sannazzaro et al. (2006)

The purpose of this chapter is to outline briefly the current state of knowledge about the effect of salinity stress on crop yield, especially in semiarid areas. This chapter focuses on interaction mechanisms which through by AM fungi ameliorate the deleterious effects of salinity, including a brief discussion on how this knowledge is currently being used and how an understanding of this could prove to be important for sustainable agriculture in the future. This chapter outlines the potential of AM fungi on sustainability crop production under saline soil in arid and semiarid areas. Methods of applying AM fungi that can overcome the challenges of salinity stresses are presented.

15.2 Mechanism on Which Salt Affects Crop

Salt stress has been shown to affect plant growth, nutrient availability, and plant physiology presses (Table 15.2).

Salinity is the concentration of dissolved mineral salts present in the soils (soil solution) and waters. The dissolved mineral salts consist of the electrolytes of cations and anions. The major cations in saline soil solutions consist of Na^+, Ca^{2+},

Table 15.2 Mechanism which through by salinity stress affect plant growth parameters

Mechanism	References
Growth processes	
<yield quality	Martínez et al. (1996) and Pardossi et al. (1999)
<root elongation <mature vegetative	Kaya et al. (2009)
<water	Munns (2005)
<cell expansion	Greenway and Munns (1980) and Nieman (1965)
<leaf expansion	Neumann et al. (1988)
<cell expansion	
<relative water content	Kabir et al. (2004) and Ahmad and Jhon 2005
<germination rate	Kaya et al. (2002), Katembe et al. (1998), Pujol et al. (2000), and Okçu et al. (2005)
<harvest index	Sibole et al. (2000)
Nutrients processes	
>concentration of ions in tissue	Demir and Kocakalikan (2002) and Francois et al. (1990)
>Na^+/Ca^{2+}, >Na^+/K^+	Sivritepe et al. (2003)
<Ca^{2+}, <K^+	Kohler et al. (2009)
<leaf water content, <root biomass	Kohler et al. (2009)
<P uptake, <P translocation	Martínez and Lauchli (1991) and Munns (1993)
>root proline accumulation	Sudhakar et al. (1993) and Cusido et al. (1987)
<chlorophyll concentration	Yeo et al. (1990), Belkhodja et al. (1994), and Kaya et al. (2001)
<photosynthesis	Downton (1977), Ball and Farquhar (1984), Behboudian et al. (1986), and Shi and Guo (2006)

Mg^{2+}, and K^+, and the major anions are Cl^-, SO_4^{2-}, HCO_3^-, CO_3^{2-}, and NO_3^- (Manchanda and Garg 2008). Other constituents contributing to salinity in hypersaline soils and waters include B, Sr_2^+, SiO_2, Mo, Ba^{2+}, and Al^{3+} (Hu and Schmidhalter 2002). Water-soluble salts accumulate in the soil solum the upper part of the soil profile (including the A and B horizons) or regolith (the layer or mantle of fragmental and unconsolidated rock material, whether residual or transported) to a level that impacts on agricultural production, environmental health, and economic welfare (United States Salinity Laboratory Staff 1954; Rengasamy 2006).The composition and concentration of soluble salts in root-zone medium solution are known to influence plant growth, both by creating osmotic imbalance and via specific physiological toxicity of ions. Osmotic stress lowers the potential energy of the solution and causes reduced growth due to the additional energy required by plants to take up water (Ben-Gal et al. 2009).

Salinity dominated by Na^+ and Cl^- not only reduces Ca^{2+} and K^+ availability but also reduces Ca^{2+} and K^+ mobility and transport to the growing parts of plants, affecting the quality of both vegetative and reproductive organs (Kohler et al. 2009). The decline of K concentration under salinity conditions has been reported by Greenway and Munns (1980) and Devitt et al. (1981). Moreover, many studies have shown that high concentrations of NaCl in the soil solution may increase the ratios of Na^+/Ca^{2+} and Na^+/K^+ ratios in plants, which would then be more susceptible to osmotic and specific ion injury as well as to nutritional disorders that result in reduced yield and quality (Sivritepe et al. 2003). Saline stress induces P deficiency by reducing P uptake or translocation (Martínez and Lauchli 1991). Saline stress decreased nitrogen concentration in pepper plants (Kaya et al. 2009).

Salt stress has been shown to affect carbohydrate partitioning and metabolism, leading to the synthesis of new compounds (Sharma et al. 1990). In plants exposed to salinity, the total nonstructural carbohydrates content in the leaves was reduced significantly compared with plants not exposed to salinity (Kohler et al. 2009). The decrease in total soluble carbohydrates due to salinity could be related also to limited carbohydrate availability, as a consequence of a decline in photosynthesis (Goicoechea et al. 2005).

Increasing salinity stress significantly increased the antioxidant enzyme activities of lettuce leaves, including those of peroxidases and catalase, compared to their respective non-stressed controls (Kohler et al. 2009). Salt stress may induce a combination of negative effects on salt-sensitive plants including osmotic stress, ion toxicity, and oxidative stress(Kohler et al. 2009). High salinity may induce imbalances in the soil, plant osmotic relationships (Wyn Jones and Gorham 1983), and in plant metabolism (Singh and Jain 1982).

Increasing salinity stress significantly decreased germination percentage (Barassi et al. 2006) High (60 mM) NaCl in nutrient solution strongly affected root elongation and mature vegetative growth of both spinach and lettuce, but especially in vegetables (Kaya et al. 2002; Barassi et al. 2006). Shoot growth is affected more than root growth in lettuce exposed to low salinities (Shannon 1997; Barassi et al. 2006).

The reduction in photosynthesis in the saline-treated plants was reported by many researchers (Downton 1977; Ball and Farquhar 1984; Behboudian et al. 1986).The adverse effects of high NaCl on chlorophyll concentration have previously been shown in rice (Yeo et al. 1990), barley (Belkhodja et al. 1994), tomato (Kaya et al. 2001), and pepper (Kaya et al. 2009). Kaya et al. (2009) reported that salt stressed caused a significant increase in electrolyte leakage compared to that in the non-stressed pepper plants. Similar results were obtained by Lutts et al. (1996) for NaCl-sensitive rice varieties wherein high salt concentration increased membrane permeability.

15.3 Mechanism on Which Plant Ameliorates the Deleterious Effects of Salinity

Many effective protection systems exist in plants that enable them to perceive, respond to, and properly adapt to various stress signals (Chen et al. 2009), and a variety of genes and gene products have been identified that involve responses to drought and high salinity stress (Chen et al. 2009).

An increase in concentration of K^+ in plants under salt stress could ameliorate the deleterious effects of salinity on growth and yield (Giri et al. 2007). Potassium plays a key role in plant water stress tolerance and has been found to be the cationic solute responsible for stomatal movements in response to changes in bulk leaf water status (Caravaca et al. 2004). Tattini (1994) reported that the resistance mechanism of salt-tolerant olive cultivars is probably related to the ability to maintain an appropriate K/Na ratio in actively growing tissue. Salt stresses increase acid phosphatase activity (Stephen et al. 1994; Ehsanpour and Amini (2003). The induction of antioxidant enzyme such as catalase and peroxidase can be considered as one mechanism of salt tolerance in plants (Hernández et al. 2003). Antioxidant enzymes are involved in eliminating H_2O_2 from salt-stressed roots (Kim et al. 2005). Salt stress induces proline accumulation in legumes (Ashraf 1989; Sharma et al. 1990; Rabie and Almadini 2005) which is thought to be involved in osmotic adjustment of stressed tissues (Delauney and Verma 1993; Ashraf and Foolad 2007). Root proline accumulation under salinity is reported in other plant species (Sudhakar et al. 1993; Cusido et al. 1987; Sharifi et al. 2007). Proline acts as a major reservoir of energy and nitrogen for utilization during salinity stress (Goas et al. 1982). Changes in the composition of carbohydrates of the host plant may play a role in increasing salt tolerance (Rosendahl and Rosendahl 1991). Salinity reduced soluble carbohydrates (Sharma et al. 1990).

15.4 Mechanism on Which AM Fungi Affect Salinity Stress

Salt resistance was improved by AM colonization in various plants (Table 15.3). While many studies have demonstrated that inoculation with AM fungi improves growth of plants under salt stress (Cho et al. 2006), relatively few mechanisms have been demonstrated that explain the increased resistance to salt stress of plants treated with AM fungi. The improved growth of AM plants has been attributed to enhanced nutrient uptake, particularly of N and P and subsequent increased growth (Jeffries et al. 2003). However, in some cases, plant salt tolerance was not related to P concentration (Ruíz-Lozano and Azcón 2000). Thus, it has been proposed that salt-tolerance mechanisms, such as enhanced osmotic adjustment and leaf hydration, increased intrinsic water use efficiency, reduced oxidative damage, or improved nutritional status, can explain the contribution of AM symbioses to the salinity resistance of host plants (Augé 2001).

Salt in soil water inhibits plants' ability to take up water, and this leads to slower growth. This is the osmotic or water-deficit effect of salinity (Munns 2005). Jimenez et al. (2003) reported that salinity decreased the leaf water and osmotic and turgor potentials in four wild and two cultivated *Phaseolus* species. Rosendahl and Rosendahl (1991) reported greater water uptake by AM plants under saline conditions. Kumar et al. (2009) also reported that relative water content in the leaves of *Jatropha* was significantly higher in AM-inoculated than in non-inoculated *Jatropha* plants under saline conditions. The beneficial effects of AM

Table 15.3 Mechanism which through by AM fungi affect salinity stress

Mechanism	References
Nutrients processes	
>P	Poss et al. (1985), Duke et al. (1986), Copeman et al. (1996), Augé (2001), Al-Karaki et al. (2001), Jeffries et al. (2003), Tian et al. (2004), Sharifi et al. (2007), Giri et al. (2007), Kohler et al. (2009), and Porras-Soriano et al. (2009)
<Na	Giri and Mukerji (2004), Ashraf et al. (2004), Ghazi and Al-Karaki (2006), Sharifi et al. (2007), Kohler et al. (2009), Kaya et al. (2009), and Porras-Soriano et al. (2009)
>Zn	Sharifi et al. (2007)
>K/Na	Kohler et al. (2009), Naidoo and Naidoo (2001), Thomas et al. (2003), Rabie and Almadini (2005), Giri et al. (2007), Porras-Soriano et al. (2009), and Sannazzaro et al. (2006)
>K	Chow et al. (1990), Graifenberg et al. (1995), Pérez-Alfocea et al. (1996), Botella et al. (1997), Giri et al. (2007), and Porras-Soriano et al. 2009
>N	Kaya et al. (2009) and Porras-Soriano et al. (2009)
Physiological processes	
>carbon dioxide exchange rate, >transpiration, >stomatal conductance, >water use efficiency	Ruíz-Lozano et al. (1996)
>soluble sugar	Feng et al. (2002)
>electrolyte concentrations in maize roots	Feng et al. (2002)
>root proline accumulation	Sharifi et al. (2007)
<shoot proline accumulation	Kaya et al. (2009) and Sharifi et al. (2007)
>chlorophyll content	Goss and de Varennes (2002), Giri and Mukerji (2004), Rabie (2005), and Kaya et al. (2009)
<electrolyte leakage	Kaya et al. (2009)
<membrane permeability	Zhongoun et al. (2007)
>water content	Colla et al. (2008)

on growth may be related to mycorrhiza-mediated effects on water absorption, nutrient uptake, and increased photosynthetic activity under salinity stress (Mukerji and Chamol 2003; Al-Karaki 2006; Miransari et al. 2008). AM mediated modifications in root morphology and biomass which may be important for water balance in the host plant under salinity stress (Giri et al. 2003).

Reduction in both K and K/Na at high salinity is another opposing effect of salinity, which impairs the function of K in the salinized plants (Tabatabaei 2006). Na can easily enter plant cells via nonselective K^+ channels (Demidchik et al. 2002), but generally, K is the preferred ion in the cytoplasm, providing a reactive surrounding for enzymes, being an osmotic equivalent under conditions of water stress and high salinity (Hammer et al. 2010). A reduction in K concentration and K/Na ratio in saline conditions was reported by Rush and Epstein (1978); Devitt et al. (1981) and Jackson and Volk (1997). Reduced the Na^+ uptake of plants and/or increased the K^+ uptake, compared with the control plants under salt stress, thus increasing the K^+/Na^+ ratio (Kohler et al. 2009). In higher plants, K^+ affects photosynthesis at various levels (Porras-Soriano et al. 2009). The role of K^+ in CO_2 fixation has been demonstrated, and an increase in the leaf potassium content is accompanied by increased rates of photosynthesis, photorespiration, RuBP carboxylase activity, and a concomitant decrease in dark respiration (Porras-Soriano et al. 2009). Loreto et al. (2002) demonstrated the limitation of photosynthesis in salinity conditions in olive cultivars as a rustle of the low chloroplast CO_2 concentration, caused by both stomatal and mesophyll conductance reduction. It has been proposed that K^+ nutrition is essential for salt tolerance in *Arabidopsis* (Zhu et al. 1998). Therefore, prevention of Na^+ accumulation in the plant and enhancement of K^+ concentrations could be part of the general mechanism of salt stress alleviation of plant by AM fungi. An improvement of K^+ uptake by AM fungi was reported in the early work by Powell (1975). A similar decrease in Na^+ accumulation under salt stress conditions has been formerly observed in *Sesbania* spp. (Giri et al. 2003; Giri and Mukerji 2004) and in *Lotus glaber* (Sannazzaro et al. 2006). Although, Rabie and Almadini (2005) observed an enhancement of K uptake by AM plants under salinity. Poss et al. (1985) and Porras-Soriano et al. (2009) reported that K uptake was affected little by AM colonization in plants grown under saline conditions.

Plant growth is affected by interactions of Na^+ or Cl^- and many mineral nutrients, causing imbalances in the nutrient availability, uptake, or distribution within plants and also increasing the plant's requirement for essential elements (Greenway and Munns 1980; Grattan and Grieve 1992). Reduced Na concentration in lettuce plants exposed to salinity, due to AM inoculation, may have helped the plants prevent accumulation of cellular Na to a toxic concentration (Kohler et al. 2009). There are several reports of lower Na^+ concentrations in AM plants, compared to non-AM plants, under salinity (Giri and Mukerji 2004; Ashraf et al. 2004; Ghazi and Al-Karaki 2006; Sharifi et al. 2007; Kohler et al. 2009; Kaya et al. 2009; Porras-Soriano et al. 2009).

Interaction between salinity and N affects growth and metabolism of plants in order to cope with the changes taking place in their environment (Papadopoulos and Rendig 1983; Shenker et al. 2003). Kaya et al. (2009) reported that mycorrhizal inoculation increased the concentration of K^+ and N in pepper plants. Ruíz-Lozano and Azcón (2000) showed that plants colonized by *Glomus* sp. accumulated N and P contents in shoots of lettuce more efficiently at the highest salt level than under the lowest salt level. Porras-Soriano et al. (2009) reported that the most active fungus in increasing NaCl tolerance in olive plants was *G. mosseae*, and it increased the N, P, and K Contents by 363 %, 552 %, and 636 %, respectively, over those values recorded in non-colonized olive. An apparent increase in salt tolerance has been noted when N levels supplied under saline conditions exceeded those that were optimum under nonsaline conditions (Papadopoulos and Rendig 1983), implying that increased fertilization, especially N, may ameliorate the deleterious effects of salinity (Ravikovitch and Porath 1967). This may be caused by ion accumulation in the leaves, particularly old leaves (Greenway and Munns 1980). The reduction in plant growth was due to the reduced leaf growth, which agrees with finding of Cramer (2002a). At the highest concentration of salinity, however, N concentration not only had no effects on growth promotion but reduced the plants' growth. It implies that other factors (such as osmotic pressure) involve in plant growth (Tabatabaei 2006). The direct factor might be salinity (such as osmotic effect, Cl, or Na toxicity) as reported by Bongi and Loreto (1989) and Xu et al. (2000). Tabatabaei (2006) concluded that further increase of N concentration at the increased salinity concentration in salt-sensitive cultivars is unlikely to improve growth and nutrition; however, in salt-tolerant cultivars, N concentration in the root zone should be increased in order to improve the plant growth. This author added that at high salinity concentration, both high and low N concentrations in the root zone have opposing effects on olive trees' growth. Therefore, use of nutrients at saline conditions for the both salt-sensitive and salt-tolerant cultivars should be carefully managed.

P availability is reduced in saline soil not only because of ion competition that reduces the activity of P but also because P concentrations in soil solution are tightly controlled by sorption processes and by the low solubility of Al–P or Fe–P precipitates (Grattan and Grieve 1999). Jakobsen et al. (1992) reported that the efficiency of P uptake by an AMF was strongly affected by the spatial distribution of its hyphae in the soil and possibly also by the differences in the capacity for uptake per unit length of hyphae. Mycorrhizal inoculation improves P nutrition of plants under salinity stress and reduces the negative effects of Na^+ by maintaining vacuolar membrane integrity, which prevents this ion from interfering in growth metabolic pathways (Rinaldelli and Mancuso 1996). Previous research has shown that the main mechanism for enhanced salinity tolerance in mycorrhizal plant was the improvement of P nutrition (Copeman et al. 1996; Al-Karaki et al. 2001). AM fungi increased P uptake, and saline stress in plants was thereby alleviated (Tian et al. 2004). Tian et al. (2004) showed that AM fungi from saline soil promoted the growth of cotton plants under saline stress by increasing P concentrations without affecting sodium and chloride concentrations. Bernstein (1975) reported that plant growth was more sensitive to excess Na than to excess Ca, and the effect of combined Na and Ca was intermediate. Then, it is possible to ameliorate Cl toxicity by increasing the NO_3 concentration in the soil solution (Xu et al. 2000). Adding Ca has been observed experimentally to relieve ionic Na toxicity and increase root elongation (Yermiyahu et al. 1997; Reid and Smith 2000; Cramer 2002b). Reid and Smith (2000) and Yermiyahu et al. (1997) reported that while Ca does allow amelioration of the toxicity caused by high intracellular Na, it cannot overcome the osmotic effects associated with high salinity. Early work by Magistad et al.

(1943) indicated that salts with different cations (Na, Ca, and Mg) and anions (Cl, SO_4) do not cause differences in plant growth when compared on an equal osmotic basis. Additionally, work on physiological aspects of Na:Ca relationships has also recognized the need to consider the osmolarity or osmotic strength of solutions with mixed ions (Yermiyahu et al. 1997; Kinraide 1999; Cramer 2002b). The decline in leaf growth is an earliest response of the plants to salinity (Munns and Termaat 1986). Neumann et al. (1988) reported that salt stress initially inhibits leaf expansion through reduced turgor. Tsang and Maum (1999) and Cantrell and Linderman (2001) reported that AM lettuce plants had greener leaves than non-AM plants at the highest salt level.

It has been proposed that salt-tolerance mechanisms, such as enhanced osmotic adjustment and leaf hydration, increased intrinsic water use efficiency, reduced oxidative damage, or improved nutritional status, can explain the contribution of AM symbioses to the salinity resistance of host plants (Augé 2001). It is reported that proline protects higher plants against salt/osmotic stresses, not only by adjusting osmotic pressure but also by stabilizing many functional units such as complex II electron transport, membranes, and proteins and enzymes such as ruBisCo (Mäkelä et al. 2000).

For plants to survive under salt stress conditions, adjustment of leaf osmotic potential is very important, and it requires intracellular osmotic balance. Under salt stress, plants accumulate some organic solutes (proline, soluble sugars, and so on) and inorganic ions to maintain higher osmotic adjustment (Yang et al. 2009). Free amino acids are important osmolytes contributing to osmotic adjustment in plants (Hajlaoui et al. 2010). With increasing external salt concentration, free amino acids accumulate in the leaves and roots of maize (Abd-El Baki et al. 2000; Neto et al. 2009; Hajlaoui et al. 2010). Sheng et al. (2011) also observed the increase of free amino acid levels in maize leaves under salt stress but to a lesser extent in AM corn plants. Arduous research has been conducted on physiological processes related to salinity and/or to specific ions, and much knowledge has been acquired (Ben-Gal et al. 2009). Ruíz-Lozano et al. (1996) concluded that the mechanisms underlying AM plant growth improvement in *Lactuca sativa* under saline conditions were based on physiological rather than on nutrient uptake (N or P). Among free amino acids, proline is a contributor to osmotic adjustment in salt-stressed maize plants (Hajlaoui et al. 2010). It appears that the presence of the AM fungi in the roots may modify the osmotic potential of the leaves as they have been shown to influence the composition of carbohydrates (Augé et al. 1987) and the level of proline (Ruíz-Lozano and Azcón 1995). Proline accumulation is thought an adaptive feature under salinity stress in AM (Jindal et al. 1993). Results show that the accumulation of proline in plant is increased by AM inoculation. Proline accumulation is thought to be an adaptive feature under salinity stress in AM (Jindal et al. 1993) and non-AM (Ashraf 1989; Sharma et al. 1990) legumes. The high level of proline enables the plants to maintain osmotic balance when growing under low water potentials (Stewart and Lee 1974). Proline acts as a major reservoir of energy and nitrogen for utilization by plants subjected to salinity stress (Goas et al. 1982; Ashraf and Foolad 2007). Reports on the effect of AM symbiosis on proline accumulation are

somewhat contradictory. Some studies have shown an increase in proline accumulation in mycorrhizal plants subjected to salt stress (Ben Khaled et al. 2003; Sharifi
et al. 2007). Enhanced proline accumulation in plant cells can increase plant
osmotic potentials (Hajlaoui et al. 2010) and abscisic acid level (Ober and Sharp
1994), thereby improving the tolerance of mycorrhizal plants to salinity. On the
contrary, some studies have shown a reduction of proline levels in AM plants under
salt stress (Duke et al. 1986; Ruíz-Lozano et al. 1996; Jahromi et al. 2008; Sheng
et al. 2011), and Kaya et al. 2009 showed that proline was significantly lower in
mycorrhizal than in non-mycorrhizal pepper plants under salinity. In plants exposed
to salinity, the total nonstructural carbohydrates content in the leaves was reduced
significantly compared with plants not exposed to salinity (Kohler et al. 2009).
Salinity reduced soluble carbohydrates (Sharma et al. 1990). Sheng et al. (2011)
reported that sugar (soluble sugars and reducing sugars) accumulation in maize
leaves decreased when salinity increased, but at the same NaCl level, the AM
symbiosis favored sugar accumulation. Similar results were observed in the shoots
of *V. radiata* (Rabie 2005) and in roots and shoots of maize (Feng et al. 2002). The
accumulation of sugars induced by the AM symbiosis is a positive response to salt
stress since sugars can prevent structural changes in soluble protein, maintain the
osmotic equilibrium in plant cells, and protect membrane integrity (Abd-El Baki
et al. 2000). The high levels of sugars in mycorrhizal plants may be the result of an
increase in photosynthetic capacity (Sheng et al. 2008; Wu et al. 2009). Thomson
et al. (1990) found a positive relationship between carbohydrate concentration in
the roots and percentage of root colonization. The change in the composition of
carbohydrates of AM plants may play a role in the salt tolerance of plants
(Rosendahl and Rosendahl 1991). Feng et al. (2002) related the mycorrhizal
tolerance to salt stress to higher accumulation of soluble sugars in plant roots.
However, Sharifi et al. (2007) showed that root soluble carbohydrates did not play a
role in responses of AM and non-AM soybeans to salinity.

The induction of antioxidant enzyme such as catalase and peroxidase can be
considered as one mechanism of salt tolerance in plants (Hernández et al. 2003).
Kohler et al. (2009) reported that the peroxidases and catalase activities of non-
mycorrhizal plants did not experience changes under moderate salinity, whereas the
mycorrhizal plants showed a suppression of such antioxidant enzymes possibly due
to lower levels of accumulated proline. These authors concluded that this decrease
in antioxidant enzymes in mycorrhizal plants could be explained partially by the
fact that these plants may have been submitted to a lower oxidative stress under
moderately saline conditions. It is worth noting that this behavior was also observed
in the peroxidase activity of fertilized plants, which probably support less stress.

Acid phosphatase is known to act under stress by maintaining a certain level of
inorganic phosphate in plant cells (Olmos and Hellin 1997).However, Kohler et al.
(2009) show that acid phosphatase did not play a role in responses of inoculated and
non-inoculated lettuce to severe salinity. Nitrate reductase (NR) was slightly
inhibited by salinity in tomato roots, while leaf NR decreased sharply (Cramer
and Lips 1995). In the leaves of tomatoes and cucumbers, NR activity increased
with exogenous NO_3 concentration (Maritinez and Cerda 1989), as NR is a

substrate-inducible enzyme (Marschner 1995), and its decreased activity under salinization has been attributed by some researchers to decreased NO_3 uptake by plants under salt stress (Lacuesta et al. 1990; Abdelbaki et al. 2000; Tabatabaei 2006). The decreased of NO_3 is accompanied by a high Cl^- uptake (Parida et al. 2004) and low rate of xylem exudation in high osmotic conditions either by NaCl or other nutrients (Tabatabaei et al. 2004). Either the reduced NO_3 uptake or translocation leads to lower NO_3 concentration in the leaves, consequently reducing NR activity of leaves under salinity conditions (Tabatabaei 2006). This finding agrees with Cramer and Lips (1995), who indicated that salinity may control NR activity through NO_3 uptake since NR activity is largely determined by NO_3 flux into the metabolic pool rather than by tissue NO_3 content itself.

Under saline condition, mycorrhizal colonization increased chlorophyll content in mung bean (Rabie 2005), pepper (Kaya et al. 2009), and in woody species, *Sesbania aegyptiaca* and *Sesbania grandiflora* (Giri and Mukerji 2004).The AM plants had greener leaves than non-AM plants under saline conditions (Kumar et al. 2009).

Salt stressed caused a significant increase in electrolyte leakage in pepper (Kaya et al. 2009), rice (Lutts et al. 1996), and lentil (Bandeoglu et al. 2004). Mycorrhizal inoculation significantly reduced the electrolyte leakage in the salt-stressed plants of pepper (Kaya et al. 2009). It was shown that salt-stressed tomato plants inoculated with mycorrhizae had lower membrane permeability than non-inoculated plants (Zhongoun et al. 2007).

High salinity levels can damage soil structure. In semiarid environments, soil aggregate stability is one of the most important properties controlling the growth of plants which, in turn, protects the soil against water erosion (Kohler et al. 2010). High salinity levels can damage soil structure. The action of Na^+ ions, when they occupy the cation exchange complex of clay particles, makes the soil more compact, thereby hampering soil aeration (Manchanda and Garg 2008). As a result, plants in saline soils not only suffer from high Na levels but are also affected by some degree of hypoxia. In semiarid and arid areas of the world, the scarcity, variability, and unreliability of rainfall and high potential evapotranspiration affect the water and soil balance of the soil. Low atmospheric humidity, high temperature, and wind velocity promote the upward movement of the soil solution and the precipitation and concentration of the salts in the surface horizons (Manchanda and Garg 2008). In arid regions, mainly chloride and sulfate types of Na, Mg, and Ca salts are concentrated (FAO 2005). The improvement of soil structural stability is of great importance in rendering these degraded, saline soils suitable for agriculture (Kohler et al. 2010). The action of Na^+ ions, when they occupy the cation exchange complex of clay particles, makes the soil more compact, thereby hampering soil aeration (Manchanda and Garg 2008). Recent discoveries suggest that glomalin, a glycoprotein (Wright et al. 1998) produced in copious amounts by AM fungal hyphae and related to soil aggregate stability (Rillig 2004), can influence soil carbon storage indirectly by stabilizing soil aggregates (Zhu and Miller 2003) and soil stability. Soil aggregate stability is one of the most important properties controlling plant growth in arid and semiarid environments by controlling soil–plant water status. AM produces glomalin, a glycoprotein which binds soil particles and hence improves the soil structure and stability (Rillig and Mummey 2006). In particular, the symbiosis between AM fungi and plants has been shown to contribute to the stability of soil aggregates, including soils of high salinity such as salt marshes (Caravaca et al. 2005). AM fungi primarily influence the stability of macroaggregates (>250 mm), which they are hypothesized to help stabilize via hyphal enmeshment aggregates (Miller and Jastrow 2000) and by deposition of organic substances (Bearden and Petersen 2000). A key factor in the contribution of AM fungi to soil aggregation is the production of the glycoprotein glomalin, which acts as an insoluble glue to stabilize aggregates (Gadkar and Rillig 2006). Operationally defined by the extraction and detection conditions (Wright and Upadhyaya 1996), it is detected in large amounts in diverse soils as glomalin-related soil protein (Rillig 2004), although the role of glomalin-related soil protein in the stabilization of saline soils has not been confirmed. Sodium is a highly dispersive agent, causing the direct breakup of aggregates and indirectly affecting aggregation through decreased plant productivity (Bronick and Lal 2005). Previous studies have described a negative relationship between soil aggregation and the percentage of Na saturation in the exchange complex (Lax et al. 1994). Kohler et al. (2010) reported that aggregate stability of soils inoculated with *G. mosseae* significantly decreased with increasing saline stress. Kohler et al. (2010) concluded that *G. mosseae* treatments increased the concentration of soil Na compared with the non-inoculated control soil under severe salinity. This suggests that the decrease in structural stability of the inoculated soils could be related to the increased concentration of soil Na (Kohler et al. 2010). Giri et al. (2007) reported that an AM fungus in *Acacia nilotica* accumulated salt and thus prevented transport of Na to shoot tissues. Cantrell and Linderman (2001) suggested that Na might be retained in intraradical AM fungal hyphae. Kohler et al. (2010) reported that the excess Na resulting from the increasing levels of salinity was retained in the soil of the plants inoculated with the AM fungus.

15.5 Would AM Fungi Isolate from Saline Soil Have a Higher Capacity to Ameliorate NaCl Effects?

As mycorrhizal fungi can adapt to edaphic conditions (Brundrett 1991), they therefore promote plant growth under saline stress, and differences in fungal behavior and efficiency may be due to the origin of the AMF (Copeman et al. 1996).

Salinity may reduce mycorrhizal colonization by inhibiting the germination of spores (Hirrel 1981), inhibiting growth of hyphae in soil and hyphal spreading after initial infection had occurred (McMillen et al. 1998), and reducing the number of arbuscules (Pfeiffer and Bloss 1988). It has been widely accepted that mycorrhizal fungi are able to adapt to edaphic conditions (Tian et al. 2004). It might be expected that an isolate from saline soil would have a higher capacity to promote plant growth under saline stress. Differences among AMFs with respect to the plant protection offered against salinity have been reported by Cantrell and Linderman (2001), Al-Karaki (2006), and Porras-Soriano et al. (2009). Ruíz-Lozano and Azcón (2000) reported AM fungal which isolate from saline soils protected lettuce plants by stimulating root development, while the effects of an isolate of *Glomus deserticola* from nonsaline soil were based on improving plant nutrition. Copeman et al. (1996) found that fungi from nonsaline soil acted as shoot growth promoters but tended to increase leaf sequestration of Cl^-. Conversely, AMF from saline soil suppressed plant growth but decreased leaf sequestration of Cl^-. They considered that this mechanism could be advantageous for long-term plant survival under salt stress. However, Tian et al. (2004) reported that the *Glomus mosseae* isolates collected from nonsaline soil (GM1) and from saline soil (GM2) did not differed in their capacity to alleviate salinity stress due to their capability to take up sodium and chloride under higher salt level. These authors suggest that GM2 had a different mechanism for improving the salinity tolerance of cotton plants at higher saline levels. Copeman et al. (1996) found that an AM fungal which isolated from nonsaline soil promoted shoot growth but tended to increase leaf sequestration of Cl. A similar result was also observed by Tian et al. (2004). Cantrell and Linderman (2001) reported that AM fungi from the saline soil were not more effective in reducing growth inhibition of lettuce (*Lactuca sativa* L.) and onion (*Allium cepa* L.) by salt than those from the nonsaline site. Copeman et al. (1996) showed that tomato shoot growth was enhanced by inoculation with AM fungi from a nonsaline soil and was inhibited by inoculation with AM fungi from a saline soil.

15.6 Conclusions

One of the most widespread agricultural problems in arid and semiarid regions is soil salinity, which adversely affect plant growth (Sharifi et al. 2007; Cerda and Martinez 1988). Salt-affected lands occur in practically all climatic regions, from the humid tropics to the polar regions (Manchanda and Garg 2008). Saline soils can be found at different altitudes, from below sea level (e.g., around the Dead Sea) to mountains rising above 5,000 m, such as the Tibetan Plateau or the Rocky Mountains (Manchanda and Garg 2008). Of nearly 160 million hectare of cultivated land under irrigation worldwide, about one-third is already affected by salt, which makes salinity a major constraint to food production (Manchanda and Garg 2008). It is the single largest soil toxicity problem in tropical Asia (Greenland 1984). The salinity of soils is a considerable problem in many parts of the world. This is particularly the case in regions with high rates of evaporation, where salts are easily accumulated in the topsoil (Hammer et al. 2010). Increased irrigation is needed to combat the spread of deserts and to meet the greater demand for food of a growing world population, but at the same time, inappropriate irrigation management leads to the accumulation of salt in poorly drained soils (Hammer et al. 2010). Important morphological barriers for ion selection in plants are the root hair membrane,

Mechanism which through by AM fungi affect crop under salt stress
- enhanced Ca²+ and K⁺ availability
- decrease Na+/Ca₂+ and Na+/K+ ratios
- enhance P uptake
- enhance N uptake
- increased Carbohydrate partitioning
- higher accumulation total soluble carbohydrates
- Increase Proline accumulation
- increase photosynthesis
- induced osmotic balance
- reduced growth
- lower energy required by plants to take up water
- increase total non-structural carbohydrates
- increase carbohydrate availability
- increased photosynthesis
- increased chlorophyll content

Mechanism which through by salt affect crop under salt stress
- reduces Ca²+ and K⁺ availability
- increase Na+/Ca₂+ and Na+/K+ ratios
- P deficiency
- decreased N
- decreased Carbohydrate partitioning
- Lower accumulation of total soluble carbohydrates
- Increase Proline accumulation
- reduction in photosynthesis
- induced osmotic imbalance
- higher energy required by plants to take up water
- reduce total non-structural carbohydrates
- reduce carbohydrate availability
- reduce photosynthesis
- reduce chlorophyll concentration

Fig. 15.1 Mechanisms which salt affect crop and mechanisms which AM fungi ameliorate the deleterious effects of salinity. Photograph effects of wheat colonized or non-colonized by endophytic fungus *Piriformospora indica* and salt-adapted *Azospirillum* (from arid area of Iran) in compared with non-inoculated (control) on 100-day-old wheat growth under water salinity (ECw = 12 dS m⁻¹). Wheat colonized by *P. indica* accelerated reproductively and growth rate of wheat (unpublished data)

the Casparian, strip and, before transfer to the shoot, the xylem membrane (Kramer 1983; Tester and Davenport 2003). In arid regions, mainly chloride and sulfate types of Na, Mg, and Ca salts are concentrated (FAO 2005). Few crop species are adapted to saline conditions (Hu and Schmidhalter 2002). AMF widely exist in salt-affected soils (Juniper and Abbott 1993). AM fungi have various mechanisms which mitigate NaCl effects and hence increase plant resistance to salt stress (Fig. 15.1). To promote plant growth under saline stress, AM fungi isolated from saline soil would have a higher capacity to promote plant growth under saline stress.

References

Abd-El Baki BGK, Siefritz F, Man HM, Weiner H, Kaldenhoff R, Kaiser WM (2000) Nitrate reductase in Zea mays L. under salinity. Plant Cell Environ 23:515–521

Abdelbaki GK, Siefritz F, Man HM, Welner H, Kaldenhoff R, Kaiser WM (2000) Nitrate reductase in *Zea mays* L under salinity. Plant Cell Environ 23:15–521

Ahmad P, Jhon R (2005) Effect of salt stress on growth and biochemical parameters of Pisum sativum L. Arch Agron Soil Sci 51:665–672

Al-Karaki GN (2006) Nursery inoculation of tomato with arbuscular mycorrhizal fungi and subsequent performance under irrigation with saline water. Sci Hortic 109:1–7

Al-Karaki GN, Hammad R, Rusan M (2001) Response of two tomato cultivars differing in salt tolerance to inoculation with mycorrhizal fungi under salt stress. Mycorrhiza 11:41–47

Apse MP, Aharon GS, Snedden WA, Bumwald E (1999) Salt tolerance conferred by overexpression of a vacuolar Na+/H+ antiport in Arabidopsis. Science 285:1256–1258

Aronson JA (1985) Economic halophytes: a global view. In: Wickens GE, Gooding JR, Field DV (eds) Plants for arid lands. George Allen and Unwin, London, pp 177–188

Ashraf M (1989) The effect of NaCl on water relations, chlorophyll and protein and proline contents of two cultivars of blackgram (*Vigna mungo* L.). Plant Soil 129:205–210

Ashraf M, Foolad MR (2007) Roles of glycine betaine and proline in improving plant abiotic stress resistance. Environ Exp Bot 59:206–216

Ashraf M, Berge SH, Mahmood OT (2004) Inoculating wheat seedlings with exopolysaccharide-producing bacteria restricts sodium uptake and stimulates plant growth under salt stress. Biol Fertil Soils 40:157–162

Augé RM (2001) Water relations, drought and vesicular-arbuscular mycorrhizal symbiosis. Mycorrhiza 11:3–42

Augé RM, Schekel KA, Wample RL (1987) Rose leaf elasticity changes in response to mycorrhizal colonization and drought acclimation. Physiol Plant 70:175–182

Azcón-Aguilar C, Barea JM (1997) Applying mycorrhiza biotechnology to horticulture: significance and potentials. Sci Hortic 68:1–24

Bacilio M, Rodríguez H, Moreno M, Hernández JP, Bashan Y (2004) Mitigation of salt stress in wheat seedlings by a gfp-tagged Azospirillum lipoferum. Biol Fertil Soils 40:188–193

Ball MC, Farquhar GD (1984) Photosynthetic and stomatal responses of two mangrove species, Avicennia marina and Aegiceras corniculatum, to long term salinity and humidity conditions. Plant Physiol 1:1–6

Bandeoglu E, Eyidogan F, Yucel M, Oktem HA (2004) Antioxidant responses of shoots and roots of lentil to NaCl-salinity stress. Plant Growth Regul 42:69–77

Barassi CA, Ayrault G, Creus CM, Sueldo RJ, Sobrero MT (2006) Seed inoculation with Azospirillum mitigates NaCl effects on lettuce. Sci Hortic 109:8–14

Bearden BN, Petersen L (2000) Influence of arbuscular mycorrhizal fungi on soil structure and aggregate stability of vertisols. Plant Soil 218:173–183

Behboudian MH, Torokfalvy E, Walker RR (1986) Effects of salinity on ionic content, water relations and gas exchange parameters in some citrus scion rootstock combinations. Sci Hortic 28:105–116

Belkhodja R, Morales F, Abadia A, Gomez-Aparisi J, Abadia J (1994) Chlorophyll fluorescence as a possible tool for salinity tolerance screening in barley (*Hordeum vulgare* L.). Plant Physiol 104:667–673

Ben Khaled L, Gomez AM, Ouarraqi EM, Oihabi A (2003) Physiological and biochemical responses to salt stress of mycorrhized and/or nodulated clover seedlings (*Trifolium alexandrinum* L.). Agronomie 23:571–580

Ben-Gal A, Borochov-Neori H, Yermiyahu U, Shani U (2009) Is osmotic potential a more appropriate property than electrical conductivity for evaluating whole-plant response to salinity? Environ Exp Bot 65:232–237

Bernstein L (1975) Effects of salinity and sodicity on plant growth. Annu Rev Phytopathol 13:295–312

Bongi G, Loreto F (1989) Gas exchange properties of salt-stressed olive (*Olea europaea* L.) leaves. Plant Physiol 90:1408–1416

Botella MA, Martinez V, Pardines J, Cerda A (1997) Salinity induced potassium deficiency in maize plants. J Plant Physiol 150:200–205

Bronick CJ, Lal R (2005) Soil structure and management: a review. Geoderma 124:3–22

Brundrett M (1991) Mycorrhizas in natural ecosystems. Adv Ecol Res 21:300–313

Cantrell IC, Linderman RG (2001) Preinoculation of lettuce and onion with VA mycorrhizal fungi reduces deleterious effects of soil salinity. Plant Soil 233:269–281

Caravaca F, Figueroa D, Barea JM, Azcón-Aguilar C, Roldán A (2004) Effect of mycorrhizal inoculation on the nutrient content, gas exchange and nitrate reductase activity of Retama sphaerocarpa and Olea europaea subsp. sylvestris under drought stress. J Plant Nutr 27:57–74

Caravaca F, Alguacil MM, Torres P, Roldán A (2005) Plant type mediates rhizospheric microbial activities and soil aggregation in a semiarid Mediterranean salt marsh. Geoderma 124:375–382

Cerda A, Martinez V (1988) Nitrogen fertilization under saline conditions in tomato and cucumber plants. J Hortic Sci 63:451–458

Chen N, Liu Y, Liu X, Chai J, Hu Z, Guo G, Liu H (2009) Enhanced tolerance to water deficit and salinity stress in transgenic *Lycium barbarum* L. plants ectopically expressing ATHK1, an Arabidopsis thaliana histidine kinase gene. Plant Mol Biol Rep 27:321–333

Cho K, Toler H, Lee J, Owenley B, Stutz JC, Moore JL, Augé RM (2006) Mycorrhizal symbiosis and response of sorghum plants to combined drought and salinity stresses. J Plant Physiol 163:517–528

Chow WS, Ball MC, Anderson JM (1990) Growth and photosynthetic responses of spinach to salinity: implication of K nutrition for salt tolerance. Aust J Plant Physiol 17:563–578

Colla G, Rouphael Y, Cardarelli M, Tullio M, Rivera CM, Rea E (2008) Alleviation of salt stress by arbuscular mycorrhizal in zucchini plants grown at low and high phosphorus concentration. Biol Fertil Soils 44:501–509

Copeman RH, Martin CA, Stutz JC (1996) Tomato growth in response to salinity and mycorrhizal fungi from saline or nonsaline soils. HortScience 31:341–344

Cramer GR (2002a) Deferential effects of salinity on leaf elongation kinetics of three grass species. Plant Soil 253:233–244

Cramer GR (2002b) Sodium–calcium interactions under salinity stress. In: Läuchli A, Luttge U (eds) Salinity: environment, plants, molecules. Kluwer, Dordrecht, pp 205–227

Cramer MD, Lips SH (1995) Enriched rhizosphere CO2 concentration can ameliorate the influence of salinity on hydroponically grown tomato plants. Plant Physiol 94:425–433

Cusido RM, Papazon J, Altabella T, Morales C (1987) Effects of salinity on soluble protein, free amino acids and nicotine contents in *Nicotiana rustica* L. Plant Soil 102:55–60

Delauney AJ, Verma DPS (1993) Proline biosynthesis and osmoregulation in plants. Plant J 4:215–223

Demidchik V, Davenport RJ, Tester M (2002) Nonselective cation channels in plants. Annu Rev Plant Biol 53:67–107

Demir Y, Kocakalikan I (2002) Effect of NaCl and proline on bean seedlings cultured in vitro. Biol Plant 45:597–599

Devitt D, Jarrell WM, Steven KL (1981) Sodium–potassium ratios in soil solution and plant response under saline conditions. Soil Sci Soc Am J 34:80–86

Downton WJS (1977) Photosynthesis in salt stressed grapevines. Aust Plant Physiol 4:183–192

Duke ER, Johnson CR, Koch KE (1986) Accumulation of phosphorus, dry matter and betaine during NaCl stress of split-root citrus seedlings colonized with vesicular-arbuscular mycorrhizal fungi on zero, one or two halves. New Phytol 104:583–590

Ehsanpour AA, Amini F (2003) Effect of salt and drought stress on acid phosphatase activities in alfalfa (*Medicago sativa* L.) explants under in vitro culture. Afr J Biotechnol 2:133–135

FAO (2005) Global network on integrated soil management for sustainable use of salt-affected soils. FAO Land and Plant Nutrition Management Service, Rome

Feng G, Zhang FS, Li XL, Tian CY, Tang C, Rengel Z (2002) Improved tolerance of maize plants to salt stress by arbuscular mycorrhiza is related to higher accumulation of soluble sugars in roots. Mycorrhiza 12:185–190

Francois LE, Donovan TJ, Maas EV (1990) Salinity effects on emergence, vegetative growth and seed yield of guar. Agron J 82:587–591

Gadkar V, Rillig M (2006) The arbuscular mycorrhizal fungal protein glomalin is a putative homolog of heat shock protein 60. FEMS Microbiol Lett 263:93–101

Ghazi N, Al-Karaki GN (2006) Nursery inoculation of tomato with arbuscular mycorrhizal fungi and subsequent performance under irrigation with saline water. Sci Hortic 109:1–7

Giri B, Mukerji KG (2004) Mycorrhizal inoculant alleviates salt stress in *Sesbania aegyptiaca* and Sesbania grandiflora under field conditions: evidence for reduced sodium and improved magnesium uptake. Mycorrhiza 14:307–312

Giri B, Kapoor R, Mukerji KG (2003) Influence of arbuscular mycorrhizal fungi and salinity on growth, biomass, and mineral nutrition of *Acacia auriculiformis*. Biol Fertil Soils 38:170–175

Giri B, Kapoor R, Mukerji KG (2007) Improved tolerance of *Acacia nilotica* to salt stress by Arbuscular mycorrhiza, Glomus fasciculatum may be partly related to elevated K/Na ratios in root and shoot tissues volume. Microb Ecol 54:753–760

Glenn EP, O'Leary JW (1985) Productivity and irrigation requirements of halophytes grown with seawater in the Sonoran Desert. J Arid Environ 9:81–91

Goas G, Goas M, Larher F (1982) Accumulation of free proline and glycine betaine in Aster tripolium subjected to a saline shock: a kinetic study related to light period. Physiol Plant 55:383–388

Goicoechea N, Merino S, Sánchez-Díaz M (2005) Arbuscular mycorrhizal fungi can contribute to maintain antioxidant and carbon metabolism in nodules of *Anthyllis cytisoides* L. subjected to drought. J Plant Physiol 162:27–35

Goss MJ, de Varennes A (2002) Soil disturbance reduces the efficacy of mycorrhizal associations for early soybean growth and N_2 fixation. Soil Biol Biochem 34:1167–1173

Graifenberg A, Giustiniani L, Temperini O, Lipucci di Paola M (1995) Allocation of Na, Cl, K and Ca within plant tissues in globe artichoke (*Cynara scolymus* L.) under saline-sodic conditions. Sci Hortic 63:1–10

Grattan SR, Grieve CM (1992) Mineral element acquisition and growth response of plants grown in saline environments. Agric Ecosyst Environ 38:275–300

Grattan SR, Grieve CM (1999) Salinity-mineral nutrient relations in horticultural crops. Sci Hortic 78:127–157

Greenland DJ (1984) Exploited plants: rice. Biologist 31:291–325

Greenway H, Munns R (1980) Mechanisms of salt tolerance in nonhalophytes. Annu Rev Plant Physiol 31:149–190

Hajlaoui H, Ayeb NE, Garrec JP, Denden M (2010) Differential effects of salt stress on osmotic adjustment and solutes allocation on the basis of root and leaf tissue senescence of two silage maize (*Zea mays* L.) varieties. Ind Crops Prod 31:122–130

Hamdy A (1990a) Management practices under saline water irrigation. In Symposium on Scheduling of Irrigation for Vegetable Crops Under Field Conditions. Acta Hortic 278:745–754

Hamdy A (1990b) Saline irrigation practices: leaching management. In: Proceedings of the water and wastewater '90' conference, Barcelona, 10 pp

Hammer EC, Nasr H, Pallon J, Olsson PA, Wallander H (2010) Elemental composition of arbuscular mycorrhizal fungi at high salinity. Mycorrhiza 21:117–129

Hernández JA, Aguilar A, Portillo B, López-Gómez E, Mataix Beneyto J, García-Legaz MF (2003) The effect of calcium on the antioxidant enzymes from salt treated loquat and anger plants. Funct Plant Biol 30:1127–1137

Hirrel MC (1981) The effect of sodium and chloride salts on the germination of *Gigaspora margarita*. Mycology 43:610–617

Hu Y, Schmidhalter U (2002) Limitation of salt stress to plant growth. In: Hock B, Elstner CF (eds) Plant toxicology. Marcel Dekker, New York, pp 91–224

Jackson WA, Volk RJ (1997) Role of potassium in photosynthesis and respiration. In: Madison WS (ed) The role of potassium in agriculture. American Society of Agronomy, Madison, WI, pp 109–188

Jahromi F, Aroca R, Porcel R, Ruiz-Lozano JM (2008) Influence of salinity on the in vitro development of *Glomus intraradices* and on the in vivo physiological and molecular responses of mycorrhizal lettuce plants. Microb Ecol 55:45–53

Jakobsen I, Abbott LK, Robson AD (1992) External hyphae of vesicular–arbuscular mycorrhizal fungi associated with *Trifolium subterraneum* L. New Phytol 120:373–379

Jeffries P, Gianinazzi S, Perotto S, Turnau K, Barea JM (2003) The contribution of arbuscular mycorrhizal fungi in sustainable maintenance of plant health and soil fertility. Biol Fertil Soils 37:1–16

Jimenez JS, Debouck DG, Lynch JP (2003) Growth, gas exchange, water relations, and ion composition of Phaseolus species grown under saline conditions. Field Crop Res 80:207–222

Jindal V, Atwal A, Sekhon BS, Singh R (1993) Effect of vesicular–arbuscular mycorrhizae on metabolism of moong plants under NaCl Salinity. Plant Physiol Biochem 3:475–481

Juniper S, Abbott LK (1993) Vesicular-arbuscular mycorrhizas and soil salinity. Mycorrhiza 4:45–57

Kabir ME, Karim MA, Azad MAK (2004) Effect of potassium on salinity tolerance of mung bean (*Vigna radiata* L. Wilczek). J Biol Sci 4:103–110

Katembe WJ, Ungar IA, Mitchell J (1998) Effect of salinity on germination and seedling growth of two Atriplex species (Chenopodiaceae). Ann Bot 82:167–175

Kaya C, Kirnak H, Higgs D (2001) Enhancement of growth and normal growth parameters by foliar application of potassium and phosphorus in tomato cultivars grown at high (NaCl) salinity. J Plant Nutr 24:357–367

Kaya C, Higgs D, Sakar E (2002) Response of two leafy vegetables grown at high salinity to supplementary potassium and phosphorus during different growth stages. J Plant Nutr 25:2663–2676

Kaya C, Ashraf M, Sonmez O, Aydemir S, Levent Tuna A, Cullu AM (2009) The influence of arbuscular mycorrhizal colonisation on key growth parameters and fruit yield of pepper plants grown at high salinity. Sci Hortic 121:1–6

Kim SY, Lim JH, ParkMR KYJ, Park TII, Seo YW, Choi KG, Yun SJ (2005) Enhanced antioxidant enzymes are associated with reduced hydrogen peroxide in barley roots under salt stress. J Biochem Mol Biol 38:218–224

Kinraide TB (1999) Interactions among Ca^{2+}, Na^+ and K^+ in salinity toxicity: quantitative resolution of multiple toxic and ameliorative effects. J Exp Bot 50:1495–1505

Kohler J, Hernández JA, Caravaca F, Roldána A (2009) Induction of antioxidant enzymes is involved in the greater effectiveness of a PGPR versus AM fungi with respect to increasing the tolerance of lettuce to severe salt stress. Environ Exp Bot 65:245–252

Kohler J, Caravaca F, Roldán A (2010) An AM fungus and a PGPR intensify the adverse effects of salinity on the stability of rhizosphere soil aggregates of Lactuca sativa. Soil Biol Biochem 42:429–434

Kramer D (1983) Genetically determined adaptations in roots to nutritional stress: correlation of structure and function. Plant Soil 72:167–173

Kumar A, Sharma S, Mishra S (2009) Influence of arbuscular mycorrhizal (AM) fungi and salinity on seedling growth, solute accumulation, and mycorrhizal dependency of Jatropha curcas L. J Plant Growth Regul 29:297–306

Lacuesta M, Gonzalez-Maro B, Gonzale-Murua C, Munoz-Rueda A (1990) Temporal study of the effect of phosphinothricin on the activity of glutamine synthetase, glutamate dehydrogenase and nitrate reductase in *Medicago sativa* L. Plant Physiol 136:410–414

Lax A, Díaz E, Castillo V, Albaladejo J (1994) Reclamation of physical and chemical properties of a salinized soil by organic amendment. Arid Soil Res Rehabil 8:9–17

Loreto F, Centritto M, Chartzoulakis K (2002) Photosynthetic limitations in olive cultivars with different sensitivity to salt stress. Plant Cell Environ 26:495–601

Lutts S, Kinet JM, Bouharmont J (1996) NaCl-induced senescence in leaves of rice (*Oryza sativa* L.) cultivars differing in salinity resistance. Ann Bot 78:389–398

Magistad OC, Ayers AD, Wadleigh CH, Gauch HG (1943) Effect of salt concentration, kind of salt, and climate on plant growth in sand cultures. Plant Physiol 18:151–166

Mäkelä P, Kärkkäinen J, Somersalo S (2000) Effect of glycinebetaine on chloroplast ultrastructure, chlorophyll and protein content, and RuBPCO activities in tomato grown under drought or salinity. Biol Plant 43:471–475

Manchanda G, Garg N (2008) Salinity and its effects on the functional biology of legumes. Acta Physiol Plant 30:595–618

Maritinez V, Cerda A (1989) Nitrate reductase activity in tomato and cucumber leaves as influenced by NaCl and N source. J Plant Nutr 12:1335–1350

Marschner H (1995) Mineral nutrition of higher plants, 2nd edn. Academic, London

Martínez V, Lauchli A (1991) Phosphorus translocation in salt-stressed cotton. Physiol Plant 83:627–632

Martínez V, Bernstein N, Läuchli A (1996) Salt-induced inhibition of phosphorus transport in lettuce plants. Physiol Plant 97:118–122

McMillen BG, Juniper S, Abbott LK (1998) Inhibition of hyphal growth of a vesicular–arbuscular mycorrhizal fungus in soil containing sodium chloride limits the spread of infection from spores. Soil Biol Biochem 30:1639–1646

Miller RM, Jastrow JD (2000) Mycorrhizal fungi influence soil structure. In: Kapulnik Y, Douds DD (eds) Arbuscular mycorrhizas: molecular biology and physiology. Kluwer, Dordrecht, pp 3–18

Miransari M, Bahrami HA, Rejali F, Malakouti MJ (2008) Using arbuscular mycorrhiza to reduce the stressful effects of soil compaction on wheat (*Triticum aestivum* L.) growth. Soil Biol Biochem 40:1197–1206

Mukerji KG, Chamol BP (2003) Compendium of mycorrhizal research. A. P. H., New Delhi, 310 pp

Munns R (1993) Physiological processes limiting plant growth in saline soils: some dogmas and hypotheses. Plant Cell Environ 16:15–24

Munns R (2005) Genes and salt tolerance: bringing them together. New Phytol 167:645–663

Munns R, Termaat A (1986) Whole plant responses to salinity. Aust J Plant Physiol 13:143–160

Muralev E, Nazarenko PI, Poplavskij VM, Kuznetsov IA (1997) Seawater desalination. In: Nuclear desalinization of seawater. Proceedings of a symposium in Taejon, Republic of Korea. International Atomic Energy Agency, Vienna, pp 355–366

Naidoo G, Naidoo Y (2001) Effects of salinity and nitrogen on growth, ion relations and proline accumulation in Triglochin bulbosa. Wetlands Ecol Manage 9:491–497

Neto ADA, Prisco JT, Gomes-Filho E (2009) Changes in soluble amino-N, soluble proteins and free amino acids in leaves and roots of salt-stressed maize genotypes. J Plant Interact 4:137–144

Neumann PM, Van Volkenburgn E, Cleland RE (1988) Salinity stress inhibits bean leaf expansion by reducing turgor, not wall extensibility. Plant Physiol 85:233–237

Nieman RH (1965) Expansion of bean leaves and its suppression by salinity. Plant Physiol 40:156–161

Ober ES, Sharp RE (1994) Proline accumulation in maize (*Zea mays* L.) primary roots at low water potentials (I. Requirement for increased levels of abscisic acid). Plant Physiol 105:981–987

Okçu G, Kaya MD, Atak M (2005) Effects of salt and drought stresses on germination and seedling growth of pea (*Pisum sativum* L). Turk J Agric Forest 29:237–242

Olmos E, Hellin E (1997) Cytochemical localization of ATPase plasma membrane and acid phosphatase by cerium based in a salt-adapted cell line of *Pisum sativum*. J Exp Bot 48:1529–1535

Papadopoulos I, Rendig VV (1983) Interactive effects of salinity and nitrogen on growth and yield of tomato plants. Plant Soil 73:47–57

Pardossi A, Bagnoli G, Malorgio F, Campiotti CA, Tognoni F (1999) NaCl effects on celery (*Apium graveolens* L.) grown in NFT. Sci Hort 81:229–242

Parida AK, Das AB (2005) Salt tolerance and salinity effects on plants: a review. Ecotoxicol Environ Saf 60:324–349

Parida AK, Das AB, Mittra B (2004) Effects of salt on growth, ion accumulation, photosynthesis and leaf anatomy of the mangrove (*Bruguiera parviflora*). Trees 18:167–174

Pérez-Alfocea F, Balibrea ME, Santa Cruz A, Estan MT (1996) Agronomical and physiological characterization of salinity tolerance in a commercial tomato hybrid. Plant Soil 180:251–257

Pfeiffer CM, Bloss HE (1988) Growth and nutrition of guayule (*Parthenium argentatum*) in a saline soil as influenced by vesicular–arbuscular mycorrhiza and phosphorus fertilization. New Phytol 108:315–321

Plenchette C, Duponnois R (2005) Growth response of the saltbush *Atriplex nummularia* L. to inoculation with the arbuscular mycorrhizal fungus *Glomus intraradices*. J Arid Environ 61:535–540

Porras-Soriano A, Soriano-Martın ML, Porras-Piedra A, Azcón R (2009) Arbuscular mycorrhizal fungi increased growth, nutrient uptake and tolerance to salinity in olive trees under nursery conditions. J Plant Physiol 166:1350–1359

Poss JA, Pond E, Menge JA, Jarrell WM (1985) Effect of salinity on mycorrhizal onion and tomato in soil with and without additional phosphate. Plant Soil 88:307–319

Powell CL (1975) Potassium uptake by endotrophic mycorrhizas (*Griselinia littoralis*, *Glomus microcarpus*, Fungi). In: Endomycorrhizas; Proceedings of a symposium, pp 461–468

Pujol JA, Calvo JF, Ramirez-Diaz L (2000) Recovery of germination from different osmotic conditions by four halophytes from Southeastern Spain. Ann Bot 85:279–286

Qadir M, Ghafoor A, Murtaza G (2000) Amelioration strategies for saline soils: a review. Land Degrad Dev 11:501–521

Rabie GH (2005) Influence of arbuscular mycorrhizal fungi and kinetin on the response of mung bean plants to irrigation with seawater. Mycorrhiza 15:225–230

Rabie GH, Almadini AM (2005) Role of bioinoculants in development of salt-tolerance of *Vicia faba* plants. Afr J Biotechnol 4:210–222

Rao DLN (1998) Biological amelioration of salt-affected soils. In: Subba Rao NS, Dommergues YR (eds) Microbial interactions in agriculture and forestry, vol 1. Science Publishers, Enfield, CT, pp 21–238

Ravikovitch S, Porath A (1967) The effect of nutrients on the salt tolerance of crops. Plant Soil 26:49–71

Reid RJ, Smith FA (2000) The limits of sodium/calcium interactions in plant growth. Aust J Plant Physiol 27:709–715

Rengasamy P (2006) World salinization with emphasis on Australia. J Exp Bot 57(5):1017–1023

Rillig MC (2004) Arbuscular mycorrhizae, glomalin and soil quality. Can J Soil Sci 84:355–363

Rillig MC, Mummey DL (2006) Mycorrhizas and soil structure. New Phytol 171:41–53

Rinaldelli E, Mancuso S (1996) Response of young mycorrhizal and non-mycorrhizal plants of olive tree (*Olea europaea* L.) to saline conditions. I. Short-term electrophysiological and long term vegetative salt effects. Adv Hortic Sci 10:126–134

Rosendahl CN, Rosendahl S (1991) Influence of vesicular arbuscular mycorrhizal fungi (*Glomus spp.*) on the response of cucumber (*Cucumis sativus* L.) to salt stress. Environ Exp Bot 31:313–318

Ruíz-Lozano JM, Azcón R (1995) Hyphal contribution to water uptake in mycorrhizal plants as affected by the fungal species and water status. Physiol Plant 95:472–478

Ruíz-Lozano JM, Azcón R (2000) Symbiotic efficiency and infectivity of an autochthonous arbuscular mycorrhizal *Glomus sp.* from saline soils and *Glomus deserticola* under salinity. Mycorrhiza 10:137–143

Ruíz-Lozano JM, Azcón R, Gómez M (1996) Alleviation of salt stress by arbuscular-mycorrhizal *Glomus species* in Lactuca sativa plants. Physiol Plant 98:767–772

Rush DW, Epstein E (1978) Genotypic response to salinity difference between salt-sensitive and salt tolerance genotypes of tomato. Plant Physiol 57:162–166

Sannazzaro AI, Ruiz OA, Albertó EO, Menéndez AB (2006) Alleviation of salt stress in Lotus glaber by Glomus intraradices. Plant Soil 285:279–287

Shannon MC (1984) Breeding, selection, and the genetics of salt tolerance. In: Staples RC, Toenniessen GH (eds) Salinity tolerance in plants: strategies for crop improvement. Wiley, New York, pp 231–254

Shannon MC (1997) Adaptation of plants to salinity. Adv Agron 60:75–120

Sharifi M, Ghorbanli M, Ebrahimzadeh H (2007) Improved growth of salinity-stressed soybean after inoculation with salt pre-treated mycorrhizal fungi. J Plant Physiol 164:1144–1151

Sharma KD, Datta KS, Verma SK (1990) Effect of chloride and sulphate type of salinity on some metabolic drifts in chickpea (*Cicer arietinum* L). Indian J Exp Biol 28:890–892

Sheng M, Tang M, Chen H, Yang B, Zhang F, Huang Y (2008) Influence of arbuscular mycorrhizae on photosynthesis and water status of maize plants under salt stress. Mycorrhiza 18:287–296

Sheng M, Tang M, Zhang F, Huang Y (2011) Influence of arbuscular mycorrhiza on organic solutes in maize leaves under salt stress. Mycorrhiza 21:423–430

Shenker M, Bell GA, Shani U (2003) Sweet corn response to combined nitrogen and salinity environmental stresses. Plant Soil 256:139–147

Shi LX, Guo JX (2006) Changes in photosynthetic and growth characteristics of Leymus chinensis community along the retrogression on the Songnen grassland in northeastern China. Photosynthetica 44:542–547

Sibole JV, Montero E, Cabot C, Poschenrieder C, Barceló J (2000) Relationship between carbon partitioning and Na+, Cl- and ABA allocation in fruits of salt-stressed bean. J Plant Physiol 157:637–642

Singh G, Jain S (1982) Effect of some growth regulators on certain biochemical parameters during seed development in chickpea under salinity. Indian J Plant Physiol 25:167–179

Singh RP, Choudhary A, Gulati A, Dahiya HC, Jaiwal PK, Sengar RS (1997) Response of plants to salinity in interaction with other abiotic and factors. In: Jaiwal PK, Singh RP, Gulati A (eds) Strategies for improving salt tolerance in higher plants. Science, Enfield, CT, pp 25–39

Sivritepe N, Sivritepe HO, Eris A (2003) The effects of NaCl priming on salt tolerance in melon seedlings rown under saline conditions. Sci Hortic 97:229–237

Stephen MG, Duff SMG, Plaxton WC (1994) The role of acid phosphatases in plant phosphorus metabolism. Physiol Plant 90:791–800

Stewart CR, Lee JA (1974) The rate of proline accumulation in halophytes. Planta 120:279–289

Sudhakar C, Reddy PS, Veeranjaneyulu K (1993) Effect of salt stress on the enzymes of proline synthesis and oxidation in greengram (*Phaseolus aureus* Roxb.) seedlings. J Plant Physiol 14:621–623

Tabatabaei SJ (2006) Effects of salinity and N on the growth, photosynthesis and N status of olive (*Olea europaea* L.) trees. Sci Hortic 108:432–438

Tabatabaei SJ, Gregory P, Hadley P (2004) Uneven distribution of nutrients in the root zone affects the incidence of blossom end rot and concentration of calcium and potassium in fruits of tomato. Plant Soil 258:169–178

Tattini M (1994) Ionic relations of aeroponically-grown olive plants during salt stress. Plant Soil 161:251–256

Tester M, Davenport R (2003) Na tolerance and Na transport in higher plants. Ann Bot (Lond) 91:503–527

Thomas HM, Morgan WG, Humphreys MW (2003) Designing grasses with a future-combining the attributes of *Lolium* and *Festuca*. Euphytica 133:19–26

Thomson BD, Robson AD, Abbott LK (1990) Mycorrhizas formed by Gigaspora calospora and Glomus fasciculatum on subterranean clover in relation to soluble carbohydrate concentrations in roots. New Phytol 114:217–225

Tian CY, Feng G, Li XL, Zhang FS (2004) Different effects of arbuscular mycorrhizal fungal isolates from saline or non-saline soil on salinity tolerance of plants. Appl Soil Ecol 26:143–148

Tsang A, Maum MA (1999) Mycorrhizal fungi increase salt tolerance of Strophostyles helvola in coastal foredunes. Plant Ecol 144:159–166

United States Salinity Laboratory Staff (1954) Diagnosis and improvement of saline and alkali soils. US Department of Agriculture, Agricultural Handbook No. 60. US Government Printer, Washington, DC

Wright SF, Upadhyaya A (1996) Extraction of an abundant and unusual protein from soil and comparison with hyphal protein of arbuscular mycorrhizal fungi. Soil Sci 161:575–585

Wright SF, Upadhyaya A, Buyer JS (1998) Comparison of N linked oligosaccharides of glomalin from arbuscular mycorrhizal fungi and soils by capillary electrophoresis. Soil Biol Biochem 30:1853–1857

Wu QS, Zou YN, He XH (2009) Contributions of arbuscular mycorrhizal fungi to growth, photosynthesis, root morphology and ionic balance of citrus seedlings under salt stress. Acta Physiol Plant 32:297–304

Wyn Jones RG, Gorham J (1983) Osmoregulation. In: Lange OL, Noble PS, Osmond CB, Zeiger H (eds) Physiological plant ecology. III. Responses to chemical and biological environments. Springer, Berlin, pp 35–56

Xu G, Magen H, Tarchitzky J, Kafkafi U (2000) Advances in chloride nutrition of plants. Adv Agron 68:97–150

Yang CW, Xu HH, Wang LL, Liu J, Shi DC, Wang GD (2009) Comparative effects of salt-stress and alkali-stress on the growth, photosynthesis, solute accumulation, and ion balance of barley plants. Photosynthetica 47:79–86

Yano-Melo AM, Saggin OJ, Maia LC (2003) Tolerance of mycorrhized banana (*Musa sp.* cv. Pacovan) plantlets to saline stress. Agriculture. Ecosyst Environ 95:343–348

Yeo AR, Yeo ME, Flowers SA, Flowers TJ (1990) Screening of rice (*Oryza sativa* L.) genotypes for physiological characters contributing to salinity resistance, and their relationship to overall performance. Theor Appl Genet 79:377–384

Yermiyahu U, Nir S, Ben-Hayyim G, Kafkafi U, Kinraide TB (1997) Root elongation in saline solution related to calcium binding to root cell plasma membranes. Plant Soil 191:67–76

Zarea MJ (2010) Conservation tillage and Sustainable Agriculture in Semi-arid Dryland Farming. In: Lichtfouse E (ed) Biodiversity, biofules, agroforestry and conservation agriculture. Springer, Dordrecht, pp 195–232, 375 pp

Zhongoun H, Chaoxing H, Zhibin Z, Zhirong Z, Huaisong W (2007) Changes of antioxidative enzymes and cell membrane osmosis in tomato colonized by arbuscular mycorrhizae under NaCl stress. Colloids Surf B Biointerfaces 59:128–133

Zhu YG, Miller RM (2003) Carbon cycling by arbuscular mycorrhizal fungi in soil-plant systems. Trends Plant Sci 8:407–409

Zhu JK, Liu JP, Xiong LM (1998) Genetic analysis of salt tolerance in Arabidopsis: evidence for a critical role of potassium nutrition. Plant Cell 10:1181–1192

Chapter 16
White-Rot Fungi in Bioremediation

Safiye Elif Korcan, İbrahim Hakkı Ciğerci, and Muhsin Konuk

16.1 Introduction to Bioremediation

Bioremediation aims to solve an environmental problem like contaminated soil or groundwater by generally using biological organisms and also sometimes pure enzyme preparations. Bioremediation technologies can be used for contaminated wastewater, ground or surface waters, soils, sediments, and air where release of pollutants or chemicals that pose a risk to human, animal, or ecosystem health has been either accidental or intentional. The aim of bioremediation is to reduce pollutant levels to undetectable, nontoxic, or acceptable levels. In other words, organopollutants are completely mineralized to CO_2, or in the case of metals, the pollutants are removed by sorption or transformation to a less toxic form.

As Alexander (1994) points out, the major requirements for successful bioremediation are that microorganisms must have catabolic activity and they must not generate toxic products during the remediation. Conditions at the site or in a bioreactor must be made conducive to microbial growth or activity. For example, an adequate supply of inorganic nutrients, sufficient O_2, or some other electron acceptor; favorable moisture content; suitable temperature; and a source of C and energy for growth are needed if the pollutant is to be cometabolized. The target compound must be available to the microorganisms. The site must not contain concentrations or combinations of chemicals that are markedly inhibitory to the biodegrading species, or means must exist to dilute otherwise render innocuous the inhibitors. Such organisms must be able to transform the compound at reasonable

S.E. Korcan • İ.H. Ciğerci
Biology Department, Faculty of Science and Literature, Afyon Kocatepe University,
ANS Campus, Gazligol Yolu, 03200 Afyonkarahisar, Turkey
e-mail: ekorcan@aku.edu.tr; cigerci@aku.edu.tr

M. Konuk (✉)
Molecular Biology and Genetics Department, Faculty of Engineering and Natural Sciences,
Üsküdar University, Altunizade, 34662 Istanbul, Turkey
e-mail: muhsin.konuk@uskudar.edu.tr

E.M. Goltapeh et al. (eds.), *Fungi as Bioremediators*, Soil Biology 32,
DOI 10.1007/978-3-642-33811-3_16, © Springer-Verlag Berlin Heidelberg 2013

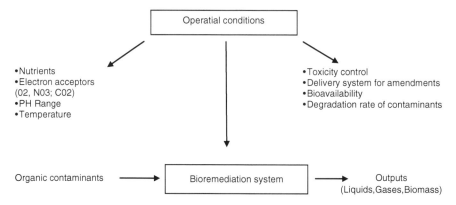

Fig. 16.1 Bioremedial Systems (Adapted from Anderson 1995)

rates and bring the concentration to levels that meet regulatory standards. Finally, the cost of the technology must be cheaper than that of other technologies that can also destroy the chemicals (Pointing 2001).

Bioremediation technology can generally be classified as in situ and bioreactors. Bioremediation of organopollutants in situ generally applies to contaminated soils with two approaches: biorestoration and bioaugmentation. In biorestoration, the physicochemical nature of the soil is changed so that indigenous microorganisms are encouraged to degrade the pollutant. On the other hand, in bioaugmentation, a known pollutant-degrading microorganism is introduced to the contaminated soil (Pointing 2001). Bioreactors are widely used for the treatment of waste before being discharged and in bioremediation, with aerobic and anaerobic treatments for the processing of solid, liquid, and slurry wastes established. Most of these utilize indigenous microorganisms (Boopathy 2000; Pointing 2001), although some of the organopollutants treated are also known as substrates for white-rot fungi (Pointing 2001).

Though many different bioremediation processes exist, there are common factors which are vital in all bioremediation processes. All bioremediation systems consist of two major things: inputs and output. The inputs include organic contaminants and operational conditions, which are composed of all the environmental conditions and other factors such as toxicity, bioavailability, and degradation rate of contaminants, and also the delivery system for amendments (Fig. 16.1).

16.2 The Parameters of Bioremediation Processes

Environmental conditions are different conditions that can affect microbial activity such as oxygen, pH, redox potential, presence of nutrients, environment temperature, and moisture content in the remediation environment. The presence of all these factors at optimum level leads to a successful bioremediation process (Meysami 2001).

16.2.1 Requirement of Oxygen and Inorganic Nutrients

Lignin-degrading fungi are aerobic microorganisms, which need certain level of oxygen to survive, grow, or metabolize (Azadpour et al. 1997). For instance, white-rot fungi and bacteria that degrade petroleum hydrocarbons require free or dissolved oxygen (Odu 1981). In addition, oil degradation requires mineral elements such as C, Ca, Mg, K, S, Fe, N, and P and various trace elements (Odu 1978). Inorganic nutrients mainly nitrogen and phosphorus as well as carbon, which is the main nutrient, are essential for all biological processes. While nitrogen can be provided in a variety of forms such as nitrate, ammonium salts, and organic compounds like urea, phosphorus can also be provided in several inorganic forms. Treatability studies are usually conducted to determine nutrient requirements, and results are specific to the particular sites and processes (Meysami 2001).

16.2.2 pH

As most natural environments possess pH values in the range between 5 and 9, it is not surprising that most microorganisms have evolved with pH tolerance within this range. For bioremediation processes, the optimum pH is site and process specific and must be determined during feasibility studies. Most microorganisms tolerate pH 5 up to 9 but prefer pH 6.5–7.5 (Meysami 2001). The optimum pH for biodegradation of hydrocarbons is around pH 6–8 (Mentzer and Ebere 1996). Biodegradation of crude petroleum in an acid soil (pH 4.5) could be doubled by liming to pH 7.4. The optimum pH for white-rot fungi is acidic and can vary from pH of 4 to 7.0 (Baker Lee et al. 1995).

16.2.3 Temperature

Temperature plays the primary role in lignin degradation by white-rot fungi. The optimum temperature range for these fungi is from 20 °C to 40 °C (Azadpour et al. 1997). In bioremediation reactions, due to the bioremedial agent's nature of being a complex community of microorganisms, a few-degree shift in temperature may result in dramatic changes in the composition and function of the community. The specific population, which has the ability to degrade the contaminant, may function just over a narrow range of temperatures or may be replaced by populations with different degradation kinetics or mechanisms (Meysami 2001). Generally, hydrocarbon biodegradation increases with temperature and peaks around 30–40 °C (Mentzer and Ebere 1996). Temperature as a limiting factor does not seem to be a problem in tropical and temperate zones. Disappearance of hydrocarbon contaminant from agricultural land can be correlated with monthly temperature averages (Dibble and Bartha 1979).

16.2.4   Water Availability

Moisture plays a crucial role in bioremediation processes. The moisture content of soil affects many other parameters in bioremediation systems such as the bioavailability of contaminants, the effective toxicity level of contaminants, the transfer of gases, species distribution, and the movement and growth of microorganism (Meysami 2001). Soil that is hydrated with 50–80 % of the maximum water-holding capacity has the greatest microbial activity (Mentzer and Ebere 1996). Below that level, osmotic and matrix forces limit the availability of water to microbes, and above that level, the reduction of air space and oxygen decreases microbial activity.

16.2.5   Adsorption Effects

Organic matters are less susceptible to microbial attack. Indeed, the rate-limiting process in biodegradation may be the desorption of contaminants (Mentzer and Ebere 1996).

16.2.6   Electron Acceptors

Carbon dioxide, sulfate, nitrate, and oxygen are among the common electron acceptors. In some cases, when they are used as substrates for reductive dehalogenation, the halogenated organic contaminants can serve as electron acceptors. Much of the energy for the growth of microorganisms is obtained during the transfer of electrons from organic substrates to inorganic electron acceptors. Therefore, appropriate electron acceptors are certainly needed for biodegradation, and the provision of these electron acceptors often constitutes the greatest challenge in the design of in situ bioremediation systems. While oxygen or nitrate can be used by some bacteria as the terminal electron acceptor, the microbial communities are distinctly different for each type of electron acceptors. Their metabolic processes and their potentials for biodegradation of pollutants are also very different (Meysami 2001).

16.3 White-Rot Fungi

The fungi family is divided into many subunits, each of which is more restrictive than the next higher level. Within this big family, the white-rot fungi (WRF) are in the division Eumycota (true fungi), subdivision Basidiomycotina, class Hymenomycetes, and subclass Holobasidiomycetidae (Hawksworth et al. 1995). The mushrooms, puffballs, conks, and crust-like fungi are varieties of white-rot fungi, which is a physiological grouping of fungi that can degrade lignin and lignin-like substances. The name white rot derives from the appearance of wood attacked by these fungi, in which lignin removal results in a bleached appearance to the substrate. White-rot fungi attack the lignin component of wood and leave the cellulose and hemicellulose less affected. The white-rot fungi that degrade lignin rather than cellulose are called selective degraders. These degraders are especially interesting in biotechnological applications since they remove lignin and leave the valuable cellulose intact. Lignin degradation by these fungi is thought to occur during secondary metabolism and typically under nitrogen starvation. However, a wide variety of lignin degradation efficiency and selectivity abilities, enzyme patterns, and substrates enhancing lignin degradation are reported for these fungi (reviewed by Hatakka 2001; Hofrichter 2002).

Most researches within the field of bioremediation have focused on bacteria with fungal bioremediation (mycoremediation), and such researches have attracted interest just within the past two decades. The toxicity of many pollutants limits natural attenuation by bacteria, but white-rot fungi can withstand toxic levels of most organopollutants (Aust et al. 2003). White-rot fungi are known for their variety and their remarkable ability to degrade complex and persistent natural materials. They produce a nonspecific, extracellular enzymatic system, which is capable of degrading lignin, one of the most resistant materials found in nature (Aust and Barr 1994). Four main genera of white-rot fungi have shown potential for bioremediation: *Phanerochaete*, *Trametes*, *Bjerkandera*, and *Pleurotus* (Hestbjerg et al. 2003). These fungi have been used widely as pollutant degraders. In contrast to bacteria, fungi are capable of extending the location of their biomass through hyphal growth. These features distinguish fungi as organisms having great potential for use in bioremediation of soils contaminated with some of persistent organic pollutants (Aust and Barr 1994). The potential targets for white-rot fungi include sorbed contaminants, high molecular weight contaminants, and complex mixtures of chemicals typical of a contaminated site (Azadpour et al. 1997).

However, white-rot fungi have advantages over bacteria in the diversity of compounds, and they are able to oxidize, notably the larger polycyclic aromatic hydrocarbons (PAHs). The extracellular nature of white-rot fungal laccase-mediator systems (LMEs) and low-molecular-mass mediators may also enhance bioavailability of pollutants to white-rot fungi in situations where bacteria, with their cell-associated pollutant catabolism, may not have access. The most significant difference between white-rot fungal and bacterial pollutant catabolism relates to their biochemistry. White-rot fungi are obligate aerobes, whereas bacteria can transform and mineralize certain pollutants under aerobic, microaerophilic, and anaerobic conditions. Bacteria generally utilize organopollutants as a nutritional C or N source, whereas substrate oxidation by LMEs of white-rot fungi does not yield any net energy gain. An additional C and N source is therefore required for primary metabolism by white-rot fungi (Pointing 2001).

16.4   Ligninolytic Enzymes

The white-rot basidiomycetes are the most efficient degraders of lignin and also the most widely studied. The enzymes implicated in lignin degradation are lignin peroxidase, which catalyzes the oxidation of both phenolic and nonphenolic units; manganese-dependent peroxidase and laccase, which oxidize phenolic compounds to give phenoxy radicals and quinines; glucose oxidase and glyoxal oxidase for H_2O_2 production; and cellobiose–quinone oxidoreductase for quinone reduction (Kirk and Farrell 1987; Thakker et al. 1992). The ability of white-rot fungi to degrade a wide number of organopollutants is in part due to the action of nonspecific system (Paszczynski and Crawford 1995). Extracellular enzymes involved in the degradation of lignin and xenobiotics by white-rot fungi include several kinds of laccases (Leontievsky et al. 1997; Thurston 1994), peroxidases (Camarero et al. 1999), and oxidases producing H_2O_2 (Guillén et al. 1992; Volc et al. 1996).

The different degrees of degradation of lignin with respect to other wood components depend on the environmental conditions and the fungal species involved (Palmieri et al. 1997). Ligninolytic enzymes, such as manganese peroxidase (MnP), lignin peroxidase, and laccase, are mainly secreted by white-rot fungi and these degrade lignin from wood in the natural environment. These enzymes are able to degrade a variety of pollutants, including PAHs, polychlorinated biphenyls (PCBs), and synthetic dyes due to their low substrate specificity (Levin et al. 2008). Based on their ligninolytic enzyme patterns, wood-rotting fungi can be divided into three groups (Hatakka 1994; Lankinen 2004): (1) LiP-, MnP-, and laccase-producing fungi; (2) MnP- and laccase-producing fungi; and (3) LiP- and laccase-producing fungi.

The most common group among the white-rot fungi is the MnP- and laccase-producing group (Hatakka 2001; Lankinen 2004). A total of six LiP and four MnP isoenzymes have been characterized, some only on lignocelluloses containing media, from the best studied selective lignin degrader, the white-rot fungus *Phanerochaete chrysosporium* (Stewart and Cullen 1999; Lankinen 2004). It has been demonstrated that there is no unique mechanism to achieve the process of lignin degradation and that the enzymatic machinery of the various microorganisms differs. *Pleurotus ostreatus*, for instance, belongs to a subclass of lignin-degrading microorganisms that produce laccase, manganese peroxidase, and veratryl alcohol oxidase but no lignin peroxidase (Palmieri et al. 1997). *P. chrysosporium* has not been reported to produce laccase, although other selective degraders of wood produce a combination of MnP and laccase—*Pleurotus ostreatus* (Giardina et al. 1996; Lankinen 2004), *Pleurotus eryngii* (Martínez et al. 1996; Muñoz et al. 1997), and *Dichomitus squalens*

(Eriksson et al. 1990; Lankinen 2004); a combination of MnP, LiP, and laccase—
Phlebia radiata (Lundell and Hatakka 1994; Lankinen 2004); or only laccase—
Pycnoporus cinnabarinus (Eggert et al. 1996; Lankinen 2004). *Pycnoporus cinnabarinus* has been shown to produce laccase as the only ligninolytic enzyme (Eggert et al. 1996). *P. radiata* produces at least three LiPs, three MnPs, and one laccase (Moilanen et al. 1996; Lankinen 2004), and *P. sanguineus* produces laccase as the sole phenol oxidase (Pointing and Vrijmoed 2000).

16.4.1 Laccase

Laccase (EC 1.10.3.2, *p*-diphenol oxidase) is one of a few enzymes that have been studied since the nineteenth century. Yoshida first described laccase in 1883 when he extracted it from the exudates of the Japanese lacquer tree, *Rhus vernicifera*. In 1896, laccase was demonstrated to be a fungal enzyme for the first time by both Bertrand and Laborde (Thurston 1994; Levine 1965). Laccases are extracellular glycoproteins containing four atoms of copper, which are distributed into three sites (T1, T2, T3) according to their spectroscopic properties. The T1 site contains the type 1 blue copper (Cu1), which coordinates a cysteine, and is responsible for the blue color of the enzyme. The T2 site contains a type 2 copper (Cu2). In the T3 site, Cu3a and Cu3b are strongly coupled (Christian et al. 2003). Not all laccases are reported to possess four copper atoms (Thurston 1994) per monomeric molecule. One of the laccases from *Pleurotus ostreatus* is said to confer no blue color and was described by the author to be a white laccase (Palmieri et al. 1997). It was determined by atomic absorption that the laccase consisted of one copper atom, one zinc atom, and two iron atoms instead of the typical four coppers. Under different cultivation conditions, fungal laccases often occur as multiple isoenzymes expressed. Laccases are either mono- or multimeric copper-containing oxidases that catalyze the one-electron oxidation of a vast amount of phenolic substrates. Molecular oxygen serves as the terminal electron acceptor and is reduced to two molecules of water (Ducros et al. 1998). Laccase differs from other peroxidases because it does not require H_2O_2 to oxidize substrates; the electrons are transferred from molecular O_2 through water. In the presence of additional cosubstrates or proper redox mediators such as 2,2′-azino-bis(3-ethylbenzothiazoline-6-sulfonate) (ABTS) or hydroxybenzotriazole (HBT), laccase degrades also less reactive, relatively recalcitrant phenols in an oxidative process involving the mediator and the substrate (Rodakiewicz-Nowak et al. 2000). The ability of laccases to oxidize phenolic compounds as well as their ability to reduce molecular oxygen to water has led to intensive studies of these enzymes (Jolivalt et al. 1999; Xu et al. 1996).

A vast amount of industrial applications for laccases have been proposed, and they include paper processing, prevention of wine discoloration, detoxification of environmental pollutants, oxidation of dye and dye precursors, enzymatic conversion of chemical intermediates, and production of chemicals from lignin. Before laccases can be commercially implemented for potential applications, however, an inexpensive enzyme source needs to be made available (Yaver et al. 2001). Two of the most intensively studied areas in the potential industrial application of laccase are the delignification or biobleaching pulp and the bioremediation of contaminating environmental pollutants (Schlosser et al. 1997).

16.4.2   Lignin Peroxidases

Lignin peroxidases (LiPs) were the first ligninolytic enzymes to be discovered (Glenn et al. 1983; Tien and Kirk 1983). LiP is considered an important ligninolytic agent, but it may act in concert with other smaller oxidants that can penetrate and open up the wood cell wall. The LiP-catalyzed oxidation of a lignin substructure begins with the extraction of one electron from the aromatic ring of the donor substrate, and the resulting species, an aryl cation radical, then undergoes a variety of postenzymatic reactions (Kirk and Farrell 1987). LiPs resemble other peroxidases such as the classical, extensively studied enzyme from horseradish, in that they contain ferric heme and operate via a typical peroxidase catalytic cycle (Kirk and Farrell 1987; Gold et al. 1989). That is, LiP is oxidized by H_2O_2 to a two-electron deficient intermediate, which returns to its resting state by performing two one-electron oxidations of donor substrates. However, LiPs are more powerful oxidants than typical peroxidases and consequently oxidize not only the usual peroxidase substrates such as phenols and anilines but also a variety of nonphenolic lignin structures and other aromatic ethers that resemble the basic structural unit of lignin (Kersten et al. 1990).

Lignin peroxidases, which are heme glycoproteins, were discovered in the extracellular broth of secondary metabolic cultures of the basidiomycete *Phanerochaete chrysosporium* (Tien and Kirk 1983) from which various isoenzyme forms have been purified and studied. Lignin peroxidase has also been isolated from the lignin-degrading basidiomycetes *Phlebia radiata* (Kantelinen et al. 1989) and *Trametes versicolor* (Jönsson et al. 1987). LiPs occur in some frequently studied white-rot fungi, e.g., *Phanerochaete chrysosporium*, *Trametes versicolor*, and *Bjerkandera* sp. (Kirk and Farrell 1987 ; Kaal et al. 1993; Orth et al. 1993), but are evidently absent in others, e.g., *Dichomitus squalens*, *Ceriporiopsis subvermispora*, and *Pleurotus ostreatus* (Perie and Gold 1991; Orth et al. 1993).

LiP is generally regarded as an important enzyme involved in the oxidative depolymerization of lignin by white-rot fungi. LiP is capable of oxidizing a variety of xenobiotic compounds, including polycyclic aromatic hydrocarbons, polychlorinated phenols, nitroaromatics, azo dyes (Paszczynski and Crawford 1991; Hammel 1995; Hammel et al. 1986), and chlorophenols (Hammel and Tardone 1988). This enzyme can also be oxidized by various polycyclic aromatic hydrocarbons and related structures, including pyrene, anthracene, benzo[*a*]pyrene, dibenzo[*p*]dioxin, and thianthrene. Certain halogenated aromatics are also oxidized, including 2-chlorodibenzo[*p*]dioxin and 2,4,6-trichlorophenol (Hammel and Tardone 1988).

16.4.3   Manganese Peroxidase

MnP (EC 1.11.1.13) is a heme glycoprotein with a molecular weight of approximately 46,000 and occurs as a family of isozymes (Kirk and Farrell 1987). The catalytic cycle of MnP is similar to that of LiP and horseradish peroxidase (Tien et al. 1986; Wariishi et al. 1988); MnPs occur in most white-rot fungi and are similar to conventional peroxidases (e.g., horseradish peroxidase), except that Mn (II) is the obligatory electron donor for reduction of the one-electron deficient enzyme to its resting state, and as a result, Mn(III) is produced (Rodakiewicz-Nowak ct al. 2000). This reaction requires the presence of organic acids such as oxalate, glyoxylate, and lactate that were shown to have an important role in the mechanism of MnP and lignin degradation (Shimada et al. 1994). Organic acids such as malate and tartrate stimulate the catalytic activity of MnP by chelating Mn (III) (Wariishi et al. 1989, 1992). The resulting Mn(III) chelates are small, diffusible oxidants that can act at a distance from the MnP active site. They are not strongly oxidizing intermediates and are consequently unable to attack the recalcitrant nonphenolic structures that predominate in lignin. However, Mn (III) chelates do oxidize the more reactive phenolic structures that make up approximately 10 % of lignin. These reactions result in a limited degree of ligninolysis and other degradative reactions (Tuor et al. 1992). MnP isozymes are encoded by several different genes in *P. chrysosporium*. Several cDNAs (MnP1, MnP2a, and MnP2b (Pribnow et al. 1989) and MP-1 (Pease et al. 1989)) and genomic DNAs (MnP1 (Godfrey et al. 1990) and MnP2 (Mayfield et al. 1994)) of MnP isozymes have been isolated and characterized. Recently, the crystal structure of MnP from *P. chrysosporium* has been reported (Sundaramoorthy et al. 1994). The analysis showed that the overall structure of MnP is closely similar to that of LiP and that MnP has two structural calcium ions and two *N*-acetylglucosamine residues. Moreover, a manganese-binding site was identified in the crystal structure.

16.5   Degraded Compounds by White-Rot Fungi

16.5.1   Textile Dyes

Michaels and Lewis (1985) show that synthetic dyes share a common feature in that they are not readily biodegradable when discharged to the environment. Dye-containing effluents are hardly decolorized by conventional biological wastewater treatments. In addition to their visual effect and their adverse impact in terms of chemical oxygen demand, many synthetic dyes are toxic, mutagenic, and carcinogenic. Researchers have showed that a lot of WRF are able to decolorize synthetic (textile) dyes: *Bjerkandera* sp., *Ceriporia metamorphosa*, *Daedalea flavida*, *Daedalea confragosa*, *Lentinus tigrinus*, *Mycoacia nothofagi*, *Phanerochaete chrysosporium*, *Phanerochaete sordida*, *Phellinus pseudopunctatus*, *Phlebia brevispora*, *Phlebia (Merulius) tremellosa*, *Phlebia fascicularia*, *Phlebia floridensis*, *Phlebia radiata*, *Piptoporus betulinus*, *Pleurotus eryngii*, *Pleurotus ostreatus*, *Pleurotus sajor-caju*, *Polyporus ciliatus*, *Polyporus sanguineus*, *Pycnoporus sanguineus*, *Stereum hirsutum*, *Stereum rugosum*, *Trametes (Coriolus) versicolor*, *Irpex lacteus*, *Geotrichum candidum*, and *Dichomitus squalens*. Paszczynski et al. (1992) confirmed the ability of ligninolytic cultures of *P. chrysosporium* to mineralize azo dyes but have not found correlation between aromatic ring substitution pattern and mineralization rates. WRF are most efficient in breaking down synthetic dyes. Even the lignin-transforming *Streptomyces chromofuscus* is a weak decolorizer compared to *P. chrysosporium*. Mineralization rates of 23.1–48.1 % for a wide range of azo dyes after 12-day incubation with *P. chrysosporium* have been recorded by Spadaro et al. (1992). Cripps et al. (1990) reported that *P. chrysosporium* was found to have 11–49 % of total azo and heterocyclic dye bound to the mycelium. In addition to sorption accounted for less than 3 % of azo- and triphenylmethane dye removal by ligninolytic cultures of *P. sanguineus* (Pointing and Vrijmoed 2000). The involvement of LMEs in the dye decolorization process has been confirmed in some studies using purified cell-free enzymes. LiP of *P. chrysosporium* has been shown to decolorize azo, triphenyl-methane, and heterocyclic dyes in the presence of veratryl alcohol and H_2O_2 (Cripps et al. 1990). Two Lac isoenzymes purified from *Trametes hispida* were able to catalyze decolorization of several synthetic dyes (Rodriguez et al. 1999). Heinfling et al. (1998a) reported that MnPs of *Bjerkandera adusta* and *Pleurotus eryngii* have also been shown to catalyze dye decolorization.

16.5.2   Polycyclic Aromatic Hydrocarbon

Polycyclic aromatic hydrocarbons (PAH) are benzene homologues formed from the fusion of four or more benzene rings. These compounds present huge problems of toxicity and persistence in the environment. These arise from natural oil deposits and vegetation decomposition, in addition to considerable anthropogenic produc-tion from the use of fossil fuels in heating and power production, transportation, waste incineration, wood burning, and industrial processes (Alloway and Ayres 1993). Gramss et al. (1998) show that WRF are the only organisms capable of significant PAH mineralization. Many studies have reported the use of WRF in PAH bioremediation such as *P. chrysosporium*, *Chrysosporium lignorum* and *C. versicolor*, *Phanerochaete sordida*, *P. ostreatus*, *Bjerkandera* sp., *Phanerochaete laevis*, *Dichomitus squalens*, *Irpex lacteus*, and *Microporus vernicipes*. Many studies have reported the use of *Pleurotus* species in bioremediation exercises (Baldrian et al. 2000). *Pleurotus tuber-regium* has been reported to ameliorate crude oil-polluted soil and the resulting soil sample supported germination and seedling of *Vigna unguiculata* (Isikhuemhen et al. 2003). Adenipekun and Fasidi (2005) reported the ability of *Lentinus subnudus* to mineralize soil contaminated with various concentrations of crude oil. Adedokun and Ataga (2007) also investigated the effects of sawdust and waste cotton as soil amendment and bioaugmentation with *Pleurotus pulmonarius* on soil polluted with crude oil,

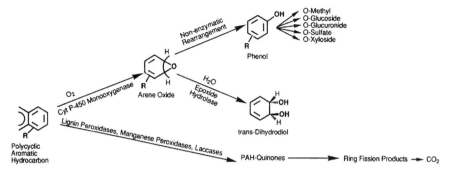

Fig. 16.2 Pathways for the fungal metabolism of polycyclic aromatic hydrocarbons

automotive gasoline oil, and spent engine oil on the growth of cowpea. There was a significant improvement on the growth of cowpea when polluted soil was amended and bioaugmented with *P. pulmonarius* after 1 month of incubation as compared with planting on polluted soil with no amendments and bioaugmentation.

Several taxa, including *P. chrysosporium* (Bumpus 1989), *Pleurotus* sp. (Bezalel et al. 1996), and *Trametes versicolor* (Morgan et al. 1991), are noteworthy for their significant PAH-mineralizing capabilities. Despite apparent non-LME-mediated PAH mineralization by some white-rot fungi grown under non-ligninolytic conditions such as *P. chrysosporium* (Dhawale et al. 1992), there is substantial and conclusive evidence that ligninolytic enzymes are involved in PAH mineralization by white-rot fungi. Several studies have shown that diverse white-rot fungi are capable of PAH mineralization and that rates of mineralization correlate with the production of LMEs (Sack et al. 1997); Bogan and Lamar (1996a) show that PAH transformation by MnP has also been shown to be limited by Mn^{2+} availability. The purified MnP of *P. chrysosporium* has been shown to oxidize twelve 3–6 rings PAH. Further evidence of lipid-peroxidation-coupled MnP-mediated PAH oxidation was observed for *Phanerochaete laevis*, which produced predominantly polar products, with no significant quinone accumulation (Bogan and Lamar 1996). By contrast, MnP of *N. forwardii* oxidized anthracene and pyrene to produce quinone products (Guenther et al. 1998).

The purified LiP of *Nematoloma forwardii* has been shown to oxidize anthracene and pyrene in the presence of the mediator veratryl alcohol (Guenther et al. 1998). Collins et al. (1996) show that purified Lacc from *Trametes versicolor* has been shown to oxidize a range of 3–5 ring PAH (in the presence of the chemical mediators). In addition, a (cell-associated) cytochrome P450 monooxygenase-mediated PAH transformation reaction is also thought to occur in certain white-rot fungi (Fig. 16.2) (Sutherland et al. 1995)

16.5.3 Chlorophenols

The various reactions in the degradation of chlorophenols (CP) produce diverse products and intermediates. The first step is catalyzed by the extracellular enzymes, followed by the action of other enzymes and the interaction of abiotic factors. Dec and Bollag (1994) have demonstrated that there are abiotic factors that also provoke dehalogenation. During the degradation of pentachlorophenol (PCP) and trichlorophenol (TCP), many more intermediary products are produced from dichlorophenol (DCP) and monochlorophenol; this is related to a higher resistance of chlorophenols to biodegradation; the greater the number of substituted chlorine atoms on the aromatic ring, the higher was the resistance to biodegradation. PCP and TCP are more resistant to degradation than mono- and dichlorophenols (Janik and Wolf 1992). Moreover, Rodakiewicz-Nowak et al. (1999) have demonstrated that chlorophenols with chlorine substitution at *meta*-positions are less metabolizable by fungal enzymes than chlorophenols in *ortho*- and *para*-positions. *Ortho*-substituted congeners are generally of lower toxicity, as the proximity of the *ortho*-substituted chlorine to the OH group on the molecule appears to shield the OH, which apparently interacts with the active site in organisms, causing the observed toxic effects. This makes 3-CP, 3,5-DCP, and 2,3,4-TCP more toxic than other mono-, di-, and trichlorophenols. Valli and Gold (1991) demonstrated the mineralization of 2,4-$[^{14}C]$dichlorophenol; after a 24-day incubation period, approximately 50 % of the substrate added to the cultures was degraded to $^{14}CO_2$. Similar results have been reported for the degradation of 2,4,5-TCP (Joshi and Gold 1993), 2,4,6-TCP (Reddy et al. 1998), and PCP (Reddy and Gold 2000). Leontievsky et al. (2000) showed that in the degradation of 2,4,6-TCP by the fungus *Panus tigrinus* (producer of the enzymes MnP and laccase) and *Coriolus versicolor* (producer of MnP, LiP and laccase), the first attack carried out by *P. tigrinus* was with MnP, while with *C. versicolor*, this was done predominantly by laccase, suggesting different methods of enzymatic regulation by the fungi in question. Valli and Gold (1991) reported that Studies on the degradation of 2,4-dichlorophenol by the fungi *P. chrysosporium* showed that manganese peroxidase degraded 2,4-DCP. Similar results have been reported by Zouari et al. (2002).

16.5.4 TNT

Fungal mineralization of 2,4,6-trinitrotoluene (TNT) has been demonstrated only for certain white-rot basidiomycetes. *Phanerochaete chrysosporium* has been selected in such studies (Hodgson et al. 2000). Transformations of TNT result in formation of the dinitrotoluenes (DNTs): 2-amino-4,6-dinitrotoluene, 2,4-diamino-6-nitrotoluene, 2,6-diamino-4-nitrotoluene, and 4-amino-2,6-dinitrotoluene. These compounds are generally not degraded further or are dimerized to even more persistent azo and azoxy dimers (Bumpus and Tatarko 1994). Donnelly et al.

1997 demonstrated that several white-rot fungi are able to transform TNT to DNTs. Jackson et al. (1999) reported degradation of TNT by non-ligninolytic strains of *P. chrysosporium*. Significantly, only white-rot fungi have been shown as capable of DNT degradation and mineralization to CO_2 (Hodgson et al. 2000).

Bending et al. (2002) showed >86 % degradation of atrazine and terbuthylazine by white-rot fungi in liquid culture and found no relationship between degradation rates and ligninolytic activity. Other studies with *P. chrysosporium* in liquid culture have reported biotransformation of the insecticide lindane independently of the production of ligninolytic enzymes (Mougin et al. 1996). These researchers ruled out the involvement of peroxidases in lindane biotransformation and mineralization, and they assessed the activity of the cytochrome P450 monooxygenase, an enzymatic system used by many organisms as a detoxification tool.

16.5.5 Pesticides

The term "pesticides" embraces an enormous diversity of products that are used in a number of different activities (Mourato et al. 2000), especially agriculture, which currently accounts for 75 % of the total use of pesticides (Buyuksonmez et al. 1999). The organophosphate insecticides are not generally persistent, and *P. chrysosporium* has been demonstrated to mineralize 12.2–27.5 % of 14C-radiolabeled chlorpyrifos, fonofos, and terbufos during an 18-day incubation (Bumpus et al. 1993). The chlorinated triazine herbicide 2-chloro-4-ethylamino-6-isopropylamino-1,3,4-triazine (atrazine) is recalcitrant in the environment, although the white-rot fungi *P. chrysosporium* (Mougin et al. 1994) and *Pleurotus pulmonarius* (Masaphy et al. 1993) have been demonstrated to transform atrazine, yielding hydroxylated and *N*-dealkylated metabolites.

16.5.6 Nylon

Nylon is a linear amide-linked polymer widely used in the textile industry. Deguchi et al. (1997) demonstrated that significant degradation of a nylon-66 membrane was observed for *P. chrysosporium* and *Trametes versicolor* under ligninolytic culture conditions. A nylon-degrading enzyme was later purified and characterized from an unnamed white-rot fungus. The characteristics of the purified protein were identical to those of MnP, although the method of catalysis was shown to involve peroxidase–oxidase reactions (as seen for horseradish peroxidase), rather than the peroxidase-Mn^{3+} type of reaction more typical of MnPs (Deguchi et al. 1998).

16.6   Conclusions

There has been extensive research to invent and improve methods for remediating polluted soils and water (Pozdnyakova et al. 2008; Reddy and Mathew 2001). Researchers determined that the fungi use lignin as a source of carbon and energy can degrade a wide range of pollutants. Their extracellular enzymes are oxidative and nonspecific and can degrade a wide range of pollutants (Pozdnyakova et al. 2008).

One of the greatest advantages of white-rot fungi is that their secreted enzymes can be purified and used as free or immobilized enzymes (Duran and Esposito 2000). Cornwell et al. (1990) reported an immobilization of LiP on porous ceramic supports that did not adversely affect LiP's stability and showed a potential for degradation of environmentally persistent aromatics. Edwards et al. (2002) demonstrated that the application of laccase and MnP from *T. versicolor* immobilized onto polysulfone ultrafiltration membranes could remove aromatic hydrocarbons from a petrochemical industrial effluent. Mielgo et al. (2003) studied the immobilization of MnP from *P. chrysosporium* and *Bjerkandera* sp. in glutaraldehyde–agarose gels, obtaining a complete immobilization in a very short period (0.5–2 h). Immobilization maintained a high percentage of MnP activity for long periods of time. On the other hand, Ahn et al. (2002) demonstrated that immobilized laccase in soil polluted by 2,4-DCP was more effective than the free enzyme in 2,4-DCP removal. Dodor et al. (2004), studying *T. versicolor* laccase immobilized on kaolinite and its potential to oxidize anthracene and benzo[a] pyrene, indicated that immobilization improved stability of laccase to temperature, pH, inhibitors, and storage time compared with the free enzyme. Oxidative enzymes have been immobilized on several natural and synthetic supports and often proposed as efficient catalytic tools to overcome several disadvantages linked to the use of free enzymes (Duran et al. 2002). To improve and enhance the applicability of isolated enzymes, research efforts have been dedicated to reduce the costs of enzymes' isolation and purification, costs that are very high and may compromise their use for practical applications. Nowadays, great progress in this area may derive from modern molecular technologies, which may provide cheaper potential sources of various enzymes by means of genetically modified microorganisms or plants. For example, several aspects of the enzymology and molecular biology of the lignin-degrading system of white-rot fungi are known (Pointing 2001) The LiPs, MnPs, and laccases genes have been cloned and sequenced, and also their expression and regulation have been studied. These advancements should lead to the successful genetic engineering of white-rot fungi and their enzyme systems. Most advantageous bioremediation strategies for treating contaminated sites can possibly be planned and applied.

References

Adedokun OM, Ataga AE (2007) Effects of amendments and bioaugmentation of soil polluted with crude oil, automotive gasoline oil, and spent engine oil on the growth of cowpea (*Vigna unguiculata* L. Walp). Sci Res Essay 2:147–149

Adenipekun CO, Fasidi IO (2005) Bioremediation of oil-polluted soil by *Lentinus subnudus*, a Nigerian white-rot fungus. Afr J Biotechnol 4:796–798

Ahn MY, Dec J, Kim J, Bollag JM (2002) Treatment of 2,4-dichlorophenol polluted soil with free and immobilized laccase. J Environ Qual 31:1509–1515

Alexander M (1994) Biodegradation and bioremediation. Academic, New York

Alloway JB, Ayres DC (1993) Chemical principles of environmental pollution. Chapman and Hall, London

Aust SD, Barr DP (1994) Mechanism white rot fungi use to degrade pollutants. Environ Sci Technnol 28:78–87

Aust SD, Swaner PR, Stahl JD (2003) Detoxification and metabolism of chemicals by white-rot fungi. In: Zhu JJPC, Aust SD, Lemley Gan AT (eds) Pesticide decontamination and detoxification. Oxford University Press, Washington, DC, pp 3–14

Azadpour A, Powell PD, Matthews J (1997) Use of lignin degrading fungi in bioremediation. Remediation 997:25–49

Baker Lee CJ, Fletcher MA, Avila OI, Callanan J, Yunker S, Munnecke DM (1995) Bioremediation of MGP soils with mixed fungal and bacterial cultures. In: Hinchee RE, Fredrickson J, Alleman B (eds) Third international in situ and on-site bioremediation symposium. pp 123–128

Baldrian PC, Der Viesche C, Gabriel S, Nerud F, Zadrazil F (2000) influence of cadmium and mercury on activities of ligninolytic enzymes and degradation of polycyclic aromatic hydrocarbons by *Pleurotus ostreatus* in soil. Appl Environ Microbiol 66:2471–2478

Bending G, Friloux M, Walker A (2002) Degradation of contrasting pesticides by white rot fungi and its relationship with ligninolytic potential. FEMS Microbiol Lett 212:59–63

Bezalel L, Hadar Y, Cerniglia CE (1996) Mineralization of polycyclic aromatic hydrocarbons by the white-rot fungus *Pleurotus ostreatus*. Appl Environ Microbiol 62:292–295

Bogan BW, Lamar RT (1996) Polycyclic aromatic hydrocarbon degrading capabilities of *Phanerochaete laevis* HHB-1625 and its extracellular ligninolytic enzymes. Appl Environ Microbiol 62:1597–1603

Boopathy R (2000) Bioremediation of explosives contaminated soil. Int Biodeter Biodegrad 46:29–36

Bumpus JA (1989) Biodegradation of polycyclic aromatic hydrocarbons by *Phanerochaete chrysosporium*. Appl Environ Microbiol 55:154–158

Bumpus JA, Tatarko M (1994) Biodegradation of 2,4,6-trinitrotoluene by *Phanerochaete chrysosporium*: identification of initial degradation products and the discovery of a TNT-metabolite that inhibits lignin peroxidase. Curr Microbiol 28:185–190

Bumpus JA, Powers RH, Sun T (1993) Biodegradation of DDE (1,1-dichloro-2,2-bis (4-chlorophenyl)ethene) by *Phanerochaete chrysosporium*. Mycol Res 97:95–98

Buyuksonmez F, Rynk R, Hess T, Bechinski E (1999) Occurrence, degradation and fate of pesticides during composting – Part I: Composting, pesticides, and pesticide degradation. Compost Sci Util 7:66–82

Camarero S, Sarkar S, Ruiz-Dueñas FJ, Martínez MJ, Martinez AT (1999) Description of a versatile peroxidase involved in natural degradation of lignin that has both Mn-peroxidase and lignin-peroxidase substrate binding sites. J Biol Chem 274:10324–10330

Christian M, Claude J, Pierre B (2003) Fungal laccases: from structure-activity studies to environmental applications. Environ Chem Lett 1:145–148

Collins PJ, Kotterman MJJ, Field JA, Dobson ADW (1996) Oxidation of anthracene and benzo[a] pyrene by laccases from *Trametes versicolor*. Appl Environ Microbiol 62:4563–4567

Cornwell KL, Tinland-Butez MF, Tardone PJ, Cabasso I, Hammel KE (1990) Lignin degradation and lignin peroxidase production in cultures of *Phanerochaete chrysosporium* immobilized on porous ceramic supports. Enzyme Microbiol Technol 12:916–920

Cripps C, Bumpus JA, Aust SD (1990) Biodegradation of azo and heterocyclic dyes by Phanerochaete chrysosporium. Appl Environ Microbiol 56:1114–1118

Dec J, Bollag JM (1994) Dehalogenation of chlorinated phenols during oxidative coupling. Environ Sci Technol 28:484–490

Deguchi T, Kakezawa M, Nishida T (1997) Nylon biodegradation by lignin-degrading fungi. Appl Environ Microbiol 63:329–331

Deguchi T, Kitaoka Y, Kakezawa M, Nishida T (1998) Purification and characterization of a nylon-degrading enzyme. Appl Environ Microbiol 64:1366–1371

Dhawale SW, Dhawale SS, Dean-Ross D (1992) Degradation of phenanthrene by *Phanerochaete chrysosporium* occurs under ligninolytic as well as non-ligninolytic conditions. Appl Environ Microbiol 58:3000–3006

Dibble JT, Bartha R (1979) Rehabilitation of oil-inundated agricultural land: a case history. Soil Sci 128:56–60

Dodor DE, Hwang HM, Ekunwe SIN (2004) Oxidation of anthracene and benzo[a]pyrene by immobilized laccase from Trametes versicolor. Enzyme Microbiol Technol 35:210–217

Donnelly KC, Chen JC, Huebner HJ, Brown KW, Autenrieth RL, Bonner JS (1997) Utility of four strains of white-rot fungi for the detoxification of 2,4,6-trinitrotoluene in liquid culture. Environ Toxicol Chem 16:1105–1110

Ducros V, Brzozowski AM, Wilson KS, Brown SH, Ostergaard P, Schneider P, Yaver DS, Pedersen AH, Davies GJ (1998) Crystal structure of the type-2 Copper depleted laccase from Coprinus cinereus at 2.2 Å resolution. Nat Struct Biol 5:310–316

Duran N, Esposito E (2000) Potential applications of oxidative enzymes and phenoloxidase-like compounds in wastewater and soil treatment: a review. Appl Catal B Environ 28:83–99

Duran N, Rosa M, D'Annibale A, Gianfreda L (2002) Applications of laccases and tyrosinases (phenoloxidases) immobilized on different supports: a review. Enzyme Microbiol Technol 31:907–931

Edwards W, Leukes WD, Bezuidenhout JJ (2002) Ultrafiltration of petrochemical industrial wastewater using immobilised manganese peroxidase and laccase: application in the defouling of polysulfone membranes. Desalination 149:275–278

Eggert C, Temp U, Eriksson KEL (1996) The ligninolytic system of the white rot fungus *Pycnoporus cinnabarinus*: purification and characterization of the laccase. Appl Environ Microbiol 62:1151–1158

Eriksson KE, Blanchette RA, Ander P (1990) Microbial and enzymatic degradation of wood and wood components. Springer, Berlin

Giardina P, Aurilia V, Cannio R, Marzullo L, Amoresano A, Siciliano R, Pucci P, Sannia G (1996) The gene, protein and glycan structures of laccase from *Pleurotus ostreatus*. Eur J Biochem 235:508–515

Glenn JK, Morgan MA, Mayfield MB, Kuwahara M, Gold ML (1983) An extracellular H2O2 requiring enzyme preparation involved in lignin biodegradation by the white rot basidiomycete Phanerochaete chrysosporium. Biochem Biophys Res Commun 144:1077–1083

Godfrey BJ, Mayfield MB, Brown JA, Gold MH (1990) Characterization of a gene encoding a manganese peroxidase from *Phanerochaete chrysosporium*. Gene 93:119–124

Gold MH, Wariishi H, Valli K (1989) Extracellular peroxidase involved in lignin degradation by the white rot basidiomycete *Phanerochaete chrysosporium*. ACS Symp Ser 389:127–140

Gramss G, Kirsche B, Viogt KD, Gunther T, Fritsche W (1998) Conversion rates of five polycyclic aromatic hydrocarbons in liquid cultures of fifty-eight fungi and the concomitant production of oxidative enzymes. Mycol Res 103:1009–1018

Guenther T, Sack U, Hofrichter M, Laetz M (1998) Oxidation of PAH and PAH-derivatives by fungal and plant oxidoreductases. J Basic Microbiol 38:113–122

Guillén F, Martinez AT, Martinez MJ (1992) Substrate specificity and properties of the aryl-alcohol oxidase from the ligninolytic fungus *Pleurotus eryngii*. Eur J Biochem 209:603–611

Hammel K (1995) Organopollutant degradation by ligninolytic fungi. In: Young LY, Cerniglia CE (eds) Microbial transformation and degradation of toxic organic chemicals. Wiley-Liss, New York, pp 331–346

Hammel KE, Tardone PJ (1988) The oxidative 4-dechlorination of polychlorinated phenols is catalyzed by extracellular fungal lignin peroxidases. Biochemistry 27:6563–6568

Hammel KE, Kalyanaraman B, Kirk TK (1986) Oxidation of polycyclic aromatic hydrocarbons and dibenzo[*p*]dioxins by *Phanerochaete chrysosporium* ligninase. J Biol Chem 261:16948–16952

Hatakka A (1994) Ligninolytic enzymes from selected white-rot fungi: production and role in lignin degradation. FEMS Microbiol Rev 13:125–135

Hatakka A (2001) Biodegradation of lignin. In: Hofrichter M, Steinbüchel A (eds) Biopolymers, vol 1. Wiley-VCH, Weinheim

Hawksworth DL, Kirk PM, Sutton BC, Pegler DN (1995) Ainsworth DN and Bisby's dictionary of the fungi, 8th edn. CAB, Oxon

Heinfling A, Martinez MJ, Martinez AT, Bergbauer M, Szewzyk U (1998) Transformation of industrial dyes by manganese peroxidases from *Bjerkandera adusta* and *Pleurotus eryngii* in a manganese-independent reaction. Appl Environ Microbiol 64:2788–2793

Hestbjerg H, Willumsen PA, Christensen M, Andersen O, Jacobsen CS (2003) Bioaugmentation of tar-contaminated soils under field conditions using Pleurotus ostreatus refuse from commercial mushroom production. Environ Toxicol Chem 22:692–698

Hodgson J, Rho D, Guiot SR, Ampleman G, Thiboutot S, Hawari J (2000) Tween 80 enhanced TNT mineralization by *Phanerochaete chrysosporium*. Can J Microbiol 46:110–118

Hofrichter M (2002) Review: lignin conversion by manganese peroxidase (MnP). Enzyme Microbiol Technol 30:454–466

Isikhuemhen O, Anoliefo G, Oghale O (2003) Bioremediation of crude oil polluted soil by the white rot fungus *Pleurotus tuber-regium* (Fr.) Sing. Environ Sci Pollut Res 10:108–112

Jackson M, Hou L, Banerjee H, Sridhar R, Dutta S (1999) Disappearance of 2,4-dinitrotoluene and 2-amino,4,6-dinitrotoluene by *Phanerochaete chrysosporium* under non-ligninolytic conditions. Bull Environ Contam Toxicol 62:390–396

Janik F, Wolf HU (1992) The Ca^{2+}-transport-ATPase of human erythrocytes as an in vitro toxicity test system-Acute effects of some chlorinated compounds. J Appl Toxicol 12:351–358

Jolivalt C, Raynal A, Caminade E, Kokel B, Le Goffic F, Mougin C (1999) Transformation of N', N'-dimethyl-N-(hydroxyphenyl)ureas by laccase from the white rot fungus Trametes versicolor. Appl Microbiol Biotechnol 51:676–681

Jönsson L, Johansson T, Sjostrom K, Nyman PO (1987) Purification of ligninase isozymes from the white-rot fungus *Trametes versicolor*. Acta Chem Scand B 41:766–769

Joshi DK, Gold MH (1993) Degradation of 2,4,5-trichlorophenol by the lignin-degrading basidiomycete Phanerochaete chrysosporium. Appl Environ Microbiol 59:1779–1785

Kaal EEJ, De Jong E, Field JA (1993) Stimulation of ligninolytic peroxidase activity by nitrogen nutrients in the white-rot fungus *Bjerkandera* sp. strain BOS 55. Appl Environ Microbiol 59:4031–4036

Kantelinen A, Hatakka A, Viikari L (1989) Production of lignin peroxidase and laccase by *Phlebia radiata*. Appl Microbiol Biotechnol 31:234–239

Kersten PJ, Kalyanaraman B, Hammel KE, Reinhammar B, Kirk TK (1990) Comparison of lignin peroxidase, horseradish peroxidase and laccase in the oxidation of methoxybenzenes. Biochem J 268:475–480

Kirk TK, Farrell RL (1987) Enzymatic combustion: the microbial degradation of lignin. Annu Rev Microbiol 41:465–505

Lankinen P (2004) Ligninolytic enzymes of the basidiomycetous fungi *Agaricus bisporus* and *Phlebia radiata* on lignocellulose-containing media. eThesis, Faculty of Agriculture and Forestry of the University of Helsinki

Leontievsky AA, Vares T, Lankinen P, Shergill JK, Pozdnyakova NN, Myasoedova NM, Kalkkinen N, Golovleva LA, Cammack R, Thurston CF, Hatakka A (1997) Blue and yellow laccases of ligninolytic fungi. FEMS Microbiol Lett 156:9–14

Leontievsky AA, Myasoedova NM, Baskunov BP, Evans CS, Golovleva LA (2000) Transformation of 2,4,6-trichlorophenol by the white rot fungi *Panus tigrinus* and *Coriolus versicolor*. Biodegradation 11:331–340

Levin L, Herrmann C, Papinutti VL (2008) Optimization of lignocellulolytic enzyme production by the white-rot fungus *Trametes trogii* in solid-state fermentation using response surface methodology. Biochem Eng J 39:207–214

Levine WG (1965) Laccase: a review. In: Peisach J (ed) The biochemistry of copper. Academic, New York, pp 371–385

Lundell T, Hatakka A (1994) Participation of Mn(II) in the catalysis of laccase, manganese peroxidase and lignin peroxidase from *Phlebia radiata*. FEBS Lett 348:291–296

Martínez MJ, Ruiz-Duenas FJ, Guillen F, Martinez AT (1996) Purification and catalytic properties of two manganese peroxidase isoenzyme from Pleurotus eryngii. Eur J Biochem 237:424–432

Masaphy S, Levanon D, Vaya J, Henis Y (1993) Isolation and characterization of a novel atrazine metabolite produced by the fungus *Pleurotus pulmonarius*, 2-chloro-4-ethylamino-6- (1-hydroxyisopropyl)amino-1,3,5-triazine. Appl Environ Microbiol 59:4342–4346

Mayfield MB, Godfrey BJ, Gold MH (1994) Characterization of the MnP2 gene encoding manganese peroxidase isozyme 2 from the basidiomycete *Phanerochaete chrysosporium*. Gene 142:231–235

Mentzer E, Ebere D (1996). Remediation of hydrocarbon contaminated sites. A paper presented at 8th Biennial International Seminar on the Petroleum Industry and the Nigerian Environment, November, Port Harcourt

Meysami P (2001) Feasibility study of fungal bioremediation of a flare pit soil using white rot fungi. Thesis, The University of Calgary

Michaels GB, Lewis DL (1985) Sorption and toxicity of azo and triphenylmethane dyes to aquatic microbial populations. Environ Toxicol Chem 4:45–50

Mielgo I, Palma C, Guisan JM, Fernandez-Lafuente R, Moreira MT, Feijoo G, Lema JM (2003) Covalent immobilisation of manganese peroxidases (MnP) from *Phanerochaete chrysosporium* and *Bjerkandera sp* BOS55. Enzyme Microbiol Technol 32:769–775

Moilanen AM, Lundell T, Vares T, Hatakka A (1996) Manganese and malonate are individual regulators for the production of lignin and manganese peroxidase isozymes and in the degradation of lignin by *Phlebia radiata*. Appl Microbiol Biotechnol 45:792–799

Morgan P, Lewis ST, Watkinson RJ (1991) Comparison of abilities of white-rot fungi to mineralize selected xenobiotic compounds. Appl Microbiol Biotechnol 34:693–696

Mougin C, Laugero C, Asther M, Dubroca J, Frasse P (1994) Biotransformation of the herbicide atrazine by the white-rot fungus *Phanerochaete chrysosporium*. Appl Environ Microbiol 60:705–708

Mougin C, Pericaud C, Malosse C, Laugero C, Asther M (1996) Biotransformation of the insecticide lindane by the white rot basidiomycete *Phanerochaete chrysosporium*. Pestic Sci 47:51–59

Mourato S, Ozdemiroglu E, Foster V (2000) Evaluating health and environmental impact of pesticide use: implications for the design of ecolabels and pesticide taxes. Environ Sci Technol 34:1456–1461

Muñoz C, Guillen F, Martinez AT, Martinez MJ (1997) Induction and characterization of laccase in the ligninolytic fungus Pleurotus eryngii. Curr Microbiol 34:1–5

Odu CTI (1978) The effect of nutrient application and aeration on oil degradation in soils. Environ Pollut 15:235–240

Odu CTI (1981) Degradation and weathering of crude oil under tropical conditions. In: Proceedings of the international seminar on the petroleum industry and the Nigerian environment. NNPC Publication

Orth AB, Royse DJ, Tien M (1993) Ubiquity of lignin degrading peroxidases among various wood degrading fungi. Appl Environ Microbiol 59:4017–4023

Palmieri G, Giardina P, Bianco C, Scaloni A, Capasso A, Sannia G (1997) A novel white laccase from *Pleurotus ostreatus*. J Biol Chem 272:31301–31307

Paszczynski A, Crawford R (1991) Degradation of azo compounds by ligninase from *Phanerochaete chrysosporium*: involvement of veratryl alcohol. Biochem Biophys Res Commun 178:1056–1063

Paszczynski A, Crawford RL (1995) Potential for bioremediation of xenobiotic compounds by the white-rot fungus *Phanerochaete chrysosporium*. Biotechnol Prog 11:368–379

Paszczynski A, Pasti-Grigsby MB, Gosczynski S, Crawford RL, Crawford DL (1992) Mineralization of sulfonated azo dyes and sulfanilic acid by *Phanerochaete chrysosporium* and *Streptomyces chromofuscus*. Appl Environ Microbiol 58:3598–3604

Pease EA, Andrawis A, Tien M (1989) Manganese-dependent peroxidase from *Phanerochaete chrysosporium*: primary structure deduced from cDNA sequence. J Biol Chem 264: 13531–13535

Perie FH, Gold MH (1991) Manganese regulation of manganese peroxidase expression and lignin degradation by the white rot fungus Dichomitus squalens. Appl Environ Microbiol 57:2240–2245

Pointing SB (2001) Feasibility of bioremediation by white-rot fungi. Appl Microbiol Biotechnol 57:20–33

Pointing SB, Vrijmoed LLP (2000) Decolorization of azo and triphenylmethane dyes by *Pycnoporus sanguineus* producing laccase as the sole phenoloxidase. World J Microbiol Biotechnol 16:317–318

Pozdnyakova NN, Nikitina VE, Turovskaya OV (2008) Bioremediation of Oil-polluted Soil with Association Including the Fungus *Pleurotus ostreatus* and Soil Microflora. Appl Biochem Microbiol 44:69–75

Pribnow D, Mayfield MB, Nipper VJ, Brown JA, Gold MH (1989) Characterization of a cDNA encoding a manganese peroxidase, from the lignin-degrading basidiomycete *Phanerochaete chrysosporium*. J Biol Chem 264:5036–5040

Reddy GVB, Gold MH (2000) Degradation of pentachlorophenol by Phanerochaete chrysosporium: intermediates and reactions involved. Microbiology 146:405–413

Reddy CA, Mathew Z (2001) Bioremediation potential of white rot fungi. In: Gadd GM (ed) Fungi in bioremediation. Cambridge University Press, Cambridge

Reddy GVB, Gelpke MDS, Gold MH (1998) Degradation of 2,4,6-trichlorophenol by Phanerochaete chrysosporium, involvement of reductive dechlorination. J Bacteriol 180:5159–5164

Rodakiewicz-Nowak J, Haber J, Pozdnyakova N, Leontievsky A, Golovleva LA (1999) Effect of ethanol on enzymatic activity of fungal laccases. Biosci Rep 19:589–600

Rodakiewicz-Nowak J, Kasture SM, Dudek B, Haber J (2000) Effect of various water-miscible solvents on enzyme activity of fungal laccases. J Mol Catal B Enzym 11:1–11

Rodríguez E, Pickard MA, Vazquez-Duhalt R (1999) Industrial dye decolorization by laccases from ligninolytic fungi. Curr Microbiol, 38:27–32

Sack U, Heinze TM, Deck J, Cerniglia CE, Martens R, Zadrazil F, Fritsche W (1997) Comparison of phenanthrene and pyrene degradation by different wood decay fungi. Appl Environ Microbiol 63:3919–3925

Schlosser D, Grey R, Fritsche W (1997) Patterns of ligninolytic enzymes in Trametes versicolor. Distribution of extra and intracellular enzyme activities during cultivation on glucose, wheat straw and beech wood. Appl Microb Biotechnol 47:412–418

Shimada M, Ma DB, Akamatsu Y, Hattori T (1994) A proposed role of oxalic acid in wood decay systems of wood rotting basidiomycetes. FEMS Microbiol Rev 13:285–296

Spadaro JT, Gold MH, Renganathan V (1992) Degradation of azo dyes by the lignin-degrading fungus Phanerochaete chrysosporium. Appl Environ Microbiol 58:2397–2401

Stewart P, Cullen D (1999) Organization and differential regulation of a cluster of lignin peroxi-dase genes of *Phanerochaete chrysosporium*. J Bacteriol 181:3427–3432

Sundaramoorthy M, Kishi K, Gold MH, Poulos TL (1994) The crystal structure of manganese peroxidase from *Phanerochaete chrysosporium* at 2.06-Å resolution. J Biol Chem 269: 32759–32767

Sutherland JB, Rafii F, Khan A, Cerniglia CE (1995) Mechanisms of polycyclic aromatic hydrocarbon degradation. In: Young LY, Cerniglia CE (eds) Microbial transformations and degradation of toxic organic chemicals. Wiley-Liss, New York, pp 269–306

Thakker GD, Evans CS, Rao KK (1992) Purification and characterisation of laccase from *Monocillium indicum* Saxena. Appl Microbiol Biotechnol 37:321–323

Thurston CF (1994) The structure and function of fungal laccases. Microbiology 140:19–26

Tien M, Kirk TK (1983) Lignin-degrading enzyme from the hymenomycete *Phanerochaete chrysosporium* Burds. Science 221:661–663

Tien M, Kirk TK, Bull C, Fee JA (1986) Steady-state and transient-state kinetic studies on the oxidation of 3,4-dimethoxybenzyl alcohol catalyzed by the ligninase of *Phanerochaete chrysosporium*. J Biol Chem 261:1687–1693

Tuor U, Wariishi H, Schoemaker HE, Gold MH (1992) Oxidation of phenolic arylglycerol beta-aryl ether lignin model compounds by manganese peroxidase from *Phanerochaete chrysosporium* oxidative cleavage of an alpha-carbonyl model-compound. Biochemistry 31:4986–4995

Valli K, Gold MH (1991) Degradation of 2,4-Dichlorophenol by the Lignin-Degrading Fungus Phanerochaete chrysosporium. J Bacteriol 1:345–352

Volc J, Kubatova E, Daniel G, Prikrylova V (1996) Only C-2 specific glucose oxidase activity is expressed in ligninolytic cultures of the white rot fungus *Phanerochaete chrysosporium*. Arch Microbiol 165:421–424

Wariishi H, Akileswaran L, Gold MH (1988) Manganese peroxidase from the basidiomycete *Phanerochaete chrysosporium*: spectral characterization of the oxidized states and the catalytic cycle. Biochemistry 27:5365–5370

Wariishi H, Dunford HB, MacDonald ID, Gold MH (1989) Manganese peroxidase from the lignin-degrading basidiomycete *Phanerochaete chrysosporium*: transient-state kinetics and reaction mechanism. J Biol Chem 264:3335–3340

Wariishi H, Valli K, Gold MH (1992) Manganese(II) oxidation by manganese peroxidase from the basidiomycete *Phanerochaete chrysosporium*: kinetic mechanism and role of chelators. J Biol Chem 267:23688–23695

Xu F, Shin W, Brown SH, Wahleitner JA, Sundaram UM, Solomon EI (1996) A study of recombinant fungal laccases and bilirubin oxidase that exhibit significant differences in redox potential, substrate specificity, and stability. Biochim Biophys Acta 1292:303–311

Yaver DS, Berka RM, Brown SH, Xu F (2001) Cloning, characterisation, expression and commercialisation of fungal laccases. In: 8th symposium on recent advances in lignin biodeg-radation and biosynthesis

Zouari H, Labat M, Sayadi S (2002) Degradation of 4-chlorophenol by the white rot fungus Phanerochaete chrysosporium in free and immobilized cultures. Bioresour Technol 84:145–150

Part V
Techniques in Mycoremediation

Chapter 17
Effect of Mobilising Agents on Mycoremediation of Soils Contaminated by Hydrophobic Persistent Pollutants

Alessandro D'Annibale, Ermanno Federici, and Maurizio Petruccioli

17.1 Introduction

In industrialised countries, soil contamination by hydrophobic persistent pollutants (HPP) is a frequently occurring scenario. Consequently, there is an understandable concern regarding the ultimate environmental fate and effects of these pollutants. Among available restoration approaches, the bioremediation encompasses a variety of techniques aimed at promoting the acceleration of natural biodegradation processes in contaminated matrices. Biological remediation techniques can be grouped into two main categories, namely, biostimulation and bioaugmentation. Biostimulation refers to a variety of in situ technical interventions, including air extraction/injection and/or addition of nitrogen and phosphorous fertilisers, or to specific matrix adjustments (i.e. moisture content and pH) assuming that resident microbiota encompass species able to bring about the breakdown of contaminants. On the contrary, bioaugmentation involves the addition of either exogenous single species or microbial consortia able to degrade the target contaminants and/or to promote their desorption.

With this regard, one of the major rate-limiting factors in the degradation of HPP is their low availability to the microbial cells. Microorganisms employ several strategies to enhance availability of HPP, such as biofilm formation and production of biopolymers with surfactant properties (Christofi and Ivshina 2002). Another option, regardless of the use of either biostimulation or bioaugmentation, involves the addition of exogenous emulsifiers and surface-active agents to enhance HPP

A. D'Annibale • M. Petruccioli (✉)
Department for Innovation in Biological, Agro-food and Forest Systems, University of Tuscia, Via San Camillo De Lellis, 01100 Viterbo, Italy
e-mail: dannib@unitus.it; petrucci@unitus.it

E. Federici
Department of Cell and Environmental Biology, University of Perugia, Via del Giochetto, 06100 Perugia, Italy
e-mail: ermanno.federici@unipg.it

E.M. Goltapeh et al. (eds.), *Fungi as Bioremediators*, Soil Biology 32, DOI 10.1007/978-3-642-33811-3_17, © Springer-Verlag Berlin Heidelberg 2013 biodegradation rate by increasing the bioavailability of the pollutant (Pannu et al. 2004; Pizzul et al. 2007).

One promising bioaugmentation approach to sites contaminated by HPP is represented by the use of species belonging to the ecological group of white-rot fungi (WRF) (Eggen 1999; Chung et al. 2000; D'Annibale et al. 2005; Ford et al. 2007). Their efficacy in bioremediation is mainly due to the presence of a non-specific, radical-based degradation machinery involved in lignin breakdown, and mainly operating in the extracellular environment (Šašek 2003). The extracellular location of lignin-modifying enzymes (LME) makes degradation ability of WRF disentangled from the need of contaminant internalisation, a rate-limiting step in bacterial breakdown of contaminants, and less dependent on solubility of the target compound. Moreover, the rather relaxed substrate specificities of LMEs of WRF, which reflect the structural heterogeneity of lignin, enable these organisms to oxidise a wide array of organic molecules (Novotny et al. 2004), including highly complex contaminant mixtures, such as Delor and creosote, that contain a variety of polychlorobiphenyls (PCB) and polycyclic aromatic hydrocarbons (PAH), respectively (Moeder et al. 2005; Covino et al. 2010a). Unlike bacteria, the frequent constitutive nature of some LME isoenzymes generally eliminates the need for these organisms to be adapted to the target chemical. In addition, lignocellulosic amendments generally used in mycoaugmentation have been shown to exert a potentially eliciting/inducing effect on LME (Crestini et al. 1996; Ford et al. 2007).

The most characterised ligninolytic enzymes of WRF include lignin peroxidase (E.C. 1.11.1.14 diarylpropane: hydrogen peroxide oxidoreductase; LiP), manganese peroxidase (E.C. 1.11.1.13 Mn(II): hydrogen peroxide oxidoreductase; MnP) and the multi-copper oxidase laccase (E.C. 1.10.3.2 $para$-benzenediol: oxygen oxidoreductase). Although both LiP and MnP are heme proteins, requiring hydrogen peroxide as the reaction oxidant, only the former enzyme is able to react directly with aromatic substrates, leading to the formation of cation radicals. By contrast, MnP oxidises Mn^{2+} to Mn^{3+} which, in turn, when stabilised by organic acids, acts as a highly diffusible mono-electronic oxidant. Laccase, instead, is able to bring about the oxygen-dependent one-electron oxidation of a wide variety of organic compounds, and its catalytic competence may be extended by the use of either natural (Böhmer et al. 1998; Johannes and Majcherczyk 2000; Cañas et al. 2007) or synthetic (Majcherczyk et al. 1998) mediators acting as electron shuttles from the enzymes to the target contaminants. Similarly, lipid peroxidation triggered by MnP has been identified as an important mechanism involved in the oxidation of PAH compounds endowed with high ionisation potentials, such as phenanthrene (8.03 eV) (Moen and Hammel 1994). It is important to realise that the fundamental steps in lignin degradation by either laccase or the ligninolytic peroxidases (i.e. LiP and MnP) involve the formation of free radical intermediates, which are formed when one electron is removed or added to the ground state of a given chemical. Such free radicals are highly reactive and rapidly able to trigger radical chain reactions. This free radical mechanism provides the basis for the non-specific nature of degradation of a variety of structurally diverse pollutants (Barr and Aust 1994; Acevedo et al. 2010). An alternative fate of HPP subjected to the action of WRF is their covalent incorporation into soil organic matter occurring *via* free radical copolymerisation of either the parent pollutant or related degradation intermediates with humic and fulvic acids precursors catalysed by LME. This mechanism leading to the so-called organic bound residue formation has been demonstrated for a variety of organic pollutants including PAH (Bogan et al. 1999), chlorophenols (Rüttimann-Johnson and Lamar 1996), trinitrotoluene (Dawel et al. 1997) and herbicides (Kim et al. 1997).

The hyphal growth of WRF enables them to readily diffuse into soil to penetrate within microaggregates, thus reaching pollutants and acting, at the same time, as effective dispersion vectors of indigenous pollutant-degrading bacteria (Kohlmeier et al. 2005). WRF, however, are generally unable to use pollutants as energy sources, and thus they require co-substrates, such as lignocellulosic substrates. With this regard, inexpensive and widely available lignocellulosic materials, such as corncobs, straw, peanut shells and sawdust, can be added as nutrients to the contaminated sites to obtain enhanced degradation of pollutants by these organisms (Boyle et al. 1998; Covino et al. 2010b).

Both properties of the ligninolytic system and mode of growth of WRF have fed the belief that the degradation capacity of these organisms is less dependent on the bioavailability of contaminants. The main objective of this chapter is to provide the basic concepts related to the action of surfactants on HPP and their use in mycoremediation processes.

17.2   Sorption Mechanisms and Bioavailability of HPP in Soil

Sorption and degradation are two key processes determining the possible persistence of a given contaminant in the soil (Haigh 1996). The bioavailability of HPP in soil is affected by both their inherent chemical properties (e.g. water solubility, acidity, molecular size and configuration, lipophilicity) and by a variety of soil physico-chemical parameters (e.g. cation exchange capacity, pH, texture, organic matter and clay contents) (Volkering et al. 1998). The term bioavailability expresses the fraction of a given chemical that is available to be taken up and subsequently transformed by soil microbiota (Juhasz 2008).

The interactions of HPP with soil are complex and include a variety of sorption mechanisms, including covalent bonding, electrostatic and other weak intermolecular interactions (Semple et al. 2003), the extent of which is largely dependent on the inherent chemical structure of the pollutant, soil's clay content and both amounts and properties of organic matter (Semple et al. 2003; Laha et al. 2009). In this respect, a(b)dsorption onto soil humic matter can significantly reduce the bioavailability of a compound, thus reducing its mobility to ground and surface waters (Kim et al. 1997). The intensity of sorption tends to increase with contact time of the pollutant with soil constituents, a phenomenon referred to as "aging", thus resulting in decreased pollutant bioavailability (Welp and Bruemmer 1999; Laha et al. 2009). Soil's pH may also affect the sorption of HPP, such as pentachlorophenol (PCP), which is readily converted into its water-soluble pheno-late anion at high pHs (Boyle 2006).

In some instances, sorption of HPP to soil may be beneficial, especially at high concentrations as their toxicity may be reduced, enabling introduced fungi to grow in the soil. It is still unclear whether fungi transform sorbed pollutants or whether they require that the pollutants be in solution. Although it has been suggested that both mechanisms might be operational in WRF (Barr and Aust 1994), the latter might be predominant. This is suggested by the high correlation found between amounts of degraded contaminants and respective bioavailabilities observed in several fungal microcosms (Bogan and Lamar 1999; Leonardi et al. 2007; Covino et al. 2010a, b). In addition, the ability of WRF to degrade PAH beyond their respective bioavailable fractions was observed for limited number of contaminants (Covino et al. 2010a) and found to be matrix dependent (Covino et al. 2010b). Thus, the use of surfactants, able to act as mobilising agents (MAs) of contaminants, appears to be a viable approach (Table 17.1).

17.3   Surfactants and Other Mobilising Agents: Structural Properties and Mode of Action

Surfactants are organic molecules comprising a hydrophilic head and a hydropho-bic tail that confer them amphiphilic properties.

As a function of the charge properties of their hydrophilic group, surfactants can be assigned to four distinct classes, namely, nonionic, anionic, cationic and zwitter-ionic ones. Surfactants can be obtained *via* either chemical means (i.e. synthetic surfactants) or can be produced by a variety of microorganisms and are thus generally referred to as biosurfactants. In synthetic ones, on the one hand, the most frequent hydrophobic moieties are polyalcohols, paraffins, olefins, alkylbenzenes and alkylphenols. On the other hand, the hydrophilic group is either a sulphate or sulphonate or carboxylate group in anionic surfactants, a quaternary ammonium group in cationic ones and polyoxyethylene, sucrose, or polypeptide in nonionic ones.

Biosurfactants, instead, can be either low molecular mass substances, such as glycolipids and lipopeptides (Rosenberg and Ron 1999; Singh et al. 2007), or high molecular weight polymers, such as emulsan, alasan or biodispersan (Calvo et al. 2009).

Above a surfactant-specific threshold concentration, referred to as critical micelle concentration (CMC), these compounds tend to form colloidal-sized clusters in water termed micelles (Volkering et al. 1998). The CMC in aqueous solutions is dependent on surfactant's chemical structure and is significantly affected by the temperature, by the ionic strength and by the presence of various organic compounds. With regard to the chemical structure, the CMC generally decreases as the hydrophobic character of the surfactant increases; anionic surfactants, however, have much higher CMCs in aqueous solution than nonionic ones with an equivalent hydrophobic group (Rosen 1989).

Table 17.1 Effect of surfactants and mobilising agents (MAs) on mycoremediation performances of contaminated soils

Surfactant or MA	Fungus	Contaminated soil typology	Bioremediation approach	Main outcome	Notes	Ref.
Nonionic surfactants						
Tween 80 (0.4 %, w/v)	*P. chrysosporium*	PAH-spiked soil	SC	Increased 4–6 ring PAH removal (15–33 %)	Increased fungal growth and MnP production	(1)
Tween 80 (2.5 %, w/w)	*I. lacteous* and *P. ostreatus*	Two aged creosote-contaminated soils	SSP with soil over wheat straw-based inoculum	Increased overall PAH removal (33 %) with *P. ostreatus* on WTP soil	Best removals for FLT, PYR, BaA, CHR and BaP	(2)
Tween 80 (5 %, w/v soil water)	*T. versicolor*	BaP-spiked soil	SSP with soil over sawdust-based inoculum	Increased BaP removal (45 %)	No interference with fungal growth	(3)
Tween 80 (0.15 %, w/v)	*P. ostreatus*	PYR-spiked soil	SSP with soil and colonised wheat seeds	Increased PYR removal (35 %)	Ready surfactant metabolisation	(4)
Tween 80 (2.5 %, w/w)	*Allescheriella* sp. and *Phlebia* sp.	PAH-spiked soil	SSP with soil over maize stalk-based inoculum	Overall increased PAH removal (16 %) with *Allescheriella*. sp. Enhanced removal of FLT, PYR and CHR (29, 26 and 22 %, respectively)	Increase of LiP activity with both fungi	(5)
Tween 80 (0.05–1.0 %, w/w)	*Polyporus* sp.	CHR-spiked soil	SSP with soil mixed with colonised wood meal	Increased CHR removal (56 %) with 0.5 % Tween	Identification of degradation intermediates	(6)
Tween 80 (0.1–0.25 %, w/w)	*Trametes versicolor* and *Trametes* sp.	PCP-contaminated soil	SSP with soil mixed with colonised pine kibbled rye	No increase in PCP removal	Increased laccase activity	(7)
Tween 80 (2.5 %, w/w)	*I. lacteus*	PAH-spiked soil	SSP with soil over colonised maize stalks	Increased overall PAH depletion (68 %)	Increased tyrosinase activity and bacterial biodiversity	(8)

(continued)

Table 17.1 (continued)

Surfactant or MA	Fungus	Contaminated soil typology	Bioremediation approach	Main outcome	Notes	Ref.
Tween 80 (2.5 %, w/w)	*P. ostreatus*	PAH-spiked soil	SSP with soil over colonised maize stalks	Increased overall PAH depletion (36 %)	Increased laccase and tyrosinase activities	(8)
Tween 40 (0.15 %, w/v)	*P. ostreatus*	PAH-spiked soil	SSP with soil over colonised wheat seeds	Increased PYR, ANT and PHE removals (25, 7 and 13 %, respectively)	Further 15 % increase of PYR removal with H_2O_2 (1 mM) addition	(4)
Tween 20 (2.5 %, w/w)	*I. lacteus* and *P. ostreatus*	PAH-spiked soil	SSP with soil over colonised maize stalks	Increased overall PAH depletion with *I. lacteus* and *P. ostreatus* (21 and 74 %, respectively)	Enhanced *P. ostreatus* laccase activity	(8)
Tween 20 (2.5 %, w/w)	*I. lacteus* and *P. ostreatus*	Two aged creosote-contaminated soils	SSP with soil over colonised straw	No increases on PAH removal in both soils	Slight increased BaA and BaP depletion with *P. ostreatus* on WTP soil	(2)
Tween 20 (2.5 %, w/w)	*Allescheriella* sp. and *Phlebia* sp.	PAH-spiked soil	SSP with soil over colonised maize stalks	Enhanced removal of FLT, PYR and CHR by *Allescheriella* sp. (43, 33 and 28 %, respectively)	Increased MnP activity with both fungi. Increases of cultivable bacteria concentration with *Phlebia* sp.	(5)
PNS (6 %, w/w)	*P. ostreatus*	Two PAH-contaminated soils	SSP with soils and colonised cottonseed hull and alder chips	Increase up to ca. 50 % of FLT and BaP removals	Increased incorporation of BaP onto humic matter	(9)
Triton X-100 (0.15 %, w/v)	*P. ostreatus*	PYR-spiked soil	SSP with soil mixed with colonised wheat seeds	Increased PYR depletion rate	None	(4)

Mobilising agent	Fungus	Soil/contaminant	Process	Result	Notes	Ref.
Triton X-100 (0.15 %, w/v)	*P. chrysosporium* and bacterial consortia	Sandy soil spiked with PCB (Aroclor 1242)	SSP	Overall 40 % PCB depletion with Di-CB being the best degraded congeners	Partial degradation of tri-, tetra-, penta- and hexa-CB also detected	(10)
Brij 30 (1–5 %, w/ w)	*P. ostreatus*	Two PAH-contaminated soils	SSP with soils over colonised lignocellulose mixture	Increased overall PAH removal by 40 % with 5 % Brij	Further improvements with a "pre-soak" phase of soil with surfactant	(11)
Trycol 6964 (1 %, w/w)	*P. ostreatus*	PAH-contaminated soil from a coking facility	SSP with soils over colonised lignocellulose mixture	Increased PAH removal up to 40–45 %	Trycol less effective than Brij and Witconol	(11)
Witconol SN-70 (1–5 %, w/w)	*P. ostreatus*	Two PAH-contaminated soils	Solid-state process with soils and colonised lignocellulose matrix	Increased overall PAH removal by approx. 40 %	ANT depletion enhanced up to 65 % with 5 % Witconol	(11)
Vegetable oils						
Soybean oil (2.5 %, w/w)	*I. lacteus*	PAH-spiked soil	SSP with soils over colonised maize stalks	Increased overall PAH depletion by 73 %	Enhanced fungal growth and decreased bacterial concentration and biodiversity	(8)
Soybean oil (2.5 %, w/w)	*P. ostreatus*	PAH-spiked soil	SSP with soils over colonised maize stalks	Increased overall PAH depletion by 84 %	Enhanced removal of CHR, BkF and BaP	(8)
Soybean oil (2.5 %, w/w)	*Allescheriella* sp. and *Phlebia* sp.	PAH-spiked soil	SSP with soil over colonised maize stalks	Increased PAH removals by 12 and 31 % with *Allescheriella* sp. and *Phlebia* sp., respectively	Growth increases for both fungi; significant drop of cultivable bacteria with *Phlebia* sp.	(5)

(continued)

Table 17.1 (continued)

Surfactant or MA	Fungus	Contaminated soil typology	Bioremediation approach	Main outcome	Notes	Ref.
Soybean oil (2.5 %, w/w)	I. lacteus and P. ostreatus	Two aged creosote-contaminated soils	SSP with soil over colonised wheat straw	Increased PAH removals by 13 and 31 % by P. ostreatus in both GS and WTP soil	Results largely affected by the contaminant bioavailability	(2)
Proprietary emulsified vegetable oil (3 %, w/w)	P. ostreatus	Creosote-contaminated soil	SSP with soil mixed with lignocellulose mixture (bench and pilot scale)	No increased overall PAH removal at both scales	Slight increases of BghiP and DBA removals at the pilot scale	(12)
Fish oil (1 %)	P. ostreatus	Creosote-contaminated soil	SSP with soil mixed or layered over spent mushroom compost	Increased overall PAH removal by 50 % increase with the layering technique	Slightly increased laccase activity	(13)
Proprietary vegetable oils (5 %, w/w)	P. ostreatus	PAH-contaminated MGP soil	SSP with soil mixed with lignocellulose mixture	Increased overall PAH removal by 40 %	BaP removal enhanced up to 50–55 %	(11)
Other mobilising agents						
OMW	I. lacteus	PAH-spiked soil	SSP with soil over colonised maize stalks	Increased overall PAH removal by 53 %	Enhanced laccase and tyrosinase activities	(8)
OMW	I. lacteus and P. ostreatus	Two aged creosote-contaminated soils	SSP with soil over colonised wheat straw	No increased overall PAH removal in both soils	None	(2)
OMW	P. ostreatus	PAH-spiked soil	SSP with soil over colonised maize stalks	Increased overall PAH removal by 25 %	Enhanced P. ostreatus growth and laccase activity	(8)

Maltosyl cyclodextrin	Absidia cylindrospora	Fluorene-spiked soil	SC	Enhanced fluorene degradation rate during the first 144 h treatment	Positive effect on indigenous microbiota	(14)
RAMEB (2.5 %, w/w)	I. lacteus	PAH-spiked soil	SSB with soil over colonised maize stalks	Increased overall PAH removal (31 %)	Increased tyrosinase activity and bacterial biodiversity	(8)
RAMEB (2.5 %, w/w)	P. ostreatus	PAH-spiked soil	SSB with soil over colonised maize stalks	No increase in PAH depletion	Increased endo-β-1,4-glucanase activity and bacterial biodiversity	(8)

PAH polycyclic aromatic hydrocarbons, *SC* slurry condition, *MnP* Mn-dependent peroxidase, *SSP* solid-state process, *WTP* wood treatment plant, *FLT* fluoranthene, *PYR* pyrene, *BaA* benzo[*a*]anthracene, *CHR* chrysene, *BaP* benzo[*a*]pyrene, *LiP* lignin peroxidase, *PCP* pentachlorophenol, *ANT* anthracene, *PHE* phenanthrene, *PNS* proprietary nonionic surfactant, *PCB* polychlorobiphenyls, *CB* chlorobiphenyls, *BkF* benzo[*k*]fluoranthene, *GS* gasholder site, *BghiP* benzo[*g,h,i*]perylene, *DBA* dibenzo[*a,h*] anthracene, *MGP* manufactured gas plant, *OMW* olive oil mill wastewater, OMW added so that to obtain a soil lipid content of 0.2 % (w/w), *RAMEB* randomly methylated β-cyclodextrins

References: (1) Zheng and Obbard (2001), (2) Leonardi et al. (2007), (3) Boyle et al. (1998), (4) Marquez-Rocha et al. (2000), (5) Giubilei et al. (2009), (6) Hadibarata and Tachibana (2009), (7) Ford et al. (2007), (8) Leonardi et al. (2008), (9) Bogan et al. (1999), (10) Viney and Bewley (1990), (11) Bogan and Lamar (1999), (12) Lamar et al. (2002), (13) Eggen (1999), (14) Garon et al. (2004)

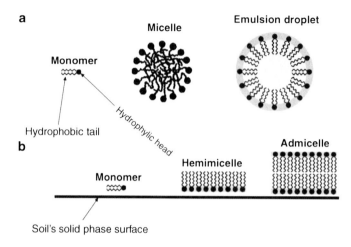

Fig. 17.1 Schematic depiction illustrating surfactant forms in both pore water (**a**) and onto the surface of soil's colloids (**b**) (Adapted from Volkering et al. 1998)

The surfactant-promoted increase in bioavailability of HPP derives from two concomitant factors. On the one hand, the pollutant solubilises within the hydrophobic core of the micelle, a phenomenon known as micellar solubilisation (Singh et al. 2007). On the other hand, the surfactant promotes a decrease in both surface and interface tensions, thus resulting in the facilitated transport of the pollutant from the solid phase to the aqueous one (Mulligan 2005; Van Hamme et al. 2006).

Interactions between HPP and surfactants in soil are strongly affected by a number of variables, mainly including surfactant concentration in soil with respect to its CMC, solubility of the pollutant and soil's physico-chemical properties (Haigh 1996; Laha et al. 2009) as well as by their inherent properties. The most important parameter that has to be considered to evaluate the mobilising ability of a given surfactant towards HPP is its CMC. In general, sub-CMC conditions in soil have little or void effect on solubilisation of HPP since the presence of surfactant can even enhance the adsorption of hydrophobic xenobiotics to soil, due to their partitioning of the pollutant into surfactant adsorbate mono- and bilayers termed hemimicelle and admicelles, respectively (Volkering et al. 1998; Mulligan 2005). Figure 17.1 shows the different forms with which surfactants are present both in pore water and on the surface of soil's colloids. This surfactant-promoted HPP absorption phenomenon was reported for a variety of mobilising agents and shown to be reversed when their concentrations were further enhanced in the system (Klummp et al. 1991; Edwards et al. 1994). Thus, solid matrices, such as soils and sediments, require much higher total surfactant concentrations (i.e. supra-CMC conditions) than clean water systems in order to achieve micellisation which is a prerequisite for pollutant desorption. In fact, although the presence of electrolytes in soil's pore water is known to reduce CMC of surfactants, their extensive adsorption onto soil surfaces requires that they be applied at concentrations largely exceeding their respective CMC (Haigh 1996; Volkering et al. 1998).

Although cyclodextrins are frequently used to mobilise contaminants, they differ from surfactants from both a structural and mode of action viewpoint (Del Valle 2004). They are cyclic oligosaccharides with the relatively less polar parts of the sugar moiety oriented towards the inner cavity. They can be divided into α-, β- and γ-cyclodextrins composed of 6, 7 and 8 glucopyranose units, respectively, with increasingly larger inner cavity (Boyle 2006). Cyclodextrins can form inclusion complexes with several pollutants, the solubility and availability of which differ from those of the noncomplexed counterparts.

17.4 Use of Mobilising Agents on Mycoremediation

Although there are contrasting reports with regard to the ability of surfactants to either stimulate or inhibit HPP biodegradation in soil, there is a general agreement that contaminant desorption often results in enhanced biodegradation.

A variety of phenomena have been invoked to explain negative impact of surfactants on microbial degradation including their preferential use as growth substrates (Tiehm 1994), a surfactant-promoted toxicity (Van Hamme et al. 2006; Singh et al. 2007), a direct effect of the micelles shielding HPP from microbial attack (Putcha and Domach 1993) and a negative effect on bacterial adhesion to nonaqueous phase liquids (Stelmack et al. 1999).

Due to their properties, surfactants exert a variety of effects on fungi, including a modification of the permeability of the plasma membranes with subsequent modification of excretion processes (Leěstan et al. 1993), influence on growth either exerted *via* their ability to act as C sources (Marquez-Rocha et al. 2000; Zheng and Obbard 2002a) or *via* their direct toxicity (Leonardi et al. 2007). Another important aspect is associated with the impact of surfactants on resident microbiota that might significantly affect the interactions between the exogenously added fungus and the indigenous microbial communities. This factor, besides affecting the competitive ability of the fungus, also modifies the outcome of the cleanup process which often arises from the synergistic interaction between indigenous and exogenous microbes (Andersson and Henrysson 1996; in der Wiesche et al. 1996; Meulenberg et al. 1997). Thus, for ease of clarity, these aspects will be discussed within separated subsections.

17.4.1 Effect of Mobilising Agents on Degradation Performances

The addition of MAs to enhance mass transfer rate and bioavailability of soil pollutants has received considerable attention in soil mycoremediation (Chung et al. 2000; Marquez-Rocha et al. 2000; Zheng and Obbard 2001; Zhou et al. 2007).

In a systematic study, Bogan and Lamar (1999) showed that the most effective surfactants on degradation of PAH in industrial soils by six WRF strains were Tween 80 and Triton X-100, belonging to the groups of alkyl ethoxylates and alkylphenol ethoxylates, respectively. Tween 80, however, was more effective than the latter in agreement with a previous study where the mineralisation of ^{14}C-labelled benzo[a]pyrene (BaP) in soil by both *Trametes versicolor* and *Trametes hirsuta* was markedly enhanced in the presence of this surfactant (Boyle et al. 1998). Similarly, Tween 80 significantly enhanced PAH degradation by white-rot fungi in both spiked (Marquez-Rocha et al. 2000) and historically contaminated soils (Leonardi et al. 2008). In another study, conversely, the use of either Tween 80 or Tween 20 did not affect the PAH-degrading ability of both *Pleurotus ostreatus* 3004 and *Irpex lacteus* 617/93 in a historically contaminated soil from a gasholder site, where the non-bioavailable fraction of contaminants was scarcely abundant (i.e. 4 % with respect to the overall PAH content) (Leonardi et al. 2007). In the same study, mycoremediation performances in a silty–loamy soil from a wood treatment plant increased for the highly condensed PAHs BaP, benzo[g,h,i]perylene (BghiP) and indeno[1,2,3-cd]pyrene (IPy), the non-bioavailable fractions of which amounted to more than 80 % of their respective content (Leonardi et al. 2007). Similar results were obtained by Zheng and Obbard (2001), who reported an enhancement in the depletion of 4- and 6-ring PAH in the presence of Tween 80 and a void effect on three-ring compounds. The same study pointed out that the surfactant-promoting effect on highly condensed PAH degradation correlated with an increased contaminant concentration in the aqueous phase (Zheng and Obbard 2001). In another study, chrysene degradation in soil by a *Polyporus* sp. isolate markedly increased in the presence of 0.5 % Tween 80 with respect to the same microcosm where the surfactant had been omitted (86 *vs.* 30 %, respectively) (Hadibarata and Tachibana 2009). Besides its susceptibility to be metabolised by fungi and its good mobilising ability, Tween 80 contains a monounsaturated acyl chain (i.e. oleic acid) which might be involved in MnP-mediated lipid peroxidation leading to radicals able to oxidise even PAH with IP higher than 7.55 eV (Moen and Hammel 1994; Yap et al. 2010).

To assess whether surfactants might affect the possible incorporation of the pollutant onto the organic matter (OM) of a soil augmented with *P. ostreatus*, a mass balance of pollutant recovery in distinct OM fractions was performed by using ^{14}C-labelled anthracene, fluoranthene and BaP (Bogan et al. 1999); among these PAH, only BaP was incorporated onto humic matter at a larger extent in the presence than in the absence of a non-specified proprietary nonionic surfactant.

The degradation ability of *I. lacteus* and *P. ostreatus* was comparatively assessed in a soil, where recent contamination was simulated by spiking with a mixture of seven PAH, and in the presence of several MAs (i.e. soybean oil, Tween 20, Tween 80, olive oil mill wastewaters and randomly methylated β-cyclodextrins) (Leonardi et al. 2008). Among them, soybean oil best affected PAH degradation, the overall residual concentrations of which were lower than those in its absence (57.7 *vs.* 201.3 and 26.3 *vs.* 160.4 mg kg^{-1} with *I. lacteus* and *P. ostreatus*, respectively) (Leonardi et al. 2008). In another study, irrespective of the fungus employed, the enhancing effect due to the use of soybean oil mainly affected the least soluble components of a PAH mixture, namely, chrysene (CHR), benzo[*k*]fluoranthene (BkFLT) and BaP (Giubilei et al. 2009). The positive effect of plant oils on fungal PAH removal from a historically contaminated manufactured gas plant soil had been first reported by Bogan and Lamar (1999); no enhancing effect was found in the non-inoculated control soil, thus suggesting that contaminant losses were not due to either resident microbiota or to abiotic mechanisms, such as auto-oxidative lipid peroxidation. With this regard, plant seed oils, including soybean oil, have molar solubility ratios for PAH similar to those of conventional MAs, thus facilitating the desorption of hydrophobic contaminants from soil organic colloids (Bogan and Lamar 1999; Pannu et al. 2004; Pizzul et al. 2007). In addition, it is worth remembering that soybean oil, due to the high content in polyunsaturated fatty acids (PUFA), is susceptible to peroxidation triggered by LME. This mechanism appears to play a relevant role in PAH oxidation by WRF (Moen and Hammel 1994; Böhmer et al. 1998). Indeed, soybean oil is characterised by a predominance of PUFA, with linoleic and linolenic acids accounting for the 50–55 and 5–9 % of the total fatty acids content, respectively (Ferrari et al. 2005). Figure 17.2 shows MnP-mediated generation of highly reactive fatty acid radicals from linoleic acid (Yap et al. 2010).

Various preparations of cyclodextrins have been used in mycoremediation with variable results. With this regards, fluorene degradation by *Absidia cylindrospora* in soil slurry system was improved in the presence of maltosyl cyclodextrins (Garon et al. 2004). Randomly methylated β-cyclodextrins (RAMEB) also exerted a positive effect on 4- and 5-ring PAH by *I. lacteus* while they did not affect *P. ostreatus* degradation performances (Leonardi et al. 2008). Although β- and γ-cyclodextrin preparations protected *T. hirsuta* from PCP toxicity, their use led to a partial and total suppression of contaminant mineralisation, respectively, since they appeared to decrease contaminant availability to the fungus (Boyle 2006). Similarly, the ability of a given surfactant to promote contaminant degradation was suggest to depend on the structure of its micelles and on the mechanisms of interaction between micellised surfactant and fungal strain (Garon et al. 2002).

Also the application mode of the surfactant might have an impact on mycoremediation. For instance, it was found that the beneficial effect of Brij 30 on mycoremediation was more pronounced when the soil was pre-soaked with surfactant and incubated for 6 days prior to fungal inoculation and allowed a fivefold reduction in the amounts of MA requirement (Bogan and Lamar 1999); when the pre-soak phase was longer, the benefits were lost likely due to the metabolisation of the surfactant by the resident microbiota.

Enhanced PAH degradation activity was observed in soil added with olive mill wastewater (OMW) and subsequently treated with *I. lacteus* and *P. ostreatus* as compared to the respective controls where OMW had been omitted (94.7 and 119.5, respectively, *vs*. 201.3 and 160.4 mg kg^{-1} soil). Thus, the rationale of using OMW as a putative MA based on its content of both residual lipids and humic-like substances (Capasso et al. 2002) was confirmed by these results. In addition, OMW contains some monomeric phenols, such as 4-hydroxybenzoic acid and

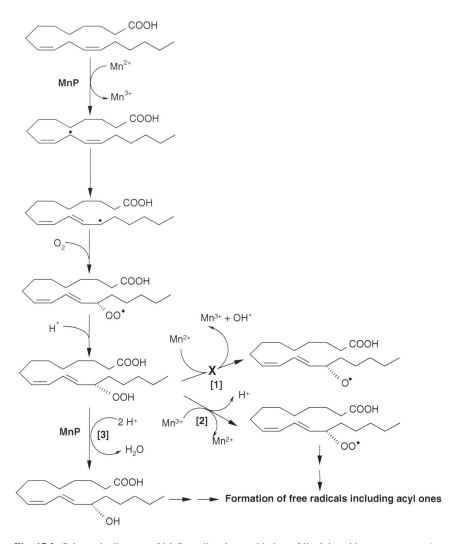

Fig. 17.2 Schematic diagram of MnP-mediated peroxidation of linoleic acid as a representative polyunsaturated fatty acid. The reaction is initiated by hydrogen abstraction at the bis-allylic position of the fatty acid by either MnP or Mn(III). Reaction [1] involving the homolytic cleavage of the O–O bond of lipid hydroperoxide by Mn(II), although theoretically possible, was not found to take place by Watanabe et al. (2001). Reaction [2] involves the homolytic cleavage of the O–H bond of the lipid hydroperoxide by Mn(III), thus generating stable acyl radicals. Free radical generation during electron transfer between MnP and lipid hydroperoxide is shown in reaction [3] (Adapted from Yap et al. 2010)

4-hydroxybenzyl alcohol that have been shown to act as valuable redox mediators in laccase-catalysed PAH oxidation (Johannes and Majcherczyk 2000). These findings might be of further support to the ever-increasing practice of OMW landspreading, currently allowed in some countries (Tomati 2001).

Although PCB are highly hydrophobic compounds, the combined use of surfactants and WRF in their degradation was mostly limited to either liquid cultures (Beaudette et al. 2000) or to washing applications (Ruiz-Aguilar et al. 2002) with a sole exception (Viney and Bewley 1990); in this last study, Triton X-100 was added to a sandy soil spiked with 1,000 mg kg^{-1} Aroclor 1242, and a 40 % PCB degradation was achieved by a mixed consortium of biphenyl-degrading bacterial isolates and the white-rot fungus *Phanerochaete chrysosporium* after 20 weeks incubation.

Another bioremediation option involves soil washing with pollutant MAs and subsequent biodegradation of leached contaminants. Several studies have employed a two-step approach based on soil washing with surfactants followed by fungal degradation of extracted contaminants in liquid cultures (Ruiz-Aguilar et al. 2002; Zheng and Obbard 2002a, b). Ruiz-Aguilar et al. (2002) reported on the degradation by three WRF (i.e. *T. versicolor*, *P. chrysosporium* and *L. edodes*) of PCB that had been leached from a historically contaminated soil by a variety of surfactants (i.e. Tween 80, Triton X-100 and Tergitol NP-10). Degradation outcomes were strongly species- and surfactant-dependent, but the combined use of Tween 80 with *P. chrysosporium* led to the highest PCB removal (70 %) at initial contaminant degradation of 1,800 mg l^{-1} in the washing solution (Ruiz-Aguilar et al. 2002). A similar approach was used by Zheng and Obbard (2002a) who employed Tween 80 to wash PAH from a 9-month aged spiked soil and investigated their degradation in a rotating biological contactor (RBC) with immobilised *P. chrysosporium*. Removal efficiencies of PAH contaminants were higher than 90 % and 76 % when the RBC was operated in batch and continuous mode, respectively.

17.4.2 Effect of Mobilising Agents on Fungal Growth

The wide biological effects exerted by surfactants require that several variables be taken into account to interpret their impact on fungal growth. Additional problems stem from the difficulty to transfer information derived from simplified model systems (i.e. liquid cultures on chemically defined media) to the soil scenery. For instance, one recurring point related to the use of MAs in soil remediation concerns their concentrations, which generally have to be well above their CMC as extensively reviewed by Haigh (1996). Consequently, fungal growth plate assays in the presence of different surfactant concentrations have been preliminarily performed with candidate strains to bioaugmentation (Garon et al. 2002; Leonardi et al. 2007); in some cases, such as with Tween 20, they were found to be predictive with respect to the occurrence of possible growth inhibitory effects on soil as observed for *P. ostreatus* (Leonardi et al. 2008). In a systematic plate-screening study conducted on a significant number of fungi, it was found that octylphenol-type (i.e. Triton X-100) and sorbitan-type (i.e. Tween 80) nonionic surfactants were tolerated at concentrations exceeding their respective CMCs, while the anionic surfactant SDS proved to be rather toxic (Garon et al. 2002). The higher tolerance of fungi towards

Tweens might be related to their ability to use this surfactant as a growth substrate as observed for *P. ostreatus* (Marquez-Rocha et al. 2000) and *P. chrysosporium* (Zheng and Obbard 2001; Ruiz-Aguilar et al. 2002), *T. versicolor* and *L. edodes* (Ruiz-Aguilar et al. 2002). Ruiz-Aguilar et al. (2002) investigated PCB degradation by three WRF (i.e. *T. versicolor*, *P. chrysosporium* and *L. edodes*) in the presence of three nonionic surfactants (Tergitol NP-10, Triton X-100 and Tween 80). In preliminary plate tests, Tween 80 had no inhibitory effect on fungal radial growth, whereas the other surfactants inhibited the growth rate by 75–95 %.

Another recurring point in mycoremediation is the frequent addition of lignocellulosic materials (LM) to contaminated soils which is done to either improve fungal colonisation (Boyle 1995; D'Annibale et al. 2005) or to formulate their inocula (Ford et al. 2007). In this respect, cellulose, which is a major component of LM, is known to have hydrophobic regions in its macromolecular structure (Pérez et al. 2002; Tengerdy and Szakacs 2003). Surfactants, the structure of which entails both hydrophobic and hydrophilic heads, are able to affect the surface properties of cellulose making it more accessible to enzymatic hydrolysis, thus leading to more effective decomposition of this macromolecule (Wu and Ju 1998; Shi et al. 2006). Moreover, the addition of a given surfactant to a solid matrix on the growth response of fungi might be significantly different in the presence and in the absence of HPP (Boyle 2006; Leonardi et al. 2008).

Among different MAs added to a PAH-spiked soil, soybean oil best supported the growth of *P. ostreatus* and *I. lacteus* that was twice that observed in soil in the absence of MAs (Leonardi et al. 2008). In another study conducted on the same contaminated soil, the biomass of *Allescheriella* sp. and *Phlebia* sp. increased 35- and about 2-fold, respectively (Giubilei et al. 2009). It is worth mentioning that soybean oil addition also resulted in a remarkable increase in biomass production in liquid cultures of other basidiomycetes including *Grifola frondosa* (Hsieh et al. 2008) and *P. ostreatus* (Belinky et al. 1994). The marked stimulatory effect of oils on growth of the white-rot fungus *Agaricus bisporus* was well above than that expected if the lipid were a mere carbon source, thus suggesting the possible presence of fungal growth factors (Wardle and Schisler 1969).

Yet, the effect of some surfactants on fungal growth appears to be species-specific and probably contaminant-specific. For instance, the presence of RAMEB did not significantly affect *P. ostreatus* and *I. lacteus* growth in a PAH-contaminated soil (Leonardi et al. 2008), while the inhibitory effect of PCP on *T. hirsuta* growth appeared to be alleviated by cyclodextrins, *via* the formation of inclusion complexes within their hydrophobic cavities (Boyle 2006).

17.4.3   Effect of Mobilising Agents on Extracellular Enzyme Production

Another aspect related to the use of MAs in soil mycoremediation that has not yet been given the right consideration is the impact of these agents on the production of lignin-modifying enzymes and extracellular glycosyl hydrolases (EGH). On the one hand, LME, due to their low substrate specificities and their ability to use small molecular weight substances, are often involved in the degradation of soil contaminants (Böhmer et al. 1998; Kotterman et al. 1998; Cañas et al. 2007). On the other hand, EGH play a fundamental role in the utilisation of LM *via* their hydrolytic action on structural polysaccharides, thus resulting in the liberation of easily assimilable compounds (Pérez et al. 2002).

With regard to LME, it has long been known that several surfactants are able to stimulate their secretion in liquid cultures of WRF (Asther and Corrieu 1987; Leěstan et al. 1993). Both Tween 80 and its acyl component, namely, oleic acid, exerted a high stimulatory activity on *P. chrysosporium* ligninase which was preceded by induction of lipase (Asther and Corrieu 1987). These results were later confirmed by Leěstan et al. (1993) who showed that, in addition to Tween 80 and Tween 20, polyoxyethylene oleate (15EO)C18:1 and the hydrophilic fraction of Tween 80 enhanced *P. chrysosporium* ligninase activity over 200-fold, indicating that the stimulatory effect was not due to the surfactants as a whole. In another study, MnP activity was fourfold stimulated by the presence of Tween 80 (Wang et al. 2008). Cultures with increased lignolytic peroxidase activities exhibited preferential enrichment of total and polar lipids and an increased unsaturation index in their plasma membrane (Leěstan et al. 1993), thus providing evidence in favour of early studies (Reese and Maguire 1969; Asther et al. 1988). The same stimulatory effects of plant oils and Tweens have been reported for laccase production by a variety of fungi (Giese et al. 2004; Yamanaka et al. 2008).

Despite these findings, suggesting a general positive effect on LME release by fungal liquid cultures, the presence of MAs in mycoremediation led to highly variable results. MnP activity from *I. lacteus* was repressed by the presence of surfactants (Leonardi et al. 2008). Addition of either Tween 80, Brij 35 or Triton X-100 to *P. chrysosporium* soil slurry cultures led to both suppression and marked reduction of LiP and MnP productions, respectively (Zheng and Obbard 2001). Conversely, LiP and MnP activities of *Phlebia* sp. grown in a PAH-spiked soil markedly increased in the presence of either soybean oil or Tween 80 (Giubilei et al. 2009); in the same study, Tween 20 led to a twofold stimulation of *Allescheriella* sp. MnP activity. The discrepant effects of surfactants frequently observed in contaminated soils with respect to liquid cultures of the same species might be partially explained by the fact that these MAs are used at concentrations largely exceeding their respective CMC. This might explain, for instance, why the highly stimulatory effect of Tween 80 on MnP activity in both stationary and shaken cultures of *I. lacteus* (Novotny et al. 2004) was not found in a PAH-contaminated soil (Leonardi et al. 2008).

With regard to the effect of MAs on EGH, OMW and RAMEB had a stimulatory effect on the activities of *P. ostreatus* endo-β-1,4-xylanase, cellobiohydrolase and endo-β-1,4-glucanase grown in a PAH-contaminated soil (Leonardi et al. 2008). An approximately twofold increase in both *Phlebia* sp. endo-β-1,4-xylanase and cellobiohydrolase activities was observed during augmentation of a PAH-spiked soil (Giubilei et al. 2009). It is worth noting that MAs may affect the hydrolases' depolymerising efficiency: in fact, nonionic surfactants such as Tween 20 and 80 are capable of modifying cellulose surface properties and alleviate enzyme deactivation by minimising the irreversible binding of the enzyme to the substrate (Converse et al. 1988; Wu and Ju 1998). This results in enhanced cellulose hydrolysis with consequent higher supply of readily available carbon sources (Shi et al. 2006). In addition, and similarly to LME, an enhancing effect of nonionic and anionic surfactants on EGH secretion has been reported in submerged cultures of fungi (Reese and Maguire 1969; Goes and Sheppard 1999; Tribak et al. 2002).

17.4.4 Effect of Mobilising Agents on Resident Microbial Communities

The impact of both mycoremediation and other associated interventions on the resident microbiota might not be ignored in order to deeply understand their failure or success. For instance, mycoremediation of PAH-contaminated soils often resulted in the accumulation of dead-end metabolites, such as PAH diones (Andersson and Henrysson 1996), and led to negligible mineralisation of contaminants (Andersson et al. 2003), thus requiring the action of the indigenous microbiota to complete the degradation process (in der Wiesche et al. 1996; Meulenberg et al. 1997). The concomitant use of surfactants might pose further problems either due to their intrinsic toxicity or to their ability to mobilise contaminants at levels that can be harmful to microbiota (Singh et al. 2007). Conventionally, enumeration of heterotrophic and/or specialised cultivable bacteria has been used to assess the impact of the combined use of WRF and MAs (Pizzul et al. 2007; Leonardi et al. 2008; Giubilei et al. 2009). Cultivable bacteria, however, represent only a minor fraction of the total microbial community (Tyson and Banfield 2005). Consequently, cultivation-independent approaches, such as denaturing gradient gel electrophoresis (DGGE) or terminal restriction fragments length polymorphism (t-RFLP) analyses of PCR-amplified 16S rRNA genes, have been adopted to gain further insights into these issues (Fedi et al. 2005; Giubilei et al. 2009).

In some cases, however, cultivation-dependent and cultivation-independent approaches provided with results that went to the same direction. For instance, microbial densities in non-inoculated and *Phlebia*-augmented PAH-contaminated soil in the presence of both soybean oil and Tween 80 were significantly lower than without these surfactants, while those in *Allescheriella* sp. microcosms were not negatively affected (Giubilei et al. 2009). DGGE analysis of PCR-amplified 16S rRNA genes showed that in *Phlebia* sp.-augmented soil, richness (S) and Shannon–Weaver index (H) were negatively affected by soybean oil, while the same parameters were generally improved in soil augmented with *Allescheriella* sp. (Giubilei et al. 2009) (Fig. 17.3). The negative effect of soybean oil on cultivable bacteria was suggested to be either due to an imbalance between the added carbon

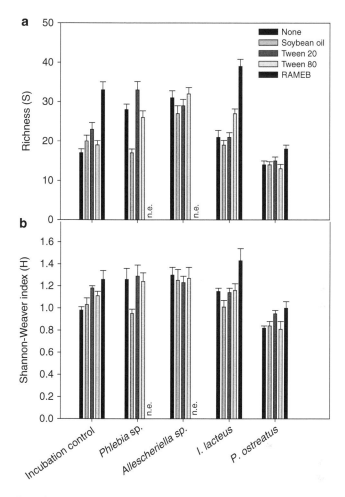

Fig. 17.3 Effect of soybean oil, Tween 20, Tween 80 and randomly methylated β-cyclodextrins (RAMEB) additions (2.5 %, w/w) on bacterial population biodiversity (**a**, richness and, **b**, Shannon–Weaver index) during mycoremediation of a PAH-spiked soil incubated for 6 weeks at 28 °C (Adapted from Leonardi et al. 2008; Giubilei et al. 2009). *n.e.* not evaluated

and soil nutrients (Pizzul et al. 2007) or to the likely occurrence of a physical barrier impairing gaseous exchanges (Borden 2007). The discrepancies in the responses of bacterial biota in *Allescheriella* sp. and *Phlebia* sp. microcosms in the presence of soybean oil might be due to the better capacity of the former fungus to catabolise the MA. In this respect, several studies have shown that Tween 80 is readily catabolised by WRF during PAH degradation (Marquez-Rocha et al. 2000; Zheng and Obbard 2001).

In another study, the use of randomly methylated β-cyclodextrins (RAMEB) enhanced both microbial densities and biodiversity of non-inoculated and *I. lacteus*- and *P. ostreatus*-augmented soils (Leonardi et al. 2008) (Fig. 17.3). In this respect, several investigators observed a generalised and marked stimulation of cultivable heterotrophic and specialised bacteria in the presence of different cyclodextrins in PCB-contaminated matrices (Fava et al. 2003; Fedi et al. 2005). However, in off-site biostimulation trials of contaminated soils, a reduction in biodiversity, expressed as the number of terminal restriction fragments, was evident (Fedi et al. 2005).

17.5 Conclusion and Perspectives

Although the combined use of surfactants with WRF has often been successfully employed in the remediation of soils contaminated by HPP even at the field scale, the variability of results suggests that their application requires substantial research on a case-by-case basis. This uncertainty stems from both the complexity and variability of interactions between surfactants and soil components and on the array of species-dependent biological effects of surfactants. Thus, the required surfactants loads onto soil are high and well above the CMC due to their high tendency to interact with soil colloids and are difficult to be estimated. This literature survey clearly shows that the large majority of surfactant-assisted mycoremediation studies have been addressed towards the use of nonionic ones. Among them, Tween 80 has been shown to be particularly promising due its good HPP-mobilising activity and high biodegradability. In addition, plant oils have been also shown to be effective in mycoremediation due to their high molar solubilisation ratios, their ability to act as co-substrates and their interaction with LME. Thus, further elucidation of both basic and applied aspects related to the use of plant oils will undoubtedly be a productive research line. Intensive researches have found that the surfactants have marked stimulatory effects on the production of fungal glycosyl hydrolases and LME and thus they may improve both decomposition of added lignocellulosic amendments and contaminant degradation (Liu et al. 2008; Liang et al. 2010). Due to the high effectiveness and low toxicity to microorganisms and environmental compatibility, the unique ability of biosurfactants to enhance the organic contaminant biodegradation and bioremediation has gained more and more attention and will undoubtedly lead to application in pollution control (Calvo et al. 2009). Despite this, biosurfactants have not been so far employed in mycoremediation techniques, and either their direct use or the combined inoculation of biosurfactant-producing strains with WRF might deserve attention. In this respect, it is interesting to note that the interspecific interactions between WRF and bacteria able to produce surfactants have resulted in increased production of relevant enzymes in bioremediation, such as LME (Crowe and Olsson 2001; Baldrian 2004). Another aspect that has been scarcely faced is the impact of the combined use of both WRF and surfactants on resident communities in contaminated sites. This is made necessary by the increasing evidence that HPP degradation derives from the cooperative action between added fungi and autochthonous communities. This interplay often involves a modification of contaminant availability exerted by the former leading to the formation of more polar metabolites the mineralisation of which is sequentially performed by resident bacteria.

Acknowledgment Financial support of this work was partially provided by the Italian Ministry of Education and University (MIUR) within the Project PRIN 2008 "Response analysis to heavy metals and polychlorobiphenyls of *Pleurotus ostreatus* planktonic cultures and its mono-specific and mixed biofilms". The authors wish to thank Prof. Federico Federici for helpful suggestions and the critical reading of the manuscript.

References

Acevedo F, Pizzul L, Castillo MP, González ME, Cea M, Gianfreda L, Diez MC (2010) Degradation of polycyclic aromatic hydrocarbons by free and nanoclay-immobilized manganese peroxidase from *Anthracophyllum discolor*. Chemosphere 80:271–278

Andersson BE, Henrysson T (1996) Accumulation and degradation of dead-end metabolites during treatment of soil contaminated with polycyclic aromatic hydrocarbons with five strains of white-rot fungi. Appl Microbiol Biotechnol 46:647–652

Andersson BE, Lundstedt S, Tornberg K, Schnuerer Y, Oeberg LG, Mattiasson B (2003) Incomplete degradation of polycyclic aromatic hydrocarbons in soil inoculated with wood-rotting fungi and their effect on the indigenous soil bacteria. Environ Toxicol Chem 22:1238–1243

Asther M, Corrieu G (1987) Effect of Tween 80 and oleic acid on ligninase production by *Phanerochaete chrysosporium* INA-12. Enzyme Microb Technol 9:245–249

Asther M, Lesage L, Drapron R, Corrieu G, Odier E (1988) Phospholipid and fatty acid enrichment of *Phanerochaete chrysosporium* INA-12 in relation to ligninase production. Appl Environ Microbiol 27:393–398

Baldrian P (2004) Increase of laccase activity during interspecific interactions of white-rot fungi. FEMS Microbiol Ecol 50:245–253

Barr DP, Aust SD (1994) Pollutant degradation by white rot fungi. Rev Environ Contam Toxicol 138:49–72

Beaudette LA, Ward OP, Pickard MA, Fedorak PM (2000) Low surfactant concentration increases fungal mineralization of a polychlorinated biphenyl congener but has no effect on overall metabolism. Lett Appl Microbiol 30:155–160

Belinky PA, Masaphy S, Levanon D, Hadar Y, Dosoretz CG (1994) Effect of medium composition on 1-octen-3-ol formation in submerged cultures of *Pleurotus pulmonarius*. Appl Microbiol Biotechnol 40:629–633

Bogan BW, Lamar RT (1999) Surfactant enhancement of the white-rot fungal PAH soil remediation. In: Bioremediation technologies for polycyclic aromatic hydrocarbons: in-situ and on-site bioremediation. Battelle, Columbus, OH, pp 81–86

Bogan BW, Lamar RT, Burgos WD, Tien M (1999) Extent of humification of anthracene, fluoranthene, and benzo[α]pyrene by *Pleurotus ostreatus* during growth in PAH-contaminated soils. Lett Appl Microbiol 28:250–254

Böhmer S, Messner K, Srebotnik E (1998) Oxidation of phenanthrene by a fungal laccase in the presence of 1-hydroxybenzotriazole and unsaturated lipids. Biochem Biophys Res Commun 244:233–238

Borden RC (2007) Effective distribution of emulsified edible oil for enhanced anaerobic bioremediation. J Contam Hydrol 94:1–12

Boyle CD (1995) Development of a practical method for inducing white-rot fungi to grow into and degrade organopollutants in soil. Can J Microbiol 41:345–353

Boyle D (2006) Effects of pH and cyclodextrins on pentachlorophenol degradation (mineralization) by white-rot fungi. J Environ Manage 80:380–386

Boyle D, Wiesner C, Richardson A (1998) Factors affecting the degradation of polyaromatic hydrocarbons in soil by white-rot fungi. Soil Biol Biochem 30:873–882

Calvo C, Manzanera M, Silva-Castro GA, Uad I, González-López J (2009) Application of bioemulsifiers in soil oil bioremediation processes. Future prospects. Sci Total Environ 407:3634–3640

Cañas AI, Alcalde M, Plou F, Martínez MJ, Martínez AT, Camarero S (2007) Transformation of polycyclic aromatic hydrocarbons by laccase is strongly enhanced by phenolic compounds present in soil. Environ Sci Technol 41:2964–2971

Capasso R, De Martino A, Arienzo M (2002) Recovery and characterization of the metal polymeric organic fraction (polymerin) from olive oil mill waste waters. J Agric Food Chem 50:2846–2855

Christofi N, Ivshina IB (2002) Microbial surfactants and their use in field studies of soil remediation. J Appl Microbiol 93:915–929

Chung NH, Lee IS, Song HS, Bang WG (2000) Mechanisms used by a white rot fungus to degrade lignin and toxic chemicals. J Microbiol Biotechnol 10:737–752

Converse AO, Matsuno R, Tanaka M, Taniguchi M (1988) A model for enzyme adsorption and hydrolysis of microcrystalline cellulose with low deactivation of the adsorbed enzyme. Biotechnol Bioeng 32:38–45

Covino S, Svobodová K, Čvančarová M, D'Annibale A, Petruccioli M, Federici F, Křesinová Z, Galli E, Cajthaml T (2010a) Inoculum carrier and contaminant bioavailability affect fungal degradation performances of PAH-contaminated solid matrices from a wood preservation plant. Chemosphere 79:855–864

Covino S, Čvančarová M, Muzikář M, Svobodová K, D'Annibale A, Petruccioli M, Federici F, Křesinová Z, Cajthaml T (2010b) An efficient PAH-degrading *Lentinus* (*Panus*) *tigrinus* strain: effect of inoculum formulation and pollutant bioavailability in solid matrices. J Hazard Mater 183:669–676

Crestini C, D'Annibale A, Giovannozzi-Sermanni G (1996) Aqueous plant extracts as stimulators of laccase production in liquid cultures of *Lentinus edodes*. Biotechnol Tech 10:243–248

Crowe JD, Olsson S (2001) Induction of laccase activity in *Rhizoctonia solani* by antagonistic *Pseudomonas fluorescens*. Appl Environ Microbiol 67:2088–2094

D'Annibale A, Ricci M, Leonardi V, Quaratino D, Mincione E, Petruccioli M (2005) Degradation of aromatic hydrocarbons by white-rot fungi in a historically contaminated soil. Biotechnol Bioeng 90:723–731

Dawel G, Kaestner M, Michels J, Poppitz W, Guenther W, Fritsche W (1997) Structure of a laccase-mediated product of coupling of 2,4-diamino-6-nitrotoluene to guaiacol, a model for coupling of 2,4,6-trinitrotoluene metabolites to a humic organic soil matrix. Appl Environ Microbiol 63:2560–2565

Del Valle EMM (2004) Cyclodextrins and their uses: a review. Process Biochem 39:1033–1046

Edwards DA, Liu Z, Luthy RG (1994) Surfactant solubilisation of organic compounds in soil/aqueous systems. J Environ Eng 120:5–22

Eggen T (1999) Application of fungal substrate from commercial mushroom production – *Pleurotus ostreatus* – for bioremediation of creosote contaminated soil. Int Biodeter Biodegr 44:117–126

Fava F, Bertin L, Fedi S, Zannoni D (2003) Methyl-β-cyclodextrin-enhanced solubilization and aerobic biodegradation of polychlorinated biphenyls in two aged-contaminated soils. Biotechnol Bioeng 81:384–390

Fedi S, Tremaroli V, Scala D, Perez-Jimenez JR, Fava F, Young L, Zannoni D (2005) T-RFLP analysis of bacterial communities in cyclodextrin-amended bioreactors development for biodegradation of polychlorinated biphenyls. Res Microbiol 156:201–210

Ferrari RA, da Silva OV, Scabio A (2005) Oxidative stability of biodiesel from soybean oil fatty acid ethyl esters. Sci Agric 62:291–295

Ford CI, Walter M, Northcott GL, Di HJ, Kameron KC, Trower T (2007) Fungal inoculum properties: extracellular enzyme expression and pentachlorophenol removal by New Zealand *Trametes* species in contaminated field soils. J Environ Qual 36:1749–1759

Garon D, Krivobok S, Wouessidjewe F, Seigle-Murandi F (2002) Influence of surfactants on solubilization and degradation of fluorene. Chemosphere 47:303–309

Garon D, Sage L, Wouessidjewe F, Seigle-Murandi F (2004) Enhanced degradation of fluorene in soil slurry by Absidia cylindrospora and maltosyl cyclodextrins. Chemosphere 56:159–166

Giese EC, Covizzi LG, Dekker RFH, Barbosa AM (2004) Influência de Tween na produção de lacases constitutivas e indutivas pelo *Botryosphaeria* sp. Acta Sci Biol Sci 26:463–470

Giubilei MA, Leopardi V, Federici E, Covino S, Šašek V, Novotny C, Federici F, D'Annibale A, Petruccioli M (2009) Effect of mobilizing agents on mycoremediation and impact on the indigenous microbiota. J Chem Technol Biotechnol 84:836–844

Goes AP, Sheppard JD (1999) Effect of surfactants on a-amylase production in a solid substrate fermentation process. J Chem Technol Biotechnol 74:709–712

Hadibarata T, Tachibana S (2009) Enhanced Chrysene Biodegradation in Presence of a Synthetic Surfactant. In: Obayashi Y, Isobe T, Subramanian A, Suzuki S, Tanabe S (eds) Interdisciplinary studies on environmental chemistry – environmental research in Asia. Terrapub, Tokyo, pp 301–308

Haigh S (1996) A review of interaction of surfactants with organic contaminants in soil. Sci Total Environ 185:161–170

Hsieh C, Wang H-L, Chen C-C, Hsu T-H, Tseng MH (2008) Effect of plant oil and surfactant on the production of mycelial biomass and polysaccharides in submerged culture of *Grifola frondosa*. Biochem Eng J 38:198–205

in der Wiesche C, Martens R, Zadrazil F (1996) Two-step degradation of pyrene by white rot fungi and soil microorganisms. Appl Microbiol Biotechnol 46:653–659

Johannes C, Majcherczyk A (2000) Natural mediators in the oxidation of polycyclic aromatic hydrocarbons by laccase mediator systems. Appl Environ Microbiol 66:524–528

Juhasz AL (2008) Can bioavailability assays predict the efficacy of PAH bioremediation? Dev Soil Sci 32:569–587

Kim J-E, Fernandes E, Bollag J-M (1997) Enzymatic coupling of the herbicide bentazon with humus monomers and characterization of reaction products. Environ Sci Technol 31:2392–2398

Klummp E, Heitman H, Schwuger MJ (1991) Interactions in surfactant/pollutant/s/soil mineral systems. Tens Surfact Deterg 28:441–446

Kohlmeier S, Smits TMH, Ford RM, Keel C, Harms H, Lukas YW (2005) Taking the fungal highway: mobilization of pollutant-degrading bacteria by fungi. Environ Sci Technol 39:4640–4646

Kotterman MJJ, Rietberg HJ, Hage A, Field JA (1998) Polycyclic aromatic hydrocarbon oxidation by the white-rot fungus *Bjerkandera* sp. strain BOS55 in the presence of non-ionic surfactants. Biotechnol Bioeng 57:220–227

Laha S, Tansel B, Ussawarujikulchai A (2009) Surfactant–soil interactions during surfactant-amended remediation of contaminated soils by hydrophobic organic compounds: a review. J Environ Manage 90:95–100

Lamar RT, White RB, Ashley KC (2002) Evaluation of white-rot fungi for the remediation of creosote-contaminated soil. Remed J 12:97–106

Leěstan D, Černileca M, Štrancarb A, Perdiha A (1993) Influence of some surfactants and related compounds on ligninolytic activity of *Phanerochaete chrysosporium*. FEMS Microbiol Lett 106:17–21

Leonardi V, Sasek V, Petruccioli M, D'Annibale A, Erbanova P, Cajthaml T (2007) Bioavailability modification and fungal biodegradation of PAHs in aged industrial soils. Int Biodeterior Biodegradation 60:165–170

Leonardi V, Giubilei MA, Federici E, Spaccapelo R, Šašek V, Novotny C, Petruccioli M, D'Annibale A (2008) Mobilizing agents enhance fungal degradation of polycyclic aromatic hydrocarbons and affect diversity of indigenous bacteria in soil. Biotechnol Bioeng 101:273–285

Liang Y-S, Yuan X-Z, Zeng G-M, Hu C-L, Zhong H, Huang D-L, Tang L, Zhao J-J (2010) Biodelignification of rice straw by *Phanerochaete chrysosporium* in the presence of dirhamnolipid. Biodegradation 21:615–624

Liu XL, Zeng GM, Tang L, Zhong H, Wang RY, Fu HY, Liu ZF, Huang HL, Zhang JC (2008) Effects of dirhamnolipid and SDS on enzyme production from *Phanerochaete chrysosporium* in submerged fermentation. Process Biochem 43:1300–1303

Majcherczyk A, Johannes C, Huttermann A (1998) Oxidation of polycyclic aromatic hydrocarbons (PAH) by laccase of *Trametes versicolor*. Enzyme Microb Technol 22:335–341

Marquez-Rocha FJ, Hernandez-Rodriguez VZ, Vazquez-Duhalt R (2000) Biodegradation of soil-adsorbed polycyclic aromatic hydrocarbons by the white rot fungus *Pleurotus ostreatus*. Biotechnol Lett 22:469–472

Meulenberg R, Rijnaarts HHM, Doddema HJ, Field JA (1997) Partially oxidized polycyclic aromatic hydrocarbons show an increased bioavailability and biodegradability. FEMS Microbiol Lett 152:45–49

Moeder M, Cajthaml T, Koeller G, Erbanová P, Šašek V (2005) Structure selectivity in degradation and translocation of polychlorinated biphenyls (Delor 103) with a *Pleurotus ostreatus* (oyster mushroom) culture. Chemosphere 61:1370–1378

Moen MA, Hammel KE (1994) Lipid peroxidation by the manganese peroxidase of *Phanerochaete chrysosporium* is the basis for phenanthrene oxidation by the intact fungus. Appl Environ Microbiol 60:1956–1961

Mulligan CN (2005) Environmental applications for biosurfactants. Environ Pollut 133:183–198

Novotny C, Svobodová K, Erbanova P, Cajthaml T, Kasinath A, Lang E, Sasek V (2004) Ligninolytic fungi in bioremediation: extracellular enzyme production and degradation rate. Soil Biol Biochem 36:1545–1551

Pannu JK, Singh A, Ward OP (2004) Vegetable oil as a contaminated remediation amendment: application of peanut oil for extraction of polycyclic aromatic hydrocarbons. Process Biochem 39:1211–1216

Pérez J, Muñoz-Dorado J, Rubia T, Martínez J (2002) Biodegradation and biological treatments of cellulose, hemicellulose and lignin: an overview. Int Microbiol 5:53–63

Pizzul L, del Pilar CM, Stenstrom J (2007) Effect of rapeseed oil on the degradation of polycyclic aromatic hydrocarbons in soil by *Rhodococcus wratislaviensis*. Int Biodeter Biodegr 59:111–118

Putcha RV, Domach MM (1993) Fluorescence monitoring of polycyclic aromatic hydrocarbon biodegradation and effect of surfactants. Environ Prog 12:81–85

Reese ET, Maguire A (1969) Surfactants as stimulants of enzyme production by microorganisms. Appl Microbiol 17:242–245

Rosen MJ (1989) Surfactants and interfacial phenomena, 2nd edn. Wiley, New York

Rosenberg E, Ron EZ (1999) High- and low-molecular-mass microbial surfactants. Appl Microbiol Biotechnol 52:154–162

Ruiz-Aguilar GML, Fernández-Sánchez JM, Rodríguez-Vázquez R, Poggi-Varaldo H (2002) Degradation by white-rot fungi of high concentrations of PCB extracted from a contaminated soil. Adv Environ Res 6:559–568

Rüttimann-Johnson C, Lamar RT (1996) Binding of pentachlorophenol to humic substances in soil by the action of white rot fungi. Soil Biol Biochem 29:1143–1148

Šašek V (2003) Why mycoremediations have not yet come into practice. In: Šašek V et al (eds) The utilization of bioremediation to reduce soil contamination: problems and solutions. Kluwer, Amsterdam, pp 247–266

Semple KT, Morriss AWJ, Paton GI (2003) Bioavailability of hydrophobic organic contaminants in soils: fundamental concepts and techniques for analysis. Eur J Soil Sci 54:809–818

Shi J-G, Zeng G-M, Yuan X-Z, Dai F, Liu J, Wu X-H (2006) The stimulatory effects of surfactants on composting of waste rich in cellulose. World J Microbiol Biotechnol 22:1121–1127

Singh A, Van Hamme JD, Ward OP (2007) Surfactants in microbiology and biotechnology: Part 2. Application aspects. Biotechnol Adv 25:99–121

Stelmack PL, Gray MR, Pickard MA (1999) Bacterial adhesion to soil contaminants in the presence of surfactants. Appl Environ Microbiol 65:163–168

Tengerdy RP, Szakacs G (2003) Bioconversion of lignocellulose in solid substrate fermentation. Biochem Eng J 13:169–179

Tiehm A (1994) Degradation of polycyclic aromatic hydrocarbons in the presence of synthetic surfactants. Appl Environ Microbiol 60:258–263

Tomati U (2001) A European regulation about olive mill waste industry. In: Proceedings of 11th international symposium on environmental pollution and its impact in the Mediterranean region, Cyprus, 6–10 October 2001, p C5

Tribak M, Ocampo JA, García-Romera I (2002) Production of xyloglucanolytic enzymes by *Trichoderma viride*, *Paecilomyces farinosus*, *Wardomyces inflatus*, and *Pleurotus ostreatus*. Mycologia 94:404–410

Tyson GW, Banfield JF (2005) Cultivating the uncultivated: a community genomics perspective. Trends Biotechnol 13:411–415

Van Hamme JD, Singh A, Ward OP (2006) Physiological aspects. Part 1 in a series of papers devoted to surfactants in microbiology and biotechnology. Biotechnol Adv 24:604–620

Viney I, Bewley RJF (1990) Preliminary studies on the development of a microbiological treatment for polychlorinated biphenyls. Arch Environ Contam Toxicol 19:789–796

Volkering F, Breure AM, Rulkens WH (1998) Microbiological aspects of surfactant use for biological soil remediation. Biodegradation 8:401–417

Wang P, Hu X, Cook S, Begonia M, Lee KS, Hwang H-M (2008) Effect of culture conditions on the production of ligninolytic enzymes by white rot fungi *Phanerochaete chrysosporium* (ATCC 20696) and separation of its lignin peroxidase. World J Microbiol Biotechnol 24:2205–2212

Wardle KS, Schisler LC (1969) The effect of various lipids on growth of mycelium of *Agaricus bisporus*. Mycologia 61:305–314

Watanabe T, Katayama S, Enoki M, Honda Y, Kuwahara M (2001) Formation of acyl radical in lipid peroxidation of linoleic acid by manganese-dependent peroxidase from *Ceriporiopsis subvermispora* and *Bjerkandera adusta*. Eur J Biochem 267:4222–4231

Welp G, Bruemmer GW (1999) Effects of organic pollutants on soil microbial activity: the influence of sorption, solubility and speciation. Ecotoxicol Environ Safety 43:83–90

Wu J, Ju LK (1998) Enhancing enzymatic saccharification of waste newsprint by surfactant addition. Biotechnol Prog 14:649–652

Yamanaka R, Soares CF, Matheus DR, Machado KMG (2008) Lignolytic enzymes produced by *Trametes villosa* CCB176 under different culture conditions. Braz J Microbiol 39:78–84

Yap CL, Gan S, Ng HK (2010) Application of vegetable oils in the treatment of polycyclic aromatic hydrocarbons-contaminated soils. J Hazard Mater 177:28–41

Zheng Z, Obbard JP (2001) Effect of non-ionic surfactants on elimination of polycyclic aromatic hydrocarbons (PAHs) in soil-slurry by *Phanerochaete chrysosporium*. J Chem Technol Biotechnol 76:423–429

Zheng Z, Obbard JP (2002a) Removal of polycyclic aromatic hydrocarbons from soil using surfactant and the white rot fungus *Phanerochaete chrysosporium* in a rotating biological contactor. J Biotechnol 96:241–249

Zheng Z, Obbard JP (2002b) Polycyclic aromatic hydrocarbon removal from soil by surfactant Solubilization and *Phanerochaete chrysosporium* oxidation. J Environ Qual 31:1842–1847

Zhou J, Weiying J, Juan D, Xingding Z, Shixiang G (2007) Effect of Tween 80 and β-cyclodextrin on degradation of decabromodiphenyl ether (BDE-209) by white rot fungi. Chemosphere 70:172–177

Chapter 18
New Insights into the Use of Filamentous Fungi and Their Degradative Enzymes as Tools for Assessing the Ecotoxicity of Contaminated Soils During Bioremediation Processes

Christian Mougin, Nathalie Cheviron, Marc Pinheiro, Jérémie D. Lebrun, and Hassan Boukcim

18.1 Introduction

Among available processes allowing the reuse of polluted soils, bioremediation is of priority interest. It exploits the capability of microorganisms, mainly bacteria or fungi, to transform pollutants, thus offering permanent solutions such as immobilization or degradation of the contaminants. For more than 2 decades, powerful capabilities of filamentous fungi, and especially those of ligninolytic white-rot basidiomycetes, have been studied and used to target specific pollutant in waste and soils (Mougin et al. 2003; Asgher et al. 2008; Novotny et al. 2009). The use of bioremediation, however, is not lacking of problems. The possible accumulation in the environment of toxic pollutants or transformation products emphasizes the fact that microorganisms, by themselves, can be insufficient to protect the biosphere from adverse toxic effects.

Here, we would like to demonstrate that fungal enzymes appear to be promising tools either for remediating polluted soil or for the assessment of its possible ecotoxicity during the bioremediation process. We illustrate in this chapter the

C. Mougin (✉) • N. Cheviron • M. Pinheiro
INRA, UR 251 PESSAC, Route de St-Cyr, 78026 Versailles Cedex, France
e-mail: christian.mougin@versailles.inra.fr; Nathalie.Cheviron@versailles.inra.fr;
Marc.Pinheiro@grignon.inra.fr

J.D. Lebrun
INRA, UR 251 PESSAC, Route de St-Cyr, 78026 Versailles Cedex, France

ESTIPA, Laboratoire Biosol, , 3 rue du Tronquet, BP 40118, 76134 Mont-Saint-Aignan
Cedex, France
e-mail: jeremie.lebrun@irstea.fr

H. Boukcim
Valorhiz, Bat 6, Parc Scientifique Agropolis II, 2196 Boulevard de la Lironde, 34980 Montferrier
le Lez, France
e-mail: hassan.boukcim@valorhiz.com

E.M. Goltapeh et al. (eds.), *Fungi as Bioremediators*, Soil Biology 32,   419
DOI 10.1007/978-3-642-33811-3_18, © Springer-Verlag Berlin Heidelberg 2013

relationship between fungal enzymes involved in polycyclic aromatic hydrocarbons (PAHs) and their ecotoxicological assessment when also considered as biomarkers. Researches on the development of these biomarkers are also presented considering model pollutants.

18.2 From Soil Bioremediation to Ecotoxicological Risk Assessment, the Example of Polycyclic Aromatic Hydrocarbons

PAHs constitute one of the most ubiquitous families of organic pollutants found in the environment. Extensive efforts of research have been developed in the field of soil bioremediation since many PAHs and their transformation products are known to be toxic, mutagenic, and carcinogenic (Penning et al. 1999; Bolton et al. 2000). Yet remediation strategies have been reviewed by Gan et al. (2009). Fungal metabolism of PAHs has been and remains extensively studied because of the increasing development of remediation processes using filamentous fungi (Mougin 2002). Numerous fungi among zygomycetes, deuteromycetes, ascomycetes, and basidiomycetes metabolize PAHs. Bacteria and fungi exhibit distinct pathways for PAH transformation. Non-ligninolytic fungi produce metabolites including trans-dihydrodiols, phenols, quinones, tetralones, and dihydrodiol epoxides. In addition, ligninolytic basidiomycetes are able to cleave the aromatic rings, mineralize them, and also produce quinones. All of these fungal metabolites are produced during phase 1 reactions. During phase 2 metabolism, some of them are conjugated with hydrophilic moieties. Then, fungal metabolites can be retained within the cells or released in their environment.

18.2.1 Fungal Biotransformation of Polycyclic Aromatic Hydrocarbons and Consequences upon Their Ecotoxicity

The biotransformation of PAHs, as well as other organic pollutants, can be due to direct metabolism or indirect effect of organisms on the environment. Three processes are typically involved in direct metabolism, namely, biodegradation, cometabolism, and synthesis (Mougin 2002).

During biodegradation, one or several interacting organisms metabolize a given PAH into carbon dioxide and other inorganic components. In this way, the organisms obtain their requirements for growth and energy from the molecule. From an environmental point of view, biodegradation is the most interesting and valuable process, because it leads to the complete breakdown of a molecule without the generation of accumulating intermediates. The prevalent form of PAH metabolism in the environment is cometabolism, in which organisms grow at the expense of a cosubstrate to transform the chemical without deriving any nutrient or energy for growth from the process.

Cometabolism is a partial and fortuitous metabolism, and enzymes involved in the initial reaction lack substrate specificity. Generally, cometabolism results only in minor modifications of the structure of the PAH, but different organisms can transform a molecule by sequential cometabolic attacks, or another can use cometabolic products of one organism as a growth substrate. Intermediate products with their own bio- and physicochemical properties can accumulate, thus causing some adverse effects on the environment.

Synthesis includes conjugation and oligomerization. In that case, xenobiotics are transformed into compounds with chemical structures more complex than those of the parent compounds. During conjugation, a xenobiotic (or one of its transformation products) is linked to hydrophilic endogenous substrates, resulting in the formation of methylated, acetylated, or alkylated compounds, glycosides, or amino acid conjugates. These compounds can be excreted from the living cells or stored. During oligomerization (or oxidative coupling), a xenobiotic combines with itself or with other xenobiotic residues (proteins, soil organic residues). Consequently, they produce high molecular weight compounds, which are stable and often incorporated into cellular components (cell wall) or soil constituents (soil organic matter). This biochemical process not only affects the activity and the biodegradability of a compound in limiting its bioavailability but also raises concern about the environmental impact of the bound residues. However, because of their chemical structure, PAHs can be conjugated, but they are poorly subjected to oligomerization.

Fungal metabolism of PAHs, as described below, has been extensively described in the past by Cerniglia and coworkers (i.e., Cerniglia and Sutherland 2001). Biodegradation aspects of PAHs have been reviewed by Haritash and Kaushik and Mougin et al. in 2009. Often, the first step of PAH metabolism consists in ring epoxidation by a cytochrome P450 monooxygenase, leading to an unstable arene oxide in animals. Arene oxides are immediately hydrated to trans-dihydrodiols by epoxide hydrolase or rearranged nonenzymatically to phenols. The carcinogenicity of trans-dihydrodiols can be lower than that of the parent compound (benzo[a] pyrene trans-9,10- and 4,5-dihydrodiols) or higher (benzo[a]pyrene trans-7, 8-dihydrodiols). The nonenzymatic rearrangement of a PAH arene oxide in solution produces phenols.

When a monooxygenase catalyzes the second oxidation of a PAH trans-dihydrodiol, the result is a dihydrodiol epoxide. Benzo[a]pyrene trans-7,8-dihydrodiol 9,10-oxide, produced by *Cunninghamella elegans*, is the ultimate carcinogenic and mutagenic metabolite of benzo[a]pyrene in mammals. The fungus also produces benzo[a]pyrene trans-9,10-dihydrodiol 7,8-oxide, which is less mutagenic. Dihydrodiol epoxides can be metabolized further by epoxide hydrolase to tetrahydrotetraols.

Phenols and trans-dihydrodiols derived from PAHs are detoxified during phase 2 reactions by alkylation or conjugation with another molecule, including sulfate, glucosides, glucuronides, and xylosides. Phenanthrene and pyrene have been shown to be converted into methoxylated compounds by *Aspergillus niger*. Sulfate conjugation, a common mammalian detoxification reaction, is also performed by fungi such as *C. elegans*. Glucuronic acid conjugates of PAHs are detoxification products in fungi. A soluble UDP-glucosyltransferase from *C. elegans* catalyzes the conjugation of several PAHs. Fungi produce also glucose or xylose conjugates from PAHs, which are no more toxic.

Finally, several strains of fungi produce quinones. Non-ligninolytic fungi formed quinones from trans-dihydrodiols. By contrast, white-rot fungi produce extracellular enzymes (mainly lignin peroxidase—LIP, manganese-dependent peroxidase—MNP, and laccase—LAC) that oxidize PAHs to form quinones. These fungi also partly metabolize PAHs to CO_2 and unidentified minor products.

In addition to various diseases and endocrine disorders, several PAHs induce mutagenic, teratogenic, and carcinogenic effects. Bioremediations processes generate compounds more polar than the parent molecules, PAHs hydroxides, ketones, and quinones, through atmospheric and metabolic reactions. Despite this fact, these polar compounds induce adverse effects on the health of environment, as well as the health of humans. Yet, in humans, lung cancer and bronchitis are possible consequences. Toxicity in animals is mainly associated to the binding of PAHs compounds on aryl hydrocarbon receptor (AhR) and thyroid hormone-related endpoints (Bekki et al. 2009). PAH quinones and ketones, which have functional groups with low polarity, have significant activities using AhR tests, thyroid receptor-based tests, and have estrogenic/antiestrogenic activity. Quinones also induce the generation of reactive oxygen species. In all living organisms, PAH metabolites can elicit several toxic and biochemical responses such as derivation of drug-metabolizing enzymes (cytochrome P450. . .). These results suggest that they might have various toxic activities in animals, more generally in the environment. In all cases, further studies on the toxicity and ecotoxicity mechanisms are necessary. Then, methyl-substituted PAHs are often in mixture with unsubstituted PAHs, and they are known to be mutagenic and carcinogenic.

18.2.2   Ecotoxicological Assessment During PAH-Contaminated Soil Bioremediation

According to the previous knowledge, the possible ecotoxicity of PAHs and related transformation products should be assessed in soils during bioremediation processes, either at the laboratory scale or in situ at the field scale.

The performance of a biological treatment of a PAH-contaminated soil has been evaluated at the field scale (Lors et al. 2009). After 6 months of incubation, the biological treatment led to a significant reduction of 2- and 3-ring PAHs and to a lesser extent to 4-ring PAHs. As a consequence, a significant decrease of the acute ecotoxicity was observed passing from highly ecotoxic before treatment to non-ecotoxic according to *Lactuca sativa* seedling and growth inhibition test and

Eisenia fetida mortality test. This could be related to the bioavailability of PAHs. Indeed, tests performed on aqueous leachates of the soil showed a strong decrease of 2- and 3-ring PAHs correlated with a significant reduction of acute and chronic ecotoxicity responses. The biological treatment led to the mutagenicity reduction and the genotoxicity disappearance in the leachate. Thus, bioassays are complementary to chemical analyses to evaluate the efficiency of a bioremediation process and to evaluate the bioavailability of the organic pollutants as the total concentration of a contaminant is not the only criterion to consider. The comparison of the ecotoxic responses allowed underlining the high sensitivity of the earthworm, Microtox, Algae, and Ames bioassays among the panel used.

In our laboratory, we performed some years ago an experiment intended to demonstrate the efficiency of the white-rot *Trametes versicolor* to decrease the amounts of PAHs in industrial soils (Rama et al. 2001). Solid lignocellulosic carriers have been developed to inoculate the fungus into a manufactured gas plant site soil. Pelleted wheat bran carriers were very efficient in stimulating the growth of fungi in the soil containing about 2,800 mg kg^{-1} PAHs. Fungal biomass and activity of extracellular LACs produced by *T. versicolor* (as markers of metabolic activity in the contaminated soil) decreased after 2 weeks of incubation. Supplementing the soil with a mixture of carbon, nitrogen, and phosphorus enhanced the fungal activity period.

In that experiment, LAC activity was measured during several weeks after inoculation of *T. versicolor*. The profile of laccase production was quite similar to that of fungal biomass production during the considered period. Laccase activity was the highest between 1 and 2 weeks after inoculation (450 nmol min^{-1} g^{-1} dry soil) and then gradually decreased. LAC identity was checked by SDS-PAGE and activity measurement. Its production seemed closely related to fungal growth. In addition, it has been shown that selected PAHs could be inducers of LACs. Yet, 9-fluorenone, as a metabolite of fluorene formed through LAC oxidation (Mougin et al. 2002a, b), increased 22-fold the activity of LAC produced by *T. versicolor*. That result demonstrated that enzymes involved in PAH metabolism can also be up- or downregulated by fungal exposure to their enzymatic substrate.

Very recently, we started an experiment intended to correlate PAH-spiked soil bioremediation by the fungus *T. versicolor* and ecotoxicological assessment of the process. An agricultural soil was spiked with phenanthrene (90 mg kg^{-1} dry soil) and benzo[*a*]pyrene (60 mg kg^{-1} dry soil), supplemented with rice bran (7 % w/w on a dry basis) and mixed. Experiments were conducted in polyethylene jars refilled with 9.0 kg dry soil during 40 weeks in the dark at room temperature. Inoculation devices have been developed in order to protect *T. versicolor*, provide nutrients, and ensure a significant fungal growth within the soil. Then, these devices should allow an easy harvesting of fungal biomass exposed to contaminants for biomarker measurements. Three situations were considered using that inoculation device (control with PAHs and without fungus, control without PAHs and with the fungus, complete assay). In a fourth situation, the fungus was inoculated after growth and coating of lignocellulosic pellets (Rama et al. 2001). All fungal inoculations were performed at the beginning of the experiments, then after 21 weeks of incubation.

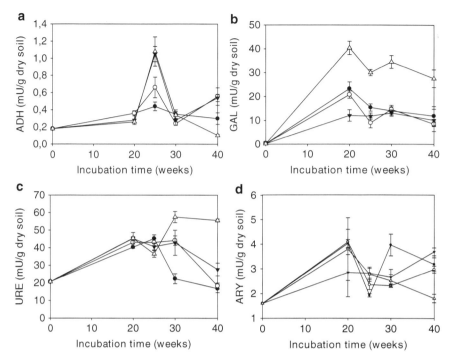

Fig. 18.1 Dehydrogenase (**a**, ADH), β-galactosidase (**b**, GAL), urease (**c**, URE), and arylsulfatase (**d**, ARS) activities in the PAHs-spiked soils during the fungal remediation process. *Symbols* refer to *filled circles*, control with PAHs and without *Trametes versicolor*; *open circles*, control without PAHs and with *Trametes versicolor*; *filled inverted triangles*, complete assay with *Trametes versicolor* in the inoculation devices; *open triangles*, complete assay with *Trametes versicolor* inoculated after growth on lignocellulosic pellets. A transient peak of ADH activity is noticed several weeks after fungal inoculation, as already reported in 2002. GAL activity was the highest in the soil inoculated with lignocellulosic pellets. There were no differences in URE activities until 25 weeks; then it decreased except in the jar inoculated with pellets. Finally, ARY activity was difficult to interpret

Whatever the inoculation process used, phenanthrene disappeared almost totally (>98 %) within the first 10 weeks in the three spiked soils. By contrast, benzo[a] pyrene was more recalcitrant to fungal attack, with a 40 % decrease in mean after 40 weeks.

As soil functioning biomarkers, global enzymatic activities have been monitored during the remediation process, namely, dehydrogenase (ADH), β-galactosidase (GAL), urease (URE), and arylsulfatase (ARS) (Fig. 18.1).

As concluding remarks of that part, we can underline that our results generally indicated a good relationship between global enzymatic activities in soils and fungal inoculation. An understanding of biochemical and physiological mechanisms involved in biological responses is necessary to understand particular situations.

Because PAH degradation profiles were quite similar in the four jars, it remains difficult to link enzymatic activities to soil contamination, including residual PAHs and their new transformation products.

18.3   Fungal Enzymes as Candidates for Biomarker Development

Filamentous fungi such as white-rot basidiomycetes are among the major decomposers of biopolymers in the environment. They have developed nonspecific and radical-based degradation mechanisms occurring in their extracellular vicinity. Numerous studies have identified the role of that enzymatic machinery (e.g., LIP and MNP, LAC...) in the transformation capacity of ligninolytic fungi towards a wide range of organic pollutants in contaminated soils (Gianfreda and Rao 2004; Baldrian 2006). In addition, these fungi possess also intracellular enzymes involved in transformation reactions, such as cytochromes P450.

18.3.1   *The Extracellular Oxidoreductases*

Lignin peroxidase (LIP, EC 1.11.1.13) and manganese peroxidase (MNP, EC 1.11.1.14) were discovered in the strain *P. chrysosporium*. LIP and MNP catalyze the oxidation of lignin units by H_2O_2. If LIP degrades non-phenolic lignin units, MNP generates Mn^{3+} which acts as a diffusible oxidizer on phenolic or non-phenolic lignin units. A versatile peroxidase (VP) has been also described in *Pleurotus* sp. and other fungi as a third type of ligninolytic peroxidase that combines the catalytic properties of LIP and MNP, being able to oxidize typical LIP and MNP substrates. In addition, a novel ligninolytic peroxidase gene (ACLnP) was cloned and characterized from the brown-rot fungus *Antrodia cinnamomea* (Huang et al. 2009). Peroxidases are glycosylated proteins with an iron protoporphyrin IX (heme) prosthetic group located at the active site. Therefore, MNPs are able to oxidize and depolymerize their natural substrate, i.e., lignin, as well as recalcitrant xenobiotics such as nitroaminotoluenes and textile dyes (Knutson et al. 2005). The use of peroxidases for soil cleaning has been studied, namely, for soils historically contaminated with aromatic hydrocarbons and detoxified by autochthonous fungi producing peroxidases (D'Annibale et al. 2006). Recently, a novel group of fungal peroxidases, known as the aromatic peroxygenases (APO), has been discovered (Pecyna et al. 2009). Members of these extracellular biocatalysts produced by agaric basidiomycetes such as *Agrocybe aegerita* or *Coprinellus radians* catalyze reactions (e.g., the peroxygenation of naphthalene, toluene, dibenzothiophene, or pyridine) which are actually attributed to cytochrome P450 monooxygenases.

Laccases (LAC, EC 1.10.3.2) belong to a large group of multicopper oxidases, which includes among others ascorbate oxidases and ceruloplasmin. They occur widely in lignin degrading filamentous fungi, including white-rot basidiomycetes. They perform the reduction of dioxygen to water while oxidizing organic substrates by a one-electron redox process. Laccases can oxidize a wide range of aromatic substrates, mainly phenolic and anilines. Their occurrence, characterization, functions, and applications have been reviewed in recent years (Baldrian 2006; Mougin et al. 2003). Laccases are therefore involved in the transformation of a wide range of phenolic compounds including natural substrates as lignin and humic substances but also xenobiotics such as trichlorophenols, pesticides, polynitrated aromatic compounds, azo dyes, and PAHs, the latter chemicals being the major source of contamination in soil. The potential use of these oxidative enzymes for the detoxification of organic pollutants has been extensively reviewed (Couto and Herrera 2006; Gianfreda and Rao 2004; Mougin et al. 2003).

18.3.2 Intracellular Cytochromes P450

Cytochromes P450 constitute a large family of heme-thiolate proteins widely distributed among living organisms. In most cases, they function as monooxygenases by binding and activating molecular oxygen, incorporating one of its atoms into an organic substrate, and reducing the second atom to form water. The result of P450 catalysis, depending on the protein and its substrate, results in most cases in hydroxylation, but epoxidation, heteroatom dealkylation, deamination, isomerization, C–C or C=N cleavage, dimerization, ring formation or extension, dehydration, dehydrogenation, or reduction has also been reported. For most eucaryotic P450s, a FAD/FMN-dependent NADPH-P450 reductase is needed to transfer the electrons used for oxygen activation from cytosolic NADPH. In filamentous fungi, P450s and reductases are usually microsomal membrane-bound proteins exposed to the cytosol. Nevertheless, soluble forms of P450s, coupling P450 and reductase in a single fusion protein, have also been found in bacteria and fungi.

P450s are encoded by a superfamily of genes. The sequences of more than 500 of them have already been recorded in all living organisms. They are named, and classified in more than 150 families, according to the identity in amino acid sequences of the deduced proteins. With a few exceptions, based on phylogenetic considerations, proteins with 40 % or less sequence identity are considered to define a new family. When two P450s are more than 55 % identical, they are designated as members of a same subfamily. Families are designated by a number, subfamilies by a letter, following the prefix CYP. Through fungal genome sequencing projects, the discovery of P450s has advanced exponentially in recent years. More than 6,000 fungal genes coding for putative P450s from 276 families have been identified.

Cytochrome P450 enzymes in the fungal kingdom have been reviewed by Cresnar and Petric in 2011. Knowledge concerning regulation of fungal P450s is only starting to accumulate. P450 expression can be regulated by fungal exposure to xenobiotics (agrochemicals, ethanol, or drugs like phenobarbital or aminopyrine). Fungal P450s can be also inhibited by mechanism-based inactivators (i.e., 1-aminobenzotriazole). Numerous environmental pollutants act as P450 effectors.

18.3.3 Other Enzymatic Systems

Although not formerly involved in xenobiotic metabolism, fungi synthetize numerous enzymatic systems. Extracellular one, synthetized by wood decaying and phytopathogenic fungi, is the flavohemoprotein cellobiose dehydrogenase (CDH, EC 1.1.3.25. It can be used as a component of amperometric biosensors for detecting quinones (Karapetyan et al. 2006).

In addition, as many organisms, fungi produce hydrolases involved in carbon, nitrogen, phosphorus, and sulfur cycles. These enzymes are commonly used for many years as sensitive indicators of soil functioning. Nevertheless, the relevant interpretation of their expression level, especially in a situation of contamination, requires a referential based on their natural spatiotemporal variability. Main hydrolases are β-glucosidase (GLU, EC 3.2.1.21), β-galactosidase (GAL, EC 3.2.1.23), N-acetyl-β-glucosaminidase (NAG, EC 3.2.1.30), urease (URE, EC 3.5.1.5), acid phosphatase (PAC, EC 3.1.3.2), alkaline phosphatase (PAL, EC, 3.1.3.1), and arylsulfatase (ARS, EC 3.1.6.1). Expression of these enzymes by fungi exposed to pollutants is not extensively studied.

All the previous paragraphs confirm that fungi possess and produce a wide range of enzymatic systems. Studying the specificity, the sensitivity, and the dose–response relationships of these systems with fungal exposure to pollutants could ensure the identification of potential biomarkers.

18.4 Ways for Research in the Field of Soil Ecotoxicology Associated to Remediation Processes

Mechanistic studies considering the regulation of fungal enzymatic systems are required in order to develop biomarkers and ecotoxicological tools. Targeted enzymes must belong to distinct families of fungal enzymes, either involved in pollutant transformation during remediation processes or involved in nutrient cycles. *T. versicolor* is retained in that preliminary approach as a fungal model efficient in degrading pollutants. In addition, the panel of environmental pollutants has been extended to metals, because organisms are exposed to mixture of inorganic and organic pollutants in soils.

18.4.1 Laccase Regulation as a Starting Point

Among fungal enzymes, laccases (LACs) have been detected and purified as constitutive or inducible isoenzymes from many strains. The stimulation of laccase production with respect to the culture medium composition has also been investigated. Metal ions and organic molecules have also been assayed for their ability to enhance the production of the inducible form of LACs. Gallic and ferulic acids were used mainly because of their structural analogy with lignin model compounds. Moreover, 2,5-xylidine (2,5-dimethylaniline) has often been used to increase enzyme production in laboratory experiments by stimulating the expression of an inducible form of LAC. The substrate range of laccase includes some potential pollutants of the environment. Unfortunately, very little data are available concerning the ability of these xenobiotics of environmental interest to interact with LAC production. We measured LAC activity in liquid cultures inoculated by *T. versicolor* (Mougin et al. 2002b). Agrochemicals, industrial compounds, and their transformation products have been assayed for their ability to enhance LAC production in liquid cultures, when added at the concentration of 0.5 mM. After 3 days of treatment, enzymatic activity in the culture medium was increased 14-fold by 4-*n*-nonylphenol and 24-fold by the aniline. LAC activity was enhanced 10-fold by oxidized derivatives of the herbicide diquat, 17-fold by *N,N'*-dimethyl-*N*-(5-chloro,4-hydroxyphenyl)urea, and 22-fold by 9-fluorenone, as presented above.

In a second time, we attempted to demonstrate LAC induction at the transcriptomic level. For that purpose, we followed the production of the gene AF414109 encoding the isoform of LAC A in *T. versicolor* cultures and the gene U44430 encoding the isoform B. RNAs were quantified by real-time PCR and their expression compared to enzymatic activity. The chemical retained as reference inducer was 2,5-dimethyaniline.

Our results showed that mRNA production and LAC activity were well correlated in mycelium of the fungus after 2 days of exposure to the inducer, with a good dose–response relationship (Fig. 18.2a). The two markers were also linked regarding the time of exposure (Fig. 18.2b).

By contrast, fungal exposure to 2,5-dimethylaniline failed to modify amounts of gene U44430, suggesting distinct regulation pathways for the two subfamilies of LACs and a specificity of responses according to chemicals.

18.4.2 Towards the Use of Profiles of Fungal Enzymes as Tools for Ecotoxicological Assessment?

In our laboratory, we developed a program of research intended to understand the physiological and physicochemical mechanisms governing the biological response of filamentous fungi to environmental pollutants. One first objective is the development of fungal biomarkers for ecotoxicity assessment. Because some

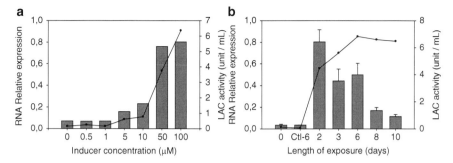

Fig. 18.2 Laccase mRNA production (*gray bars*, gene AF414109) and laccase activity (*black solid line*) in liquid cultures of *T. versicolor* spiked with 2,5-dimethylaniline, with respect to inducer concentration (**a**, after 2 days of induction) and the length of exposure (**b**, inducer concentration of 100 μM). Results exhibited good relationships of mRNA production and LAC activity with respect to inducer concentration and length of exposure

information was already available concerning organic pollutants (see above), recent experiments were developed with metals as pollutants. First experiments are performed in pure fungal liquid cultures and extended to polluted soils in a second time. The conclusions are relevant in the case of soil PAH contamination because they consider in all cases the chemical speciation of the pollutants, their bioavailability, and the exposure of organisms.

The relationship between the physiological state of fungi and the response of their functional enzymatic system has been clarified (Lebrun et al. 2010). Our study aims at establishing how the development phases modulate the secretion of enzymes in *T. versicolor* exposed to a single pollutant, copper (Cu). For that purpose, extracellular hydrolases (GLU, GAL, NAG, PAC, PAL, ARS) and oxidoreductases (LIP, MNP and LAC) were monitored in liquid cultures for 2 weeks. Cu was added either during the growing or stationary phases at 20 or 200 ppm. Our results showed that Cu at the highest concentration modifies the secretion of enzymes, whatever the development phase to which the fungus was exposed. In a general way, the production of hydrolases is decreased by Cu whereas that of oxidoreductases is highly increased. However, the sensitivity of enzyme responses to Cu depends on the phase development and the type of secreted enzyme. Furthermore, LIP, which was not measured in the control cultures, was specifically produced in the presence of Cu. Our results confirm that oxidoreductases may be appropriate biomarkers for ecotoxicity assessment. It remains now necessary to investigate the response of other types of fungal enzymes after exposure to chemicals in mixtures.

Then, the sensitivity of biological responses must be addressed regarding the bioavailability of pollutants. Lebrun et al. (2011b)aimed at enhancing the secretion of lignin-modifying oxidoreductases in *T. versicolor* by favoring the bioavailability of essential metals. For this purpose, the fungus was exposed to Cu or zinc (Zn) in liquid culture media exhibiting different complexation levels. Metal speciation was determined experimentally or theoretically to quantify free metal species, supposed to be the most bioavailable, and species complexed to ligands. Although Zn^{2+} contents were high in media, Zn had no effect on oxidoreductase production. Conversely, Cu highly induced MNP and LAC productions until 40 and 310 times when compared to unexposed controls. This inductive potential was highly correlated to Cu^{2+} contents in media. Furthermore, in lowly complexing media, the response threshold of oxidoreductases to Cu greatly decreased and an unexpected production of LIP occurred, as a confirmation of the previous result.

Moreover, Lebrun et al. (2011a) investigated the effect of Zn, Cu, lead (Pb), and cadmium (Cd), tested alone or in equimolar cocktail, on the secretion profiles at enzymatic and protein levels *in T. versicolor* cultures. They monitored extracellular hydrolases (GLU, GAL, NAG, and PAC) and ligninolytic oxidases (LAC and MNP). Fungal secretome was analyzed by electrophoresis, and LAC secretion was characterized by western blot and mass spectrometry analyses. They showed that all hydrolase activities were inhibited by the metals tested alone or in cocktail, whereas oxidase activities were specifically stimulated by Cu, Cd, and metal cocktail. At protein level, metal exposure modified the electrophoretic profiles of fungal secretome and affected the diversity of secreted proteins. The LAC isoenzymes, A and B, were differentially glycosylated according to the metal exposure. The amount of secreted LAC A and LAC B was strongly correlated with the stimulation of LAC activity by Cu, Cd, and metal cocktail. In conclusion, modification of enzyme activities can now be linked to modifications of protein secretion in relation to transcriptomic probes.

Finally, the relation between degradation of PAHs and the ligninolytic enzymes remains difficult to be established in the soil matrix. Studies in liquid medium showed the ability of these enzymes to degrade the PAHs, and principles governing the various reactions are globally established. However, the experiments of biodegradation of PAHs in soil have generally provided conflicting results. These results could be explained firstly by the difficulty of extracting these enzymes in the soil and measure of their activities and secondly by the absence of the knowledge on the interactions between soil physical properties and the pollutant bioavailability and enzymatic activities. These are new challenges for the environmental sciences, especially by the study of physical coupling of processes in soils (Braudeau and Mohtar 2009).

18.5 Conclusions and Perspectives

Through this chapter, we highlighted the following points:

– Considering the example of PAHs, we showed the complexity of the biochemical reactions catalyzed by fungi during remediation processes.
– These complex pathways resulted in the formation of numerous metabolites, exhibiting in most cases a real ecotoxicity.

- Fungal enzymatic systems, either involved in pollutant transformation or not, behave as potent biomarkers for ecotoxicological assessment.
- It remains necessary to develop additional research to understand the biochemical mechanisms responsible of fungal biological responses.
- That knowledge allowed considering profiles of protein expression as biomarkers of soil ecotoxicity.

More research remains necessary to identify the most relevant protein profiles produced by fungi exposed to soil pollutants. It is also essential to understand all the mechanisms governing protein expression in polluted soils, as well as the selectivity and sensitivity of the responses. Proteomics and transcriptomics are actual approaches to study these topics. It is also fundamental to study the role of the soil physics properties in the enzymatic reactions measured in situ in order to allow the coupling between physical and biological processes in soils. That will make it possible, through a mechanistic modeling, the generalization of the results and the scale transfer for the use of the enzymes as biomarkers of soil ecotoxicity in the environmental diagnosis.

Acknowledgement The authors thank VALORHIZ, the French Ministry for the Higher Education and Research, and French Public company OSEO for their financial contributions to this R&D project. The authors also thank Dr. Claude Jolivalt (Ecole Nationale Supérieure de Chimie, Paris, France) for performing RNA extraction and real-time PCR experiments, as well as Dr. Karine Laval and Dr. Isabelle Trinsoutrot-Gattin (ESITPA, Mont St-Aignan, France) for comanaging the Ph.D. research of Jeremie D. Lebrun.

References

Asgher M, Bhatti HN, Ashraf M, Legge R (2008) Recent developments in biodegradation of industrial pollutants by white rot fungi and their enzyme system. Biodegradation 19:771–783

Baldrian P (2006) Fungal laccases – occurrence and properties. FEMS Microbiol Rev 30:215–242

Bekki K, Takigami H, Suzuki G, Tang N, Hayakawa K (2009) Evaluation of toxic activities of polycyclic aromatic hydrocarbon derivatives using in vitro bioassays. J Health Sci 55:601–610

Bolton JL, Trush MA, Penning TM, Dryhurst G, Monks TJ (2000) Roles of quinones in toxicology. Chem Res Toxicol 13:135–160

Braudeau E, Mohtar RH (2009) Modeling the soil system: bridging the gap between pedology and soil-water physics. Global Planet Change 67:51–61

Cerniglia CE, Sutherland JB (2001) Bioremediation of polycyclic aromatic hydrocarbons by ligninolytic and non-ligninolytic fungi. In: Gadd GM (ed) Fungi in bioremediation, British Mycological Society Symposium Series. Cambridge University Press, Cambridge, pp 136–187

Couto SR, Herrera JLT (2006) Industrial and biotechnological applications of laccases: a review. Biotechnol Adv 24:500–513

Cresnar B, Petric S (2011) Cytochrome P450 enzymes in the fungal kingdom. Biochim Biophys Acta 1814:29–35

D'Annibale A, Rosetto F, Leonardi V, Federici F, Petruccioli M (2006) Role of autochthonous filamentous fungi in bioremediation of a soil historically contaminated with aromatic hydrocarbons. Appl Environ Microbiol 72:28–36

Gan S, Lau EV, Ng HK (2009) Remediation of soils contaminated with polycyclic aromatic hydrocarbons (PAHs). J Hazard Mater 172:532–549

Gianfreda L, Rao MA (2004) Potential of extra cellular enzymes in remediation of polluted soils: a review. Enzyme Microb Technol 35:339–354

Haritash AK, Kaushik CP (2009) Biodegradation aspects of polycyclic aromatic hydrocarbons (PAHs): a review. J Hazard Mater 169:1–15

Huang ST, Tzean SS, Tsai BY, Hsieh HJ (2009) Cloning and heterologous expression of a novel ligninolytic peroxidase gene from poroid brown-rot fungus *Antrodia cinnamomea*. Microbiology 155:424–433

Karapetyan KN, Fedorova TV, Vasilchenko LG, Ludwig R, Haltrich D, Rabinovich ML (2006) Properties of neutral cellobiose dehydrogenase from the ascomycete *Chaetomium* sp. INBI 2-26(-) and comparison with basidiomycetous cellobiose dehydrogenases. J Biotechnol 121:34–48

Knutson K, Kirzan S, Ragauskas A (2005) Enzymatic biobleaching of two recalcitrant paper dyes with horseradish and soybean peroxidase. Biotechnol Lett 27:753–758

Lebrun JD, Trinsoutrot-Gattin I, Laval K, Mougin C (2010) Insights into the development of fungal biomarkers for metal ecotoxicity assessment: case of *Trametes versicolor* exposed to copper. Environ Toxicol Chem 29:902–908

Lebrun JD, Demont-Caulet N, Cheviron N, Laval K, Trinsoutrot-Gattin I, Mougin C (2011a) Secretion profiles of fungi as potential tools for metal ecotoxicity assessment: a study of enzymatic system in *Trametes versicolor*. Chemosphere 82:340–345

Lebrun JD, Lamy I, Mougin C (2011b) Favouring the bioavailability of Zn and Cu to enhance the production of lignin-modifying enzymes in *Trametes versicolor* cultures. Bioresour Technol 102:3103–3109

Lors C, Perie F, Grand C, Damidot D (2009) Benefits of ecotoxicological bioassays in the evaluation of a field biotreatment of PAHs polluted soil. Global Nest J 11:251–259

Mougin C (2002) Bioremediation and phytoremediation of industrial PAH-polluted soils. Polycycl Aromat Comp 22:1011–1043

Mougin C, Jolivalt C, Malosse C, Sigoillot JC, Asther M, Chaplain V (2002a) Interference of soil contaminants with laccase activity during the transformation of complex mixtures of polycyclic aromatic hydrocarbons (PAH) in liquid media. Polycycl Aromat Comp 22:673–688

Mougin C, Kollmann A, Jolivalt C (2002b) Enhanced production of laccases by the fungus *Trametes versicolor* by the addition of xenobiotics. Biotechnol Lett 24:139–142

Mougin C, Jolivalt C, Briozzo P, Madzak C (2003) Fungal laccases: from structure-activity studies to environmental applications. Environ Chem Lett 1:145–148

Mougin C, Boukcim H, Jolivalt C (2009) Bioremediation strategies based on fungal enzymes as tools. In: Singh A, Kuhad RC, Ward OP (eds) Advances in applied bioremediation, Soil biology series. Springer, Heidelberg, pp 123–149, Chapter 7

Novotny C, Cajthaml T, Svobodova K, Susla M, Sasek V (2009) *Irpex lacteus*, a white-rot fungus with biotechnological potential – review. Folia Microbiol 54:375–390

Pecyna MJ, Ullrich R, Bittner B, Clemens A, Scheibner K, Schubert R, Hofrichter M (2009) Molecular characterization of aromatic peroxygenase from *Agrocybe aegerita*. Appl Microbiol Biotechnol 84:885–897

Penning TM, BurcZynski ME, Hung CF, McCoull KD, Palackal NT, Tsuruda LS (1999) Dihydrodiol dehydrogenases and polycyclic aromatic hydrocarbon activation: generation of reactive and redox active o-quinones. Chem Res Toxicol 12:1–18

Rama R, Sigoillot J-C, Chaplain V, Asther M, Mougin C (2001) Inoculation of filamentous fungi in manufactured gas plant site soils and PAH transformation. Polycycl Aromat Comp 18:397–414

Chapter 19
Bioremediation and Genetically Modified Organisms

Morad Jafari, Younes Rezaee Danesh, Ebrahim Mohammadi Goltapeh, and Ajit Varma

19.1 Introduction: Biotechnology: A Forceful Path for Bioremediation

The resultant accumulations of the various organic chemicals in the environment, particularly in soil, are of significant concern because of their toxicity, including their carcinogenicity, and also because of their potential to bioaccumulate in living systems. A wide variety of nitrogen-containing industrial chemicals are produced for use in petroleum products, dyes, polymers, pesticides, explosives, and pharmaceuticals. Many of these chemicals are toxic and threaten human health and are classified as hazardous by the various world organizations related to environment protection such as United States Environmental Protection Agency. The conventional remediation technologies (other than bioremediation) used for in situ and ex situ remediation are typically expensive and destructive. Biotechnological processes for the bioremediation of chemical pollutants offer the possibility of in situ treatments and are mostly based on the natural activities of microorganisms. Biotechnological processes to destroy hazardous wastes offer many advantages over physicochemical processes.

M. Jafari (✉)
Department of Agronomy and Plant Breeding, Faculty of Agriculture, Urmia University, Urmia, Iran
e-mail: m.jafari@urmia.ac.ir

Y.R. Danesh
Department of Plant Protection, Faculty of Agriculture, Urmia University, Urmia, Iran
e-mail: Y.rdanesh@urmia.ac.ir; Younes_rd@yahoo.com

E.M. Goltapeh
Department of Plant Pathology, College of Agriculture, Tarbiat Modares University, Tehran, Iran
e-mail: emgoltapeh@modares.ac.ir

A. Varma
Amity Institute of Microbial Technology (AIMT), Amity University Uttar Pradesh, Noida, Uttar Pradesh, India
e-mail: ajitvarma@amity.edu

E.M. Goltapeh et al. (eds.), *Fungi as Bioremediators*, Soil Biology 32,
DOI 10.1007/978-3-642-33811-3_19, © Springer-Verlag Berlin Heidelberg 2013

When successfully operated, biotechnological processes may achieve complete destruction of organic wastes (Iwamoto and Nasu 2001).

Interest in bioremediation of polluted soil and water has increased in the last two decades primarily because it was recognized that organisms such as microbes were able to degrade toxic xenobiotic compounds which were earlier believed to be resistant to the natural biological processes occurring in the soil. Microbial activity in soils accounts for most of the degradation of organic contaminants. However, information about chemical and physical mechanisms can also useful to identification of significant transformation pathways for these compounds (Singh et al. 2009).

Biotechnology has the potential to play an immense role in the development of treatment processes for contaminated soil. As with any microbial process, optimizing the environmental conditions in bioremediation processes is a central goal in order that the microbial, physiological, and biochemical activities are directed toward biodegradation of the target contaminants. However, an important factor limiting the bioremediation of sites contaminated with certain hazardous compounds is the slow rate of degradation (Iwamoto and Nasu 2001). This slow degradation rate often limits the practicality of using microorganisms in remediating contaminated sites. This is an area where genetic engineering can make a marked improvement.

Recombinant DNA techniques have been studied intensively to improve the degradation of hazardous wastes under laboratory conditions. With advances in biotechnology, bioremediation has become a rapidly growing area and has been commercially applied for the treatment of hazardous wastes and contaminated sites (Dua et al. 2002). A center and a database have been established on biocatalysis and biodegradation (http://umbbd.ahc.umn.edu).

19.2   Genetic Engineering of Organisms for Bioremediation

Organisms can be supplemented with additional genetic properties for the biodegradation of specific pollutants if naturally occurring organisms are not able to do that job properly or not quickly enough. The combination of microbiological and ecological knowledge, biochemical mechanisms, combining different metabolic abilities of organisms, and manipulation of pivotal genetic factors influence in biodegradation and bioprocessing of mankind pollutions, bottlenecks in environmental cleanup may be circumvented. Using genetic engineering techniques for development of a new organism with beneficial properties applicable in bioremediation can be classified into two main categories that are separately discussed as follow:

19.2.1 Genetic Engineering of Microorganisms

The key players in bioremediation are microorganisms that live virtually every-where. They are ideally suited to the task of contaminant destruction, because they possess enzymes that allow them to use as environmental contaminants as food, and because they are so small that they are able to contact contaminants easily. Genetically engineered microorganisms (GEMs) have shown potential in applications for bioremediation in soil, groundwater, and activated sludge environments, due to the enhanced degradative capabilities of a wide range of contaminants (Menn et al. 2008). Recent advances in molecular biology have opened up new perspectives to progress in engineering microorganisms with the aim of performing specific bioremediation.

19.2.1.1 Bacterial Engineering for Bioremediation Purposes

Bacteria possess a high potential force for degradation of environmental pollu-tants. Microorganisms that can degrade various pollutants (e.g., nitroaromatics, chloroaromatics, polycyclic aromatics, biphenyls, polychlorinated biphenyls (PCBs), and components of oil have been isolated with the eventual goal of exploiting their metabolic potential for the bioremediation of contaminated sites (Samanta et al. 2002; Parales and Haddock 2004). However, some of the more recalcitrant and toxic xenobi-otic compounds, such as highly nitrated and halogenated aromatic compounds, as well as some pesticides and explosives, are usually stable, chemically inert under natural conditions, and not known to be degraded efficiently by many microorganisms (Parrilli et al. 2010). Also, the toxicity of some of these organic pollutants to the existing microbial populations, coupled with complications caused by mixtures of pollutants, is a major hindrance to successful biodegradation by microbes. These limitations to bioremediation have paved the way for the development of GEMs or "designer biocatalysts" that contain artificially designed catabolic pathways (Paul et al. 2005).

There are several strategies to enhance capability of bacteria for efficient biodegradation that we described briefly some of these below:

Optimizing Biocatalysts

The construction of an optimized "biocatalyst" requires a bank of genetic modules that encode desired properties that can be combined to generate novel, improved, and efficient degradation activities. So far, several microorganisms have been modified to make them potent biocatalysts. With the aim of treating a site contaminated with various polychlorinated biphenyls, for example, genetic engineering has been used to alter the substrate specificity of a biphenyl dioxygenase enzyme involved in PCB degradation in *Pseudomonas* sp. LB400 and *Pseudomonas alcaligenes* KF707

(Kimura et al. 1997). Variants of the enzyme "biphenyl dioxygenase" were created by combining the substrate range of the enzyme obtained from both of these organisms, so that the variants could hydroxylate both double ortho- and double para-substituted PCBs. One strategy for designing superior biocatalysts is the rational combination of catabolic segments from different organisms within one recipient strain. Thereby, complete metabolic routes for xenobiotics, which are only co-metabolized, can be generated and the formation of dead-end products or even toxic metabolites can be avoided. This strategy has been applied successfully for the degradation of highly toxic trihalopropanes, for which mineralization has not yet been described (Bosma et al. 1999).

Protein Engineering

Protein engineering can be exploited to improve an enzyme's stability, substrate specificity and kinetic properties. Rational design of proteins performed by site-directed mutagenesis requires an understanding of structure–function relationships in the molecule and therefore a detailed knowledge of the three-dimensional structure of the enzyme itself (Schanstra et al. 1996), or of at least one member of the protein family, to allow the structure of the protein under study to be modeled. However, the number of degradative enzymes whose structure has been elucidated is still small, and this constitutes a major limitation for rational protein design. Where phenotypic selection of desired variants is possible, rare spontaneous or induced mutants may be readily obtained; where phenotypic selection of variants is not possible, other, more-efficient approaches are needed (Timmis and Pieper 1999).

One approach to combining the best attributes of related enzymes is to exchange subunits or subunit sequences. For example, enzyme variants with superior trichloroethylene (TCE)-transformation kinetics were obtained by exchanging subunits between the multicomponent toluene and biphenyl dioxygenases (Furukawa et al. 1994). Further experiments to exchange domains of the subunits of biphenyl or (chloro) benzene dioxygenases exhibiting different substrate specificities resulted in chimeric enzymes with broader substrate specificities than the parental enzymes 40–43. A recombinant *E. coli* strain was genetically engineered to coproduce OPH and carboxylesterase B1 for the simultaneous degradation of organophosphorus, carbamate, and pyrethroid classes of pesticides (Lan et al. 2006). However, E. coli strains are not suitable for in situ remediation since they are not adapted to these environments. A more realistic approach is to engineer soil bacteria that are known to survive in contaminated environments for an extended period. A new study for expanding the substrate range of enzymes recently was reported by Yang et al. (2010). In this research, *Stenotrophomonas* sp. strain YC-1, a native soil bacterium was genetically engineered to produce organophosphorus hydrolase (OPH) enzyme with broader substrate range for organophosphates (OPs). A mixture of six Synthetic organophosphate pesticides could be degraded completely within 5 h. The broader substrate specificity in combination with the rapid degradation rate makes this engineered strain a promising candidate for in situ remediation of OP-contaminated sites. A recently developed and powerful alternative method for obtaining proteins with new activities involves shuffling their gene sequences (Crameri et al. 1997; Harayama 1998). By random shuffling of DNA segments between the large subunit of two wild-type biphenyl dioxygenases, variants were obtained with extended substrate range of biphenyl dioxygenases toward PCBs (Kumamaru et al. 1998; Bruhlmann and Chen 1999). DNA-shuffling methods (i.e., the random fragmentation of a population of mutant genes of a certain family followed by random reassembly) have been developed, which allow the creation of a vast range of chimeric proteins and protein variants for biodegradation applications (Pieper and Reineke 2000).

Biosurfactants Production for Bioavailability of Xenobiotics

One of the main reasons for the prolonged persistence of hydrophobic organic compounds in the environment is their solubilization-limited bioavailability. Surfactants can improve the accessibility of these substrates to microbial attack. Almost all the industrially produced surfactants are chemically derived from petroleum and require both synthesis and several purification steps, rendering the process costly and liable for contamination with unknown hazards (Sitohy et al. 2010). A possible way to enhance bioavailability of xenobiotics (pesticides, pharmaceuticals, petroleum compounds, polycyclic aromatic hydrocarbons, PCBs etc.) and, thereby, their biodegradation is the application of "biosurfactants" (natural surfactants of microbial origin). These molecules consist of both a hydrophilic and hydrophobic (Pieper and Reineke 2000) and the high surface activity, heat and pH stability, low toxicity, ecological acceptability, and biodegradability of them constitute important advantages over synthetic surfactants (Timmis and Pieper 1999). So, they may be recommended to replace the presently used chemically synthesized. Despite the fact that application of biosurfactants has been shown potentially to increase the degradation rate of hydrophobic pollutants, the high cost of biosurfactant production restricts their application. Current efforts are therefore directed toward the design of recombinant biocatalysts that exhibit a desired catabolic trait and produce a suitable biosurfactant (Sullivan 1998). Also, the combination of surfactant production with degradative capabilities in a "single bacterial strain" will offer advances for in situ bioremediation, but further insights into the genetic organization and regulation of surfactant production are needed (Gallardo et al. 1997). For example, to improve the biodesulfurization process, Gallardo et al. (1997) designed a recombinant biocatalyst that combines the Dsz phenotype with potential interest for the production of biosurfactants. They developed a recombinant bacterium *Pseudomonas aeruginosa* PG201 that were able to desulfurize dibenzothiophene more efficiently than the native host. These new biocatalysts combine relevant industrial and environmental traits, such as production of biosurfactants, with the enhanced biodesulfurization phenotype.

Completing Pathways for Fully Degradation of a Substrate

In some cases, although a complete pathway for a particular substrate may not exist in a single organism, partial and complementary pathway segments may exist in different organisms. The development of an organism exhibiting a desired catabolic phenotype may therefore require the combination of determinants for complementary pathway segments in order to form a complete pathway sequence for a target substrate (Timmis and Pieper 1999). Complete metabolic pathways may be needed due to: First, co-metabolic processes need an input of energy and therefore represent a metabolic burden for the microorganism. Second, the end metabolites produced by incomplete pathways may be toxic or subject to further transformations by other microorganisms, forming reactive or toxic molecules. One example of this is found in PCBs metabolism, in which microorganisms usually metabolize only one aromatic ring and accumulate the others as the corresponding chlorobenzoates, which have been shown to be inhibitory (Stratford et al. 1996).

19.2.1.2 Genetically Engineered Fungi for Mycoremediation

Recent advances in molecular biology, biotechnology, and enzymology are the driving forces toward engineer-improved fungi and enzymes for mycoremediation. The ease of genetic engineering, transportation, and scaling-up makes fungi the organisms of choice in bioremediation (Obire et al. 2008). A number of the genetic engineering approaches that have been developed have proven beneficial in adding the desired qualities in metabolic pathways or enzymes. Strain manipulation is becoming easier with the exponential expansion of molecular tool boxes and genome sequences. However, the best source is that of the genes of fungi, where mycotransformation is well understood. Specific gene alterations can be designed and controlled via metabolic engineering. Metabolic control is shared by enzymes (i.e., enzymes are democratic). Mathematical modeling of metabolic control analysis can be used to make predictions as to how metabolic pathways will respond to manipulation. Fungal genes can be cloned to meet the objectives of mycoremediation. Fungal mutants that oversecrete specific enzymes can be produced, and various processes using such mutants may be designed and scaled up in the treatment of wastes and wastewaters. Fungal protoplasts can be exploited to enhance processes related to mycoremediation. At present, efforts to increase flux through specific pathways have met with limited success. Potentially, the future of metabolic engineering is bright, but there is still a long way to go to understand this area of the metabolic network before the introduction of bioengineered yeast or fungi in the field of mycoremediation. Recent advances in biotechnology can open the door for the development of genes responsible for the mineralization of PCBs by fungi. Genes encoding Lignin peroxidase in 30 fungal species have been screened that may open new frontiers for the degradation of PCBs. A dendrogram illustrating a sequence relationship among 32 fungal peroxidases has been presented. A great future lies in successful genetic splicing and bringing together pathway fragments with a view to constructing an entirely new white-rot fungus that can utilize PCBs as the sole source of carbon (Harbhajan 2006). The first complete eukaryotic genome belongs to the yeast *Saccharomyces cerevisiae* (Dujon 1996). The genome sequence has laid a strong foundation for work in the disciplines of agriculture, industry, medicine, and remediation. In a paper for fungal comparative genomics, the Fungal Genome Initiative (FGI) Steering Committee identified a coherent set of 44 fungi as immediate targets for sequencing (Birren et al. 2003). Several projects have released information on the genome sequences of the yeasts *Schizosaccharomyces pombe* and *Candida albicans* and the filamentous fungi *Aspergillus nidulans*, *Aspergillus fumigatus*, *Neurospora crassa*, and *Coprinus cinereus*. The 13.8 million base pair genome of *S. pombe* consists of 4,940 protein coding genes, including mitochondrial genome and genes (Wood et al. 2002). Ten thousand genes are predicted in the 40-Mb genome in the sequence of the first filamentous fungus, *N. crassa* (Galagan et al. 2003). The 30 million base pair genome of the first basidiomycete, *Phanerochaete chrysosporium* strain RP78, has been sequenced using a whole-genome shotgun approach (Martinez et al. 2006). The genome reveals genes encoding oxidases, peroxidases, and hydrolytic enzymes involved in wood decay. This opens up new horizons related to the process of biodegradation of lignin and organopullutants and in the area of mycoremediation. Recently, yeast has been engineered with a binding affinity to cellulose (Nam et al. 2002). Genes encoding the cellulose binding domain (CBD) from cellobiohydrolase I (CBHI) and cellobiohydrolase II (CBHII) of *Trichoderma reesei* have been expressed on the cell surface of *Saccharomyces cerevisiae*.

Unlike bacteria, the role of biotechnological innovations related to biodegradation by fungi is relatively less well understood. Moreover, bacteria and fungi exhibit different mechanisms in the biodegradation of pollutants such as pesticides. Significant progress has been achieved in molecular biology related to fungi, especially related to the extraction of genetic material (RNA and DNA), gene cloning, and genetic engineering of fungi. The development of biotechnology for using white rot fungi for environmental pollution control has been implemented to treat various refractory wastes and to remediate contaminated soils (Gao et al. 2010).

19.2.1.3 Genetic Engineering of Plants for Phytoremediation

When microorganisms are used for remediation of xenobiotics, both inoculation of microorganisms and nutrient application are essential for their maintenance at adequate levels over long periods (Eapen et al. 2007). Besides, the microbes which show highly efficient biodegradation capabilities under laboratory conditions may not perform equally well at actual contaminated sites (Macek et al. 2008). However, there are two main problems with the introduction of transgenic microorganisms: the bureaucratic barriers blocking their release into the environment and the poor survival rate of those engineered strains that have been introduced into the contaminated soil (Suresh and Ravishankar 2004; Abhilash et al. 2009). In comparison, phytoremediation is easier to manage, because it is an autotrophic system of large biomass that requires little nutrient input. Moreover, plants offer protection against water and wind erosion, preventing contaminants from spreading; also they are robust in growth, are a renewable resource and can be used for in situ remediation. So, phytoremediation for removal of xenobiotics can be an alternate/supplementary method (Suresh and Ravishankar 2004; Abhilash et al. 2009).

Although much research has been done to demonstrate the success of phytore-mediation, resulting in its use on many contaminated sites (Abhilash et al. 2009), the method still lacks wide application. Further, detoxification of organic pollutants by plants is often slow, leading to the accumulation of toxic compounds in plants that could be later released into the environment (Aken 2008). The question of how to dispose of plants that accumulate organic pollutants is also a serious concern. A direct method for enhancing the effectiveness of phytoremediation is to overexpress in transgenic plants the genes involved in metabolism, uptake, or transport of specific pollutants (Cherian and Oliveira 2005; Doty 2008; Aken 2008; Macek et al. 2008). Besides, being autotrophic organisms, plants do not actually use organic compounds for their energy and carbon metabolism. As a consequence, they usually lack the catabolic enzymes necessary to achieve full mineralization of organic molecules, potentially resulting in the accumulation of toxic metabolites, and they lack xenobiotic degradative capabilities of bacteria. Hence, introduction of genes for degradation of xenobiotics from microbes or other eukaryotes such as mammals, where the potential already exist to plants will further enhance their ability to degrade/mineralize the recalcitrant contaminants (Eapen et al. 2007).

Genes involved in degradation of xenobiotic pollutants can be isolated from bacteria/ fungi/animals/plants and introduced into candidate plants using *Agrobacterium* mediated or direct DNA methods of gene transfer. Transgenic plants for phytore-mediation were first developed for remediating heavy metal contaminated soil sites (Misra and Gedamu 1989; Rugh et al. 1996). The first attempt to develop engineered plants for phytoremediation of organic pollutants targeted explosives and halogenated organic compounds in tobacco plants (Doty et al. 2000). These plants have been developed to contain either transgenes responsible for the metabolization of xenobiotics or transgenes that result in the increased resistance of pollutants (Abhilash et al. 2009).

Phytoremediation is a broad term that comprises several techniques used for water and soil decontamination. Thus, we will focus on some main applications of transgenic plants for phytoremediation.

Transgenic Plants for Remediation of Toxic Explosives

Best known for their explosives properties, nitro-substituted compounds, such as 2,4,6-trinitrotoluene (TNT), hexahydro-1,3,5-trinitro-1,3,5-triazine (RDX), and glycerol trinitrate (GTN), are also toxic and persistent environmental pollutants contaminating numerous military sites. Manufacture of explosives, testing and firing on military ranges, and decommission of ammunition stocks have generated toxic wastes, leading to large-scale contamination of soils and groundwater. From laboratory studies, most nitro-substituted explosives were found to be toxic for all classes of organisms, including bacteria, algae, plants, invertebrates, and mammals (Talmage et al. 1999). Traditional remediation of explosive-contaminated sites requires soil excavation before treatment by incineration or land-filling, which is costly, damaging for the environment, and, in many cases, practically infeasible owing to the range of contamination (Hannink et al. 2002).

Although the microbial catabolic pathways leading to the complete mineralization of explosives are yet to be revealed, it is generally accepted that these compounds can be transformed into various intermediates in wide range of microorganisms by various enzymes (Ramos et al. 2005).

Despite promising observations about ability of plants for metabolization of explosive pollutants, the application, however, may be limited by the fact that the indigenous biodegradability of plants is less effective than those of adapted microorganisms. As explosives are phytotoxic, phytoremediation of these pollutants is very difficult. This limitation might be overcome by incorporating bacterial nitroreductase genes into the plant genomes (French et al. 1999; Rosser et al. 2001). French et al. (1999) introduced pentaerythritol tetranitrate (PETNr) (a monomeric flavin mononucleotide (FMN)-containing protein) reductase into *Nicotiana tabacum*, resulting in increased tolerance TNT. Furthermore, tobacco plants expressing PETNr were able to germinate and grow naturally on solid media containing 1 mM GTN, a concentration that would be lethal to non-transgenic plants. The catabolic fingerprinting in TNT degrading bacterium *Enterobacter cloacae* reveals that this step was shown to be catalyzed by an FMN containing nitroreductase enzyme (NR). This NR enzyme can transform TNT significantly faster than PETNr, and when expressed in transgenic plants, NR also confers greater tolerance to TNT than PETNr (Hannink et al. 2001; Rylott and Bruce 2009). The overexpression of this NR gene in transgenic tobacco resulted in the enhanced tolerance to TNT contamination.

Recently, Van Dillewijn et al. (2008) developed a transgenic aspen incorporated with a nitroreductase, pseudomonas nitroreductase A (pnrA), isolated from the bacterium *Pseudomonas putida* for the enhanced degradation of TNT. When compared with the non-transgenic plants, the transgenic trees were able to take up higher levels of TNT from liquid culture and soil. Latest studies revealed that overexpression of two of the uridine diphosphate (UDP) glycosyltrasferases (UGTs) (743B4 and 73C1 isolated from *Arabidopsis thaliana*) genes in Arabidopsis thaliana resulted in increased conjugate production and enhanced root growth in 74B4 overexpression seedlings grown in liquid culture containing TNT (Gandia-Herrero et al. 2008).

Besides TNT, plants were also transformed for improving performances against RDX, which is today the most widely used military explosive. *A. thaliana* plants were engineered to express a bacterial gene, *XplA*, encoding a RDX-degrading fused flavodoxin-cytochrome P450-like enzyme (Rylott et al. 2006). The donor strain, *Rhodococcus rhodochrous* strain 11Y, originally isolated from RDX-contaminated soil (Seth-Smith et al. 2002). Liquid cultures of *A. thaliana* expressing *XplA* removed 32–100 % of RDX (initial concentration 180 μM), while less than 10 % was removed by wild-type plants. Some selected examples of transgenic plants developed for degradation of xenobiotics pollutants are listed in Table 19.1.

Transgenic Plants for Remediation of Heavy Metal

Actually, heavy metal (nonradioactive As, Cd, Co, Cu, Ni, Zn, and Cr and radioactive Sr, Cs, and U) pollution has become one of the most serious environmental problems today (Alkorta et al. 2004). For instance, arsenic, a nonessential metalloid, is an environmental pollutant of prime concern which is causing a global epidemic of poisoning, with tens of thousands of people having developed skin lesions, cancers, and other symptoms (Pearce 2003; Alkorta et al. 2004).

Although many studies have been carried out to investigate the possibility of using microorganisms to aid in the remediation of metal polluted environments, microorganisms do not solve the critical problem of the removal of metals from the polluted soil. As a matter of fact, bacteria can only transform metals from one oxidation state or organic complex to another but not extract them from the polluted soil (Garbisu et al. 2002). Fortunately, the possibility of using plants that can literally extract the metals from the polluted soil was raised. From accumulating high levels of metal and translocating it to the harvestable segments of the plant, a plant suitable for phytoextraction should grow rapidly and reach a high biomass.

Over 400 hyperaccumulator plants have been reported and include members of the Asteraceae, Brassicaceae, Caryophyllaceae, Cyperaceae, Cunouniaceae, Fabaceae, Flacourtiaceae, Lamiaceae, Poaceae, Violaceae, and Euphobiaceae. The Brassicaceae is a very important hyperaccumulator group (Baker and Brooks 1989). The importance of improvement of metal uptake by breeding or genetic modification can be illustrated by the fact that more important and interesting reviews on engineering of GM plants suitable for metal accumulation appeared simultaneously (Clemens et al. 2002). Several metal homeostasis genes are constitutively expressed at very high levels in metal hyperaccumulators, when compared with closely related nonaccumulators. In *Arabidopsis halleri*, these include genes encoding several membrane transporter proteins of the ZRT-IRT-related protein (ZIP) family (zinc-regulated transporter, iron-regulated transporter) (Guerinot 2000), which are likely to mediate zinc influx into the cytoplasm and two isoforms of the enzyme nicotianamine synthase. These genes are expressed at low levels or only upregulated under conditions of zinc deficiency in *A. thaliana*. Other genes found to be constitutively expressed at high levels in the hyperaccumulator species *A. halleri* encode membrane transport proteins of the heavy metal P-type ATPase (HMA) family of P1B-type metal ATPases, which are potentially involved in metal export into the apoplast for metal detoxification or for root-to-shoot metal translocation in the xylem (Kramer 2005).

Plants have a family of metallothionein (MTs) genes encoding cysteine-rich peptides that are generally composed of 60–80 amino acids and contain 9–16 cysteine residues (Chatthai et al. 1997). MTs can protect plants from effects of toxic metal ions such as Ag, Cd, Co, Cu, Hg, and Ni. Metallothionein genes have

Gene	Gene product	Source	Plant	Target pollutant	References
NfsI	Nitroreductase	*Enterobacter cloacae*	*N. tabacum*	TNT	Hannink et al. (2001, 2007)
XplA, XplB	Cytochrome P450 monooxygenase	*Rhodococcus rhodochorus*	*A. thaliana*	RDX	Jackson et al. (2007)
Onr	Pentaerythritol tetranitrate reductase (PETNr)	*Enterobacter cloaceae*	*N. tabaccum*	GTN, TNT	French et al. (1999)
NfsA	Nitroreductase	*E. coli*	*A. thaliana*	TNT	Kurumata et al. (2005)
pnrA	Nitroreductase	*Pseudomonas putida*	*P. tremula × P. tremuloides*	TNT	Van Dillewijn et al. (2008)
743B4, 73C1	Glycosyltransferases	*A. thaliana*	*A. thaliana*	TNT	Gandia-Herrero et al. (2008)
merP	Hg^{2+}-binding protein MerP	*Bacillus megaterium*	*A. thaliana*	Hg^{2+}	Hsieh et al. (2009)
TaPCS1	Phytochelatins synthase	*T. aestivum*	*N. glauca*	Pb^{2+}, Cd^{2+}	Gisbert et al. (2003) and Martinez et al. (2006)
gshI	γ-Glutamylcysteine synthetase	*E. coli*	*Indian mustard*	Cd	Zhu et al. (1999b)
gshII	Glutathione synthetase	*Oriza sativa*	*Indian mustard*	Cd	Zhu et al. (1999a)
CAX-2	Vacuolar transporters	*A. thaliana*	*N. tabaccum*	Cd, Ca, Mn	Hirschi et al. (2000)
TnMERI1	mercuric ion binding protein (MerP)	*Bacillus megaterium* strain MB1	*A. thaliana*	Hg	Hsieh et al. (2009)
CYP1A1, CYP2B6, CYP2C19	Cytochrome P450 monooxygenase	*H. sapiens*	*Oryza sativa*	Herbicide (atrazine, metolachlor)	Kawahigashi et al. (2006)
CYP1A1, CYP2B6, CYP2C9, CYP2C19	Cytochrome P450 monooxygenase	*H. sapiens*	*Solanum tuberosum, Oryza sativa*	Sulfonylurea and other herbicides	Inui and Ohkawa (2005)
GstI-6His	Glutathione S-transferases	*Zea mays*	*N. tabaccum*	Alachlor	Karavangeli et al. (2005)
atzA	Atrazine chlorohydrolase	*Bacteria*	*Medicago sativa, N. tabaccum*	Atrazine	Wang et al. (2005)

been cloned and introduced into several plant species. Transfer of human MT-2 gene in tobacco or oil seed rape resulted in plants with enhanced Cd tolerance (Misra and Gedamu 1989) and pea MT gene in Arabidopsis thaliana enhanced Cu accumulation (Evans et al. 1992). The choice of promoter used was found to be of great importance for metallothionein genes. The ribulose biphosphate carboxylase (rbcs) promoter was repressed by high Cd concentration while mannose synthase promoter was induced by Cd (Stefanov et al. 1997).

Phytochelatins are another group of metal binding proteins and are involved in heavy metal sequestration that are non-translationally synthesized from reduced glutathione. Phytochelatins complex with metals and help in storage in vacuoles (Cobbett 2000). Genetic engineering of plants for synthesis of metal chelators will improve the capability of plant for metal uptake (Pilon-Smits and Pilon 2002; Clemens et al. 2002). Transgenic *B. juncea* overexpressing different enzymes involved in phytochelatin synthesis were shown to extract more Cd, Cr, Cu, Pb, and Zn than wild plants (Zhu et al. 1999a,b) Transgenics engineered to have higher levels of metal chelators showed enhanced cadmium and zinc accumulation in greenhouse experiments using polluted soil (Bennett et al. 2003). Also, transgenic plants engineered to have enhanced sulfate/selenate reduction showed fivefold higher selenium accumulation in the field (Banuelos et al. 2005). The constitutive overexpression of phytochelatin synthase of *Triticum aestivum* (TaPCS1) in shrub *Nicotiana glaucum* substantially increased its tolerance to Pb^{2+} and Cd^{2+} and greatly improved accumulation of Cu^{2+}, Zn^{2+}, Pb^{2+} and Cd^{2+} in shoots (Gisbert et al. 2003; Martinez et al. 2006). The overexpressed gene conferred up to 36 and 9 times more Pb^{2+} and Cd^{2+} accumulation, respectively, in shoots of the transgenic line NgTP1 under hydroponic conditions, reflected in the increased accumulation of these metals from mining soil. Hsieh et al. (2009) reported an increase in mercury (Hg) accumulation and tolerance of *A. thaliana* when mercuric ion binding protein (MerP), originated from transposon TnMERI1 of transposon TnMERI1 *Bacillus megaterium* strain MB1, was expressed in the transgenic plants. Table 19.1 shows instance of genes, which have been used for the development of transgenic plants for phytoremediation of toxic metals.

Transgenic Plants for Enhanced Remediation of Herbicides

Herbicides are economically important, because they prevent losses in crop yield due to weed infestation (Kawahigashi et al. 2008). However, the overuse and repeated use of same herbicide can lead to the development of herbicide resistant weeds. According to the Weed Science Society of America, over 310 biotypes of herbicide resistant weeds have been reported in agricultural fields and gardens worldwide. As a result of these herbicide tolerance, larger amount of herbicides are needed to kill these weeds, so that residues contaminate the soil and nearby water bodies. Plants used for decontamination of these contaminated system should be resistant to herbicides.

Among the various enzymatic groups, cytochrome P450 plays a major role in the enhanced degradation of herbicides. Cytochrome P450 enzymes comprise a super-family of heme proteins crucial for the oxidative, peroxidative, and reductive metabolism of a diverse group of compounds, including endobiotics, such as steroids, bile acids, fatty acids, prostaglandins, and leukotrienes, and xenobiotics, including most of the therapeutic drugs and environmental pollutants (Abhilash et al. 2009). In almost all living organisms, these enzymes are present in more than one form, thus forming one of the largest families of enzymes. The enzyme system is located in microsomes and consists of several cytochrome P450 isoforms. Although cytochrome P450 (P450 or CYP) monooxygenases in higher plants play an important role in the oxidative metabolism of endogenous and exogenous liphophilic compounds (Eapen et al. 2007; Doty 2008), molecular information on P450 species metabolizing xenobiotics in plants is quite limited.

Humans have been estimated to have at least 53 different CYP genes and 24 pseudogenes. So far, it has been reported that 11 P450 species (Abhilash et al. 2009). A study of 11 human P450s in the CYP1, 2, and 3 families using a recombinant yeast expressing system showed that they can metabolize 27 herbicides and 4 insecticides (Inui and Ohkawa 2005). Further, another study conducted by same research group found that human CYP1A1 metabolized 16 herbicides, including triazines, ureas, and carbamates, and CYP2B6 metabolized more than 10 herbicides, including chloroace-tanilides, oxyacetamides, and 2,6-dinitroanilines, three insecticides, and two industrial chemicals. In recent years, some crop plants were also genetically engineered with mammalian P450 cytochrome genes to confer herbicide resistance. Rice is a good candidate for metabolizing herbicides and reducing the load of herbicides in paddy fields and streams. The expression of mammalian cytochrome P450 genes in transgenic potatoes and rice plants has been used to detoxify herbicides (Inui and Ohkawa 2005). Several cytochrome P450 genes such as CYP1A1, CYP2B6 and CYP2C19, when introduced into rice plants, showed tolerance to herbicide atrazine, metolachlor, and norfluazon and could decrease the amount of herbicides, owing to increased metabolism by the introduced P450 enzymes (Kawahigashi et al. 2005).

As with P450s, overexpression of glutathione *S*-transferases (GST) genes enhances the potential for phytoremediation of herbicides. Glutathione-S-transferases catalyze nucleophilic attack of the sulfur atom of glutathione on electrophilic group of a variety of hydrophobic substrates, which include herbicides such as chloroace-tanilides and triazine. Transgenic tobacco plants overexpressing maize GST was shown to remediate chloroacetanilide herbicide—alachlor (Karavangeli et al. 2005). In addition to the approaches involving P450 and GST genes, various transgenic plants that exhibit herbicide tolerance can be used for phytoremediation. Transgenic alfalfa, tobacco, and *Arabidopsis* plants expressing a bacterial atrazine chlorohydrolase (*atzA*) gene show enhanced metabolic activity against atrazine—a widely used herbicide (Wang et al. 2005). Transgenic tobacco plants expressing the Mn peroxidase gene from *Coriolus versicolor* reduced pentachlorophenol (PCP) in the culture media with high efficiency (Iimura et al. 2002). Some selected transgenic plants developed to bioremediation of herbicides are given in Table 19.1.

19.2.2   A Glance on Anticipated Risks of Genetically Modified Organisms

First, it should be noted that from the biosafety viewpoint, not all naturally occurring soil bacteria are ideal as bioremediation agents. For example, *Burkholderia cepacia* has potential as an agent for bioremediation and for biological control of phytopathogens. However, it is a human pathogen known to be involved in cystic fibrosis and it is resistant to multiple antibiotics (Holmes et al. 1998). This has led to rejection by the US Environmental Protection Agency (EPA) of its use as an environmental agent (Davison 2005). For transgenic modified microorganisms, many authorities are particularly reluctant to authorize the release of them (Sayler and Sayre 1995). The transgene is usually derived from another soil microorganism, thus no new gene is added to the soil microbial community. It is also very probable that the introduced engineered strain will not survive for long in the soil environment; at least not long after its specific substrate is exhausted. However, several cutting-edged strategies have developed for mitigation of probably risks of genetically modified organisms (for details please see Davison 2005; Pandey et al. 2005) to achieve efficient and safer bioremediation of contaminated sites.

A great deal has been written about the potential and imagined risks of transgenic plants for agricultural use (Davison 2005; Singh et al. 2006) and much, but not all, of it applies to transgenic plants for use for phytoremediation. It seems unlikely, at least in the short term, that transgenic phytoremediation plants will contain herbicide resistance, insect resistance, and virus resistance genes, which have been major subjects of biosafety discussions. In addition, phytoremediation plants will not be intended as human or animal foods, so that food safety, allergenicity, and labeling are not relevant issues and the degraded products should be less dangerous compared to the parent pollutant. Finally, on a more optimistic note, phytoremediation is generally seen as posing fewer biosafety concerns. However, the methods for mitigation of the potential risks of transgenic phytoremediating plants have been described by several research papers (for more details see Davison 2005; Gressel and Al-Ahmad 2005; Kotrba et al. 2009).

19.3   Conclusion and Future Perspectives

Among the top ten biotechnologies for improving human health, bioremediation is recognized as one of the technologies (Eapen et al. 2007). The application of molecular-biology-based techniques in bioremediation is being increasingly used and has provided useful information for improving of bioremediation strategies. Furthermore, environmental metagenomic data from soil and sea can be a useful source of genes. Combinational approaches such as genome shuffling are also useful for generating new genes or modifying enzyme activities to allow efficient bioremediation (Kawahigashi 2009). This new biotechnology approach will open exciting new vistas for enhancing bioremediation programs in the coming years.

Whereas bioremediation using transgenic bacteria seems presently to be in the doldrums, phytoremediation using transgenic plants could offer some new answers to environmental cleanup of toxic wastes. New genetic method risk-mitigation may help ensure that neither the transgenic plants, nor the transgenes they contain, will escape into the environment (Davison 2005). The potential of engineered phytoremediation plants should be demonstrated in field trials, some of which have emerged in the last few years. The ecological impact and underlying economics of phytoremediation with transgenics should be carefully evaluated and weighted against known disadvantages of conventional remediation techniques or risks of having the recalcitrant heavy metal or metalloid species in our environment (Kotrba et al. 2009).

In addition, the combination of plants for removing or degrading toxic pollutants and rhizospheric microorganisms for enhancing the availability of hydrophobic compounds can break down many types of toxic foreign chemicals, including herbicides. In view of the importance of mycorrhizal (macro) fungi in plant growth and particularly in the mobilization and cycling of elements in the soil, the colonization of contaminated soils with the suitable fungal species would be beneficial to promote bioavailability of the environmental pollutants. Gadd (2007) further demonstrated suitability of genetic engineering approach in constructing fungi with improved metalloresistance.

References

Abhilash PC, Jamil S, Singh N (2009) Transgenic plants for enhanced biodegradation and phytoremediation of organic xenobiotics. Biotechnol Adv 27:474–488

Aken BV (2008) Transgenic plants for phytoremediation: helping nature to clean up environmental pollution. Trends Biotechnol 26:225–227

Alkorta I, Hernández-Allica J, Becerril JM, Amezaga I, Albizu I, Garbisu C (2004) Recent findings on the phytoremediation of soils contaminated with environmentally toxic heavy metals and metalloids such as zinc, cadmium, lead and arsenic. Rev Environ Sci Biotechnol 3:71–90

Baker AJM, Brooks RR (1989) Terrestrial higher plants which hyperaccumulate metallic elements – a review of their distribution, ecology and phytochemistry. Biorecovery 1:81–126

Banuelos G, Terry N, LeDuc DL, Pilon-Smits EH, Mackey B (2005) Field trial of transgenic Indian mustard plants shows enhanced phytoremediation of selenium-contaminated sediment. Environ Sci Technol 39:1771–1777

Bennett LE, Burkhead JL, Hale KL, Terry N, Pilon M, Pilon-Smits EA (2003) Analysis of transgenic Indian mustard plants for phytoremediation of metal-contaminated mine tailings. J Environ Qual 32:432–440

Birren B, Fink G, Lander E (2003) Fungal genome initiative: a white paper for fungal comparative genomics. Center for Genome Research, Cambridge, MA

Bosma T, Kruzinga E, Bruin EJD, Poelarends GJ, Janssen DB (1999) Utilization of trihalogenated propanes by *Agrobacterium radiobacter* AD1 through heterologous expression of the haloalkane dehalogenase from *Rhodococcus* sp. strain M15-3. Appl Environ Microbiol 65:4575–4581

Bruhlmann F, Chen W (1999) Tuning biphenyl dioxygenase for extended substrate specificity. Biotechnol Bioeng 63:544–551

Chatthai M, Kaukinen KH, Tranbarger TJ, Gupta PK, Misra S (1997) The isolation of a novel metallothionein related cDNA expressed in somatic and zygotic embryos of Douglas fir: regulation of ABA, osmoticum and metal ions. Plant Mol Biol 34:243–254

Cherian S, Oliveira MM (2005) Transgenic plants in phytoremediation: recent advances and new possibilities. Environ Sci Technol 39:9377–9390

Clemens S, Palmgren M, Kraemer U (2002) A long way ahead: understanding and engineering plant metal accumulation. Trends Plant Sci 7:309–315

Cobbett CS (2000) Phytochelatins and their role in heavy metal detoxification. Plant Physiol 125:825–832

Crameri A, Dawes G, Rodriguez E, Silver S, Stemmer WP (1997) Molecular evolution of an arsenate detoxification pathway by DNA shuffling. Nat Biotechnol 15:436–438

Davison J (2005) Risk mitigation of genetically modified bacteria and plants designed for bioremediation. J Ind Microbiol Biotechnol 32:639–650

Doty SL (2008) Enhancing phytoremediation through the use of transgenic plants and entophytes. New Phytol 179:318–333

Doty SL, Shang QT, Wilson AM, Moore AL, Newman LA, Strand SE, Gordon MP (2000) Enhanced metabolism of halogenated hydrocarbons in transgenic plants contain mammalian P450 2E1. Proc Natal Acad Sci USA 97:6287–6291

Dua M, Singh A, Sethunathan N, Johri AK (2002) Biotechnology and bioremediation: successes and limitations. Appl Microbiol Biotechnol 59:143–152

Dujon B (1996) The yeast genome project: What did we learn? Trends Genet 12:263–270

Eapen S, Singh S, D'Souza SF (2007) Advances in development of transgenic plants for remediation of xenobiotic pollutants. Biotechnol Adv 25:442–451

Evans KM, Gatehouse JA, Lindsay WP, Shi J, Tommey AM, Robinson NJ (1992) Expression of the pea metallothionein like gene Ps MTA in *Escherichia coli* and *Arabidopsis thaliana* and analysis of trace metal ion accumulation:implications of Ps MTA function. Plant Mol Biol 20:1019–1028

French CJ, Rosser SJ, Davies GJ, Nicklin S, Bruce NC (1999) Biodegradation of explosives by transgenic plants expressing pentaerythritol tetranitrate reductase. Nat Biotechnol 17:491–494

Furukawa K, Hirose J, Hayashida S, Nakamura K (1994) Efficient degradation of trichloroethylene by a hybrid aromatic ring dioxygenase. J Bacteriol 176:2121–2123

Gadd GM (2007) Geomycology: biogeochemical transformations of rocks, minerals, metals and radionuclides by fungi, bioweathering and bioremediation. Mycol Res 11:3–49

Galagan JE, Calvo SE, Borkovich KA et al (2003) The genome sequence of the filamentous fungus *Neurospora crassa*. Nature 422:859–868

Gallardo ME, Ferrandez A, de Lorenzo V, Garcia JL, Diaz E (1997) Designing recombinant Pseudomonas strains to enhance biodesulfurization. J Bacteriol 179:7156–7160

Gandia-Herrero F, Lorenz A, Larson T, Graham IA, Bowles J, Rylott EL (2008) Detoxification of the explosive 2,4,6- trinitrotoluene in Arabidopsis: discovery of bi-functional O and C-glucosyltransferases. Plant J 56:963–974

Gao D, Du L, Yang J, Wu WM, Liang H (2010) A critical review of the application of white rot fungus to environmental pollution control. Crit Rev Biotechnol 30:70–77

Garbisu C, Hernandez-Allica J, Barrutia O, Alkorta I, Becerril JM (2002) Phytoremediation: a technology using green plants to remove contaminants from polluted areas. Rev Environ Health 17:173–188

Gisbert C, Ros R, De Haro A, Walker DJ, Pilar Bernal M, Serrano R (2003) A plant genetically modified that accumulates Pb is especially promising for phytoremediation. Biochem Biophys Res Commun 303:440–445

Gressel J, Al-Ahmad H (2005) Assessing and managing biological risks of plants used for bioremediation, including risks of transgene flow. Z Naturforsch C 60:154–165

Guerinot ML (2000) The ZIP family of metal transporters. Biochim Biophys Acta 1465:190–198

Hannink N, Rosser SJ, French CE, Basran A, Murray JAH, Nicklin S, Bruce NC (2001) Phytodetoxification of TNT by transgenic plants expressing a bacterial nitroreductase. Nat Biotechnol 19:1168–1172

Hannink NK, Rosser SJ, Bruce NC (2002) Phytoremediation of explosives. CRC Crit Rev Plant Sci 21:511–538

Hannink NK, Subramanian M, Rosser SJ, Basran A, Murray JAH, Shanks JV, Bruce NC (2007) Enhanced transformation of TNT by tobacco plants expressing a bacterial nitroreductase. Int J Phytoremediation 9:385–401

Harayama S (1998) Artificial evolution by DNA shuffling. Trends Biotechnol 16:76–82

Harbhajan S (2006) Mycoremediation: fungal bioremediation. Wiley, Hoboken, NJ, pp 1–592

Hirschi KD, Korenkov V, Wilganowski N, Wagner G (2000) Expression of Arabidopsis *CAX2* in tobacco: altered metal accumulation and increased manganese tolerance. Plant Physiol 124:125–133

Holmes A, Govan J, Goldstein R (1998) Agricultural use of *Burkholderia* (Pseudomonas) *epacia*: a threat to human health? Emerg Infect Dis 4:221–227

Hsieh JL, Chen CY, Chiu MH, Chein MF, Chang JS, Endo G, Huang CC (2009) Expressing a bacterial mercuric ion binding protein in plant for phytoremediation of heavy metals. J Hazard Mater 161:920–925

Iimura Y, Ikeda S, Sonoki T, Hayakawa T, Kajita S, Kimbara K, Tatsumi K, Katayama Y (2002) Expression of a gene for Mn peroxidase from Coriolus versicolor in transgenic tobacco generates potential tools for phytoremediation. Appl Microbiol Biotechnol 59:246–251

Inui H, Ohkawa H (2005) Herbicide resistance in transgenic plants with mammalian P450 monooxygenase genes. Pest Manag Sci 61:286–291

Iwamoto T, Nasu M (2001) Current bioremediation practice and perspective. J Biosci Bioeng 92:1–8

Jackson EG, Rylott EL, Fournier D, Hawari J, Bruce NC (2007) Exploring the biochemical properties and remediation applications of the unusual explosive-degrading P450 system XplA/B. Proc Natl Acad Sci USA 104:16822–16827

Karavangeli M, Labrou NE, Clonis YD, Tsaftaris A (2005) Development of transgenic tobacco plants overexpressing maize glutathione S-transferase I for chloroacetanilide herbicides phytoremediation. Biomol Eng 22:121–128

Kawahigashi H (2009) Transgenic plants for phytoremediation of herbicides. Curr Opin Biotechnol 20:225–230

Kawahigashi H, Hirose S, Inui H, Ohkawa H, Ohkawa Y (2005) Enhanced herbicide cross-tolerance in transgenic rice plants coexpressing human CYP1A1, CYP2B6, and CYP2C19. Plant Sci 168:773–781

Kawahigashi H, Hirose S, Ohkawa H, Ohkawa Y (2006) Phytoremediation of herbicide atrazine and metolachlor by transgenic rice plants expressing human CYP1A1, CYP2B6 and CYP2C19. J Agric Food Chem 54:2985–2991

Kawahigashi H, Hirose S, Ohkawa H, Ohkawa Y (2008) Transgenic rice plants expressing human P450 genes involved in xenobiotic metabolism for phytoremediation. J Mol Microbiol Biotechnol 15:212–219

Kimura N, Nishi A, Goto M, Furukawa K (1997) Functional analyses of a variety of chimeric dioxygenases constructed from two biphenyl dioxygenases that are similar structurally but different functionally. J Bacteriol 179:3936–3943

Kotrba P, Najmanova J, Macek T, Ruml T, Mackova M (2009) Genetically modified plants in phytoremediation of heavy metal and metalloid soil and sediment pollution. Biotechnol Adv 27:799–810

Kramer U (2005) Phytoremediation: novel approaches to cleaning up polluted soils. Curr Opin Biotechnol 16:133–141

Kumamaru T, Suenaga H, Mitsuoka M, Watanabe T, Furukawa H (1998) Enhanced degradation of polychlorinated biphenyls by directed evolution of biphenyl dioxygenase. Nat Biotechnol 16:663–666

Kurumata M, Takahashi M, Sakamoto A, Ramos JL, Nepovim A, Vanek T, Hirata T, Morikawa H (2005) Tolerance to, and uptake and degradation of 2,4,6-trinitrotoluene (TNT) are enhanced by the expression of a bacterial nitroreductase gene in *Arabidopsis thaliana*. Z Naturforsch C 60:272–278

Lan WS, Gu JD, Zhang JL, Shen BC, Jiang H, Mulchandani A, Chen W, Qiao CL (2006) Coexpression of two detoxifying pesticide-degrading enzymes in a genetically engineered bacterium. Int Biodeterior Biodegrad 58:70–76

Macek T, Kotrba P, Svatos A, Novakova M, Demnerova K, Mackova M (2008) Novel roles for genetically modified plants in environmental protection. Trends Biotechnol 26:146–152

Martinez M, Bernal P, Almela C, Velez D, Garcia-Agustin P, Serrano R (2006) An engineered plant that accumulates higher levels of heavy metals than Thlaspi caerulescens, with yields of 100 times more biomass in mine soils. Chemosphere 64:478–485

Menn FM, Easter JP, Sayler GS (2008) Genetically engineered microorganisms and bioremediation. In: Rehm HJ, Reed B (eds) Biotechnology set. Wiley, Hoboken, NJ, pp 441–463

Misra S, Gedamu L (1989) Heavy metal tolerant transgenic *Brassica napus* L. and *Nicotiana tabacum* L. plants. Theor Appl Genet 78:161–168

Nam JM, Fujita Y, Arai T, Kondo A, Morikawa Y, Okada H, Ueda M, Tanaka A (2002) Construction of engineered yeast with the ability of binding to cellulose. J Mol Catal B Enzym 17:197–202

Obire OE, Anyanwu C, Okigbo RN (2008) Saprophytic and crude oil-degrading fungi from cow dung and poultry droppings as bioremediating agents. Int J Agric Technol 4:81–89

Pandey G, Paul D, Jain RK (2005) Conceptualizing "suicidal genetically engineered microorganisms" for bioremediation applications. Biochem Biophys Res Commun 327:637–639

Parales RE, Haddock JD (2004) Biocatalytic degradation of pollutants. Curr Opin Biotechnol 15:374–379

Parrilli E, Papa R, Tutino ML, Sannia G (2010) Engineering of a psychrophilic bacterium for the bioremediation of aromatic compounds. Bioeng Bugs 1:213–216

Paul D, Pandey G, Pandey J, Jain RK (2005) Accessing microbial diversity for bioremediation and environmental restoration. Trends Biotechnol 23:135–142

Pearce F (2003) Arsenics fatal legacy grows. New Sci 179:4–5

Pieper DH, Reineke W (2000) Engineering bacteria for bioremediation. Curr Opin Biotechnol 11:262–270

Pilon-Smits E, Pilon M (2002) Phytoremediation of metals using transgenic plants. Crit Rev Plant Sci 21:439–456

Ramos JL, Gonzalez-Perez MM, Caballero A, van Dillewijn P (2005) Bioremediation of polynitrated aromatic compounds: plants and microbes put up a fight. Curr Opin Biotechnol 16:275–281

Rosser SJ, French CE, Bruce NC (2001) Engineering plants for the phytoremediation of explosives. In Vitro Cell Dev Biol Plant 37:330–333

Rugh CL, Wilde D, Stack NM, Thompson DM, Summer AO, Meagher RB (1996) Mercuric ion reduction and resistance in transgenic Arabidopsis thaliana plants expressing a modified bacterial merA gene. Proc Natl Acad Sci USA 93:3182–3187

Rylott EL, Bruce NC (2009) Plants disarm soil: engineering plants for the phytoremediation of explosives. Trends Biotechnol 27:73–81

Rylott EL, Jackson RG, Edwards J, Womack GL, Seth-Smith HMB, Rathbone DA, Strand SE, Bruce NC (2006) An explosive-degrading cytochrome P450 activity and its targeted application for the phytoremediation of RDX. Nat Biotechnol 24:216–219

Samanta SK, Singh OV, Jain RK (2002) Polycyclic aromatic hydrocarbons: environmental pollution and bioremediation. Trends Biotechnol 20:243–248

Sayler GS, Sayre P (1995) Risk assessment for recombination Pseudomonas released into the environment for hazardous waste degradation. In: Bioremediation: The Tokyo '94 Workshop. OECD Documents, Paris, pp 263–272

Schanstra JP, Ridder IS, Heimeriks GJ, Rink R, Poelarends GJ, Kalk KH, Dijkstra BW, Janssen DB (1996) Kinetics of halide release of haloalkane dehalogenase with higher catalytic activity and modified substrate range. Biochemistry 35:13186–13195

Seth-Smith HMB, Rosser SJ, Basran A, Travis ER, Dabbs ER, Nicklin S, Bruce NC (2002) Cloning, sequencing, and characterization of the hexahydro-1, 3, 5-trinitro-1, 3, 5-triazine degradation gene cluster from *Rhodococcus rhodochrous*. Appl Environ Microbiol 68:4764–4771

Singh OV, Ghai S, Paul D, Jain RK (2006) Genetically modified crops: success, safety assessment, and public concern. Appl Microbiol Biotechnol 71:598–607

Singh A, Kuhad RC, Ward OP (2009) Biological remediation of soil: an overview of global market and available technologies. In: Singh A et al (eds) Advances in applied bioremediation. Springer, Heidelberg, pp 1–19

Sitohy MZ, Rashad MM, Sharobeem SF, Mahmoud AE, Nooman MU, Al Kashef AS (2010) Bioconversion of soy processing waste for production of surfactants. Afr J Microbiol Re 4:2811–2821

Stefanov I, Frank J, Gedamu L, Mishra S (1997) Effect of cadmium treatment on the expression of chimeric genes in transgenic seedlings and calli. Plant Cell Rep 16:291–294

Stratford J, Wright MA, Reineke W, Mokross H, Havel J, Knowles CJ, Robinson GK (1996) Influence of chlorobenzoates on the utilisation of chlorobiphenyls and chlorobenzoate mixtures by chlorobiphenyl/chlorobenzoate-mineralising hybrid bacterial strains. Arch Microbiol 165:213–218

Sullivan ER (1998) Molecular genetics of biosurfactant production. Curr Opin Biotechnol 9:263–269

Suresh B, Ravishankar GA (2004) Phytoremediation – a novel and promising approach for environmental clean-up. Crit Rev Biotechnol 24:97–124

Talmage SS, Opresko DM, Maxwell CJ, Welsh CJ, Cretella FM, Reno PH, Daniel FB (1999) Nitroaromatic munitions compounds: environmental effects and screening values. Rev Environ Contam Toxicol 161:1–156

Timmis KN, Pieper DH (1999) Bacteria designed for bioremediation. Trends Biotechnol 17:201–204

Van Dillewijn P, Couselo JL, Corredoira E, Delgado E, Wittich RM, Ballester A (2008) Bioremediation of 2, 4, 6-trinitrotoluene by bacterial nitroreductase expressing transgenic aspen. Environ Sci Technol 42:7405–7410

Wang L, Samac DA, Shapir N, Wackett LP, Vance CP, Olszewski NE, Sadowsky MJ (2005) Biodegradation of atrazine in transgenic plants expressing a modified bacterial atrazine chlorohydrolase (*atzA*) gene. Plant Biotechnol J 3:475–486

Wood V, Gwilliam R, Rajandream MA et al (2002) The genome sequence of *Schizosaccharomyces pombe*. Nature 415:871–880

Yang C, Song C, Mulchandani A, Qiao C (2010) Genetic engineering of *Stenotrophomonas* strain YC-1 to possess a broader substrate range for organophosphates. J Agric Food Chem 58:6762–6766

Zhu Y, Pilon-Smits EA, Tarun AS, Weber SU, Jouanin L, Terry N (1999a) Cadmium tolerance and accumulation in Indian mustard is enhanced by overexpressing g-glutamylcysteine synthetase. Plant Physiol 121:1169–1177

Zhu Y, Pilon-Smits EAH, Jouanin L, Terry N (1999b) Overexpression of glutathione synthetase in Brassica juncea enhances cadmium tolerance and accumulation. Plant Physiol 119:73–79

Chapter 20
Molecular Techniques in Fungal Bioremediation

Morad Jafari, Younes Rezaee Danesh, and Yubert Ghoosta

20.1 Introduction

Analytical chemical measurements do not provide direct information regarding the biological effects of toxic compounds and also about the concentrations available for microbial biodegradation.

Biological analyses based upon biochemical and molecular genetics methods should come into play and form a valuable extension and addition to chemical analytics in the assessment of polluted sites (Power et al. 1998). The needs for timely treatment of contaminated sites could be evaluated, and the remaining toxicological risks after treatment established. Similarly, genetic and biochemical techniques can contribute greatly to our knowledge about the potential activity of the microorganisms at polluted sites. The application of molecular biological techniques to detect and identify microorganisms by certain molecular markers has been more frequently used in microbial ecological studies (Iwamoto and Nasu 2001). Advances in molecular biology led to the development of culture-independent approaches for describing bacterial communities without bias, i.e., the selectivity and unrepresentativity of the total community due to cultivation (Ranjard et al. 2001). Here, we describe high-throughput molecular methods that can be utilized for characterization of fungal community dynamics in polluted environmental sites that a simplified scheme of them is shown in Fig. 20.1.

M. Jafari (✉)

Department of Agronomy and Plant Breeding, Faculty of Agriculture, Urmia University, Urmia, Iran
e-mail: m.jafari@urmia.ac.ir

Y.R. Danesh • Y. Ghoosta
Department of Plant Protection, Faculty of Agriculture, Urmia University, Urmia, Iran
e-mail: Y.rdanesh@urmia.ac.ir; y.ghoosta@urmia.ac.ir

E.M. Goltapeh et al. (eds.), *Fungi as Bioremediators*, Soil Biology 32,
DOI 10.1007/978-3-642-33811-3_20, © Springer-Verlag Berlin Heidelberg 2013

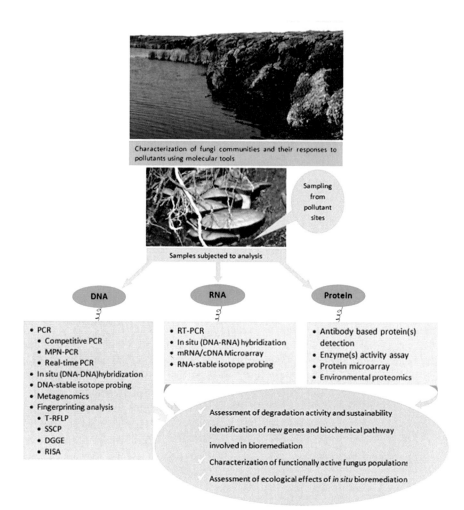

Fig. 20.1 Graphical representation of high-throughput molecular techniques used in characterization of fungi population dynamics for developing efficient bioremediation strategies

20.2 Molecular Techniques in Fungi Bioremediation

20.2.1 The Polymerase Chain Reaction

Recently, molecular strategies for the detection of fate and diversity of fungi in soils have been developed. These methods can identify fungi on a species level and are based on the use of variable internally transcribed spacer regions. A polymerase chain reaction (PCR) has been used to amplify rDNA sequences selectively to determine the total fungal diversity and taxonomy using universal primers (White

Table 20.1 The gene-specific primers that are commonly used for fingerprinting studies of fungal community dynamics in environmental samples

Primer name	Primer sequence (5′ → 3′)	Reference
2234C[a]	GTTTCCGTAGGTGAACCTGC	Ranjard et al.
3126T[a]	ATATGCTTAAGTTCAGCGGGT	(2001)
nu-SSU-0817-59[b]	TTAGCATGGAATAATRRAATAGGA	Borneman and
nu-SSU-1196-39[b]	TCTGGACCTGGTGAGTTTCC	Hartin
nu-SSU-1536-39[b]	ATTGCAATGCYCTATCCCCA	(2000)
EF4f[c]	GGAAGGG[G/A]TGTATTTATTAG	van Elsas et al.
Fung5r[c]	GTAAAAGTCCTGGTTCCC	(2000) and
NS2f[c]	GGCTGCTGGCACCAGACTTGC	De Felice
		et al. (2009)

[a]ITS1-5.8S-ITS2 region
[b]Two new PCR primer pairs designed to amplify rDNA from all major taxonomic groups of fungi
[c]Two-step, nested PCR based on the EF4f-Fung5r (first PCR) and NS2f-Fung5rC (nested PCR)

et al. 1990). Specific primers can be designed to amplify and analyze fungal DNA from cultures. Table 20.1 shows available common primers for using in PCR-based molecular methods for the detection of degradative fungi in various media. The PCR essentially enables the detection of a particular DNA fragment, and thus a particular organism or group, from within a more complex mixture by enzymatically amplifying it (Steffan and Atlas 1991). The PCR is so sensitive that detection of as few as ten copies of a specific DNA fragment present within a complex mixture of DNA is possible. Since DNA can be isolated from practically every environment microorganisms can be detected by the PCR from virtually everywhere. Examples of the broad applicability of the PCR include most probable number (MPN)-PCR (Sykes et al. 1992) and competitive PCR (Siebert and Larrick 1992), which have been used to attempt quantification of target DNAs within environmental samples.

The PCR has also been used to detect groups of related genes, related microorganisms, from either a functional or a taxonomic perspective depending on the target genes. In this instance, primers, which hybridize to conserved DNA, are used to amplify a particular gene fragment. For example, the use of conserved sites within the 18S rDNA or 28S rDNA as priming sites for the PCR is now very well established for amplification of ribosomal gene fragments (Amann et al. 1995). Subsequent DNA sequencing of these fragments gives information on many of the diversity of microorganisms from that particular environment. Similarly, these methods can be applied for conserved genes encoding pollutant degradative pathways. However, the PCR also has its limitations (For instance, inhibition DNA polymerase by humic acids in environmental samples and difficult quantification using PCR because of the nonlinearity of the amplification process) that must be taken into consideration (Siebert and Larrick 1992).

20.2.2   In Situ Hybridization

Labeled probes bind to a single-stranded target sequence within the cell. Labeled cells can then be specifically detected by microscopic techniques. Most notably, whole cell and in situ hybridizations are performed with fluorescently labeled oligonucleotides binding to target sites on the ribosomal RNAs. Ribosomal RNAs have made good targets since they are present in relatively large numbers in active cells. Therefore, the signal derived from the fluorescent probes is sufficiently strong to be detected by epifluorescence microscopy and to identify single cell. rRNAs are ubiquitous and contain both regions of highly conserved and regions of highly variable sequence. This makes it possible to create very specific species probes and more general probes for taxonomic groups of microorganisms. In situ and whole cell hybridizations have also been used to determine in situ activity of microorganisms (Prin et al. 1993; Hahn et al. 1995). In vitro transcripts, labeled with reporter molecules (such as biotin or digoxigenin) or fluorophores, can be used to detect specific mRNA sequences within microbe cells in the same manner as with the oligonucleotide probes. This method would be extremely valuable in determining the activity of microorganisms within populations.

20.2.3   Antibodies for Determination of Microbial Abundance and Enzyme Levels in Polluted Environments

The actual degradative activity for the breakdown of a specific pollutant in an environmental compartment is a function of both the number of microbial cells, which harbor this ability and the level of expression of the pollutant-degrading catabolic enzymes in these cells (Power et al. 1998). Antibodies can be used to detect specific cells or proteins in order to estimate the biodegradative potential of a specific site, instead of or in conjunction with genetic markers. Cell-specific antibodies can be a rapid and sensitive method for the detection of certain microorganisms (McDermott 1997).

   With combined use of surface-directed and enzyme-specific antibodies, the biodegradative capacities of a particular site are perceivable. Of course, the antibodies must be sufficiently specific to distinguish the target protein or cell from others present in the mixture. The ease and rapidity of antibody assays lend themselves to development for field studies Power et al. (1998) have used this approach to investigate the presence and activity of nitrilotriacetic acid (NTA) degrading bacteria in both model systems and in the environment. NTA-degrading bacteria were quantified with polyclonal antibodies raised against the cell surface of *Chelatobacter heintzii* and *Chelatococcus asaccharovorans* in wastewater treatment plants.

20.2.4 High-Throughput Techniques for Characterization of Fungal Community Structures in Contaminated Sites

A number of culture-independent molecular techniques currently used to study complex microbial communities are compatible with a high-throughput setup, such as fingerprinting techniques, real-time PCR, microarrays, metagenomics, metatranscriptomics, metaproteomics, or metabolomics.

Molecular fingerprinting techniques were introduced in soil microbial ecology in the past 15 years and allowed the study of the ecology of microorganisms, which could not be cultivated in synthetic media yet constitute the majority of soil microorganisms (Kirk et al. 2004). These techniques are based on the separation of amplicons after PCR amplification of phylogenetic (e.g., 16S rRNA for bacteria and 5.8S rRNA for fungi) or functional genes using universal or specific primers. The high specificity, sensitivity, and reproducible consistency of PCR detection of specific DNA sequences in complex environmental samples have contributed significantly to the advancement of knowledge in environmental microbiology (Plaza et al. 2001). Some of fingerprinting techniques have the potential for high-throughput design, such as terminal restriction fragment length polymorphism (T-RFLP), length heterogeneity analysis by PCR (LH-PCR), single-strand conformation polymorphism (SSCP), denaturing gradient gel electrophoresis (DGGE), or ribosomal intergenic spacer analysis (RISA).

20.2.4.1 Terminal Restriction Fragment Length Polymorphism

T-RFLP analysis is one of the most frequently used high-throughput fingerprinting methods. Because of its relative simplicity, T-RFLP analysis has been applied to the analysis of fungal ribosomal genes (Schutte et al. 2008). In this technique, target genes from whole-community DNA extracts were amplified by using specific primer pairs, one of which is fluorescently labeled. Some of the gene-specific primers that commonly used in fingerprinting studies of fungi are listed in Table 20.1. Resulting fragments was then obtained using enzymatic restriction of PCR amplicons and separated according to their size. The separated fragments are visualized by an automated DNA sequencer as a pattern of peaks on an electropherogram. Only the labeled terminal fragments (T-RFs) are detected and quantified. Individual RFs can be assigned presumptively to operational taxonomic units, which ideally correspond to phylogenetically related microorganisms, based on *in silico* search for matching restriction sites in sequences from clone libraries established in parallel from the same sample. In addition, the identification of T-RFLP peaks can be directly obtained by comparing them to databases (Marsh et al. 2000). In general, the T-RFLP technique has been proven to be reproducible as an accurate tool for community fingerprinting (Osborn et al. 2000). Advances in the use of T-RFLP technique to characterize microbial communities have been discussed by Schutte et al. (2008).

20.2.4.2 Denaturing Gradient Gel Electrophoresis

Another fingerprinting technique is denaturing gradient gel electrophoresis (DGGE), which rely on the separation of PCR amplicons based on their sequence. In this technique, an interest pattern band can be excited from the gel and further sequenced for taxonomic assignment. The high-throughput version of DGGE, denaturing high-performance liquid chromatography (D-HPLC), separates DNA fragments within minutes using fast and repeatable reverse-phase ion-pair chromatography and then the fragments can be collected at the end of the column for further automated sequencing (Barlaan et al. 2005).

Several group-specific or universal primers provide relatively large products (400–700 bp), which either could not be analyzed in DGGE (>600 bp) or provide poor resolution (400–600 bp). Thus, we commonly follow a nested or semi-nested PCR approach to obtain PCR products for DGGE analysis. In this procedure, in the first PCR round, the soil DNA sample is amplified with group-specific primers and the product obtained is subjected to a second, nested PCR with universal primers. It should be noted that in the second PCR round one of the two primers should carry a so-called GC clamp, a 40-bp G- and C-rich sequence, at its $5'$ end to prevent complete denaturation of the dsDNA fragments at high denaturant concentrations (De Felice et al. 2009; Karpouzas and Singh 2010). Thus the second PCR round produces shorter products with a GC clamp attached, which are suited for DGGE analysis. The genetic profile of the fungi community, developing in a Viagra polluted water environment, was evaluated by PCR-DGGE technique using 18S rRNA gene-specific primers (De Felice et al. 2009).

DGGE and TRFLP alone or in combination with cloning and sequencing have been used in different steps of bio-augmentation strategies. For instance, these two techniques have been used in conjunction with culture-dependent techniques in order to elucidate the phylogeny and the ecological significance of the different members of complex pesticide-degrading microbial consortia (Singh et al. 2003; Breugelmans et al. 2007) analysis in order to trace the fate of degrading microorganisms released in polluted soil ecosystems (Cunliffe and Kertesz 2006). In addition, DGGE and T-RFLP study possible perturbations in the soil microbial community, caused during bio-augmentation or bio-stimulation by the release of pollutant-degrading microorganisms (MacNaughton et al. 1999; Raina et al. 2007).

20.2.4.3 rRNA Intergenic Spacer Analysis

RISA, the rRNA intergenic transcribed spacer (ITS) regions that are located between the 18S and 28S ribosomal genes in fungi are target sites. Ranjard et al (2001) expanded the use of ARISA (automated rRNA intergenic spacer analysis) to fungal communities (F-ARISA) by designing primers that allow differentiation between fungal populations by determining the length heterogeneity of the region between the 18S and 28S rRNA genes, which includes the 5.8S rRNA gene as well

Fig. 20.2 Length distribution of ITS1-5.8S-ITS2 regions at ribosomal operon to assess the extent of the variability within the main fungal taxonomic groups

as two ITS (Fig. 20.2). ITS regions display higher heterogeneity in both length and nucleotide sequence than their flanking genes. Similarly to T-RFLP or SSCP, an automated approach has been proposed (Fisher and Triplett 1999), in which (1) a fluorescence tagged primer is used for PCR, and (2) the electrophoretic step is performed with an automated system. Serial analysis of ribosomal sequence tags/ribosomal DNA (SARST or SARD) and single-point genome signature tags (SP-GSTs) are other two tagging methods for high-throughput profiling of complex microbial communities. These techniques provide a fingerprint of microbial communities in the form of concatemers of PCR-amplified tag sequences (Ashby et al. 2007). Please refer to Neufeld et al. (2004) and van der Lelie et al. (2006) for more details.

20.2.4.4   Real-Time PCR

Real-time PCR is the continuous collection of fluorescent signals from one or more polymerase chain reactions over a range of cycles. Quantitative real-time PCR is the conversion of the fluorescent signals from each reaction into a numerical value for each sample (Dorak 2006). Real-time PCR is highly sensitive, down to a detection limit of 1–2 genome copies (Inglis and Kalischuk 2004). Real-time PCR does not require any tedious post-PCR steps for the quantification of amplicons, as their amount is monitored in real time. Therefore, this is a high-throughput technique with superior analytical sensitivity for the detection and quantification of specific genes in environmental samples (Harms et al. 2003). A quantitative fingerprinting method combining real-time PCR and T-RFLP enabled simultaneous determination of microbial abundance and diversity within a complex wastewater community (Yu et al. 2005).

20.2.4.5   Microarray

Microarray method is a high-throughput technique that compared to traditional nucleic acid membrane hybridization, offer the advantage of miniaturization (thousands of probes can be spotted on a slide), high sensitivity, and rapid detection (Eyers et al. 2004). Fluorescent dyes can be enzymatically or chemically incorporated in the sample to be hybridized; therefore, readout of the microarray is based on the detection of a fluorescence signal. The sensitivity of microarrays applied to environmental samples can be problematic, because potential contaminants (e.g., humic acids) in these samples can inhibit enzymatic reactions and generate a high signal background on the microarray (Stenuit et al. 2008). Microarrays [phylogenetic oligonucleotide- (POAs), functional gene-array (FGAs) or community genome-arrays (CGAs)] are used ever more extensively to characterize the phylogenetic and catabolic diversity of contaminated environments. In addition, they have the potential of "dynamically" monitoring changes in the phylogenetic composition of microbial communities or changes in catabolic gene expression levels during biodegradation processes.

20.2.4.6 Metagenomics: Environmental Functional Genomics

The term "metagenomics" is derived from the statistical concept of *meta*-analysis (the process of statistically combining separate analyses) and genomics (the comprehensive analysis of an organism's genetic material) (Rondon et al. 2000). Metagenomics is the culture-independent genomic analysis of microbial communities. Environmental DNA and RNA can be obtained directly from environment (Vo et al. 2007). Metagenomics enables the retrieval of unknown sequences or functions from the environment, whereas methods relying on PCR amplification or microarrays (POA or FGA) presuppose knowledge of gene sequences. Usually, the term "metagenomics" refers to the construction of metagenomic libraries: (a) generation of DNA fragments of appropriate size, (b) ligation of the fragments into an appropriate cloning vector (cosmid, fosmid, or bacterial artificial chromosome (BAC) vectors), (c) introduction of the recombinant vectors into a suitable bacterial cloning host, and (d) screening of clones harboring particular activities or containing specific sequences (Daniel 2005). Moreover, in some heavily contaminated environments harboring very low cell densities, direct extraction of metagenomic DNA does not provide enough genomic material for subsequent library construction. New PCR-independent amplification techniques are then necessary, like multiple displacement amplification (MDA) using phi29 DNA polymerase (Binga et al. 2008). While introducing an amplification bias, such a technique of whole-genome amplification of metagenomic DNA from very minute microbial sources gives genomic information that would otherwise remain inaccessible (Abulencia et al. 2006). Metagenomic analyses of a variety of environmental samples, for example, sea water, cave water estuarine, brackish sediments, freshwater sediments, peat soil, temperature forest soil, or acidic forest soil have revealed the enormous genetic diversity of complex microbial communities present in these environments (Vo et al. 2007).

20.2.4.7 Stable Isotope-Probing Techniques

Stable isotope probing (SIP) is one of the many emerging inquiry tools used by environmental microbiologists. It is a molecular technique that allows investigators to follow the flow of atoms in isotopically enriched molecules through complex microbial communities into metabolically active microorganisms (Wackett 2004). In conjunction with other tools, SIP seeks to discover the microorganisms responsible for catalyzing biogeochemical reactions in soils, sediments, and waters (Madsen 2006). SIP is based on the introduction of a substrate labeled with a stable isotope (for example, ^{13}C, ^{15}N, and ^{18}O) into a system, allowing microbes to incorporate heavy isotopes into biomarkers such as phospholipids fatty acids (PLFAs), DNA, and ribosomal RNA (rRNA). In nature, most of carbon sources are ^{12}C-based. Therefore, the isotope-labeled nucleic acids indicate the genomes or transcriptomes from microorganisms that actually mineralized and assimilated the stable-isotope substrate into their biomass. After extracting these biomaterials, it is possible to separate light and heavy fractions using ultracentrifugation, thus isolating the genomes (DNA), transcriptomes (RNA), and fatty acids of the microbial communities that have the target functions, i.e., mineralization of target carbon substrates in the environmental conditions (Vo et al. 2007). De Rito and Madsen (2009) distinguished a soil fungus involved in phenol biodegradation at an agricultural field sites using field-based DNA SIP of the 18S–28S internal transcribed spacer (ITS) region. They successfully confirmed *Trichosporon multisporum*'s ability to metabolize phenol in field-based soil experiments by this technique.

As above mentioned, metagenomic analyses of environmental samples are able to reveal enormous genetic diversity of complex microbial communities present in polluted site. Whole clones from metagenomic libraries can be sequenced and finally novel genes for enhancing biodegradation of some persistent pollutants can be discovered (Venter et al. 2004). However, this approach is known to be economically nonefficient. Most of laboratories cannot conduct that type of sequencing based metagenomics studies because of the extremely high cost. Although sequencing analysis is still needed, functional genomes isolated by SIP can be transferred into metagenomic library, sequence analysis for such selected functional genomes would be feasible, because the reduced size of clone library may become adequate for sequencing analysis. Therefore, integration of metagenomics with SIP will significantly reduce the amount and cost of sequencing (Vo et al. 2007; Kalyuzhnaya et al. 2008).

20.2.4.8 Proteomics

A variety of contaminants has been shown to be unusually recalcitrant, i.e., microorganisms either do not metabolize or transform them into certain other metabolites that again accumulate in the environment. Therefore, it may be more productive to explore new catabolic pathways that might lead towards complete mineralization of these pollutants. One of the reasons, our knowledge of microbial degradation pathways is so incomplete is the immense complexity of microbial physiology that allows response and adaptability to various internal and external stimuli (Fulekar 2007). New sophisticated techniques in medical science make it possible to explore global protein expression (proteomics) and low-molecular-weight metabolite expression (metabolomics) in environmental remediation. By applying proteome- and metabolome-based techniques to environmental samples, it is now possible to develop models that can predict microbial activities under various bioremediation strategies.

Metabolomics involves a nontargeted, holistic analysis of the set of metabolites produced by cellular proteins in response to various environmental stimuli. One particular advantage of metabolomics is that, when combined with multivariate data analysis (MVDA) tools like principal component discriminant analysis and partial least squares, it allows us to monitor changes in an organism as it is exposed to environmental pollutants (Singh 2006). While proteomics studies the global expression of proteins, metabolomics characterizes and quantifies their end products: the metabolites, produced by an organism under a given set of conditions (Whitfield et al. 2004). Unlike past studies based on predefined metabolites, metabolomics examines all the metabolites present in a biological system; thus, there is no bias associated with the choice of metabolites to be studied. However, metabolites in a site-specific organism are part of an in vivo metabolite flux that regulates entire metabolic pathways. Additionally, metabolism-based wide fluxes (fluxomes) allow us to pinpoint scenarios of physiological regulation in an organism (Singh 2006).

20.3   Conclusion

Although biochemical and genetic methods often result in clear-cut signals or bands, validation of an extrapolation from these results to overall environmental quality assessment has been limited and remains difficult. Molecular tools as described in this chapter can answer many of limitations. Nowadays, these technologies are expected to boost the discovery of new catabolic activities and to provide quantitative and timely information for the management and cleanup of contaminated sites and effluents in a perspective of sustainable development (Stenuit et al. 2008). The integrated functional genomics tools have a huge potential in applying in many areas of environmental studies for biological treatment engineering and are highly expected to provide more mechanistic and useful information on eco-physiology (systems biology) in bioprocesses in engineered and natural environmental systems (Vo et al. 2007). Continued scientific advancement will ultimately allow for comprehensive integrated approaches using gene, protein, and metabolite expression to study the functional physiology of an organism. Combined approaches will probably allow us to understand better microbial characteristics in pollutant sites, which may eventually pave the way toward effective bioremediation.

References

Abulencia CB, Wyborski DL, Garcia JA, Podar M, Chen W, Chang SH, Watson D, Brodie EL, Hazen TC, Keller M (2006) Environmental whole-genome amplification to access microbial populations in contaminated sediments. Appl Environ Microbiol 72:3291–3301

Amann RI, Ludwig W, Schleifer KH (1995) Phylogenetic identification and in situ detection of individual microbial cells without cultivation. Microbiol Rev 59:143–169

Ashby MN, Rine J, Mongodin EF, Nelson KE, Dimster-Denk D (2007) Serial analysis of rRNA genes and the unexpected dominance of rare members of microbial communities. Appl Environ Microbiol 73:4532–4542

Barlaan EA, Sugimori M, Furukawa S, Takeuchi K (2005) Profiling and monitoring of microbial populations by denaturing high-performance liquid chromatography. J Microbiol Methods 61:399–412

Binga EK, Lasken RS, Neufeld JD (2008) Mini-review: something from (almost) nothing: the impact of multiple displacement amplification on microbial ecology. ISME J 2:233–241

Borneman J, Hartin RJ (2000) PCR primers that amplify fungal rRNA genes from environmental samples. Appl Environ Microbiol 66:4356–4360

Breugelmans P, D'Huys PT, Mot RD, Springael D (2007) Characterization of novel linuron-mineralizing bacteria consortia enriched from long-term linuron-treated agricultural soils. FEMS Microbiol Ecol 62:374–385

Cunliffe M, Kertesz MA (2006) Effect of *Sphingobium yanoikuyae* B1 on bacterial community dynamics and polycyclic aromatic hydrocarbon degradation in aged and freshly PAH-contaminated soils. Environ Pollut 114:228–237

Daniel R (2005) The metagenomics of soil. Nat Rev Microbiol 3:470–478

De Felice B, Argenziano C, Guida M, Trifuoggi M, Russo F, Condorelli V, Inglese M (2009) Molecular characterization of microbial population dynamics during sildenafil citrate degradation. Mol Biotechnol 41:123–132

De Rito CM, Madsen EL (2009) Stable isotope probing reveals *Trichosporon* yeast to be active in situ in soil phenol metabolism. ISME J 3:477–485

Dorak MT (ed) (2006) Real-time PCR. Taylor & Francis, Abingdon

Eyers L, George I, Schuler L, Stenuit B, Agathos SN, El Fantroussi S (2004) Environmental genomics: exploring the unmined richness of microbes to degrade xenobiotics. Appl Microbiol Biotechnol 66:123–130

Fisher MM, Triplett EW (1999) Automated approach for ribosomal intergenic spacer analysis of microbial diversity and its application to freshwater bacterial communities. Appl Environ Microbiol 65:4630–4636

Fulekar MH (2007) Bioremediation technologies for environment. Indian J Environ Prot 27:264–271

Hahn D, Amann RI, Zeyer J (1995) Detection of mRNA in Streptomyces cells by whole-cell hybridization with digoxigenin-labeled probes. Appl Environ Microbiol 59:2753–2757

Harms G, Layton AC, Dionisi HM, Gregory IR, Garrett VM, Hawkins SA (2003) Real-time PCR quantification of nitrifying bacteria in a municipal wastewater treatment plant. Environ Sci Technol 37:343–351

Inglis GD, Kalischuk LD (2004) Direct quantification of *Campylobacter jejuni* and *Campylobacter lanienae* in feces of cattle by real-time quantitative PCR. Appl Environ Microbiol 70:2296–2306

Iwamoto T, Nasu M (2001) Current bioremediation practice and perspective. J Biosci Bioeng 92:1–8

Kalyuzhnaya MG, Lapidus A, Ivanova N, Copeland AC, McHardy AC, Szeto E et al (2008) High-resolution metagenomics targets specific functional types in complex microbial communities. Nat Biotechnol 26:1029–1034

Karpouzas DG, Singh BK (2010) Application of fingerprinting molecular methods in bioremediation studies. In: Cummings S (ed) Bioremediation – methods and protocols. Humana, New York, pp 68–88

Kirk JL, Beaudette LA, Hart M, Moutoglis P, Klironomos JN, Lee H, Trevors JT (2004) Methods of studying soil microbial diversity. J Microbiol Methods 58:169–188

MacNaughton SJ, Stephen JR, Venosa AD, Davis GA, Chang YJ, White DC (1999) Microbial population changes during bioremediation of an experimental oil spill. Appl Environ Microbiol 65:3566–3574

Madsen EL (2006) The use of stable isotope probing techniques in bioreactor and field studies on bioremediation. Curr Opin Biotechnol 17:92–97

Marsh TL, Saxman P, Cole J, Tiedje J (2000) Terminal restriction fragment length polymorphism analysis program, a web-based research tool for microbial community analysis. Appl Environ Microbiol 66:3616–3620

McDermott TR (1997) Use of fluorescent antibodies for studying the ecology of soil- and plant-associated microbes. In: Hurst CJ, Knudsen GR, McInerney MJ, Stetzenbach LD, Walter MJ (eds) Manual of environmental microbiology. American Society for Microbiology, Washington, DC, pp 473–481

Neufeld JD, Yu Z, Lam W, Mohn WW (2004) Serial analysis of ribosomal sequence tags (SARST): a high-throughput method for profiling complex microbial communities. Environ Microbiol 6:131–144

Osborn AM, Moore ERB, Timmis KN (2000) An evaluation of terminal restriction fragment length polymorphism (T-RFLP) analysis for the study of microbial community structure and dynamics. Environ Microbiol 2:39–50

Plaza G, Ulgif K, Hazen TC, Brigmon RL (2001) Use of molecular techniques in bioremediation. Acta Microbiol Pol 50:205–218

Power M, van der Meer JR, Tchelet R, Egli T, Eggen R (1998) Molecular-based methods can contribute to assessments of toxicological risks and bioremediation strategies. J Microbiol Methods 32:107–119

Prin Y, Mallein-Gerin F, Simonet P (1993) Identification and localization of Frankia strains in *Alnus* nodules by in situ hybridization of *nif H* mRNA with strain-specific oligonucleotide probes. J Exp Bot 44:815–820

Raina V, Suar M, Singh A, Prakash O, Dadhwal M, Gupta SK (2007) Enhanced biodegradation of hexachlorocyclohexane (HCH) in contaminated soils via inoculation with *Sphingobium indicum* B90A. Biodegradation 19:27–40

Ranjard L, Poly F, Lata JC, Mougel C, Thioulouse J, Nazaret S (2001) Characterization of bacterial and fungal soil communities by automated ribosomal intergenic spacer analysis fingerprints: biological and methodological variability. Appl Environ Microbiol 67:4479–4487

Rondon MR, August PR, Bettermann AD, Brady SF, Grossman TH, Liles MR, Loiacono KA, Lynch BA, MacNeil IA, Minor C (2000) Cloning the soil metagenome: a strategy for accessing the genetic and functional diversity of uncultured microorganisms. Appl Environ Microbiol 66:2541–2547

Schutte UME, Abdo Z, Bent SJ, Shyu C, Williams CJ, Pierson JD, Forney LJ (2008) Advances in the use of terminal restriction fragment length polymorphism (T-RFLP) analysis of 16S rRNA genes to characterize microbial communities. Appl Microbiol Biotechnol 80:365–380

Siebert PD, Larrick JW (1992) Competitive PCR. Nature 359:557–558

Singh OV (2006) Proteomics and metabolomics: the molecular make-up of toxic aromatic pollutant bioremediation. Proteomics 6:5481–5492

Singh BK, Walker A, Morgan JAW, Wright DJ (2003) Effects of soil PH on the biodegradation of chlorpyrifos and isolation of a chlorpyrifos-degrading bacterium. Appl Environ Microbiol 69:5198–5206

Steffan RJ, Atlas RM (1991) Polymerase chain reaction: applications in environmental microbiology. Annu Rev Microbiol 45:137–161

Stenuit B, Eyers L, Schuler L, Agathos SN, George I (2008) Emerging high-throughput approaches to analyze bioremediation of sites contaminated with hazardous and/or recalcitrant wastes. Biotechnol Adv 26:561–575

Sykes PJ, Neoh SH, Brisco MJ, Hughes E, Condon J, Morley AA (1992) Quantitation of targets for PCR by use of limiting dilution. Biotechniques 13:444–449

van der Lelie D, Lesaulnier C, McCorkle S, Geets J, Taghavi S, Dunn J (2006) Use of single-point genome signature tags as a universal tagging method for microbial genome surveys. Appl Environ Microbiol 72:2092–2101

van Elsas JD, Frois-Duarte G, Keijzer-Wolters A, Smit E (2000) Analysis of the dynamics of fungal communities in soil via fungal-specific PCR of soil DNA followed by denaturing gradient gel electrophoresis. J Microbiol Methods 43:133–151

Venter JC, Remington K, Heidelberg JF, Halpern AL, Rusch D, Eisen JA, Wu D, Paulsen I, Nelson KE, Nelson W (2004) Environmental genome shotgun sequencing of the Sargasso Sea. Science 304:66–74

Vo NXQ, Kang H, Park J (2007) Functional metagenomics using stable isotope probing: a review. Environ Eng Res 12:231–237

Wackett LP (2004) Stable isotope probing in biodegradation research. Trends Biotechnol 22:153–154

White TJ, Bruns T, Lee S, Taylor JW (1990) Amplification and direct sequencing of fungal ribosomal RNA genes for phylogenetics. In: Innis MA, Gelfand DH, Sninsky HH, White TJ (eds) PCR protocols: a guide to methods and applications. Academic, New York, pp 315–322

Whitfield PD, German AJ, Noble PJ (2004) Metabolomics: an emerging post-genomic tool for nutrition. Br J Nutr 92:549–555

Yu CP, Ahuja R, Sayler GS, Chu KH (2005) Quantitative molecular assay for fingerprinting microbial communities of wastewater and estrogen-degrading consortia. Appl Environ Microbiol 71:1433–1444

Chapter 21
Bioremediation of Heavy Metals Using Metal Hyperaccumulator Plants

Sangita Talukdar and Soniya Bhardwaj

21.1 Introduction

Contamination of soils with heavy metals, either by natural causes or due to pollution, often has pronounced effects on the vegetation, environment and human health. Due to the continual influx of heavy metal contaminants and pollutants into the biosphere from both natural and anthropogenic sources, these metals and metalloids accumulates in soil and have noxious effect on both plants and animals. The heavy metals are strongly deleterious to metal-sensitive enzymes, which result in growth inhibition and death of the cells. Some of these heavy metals are micronutrients which are necessary for plant growth, such as zinc (Zn), copper (Cu), manganese (Mn), nickel (Ni) and cobalt (Co), whereas cadmium (Cd), lead (Pb) and mercury (Hg) have no known biological function and are extremely toxic. At the outset, heavy metal pollution is of major concern due to the increasing levels of pollution and its evident impact on human health through the food chain.

Cadmium (Cd) and mercury (Hg) are common hazardous constituents of industrial Effluents. These are the non-essential elements that negatively affect plant growth and development and are toxic to living organisms even at low concentrations. Cd can be easily taken up and accumulated by plants and crops through their root systems and is present in all food (Alam et al. 2003). Research undertaken over the past 40 years has identified the irrefutable relationship between long-term consumption of Cd-contaminated rice and human Cd diseases such as itai-itai and proximal tubular renal dysfunction (Simmons et al. 2005). Due to the potential toxicity and high persistence of metals, soils polluted with heavy metals are a critical environmental issue, which requires an effective and affordable solution.

S. Talukdar (✉) • S. Bhardwaj
Amity Institute of Microbial Technology (AIMT), Amity University Uttar Pradesh, Noida, Uttar Pradesh, India
e-mail: sangita.talukdar@gmail.com; soniya.bhardwaj11@rediffmail.com

E.M. Goltapeh et al. (eds.), *Fungi as Bioremediators*, Soil Biology 32,
DOI 10.1007/978-3-642-33811-3_21, © Springer-Verlag Berlin Heidelberg 2013

Soil contaminated with Pb causes a sharp decline in crop productivity and therefore causing serious problem for agriculture (Johnson and Eaton 1980). Its increasing levels in soil environment inhibit germination of seeds and exert a wide range of adverse effects on growth and metabolism of plants. The higher uptake of lead causes toxicity symptom in plants like stunted growth, chlorosis and inhibition of root growth (Burton et al. 1984). It inhibits photosynthesis, mineral uptake and water balance; changes hormonal status and affects the membrane structure and permeability in plants. Like various other heavy metals, lead influences the activity of wide range of enzymes of different metabolic pathways. It strongly inhibits the activity of δ-amino laevulinate dehydrogenase, which is the key enzyme of chlorophyll biosynthesis (Prasad and Prasad 1987). It also inhibits the activity of enzymes of the reductive pentose pathway (Hampp et al. 1973). It also inhibits the activity of ATP synthase/ATPase (Tu Shu and Brouillette 1987).

Zinc (Zn) being the second most abundant transition metal in biological systems, including plants, is stable and inert to oxidoreduction, in contrast to the neighbouring transition elements in the periodic table (Vallee and Falchuk 1993) and plays a central role for a broad range of biological processes. However, Zn can be toxic to living organism when exposed at high concentration. Zn released from mines and industrial sources contaminate soil. This high amount of Zn can be transferred to human.

Despite a worldwide intensification of agriculture and tremendous progress towards increasing yields in major crops over the last decades, the goal to reduce the problems associated with heavy metal toxicity in plants is far from being achieved. It is because of human activity and geological origin of the soil. The metal concentrations in soil range from less than 1 mg kg^{-1} (ppm) to high as 100,000 mg kg^{-1} (Blaylock and Huang 2000). The high metal concentrations in contaminated soils result in decreased soil microbial activity, soil fertility, and yield losses (McGrath et al. 1995). For decreasing the toxicity of these heavy metals, several technologies and methods have been developed to remove them from polluted soil. These methods such as excavation and landfill, thermal treatment, acid leaching, and electroreclamation, which are being used for heavy metal remediation, are not suitable in practical applications. It is so because of many reasons like high cost, low efficiency, large destruction of soil structure and fertility and high dependence on the contaminants of concern, soil properties, site conditions and so on (Jing et al. 2007).

A cost-effective method to bioremediation of heavy metals is the usage of metal hyperaccumulator plants. What are these metal hyperaccumulator plants and how can they be used in bioremediation?

21.2 Metal Hyperaccumulating Plant Species

Requirement of the metals varies from species to species and shows large differences among the metal "non-hyperaccumulator" and the "hyperaccumulator" species. "Hyperaccumulator" species are defined as plants having a Zn concentration above 10,000 μg g^{-1} dw, Ni concentration higher than 1,000 μg g^{-1} dw or

Cd concentration above 100 µg g^{-1} dw. In comparison, a non-accumulator plant contains 30–100 µg g^{-1} dw of Zn and 1–10 µg g^{-1} dw of Ni. A Cd concentration of more than 10 µg g^{-1} dw is toxic. *Thlaspi caerulescens* (Tc), a Zn/Cd/Ni hyperaccumulator species, can accumulate up to 30,000 µg Zn g^{-1} dw foliar concentration (Brown et al. 1995), 4,000 µg Ni g^{-1} dw (Reeves and Brooks 1983; McGrath et al. 1993) and 2,700 µg Cd g^{-1} dw (Lombi et al. 2000). Like other hyperaccumulators, *T. caerulescens* exhibits enhanced metal uptake, as well as enhanced metal translocation to the shoots (Lasat et al. 1996; Shen et al. 1997; Schat et al. 2000). Approximately, 400 species have been reported as Ni, Zn, Cd, Pb, Cu, Co and Mn hyperaccumulators, which belong to a wide range of unrelated families (Baker and Brooks 1989; Baker et al. 2000). Among these 400 species, 317 are Ni hyperaccumulators and 12 are identified as Zn hyperaccumulators (Baker et al. 2000). Many Zn hyperaccumulators belong to the Brassicaceae family; 11 of them are *Thlaspi* species and one is an *Arabidopsis* species (*Arabidopsis halleri*). They are mainly found on calamine soils enriched in Zn, Pb and Cd, either naturally or due to mining or metal smelting in different parts of the world. Few species are naturalised in parts of Scandinavia.

The biggest concern now is to understand the mechanism of the plants, especially the hyperaccumulator plants, in order for better utilisation for bioremediation technique. Metal transporters play a pivotal role in metal uptake, localisation in the cell of hyperaccumulator plants as well as non-accumulator plants. An efficient root uptake system, as well as root-to-shoot translocation of metals, is among the more important characteristics of a metal hyperaccumulator. So it is vital to understand the function of individual metal transporters in connection to bioremediation.

21.3 Metal Transporters

21.3.1 Metal Uptake Proteins

A primary control point for metal homeostasis appears to be the regulation of metal uptake across the plasma membrane into the cell (Guerinot 2000). Metal ions are hydrophilic and do not cross cell membranes or other organelle membranes by passive diffusion (Cousins and McMahon 2000). Plants have evolved mechanisms that allow the transport of metal ions, through different categories of metal transporters such as the ZIP (ZRT, *I*RT-like *p*roteins) family (Guerinot 2000), cation diffusion facilitators (CDFs) (Williams et al. 2000), heavy metal (or CPx-type) ATPases, the natural resistance-associated macrophage proteins (Nramps) and the cation antiporters (Gaxiola et al. 2002), found to be located in different organelles within the cell.

21.3.2   ZIP Family Transporters

The ZIPs are involved in the transport of Fe, Zn, Mn and Cd with family members differing in their substrate range and specificity (Guerinot 2000; Mäser et al. 2001). About 85 ZIP family members have now been identified from bacteria, archea and all types of eukaryotes, including 15 genes in *A. thaliana* (Mäser et al. 2001).

The ZIP proteins are predicted to have eight transmembrane (TM) domains with the amino- and carboxyl-terminal ends situated on the outer surface of the plasma membrane (Guerinot 2000). The overall length varies widely, mostly because of a variable region between TM-3 and TM-4. This region is predicted to be on the cytoplasmic side and contains a potential metal-binding domain, rich in histidine residues. The most conserved region of these proteins lies in TM-4 and is predicted to form an amphipathic helix containing a fully conserved histidine that may form part of an intra-membranous metal binding site involved in transport (Guerinot 2000; Mäser et al. 2001). Its transport function after heterologous expression in yeast is eliminated when conserved histidines or certain adjacent residues are replaced by different amino acids (Rogers et al. 2000).

The first member ZIP family identified from plant (Eide et al. 1996) was *AtIRT1* (iron-regulated transporter 1) cloned from *A. thaliana* and identified by functional complementation of the Fe-uptake-deficient yeast double mutant *fet3 fet4*. *AtIRT1* is thought to be the major transporter for high affinity Fe uptake by roots (Connolly et al. 2002; Vert et al. 2002). Plants overexpressing *AtIRT1* also accumulate higher concentrations of Cd and Zn than wild types under Fe deficient conditions, indicating an additional role in the transport of these metals (Connolly et al. 2002). This is also supported by transport studies in yeast (Eide et al. 1996; Korshunova et al. 1999). The *irt1-1 A. thaliana* knockout mutant is chlorotic and shows severe growth defects that can be rescued by exogenous application of Fe (Vert et al. 2002). *AtIRT1* is expressed predominantly in the external layers of the root under Fe-deficient conditions and the protein is localised to the plasma membrane. Mutants of *AtIRT1* also showed significant changes in photosynthetic efficiency and developmental defects that are consistent with a deficiency in Fe transport and homeostasis (Henriques et al. 2002; Varotto et al. 2002).

Interestingly, *AtIRT2* is also expressed in root epidermal cells under Fe deficiency. However, *AtIRT2* cannot complement the loss of *AtIRT1* (Grotz and Guerinot 2002) and appears to have a greater specificity to substrates. Although the gene can complement Fe and Zn uptake mutants, the protein does not transport Cd or Mn in yeast (Vert et al. 2001). This suggests that these transporters have different functions in *A. thaliana*.

In tomato *LeIRT1* and *LeIRT2* were studied and both genes were shown to be expressed in roots (Eckhardt et al. 2001). Expression of *LeIRT1* was found to be strongly enhanced by Fe limitation, whereas this was not the case for *LeIRT2*. *LeIRT1* was also up-regulated by phosphorus (P) and potassium (K) deficiencies in the root medium. This suggests a possible co-regulation of the transporter genes for certain essential minerals (Wang et al. 2002). Studies in yeast suggest that *LeIRT1* and *LeIRT2* also have a broad range of substrate transport (Eckhardt et al. 2001). *OsIRT1* from rice, which has high homology to the *A. thaliana AtIRT1* gene, is also predominantly expressed in roots and is induced by Fe- and Cu-deficiency (Bughio et al. 2002).

Based largely on yeast complementation studies, further information is available on the functional properties of other plant ZIP transporters. The ZIP1-3 transporters from *A. thaliana* restore Zn uptake in the yeast Zn uptake double mutant *zrt1 zrt2* and were proposed to play a role in Zn transport (Grotz et al. 1998; Guerinot 2000). *ZIP1*, *ZIP3* and *ZIP4* are expressed in the roots of Zn-deficient plants, while *ZIP4* is also expressed in the shoots (Grotz et al. 1998; Guerinot 2000). Wintz et al. (2003) demonstrated that two ZIP genes, *AtZIP2* and *AtZIP4*, are involved in copper transport.

The proposed role of ZIP transporters in Zn nutrition is supported by the characterisation of homologues from other species. A member of the ZIP family, GmZIP1, has now been identified in soybean (Moreau et al. 2002). By functional complementation of the *zrt1 zrt2* yeast cells, *GmZIP1* was found to be highly selective for Zn, while yeast Zn uptake was inhibited by Cd. *GmZIP1* was specifically expressed in the nodules and not in roots, stems or leaves, and the protein was localised to the peribacteroid membrane, indicating a possible role in the symbiosis (Moreau et al. 2002).

The yeast ZIPs, *ZRT1* and *ZRT2*, were shown to be high- and low-affinity Zn transporters, respectively (Eide 1998; Guerinot 2000); *ZRT1* is glycosylated and present at the plasma membrane. A third ZIP homologue in yeast, *ZRT3*, is proposed to function in the mobilisation of stored Zn from the vacuole (MacDiarmid et al. 2000).

Why do plants need so many ZIPs? This diversity may be required for a variety of reasons: (1) to provide the high- and low-affinity systems needed to cope with varying metal availability in the soil; (2) to provide the specific requirements for transport at the different cellular and organellar membranes within the plant and (3) to respond to a variety of stress conditions. High-affinity transporters are selective for their target metals and are tightly regulated according to metal need. Low-affinity transporters are less responsive to metal need and are somewhat less selective for the metals they transport (Radisky and Kaplan 1999).

21.3.3 Metal Efflux Proteins

Plasma membrane controlled Zn regulation is not sufficient to control the free metal concentration in the cells. The vacuole, which occupies most of the plant cell volume, plays a major role in the regulation of ion homeostasis in the cell and in detoxification of the cytosol (Marschner 1995). Several CDF genes involved in sequestration of metals in the cell have been identified in different plant species.

21.3.4   CDF Family or Cation Efflux Family

Over-expression of CDF family members are described to confer Zn, Cd, Co or Ni tolerance to plants (Paulsen and Saier 1997; Eide 1998). Several evidences suggest that these transporters either sequester metal ions within vacuoles or export metal ions out of the cells (Paulsen and Saier 1997), though the mechanisms underlying the transport are not well understood for most of the CDF proteins. Proteins of the CDF family from diverse sources have the following features in common: (1) they share an N-terminal signature sequence that is specific to the family; (2) the proteins possess six trans-membrane-spanning regions; (3) they share a cation efflux domain; and (4) most of the eukaryotic members possess an intracellular histidine-rich domain that is absent from the prokaryotic members (Paulsen and Saier 1997). Several CDF family members have been studied from *Saccharomyces cerevisiae*, *Escherichia coli*, *Bacillus subtilis* and *Schizosaccharomyces pombe* and were found to contribute to the storage of Zn and other metals into the vacuole (Li and Kaplan 1998; MacDiarmid et al. 2002; Miyabe et al. 2001; Guffanti et al. 2002; Chao and Fu 2004; Li and Kaplan. 2001; Clemens et al. 2002).

The genome of *A. thaliana* encodes 12 putative CDF genes, which are highly divergent in sequence, but share some characteristics of CDF family membrane transport proteins (Blaudez et al. 2003). *AtZAT* (Zn transporter of *A. thaliana*) later renamed as *AtMTP1* (Metal Tolerance Protein 1) was the first plant CDF, identified as a cDNA (van der Zaal et al. 1999; Mäser et al. 2001). Under normal or excess Zn supply, *AtMTP1* transcripts were shown to be present at low levels in seedlings (van der Zaal et al. 1999). *35S::AtMTP1* transformed *A. thaliana* plants resulted in enhanced Zn tolerance of transgenic seedlings and slightly increased Zn accumulation in roots. Studies by Kobae et al. (2004) suggested *AtMTP1* to be a potential Zn transporter in *A. thaliana*. A further study by Desbrosses-Fonrouge et al. (2005) showed that AtMTP1 acts as a Zn transporter, localised in the vacuolar membrane, mediating Zn detoxification and leaf Zn accumulation. Bloss et al. (2002) expressed *AtMTP1* in *E. coli* and studied the purified protein in reconstituted proteoliposomes. The protein transported Zn into proteoliposomes by a mechanism that relied on the Zn gradient across the membrane and not on a proton gradient. Two of the 12 genes encoding putative CDF proteins in *A. thaliana*, *AtMTP2* and *AtMTP3*, are closely related to *AtMTP1* (64.4 % and 67.6 % identity, respectively). *AtMTP2* was shown by expression profiling to be induced in a Zn deficiency condition (van de Mortel et al. 2006). AtMTP3 was found to contribute to basic cellular Zn tolerance and was involved in the control of Zn partitioning (Arrivault et al. 2006). Silencing of *AtMTP3* causes Zn hypersensitivity and enhances Zn accumulation in above ground organs of plants exposed to excess Zn or to Fe deficiency.

Over-expression of *PtdMTP1* from hybrid poplar (*Populus trichocarpa* × *P. deltoides*) in *A. thaliana* also conferred Zn tolerance (Blaudez et al. 2003). Heterologous expression of *PtdMTP1* from poplar in various yeast mutants was shown to confer resistance specifically to Zn, possibly as a result of transport into the vacuole (Kohler et al. 2003).

Persans et al. (2001) isolated CDF member genes (*TgMTPs*) from the Ni-hyperaccumulating species *Thlaspi goesingense*. These genes conferred metal tolerance to *S. cerevisiae* mutants defective in vacuolar *COT1* and *ZRC1* Zn transporters. They suggested that TgMTP1 transports metals into the vacuole. In vivo and in vitro immunological staining of hemagglutinin (HA)-tagged *TgMTP1::HA* revealed that the protein is localised in both vacuolar and plasma membranes in *S. cerevisiae*. It was assumed that TgMTP1 functions by enhancing plasma membrane Zn efflux thereby conferring Zn resistance in *S. cerevisiae*. Transient expression in *A. thaliana* protoplasts also revealed that *TgMTP1::GFP* is localised at the plasma membrane, suggesting that *TgMTP1* may enhance the Zn efflux in plants, which is different from the other endogenous *A. thaliana* CDFs. Delhaize et al. (2003) showed that the *ShMTP1* cDNA from *Stylosnathes hamata*, a tropical legume tolerant to acid soils with high concentrations of Mn, conferred Mn tolerance to yeast and plants by a mechanism that is likely to involve the sequestration of Mn into internal organelles. Expression studies on *AhMTP1* from *A. halleri* showed substantially higher transcript levels in the leaves and up-regulation upon exposure to high Zn concentrations in the roots of *A. halleri* and vacuolar localisation leading to Zn accumulation in the plant (Dräger et al. 2005).

21.4 Metal Transport in *T. caerulescens*

Lasat et al. (1996) described the physiological characterisation of Zn uptake of *T. caerulescens* and *T. arvense* (a non-hyperaccumulator species). A concentration-dependent Zn^{2+} influx into the root was recorded, using radiotracer flux techniques. In both species, there was a saturable component following Michaelis–Menten kinetics. V_{max} was much higher in *T. caerulescens* than in *T. arvense*, whereas the K_m values appeared similar for both species. This suggests a higher expression of functionally similar Zn transporters in *T. caerulescens* compared to *T. arvense* roots (Lasat et al. 2000). Time-course analysis of Zn accumulation in roots and shoots supported this finding. The Zn content in the *T. caerulescens* roots was found to be two times higher than in the *T. arvense* roots at the start of the experiment, despite a much higher rate of Zn translocation to the shoot later on (Lasat et al. 1996). Further, Lasat et al. (2000) and Pence et al. (2000) isolated *TcZNT1*, a *ZIP* family Zn transporter gene from *T. caerulescens*, by functional complementation of the yeast *zhy3* mutant, defective in Zn uptake (Zhao and Eide 1996).

Assunção et al. (2001) cloned the *TcZNT1* gene as well as an apparent paralogue, the *TcZNT2* gene, based on the homology to the *A. thaliana AtZIP4* gene from accession La Calamine. In non-accumulator species, like *A. thaliana* or *T. arvense*, the orthologues of *TcZNT1* are mainly expressed in roots, but only under Zn deficiency conditions. At normal or elevated Zn supply, their transcription is strongly down-regulated (Grotz et al. 1998; Pence et al. 2000; Assunção et al. 2001). In contrast, in *T. caerulescens* both *TcZNT1* and *TcZNT2* are highly expressed in roots and at a lower level in shoots, not only under conditions of Zn deficiency, but also at normal or elevated Zn supply (Pence et al. 2000; Assunção et al. 2001). TcZNT1-mediated Zn uptake in yeast showed a saturable component (Pence et al. 2000) with a K_m value very similar to the one found for *T. caerulescens* (Lasat et al. 1996). Furthermore, the V_{max} of Zn influx in roots of *T. caerulescens* grown under different Zn concentrations correlated very well with the root *TcZNT1* transcript levels and the K_m values were very similar at all the Zn exposure levels tested (Pence et al. 2000). Similar experiments were done to show evidence of Cd transport of TcZNT1, concluding that TcZNT1 mediates high-affinity Zn uptake as well as low-affinity Cd uptake (Pence et al. 2000).

Why these genes are apparently over-expressed in *T. caerulescens* is still unknown. An increased expression of *TcZNT1* and *TcZNT2* may be one of the evolutionary changes on the way from non-accumulator to hyperaccumulator (Assunção et al. 2001). One possibility could be that a Zn-responsive element in the *TcZNT1* promoter has been altered, altering the Zn-imposed transcriptional down regulation (Assunção et al. 2001). However, since two genes (*TcZNT1* and *TcZNT2*) are over-expressed, an alteration in the Zn receptor and signal transduction machinery appears more plausible (Lasat et al. 2000; Pence et al. 2000). Recently, a gene encoding a basic helix-loop-helix (bHLH) transcription factor involved in the regulation of Fe status was identified in tomato. Mutation of this gene leads to much lower expression of the main root Fe transporter (Ling et al. 2002). One can envision that a similar transcription factor in *T. caerulescens* is involved in the regulation of Zn status. Modification of one transcription factor often changes the expression of several genes, as could be the case for the *TcZNT1* and *TcZNT2* paralogues (Assunção et al. 2001). Alternatively, if the Zn sequestration machinery is much more efficient in *T. caerulescens* than in non-accumulator species, this can create a state of "physiological Zn deficiency" (Assunção et al. 2001). In this situation, cellular sensing machinery does not sense Zn at appropriate levels, even though the Zn supply rates and total cellular Zn concentrations would be adequate or even toxic for non-hyperaccumulator plants (Assunção et al. 2001).

Similar to the kinetics studies with Zn, Lombi et al. (2001) and Zhao et al. (2002) also established the kinetic parameters of Cd and Zn influx into the roots of plants from two calamine *T. caerulescens* accessions, Prayon and Ganges, with different Cd accumulation capacities (much higher in Ganges). The non-saturable component of the Cd influx was the same in both accessions and the V_{max} of the saturable component was about five times higher in Ganges than in Prayon, while the maximum saturable Zn influx rates were about equal. Cd uptake in Prayon was significantly suppressed in the presence of equimolar concentrations of Zn and Mn, suggesting that in Prayon Zn transporters largely mediate Cd uptake. However, a similar treatment did not affect Cd uptake in Ganges, suggesting the existence of a transporter with high selectivity to Cd, as compared to Zn and Mn at least, which would be expressed much higher in Ganges than in Prayon (Lombi et al. 2001; Zhao et al. 2002).

Assunção et al. (2001) showed constitutively high expression of *TcZTP1* (Zn Transporter 1), a member of the CDF family in *T. caerulescens*. *TcZTP1* is very similar to *AtMTP1* of *A. thaliana*. It has 90 % identity at the DNA level and

75 % aa identity. Still the molecular basis of Zn uptake and transport in *T. caerulescens* is largely unexplored (Assunção et al. 2003). Two related ZIP genes have been cloned from Ni hyperaccumulator *T. japonicum*, *TjZNT1* and *TjZNT2*. Expressing either *TjZNT1* or *TjZNT2* in yeast shows increased resistance to Ni^{2+} (Mizuno et al. 2005), highlighting a potential role for these genes in Ni tolerance. Further studies are necessary to determine if or how these proteins function to hyperaccumulate metals in plants and to understand the differential regulation of the similar genes in the hyperaccumulator plants compared with non-accumulator plants.

Furthermore, it is fascinating that *T. caerulescens* and other hyperaccumulator plants cope with such high Zn and other heavy metal concentrations, which are very toxic to other non-accumulator plants. This is why the hyperaccumulator plants are assumed to have specific or higher activity of mechanisms for root-to-shoot transport, xylem loading and unloading (Lasat et al. 1998) and vacuolar sequestration of heavy metals, particularly in the leaf epidermal cells (Vázquez et al. 1999; Küpper et al. 1999), trichomes (Krämer et al. 1997) or stomatal guard cells (Heath et al. 1997) compared to the non-accumulating plants. Only few studies have been carried out to understand the mechanisms behind. Further research is necessary to understand the underlined mechanisms.

21.5 Bioremediation Using Metal Hyperaccumulator Plants

A great interest has developed recently in the use of terrestrial plants as a green technology for the remediation of surface soils contaminated with toxic heavy metals (Pence et al. 2000). This developed the new field of environmental biotechnology, termed phytoremediation, which uses plants to extract heavy metals from the soil and to concentrate them in harvestable shoot tissue (Salt et al. 1995). Hence, the hyperaccumulator plants have great potential for bioremediation of Zn and Cd.

Approximately, 400 species have been reported as Ni, Zn, Cd, Pb, Cu, Co and Mn hyperaccumulators, which belong to a wide range of unrelated families (Baker and Brooks 1989; Baker et al. 2000). *Brassica juncea*, *Brassica napus*, *Helianthus annuus* and *Thlaspi caerulescens* are only few potential examples. *Thlaspi caerulescens*, a Zn/Cd/Ni hyperaccumulator species, can accumulate high amount of Zn, Cd and Ni in its roots and shoots, has great potential for application in phytoextraction of the metals. *Brassica napus* and *Brassica juncea* is well known for phytoextraction for several heavy metals (Cr, Hg, Pb, Se, Zn and Ag). Sunflower or *Helianthus annuus* are also commonly known as metal accumulator for several heavy metals.

However, it is important to understand the individual hyperaccumulator species for specific metal extraction and processing. This will further improve the technique of phytoremediation. Several criteria must be met before a plant is considered to be well suited for phytoremediation (Song et al. 2003). An "ideal" phytoremediator should be fast growing, develop a large biomass, be tolerant to and accumulate high concentrations of toxic metals in the shoot and be easily cultivated and harvested. Although natural hyperaccumulators can tolerate and accumulate high concentration of toxic metals, they usually produce little biomass, they grow slowly and cannot be easily cultivated. Now with the advancement of molecular biology, scientists are performing genetic engineering to improve the metal accumulation capacity of fast-growing and high-biomass plants.

Transferring the genes responsible for the hyperaccumulating phenotype to higher shoot-biomass-producing plants has been suggested as a potential avenue for enhancing phytoremediation as a viable commercial technology (Brown et al. 1995). Several engineering approaches to enhance metal uptake can be envisioned (Clemens et al. 2002). (1) The number of uptake sites could be increased; (2) the specificity of uptake systems could be altered, so that the competition among the unwanted cations is reduced and (3) the sequestration capacity could be enhanced by increasing the number of intracellular high-affinity binding sites or the rates of transport into organelles (Clemens et al. 2002).

In summary, better understanding of the metal accumulating plant species and their engineering to cater the practical need can be a revolutionary step towards cost-effective and efficient method of metal bioremediation.

21.6   Conclusion and Future Perspectives

The mechanism of individual hyperaccumulator species for specific metal extraction and processing can help in phytoremediation, which is a cost-effective and efficient method of excavating heavy metals from the environment. To further understand the role of the heavy metal transporters in metal homeostasis in plants a more detailed analysis of the different members is required to study the mechanism of metal hyperaccumulation in the hyperaccumulator plants as compared to the non-accumulating plants. This analysis should include determination of the proteins expression patterns, membrane localisation, metal specificity and transport mechanisms, including structure/function analyses. An improved understanding of the genetic and biochemical basis of metal hyperaccumulation in plants can help both in development of plants ideally suited for phytoremediation of pollutant metals and enhancing the micronutrient quality of staple food crop.

References

Alam MZ, Fakhrul-Razi A, Molla AH (2003) Biosolids accumulation and biodegradation of domestic wastewater treatment plant sludge by developed liquid state bioconversion process using a batch fermenter. Water Res 37:3569–3578

Arrivault S, Senger T, Kramer U (2006) The Arabidopsis metal tolerance protein AtMTP3 maintains metal homeostasis by mediating Zn exclusion from the shoot under Fe deficiency and Zn oversupply. Plant J 46(5):861–879

Assunção AGL, Martins PDC, De Folter S, Vooijs R, Schat H, Aarts MGM (2001) Elevated expression of metal transporter genes in three accessions of the metal hyperaccumulator Thlaspi caerulescens. Plant Cell Environ 24(2):217–226

Assunção AGL, Schat H, Aarts MGM (2003) Thlaspi caerulescens, an attractive model species to study heavy metal hyperaccumulation in plants. New Phytol 159(2):351–360

Baker AJM, Brooks RR (1989) Terrestrial higher plants which hyperaccumulate metallic elements – a review of their distribution, ecology and phytochemistry. Biorecovery 1:81–126

Baker AJM, McGrath SP, Reeves DR, Smith JAC (2000) Metal hyperaccumulator plants: a review of the Ecology and Physiology of the Biological Resource for Phytoremediation of Metal-Polluted Soils. In: Terry N, Banuelos G (eds) Phytoremediation of contaminated soils & water. CRC, Boca Raton, FL, pp 171–188

Blaudez D, Kohler A, Martin F, Sanders D, Chalot M (2003) Poplar metal tolerance protein 1 confers zinc tolerance and is an oligomeric vacuolar zinc transporter with an essential leucine zipper motif. Plant Cell 15(12):2911–2928

Blaylock MJ, Huang JW (2000) Phytoextraction of metals. In: Raskin I, Ensley BD (eds) Phytoremediation of toxic metals: using plants to clean up the environment. Wiley, Toronto, p 303

Bloss T, Clemens S, Nies DH (2002) Characterization of the ZAT1p zinc transporter from Arabidopsis thaliana in microbial model organisms and reconstituted proteoliposomes. Planta 214(5):783–791

Brown S, Chaney R, Angle J, Baker A (1995) Zinc and cadmium uptake by hyperaccumulator Thlaspi caerulescens grown in nutrient solution. Soil Sci Soc Am J 59:125–133

Bughio HR, Soomro AM, Baloch AW, Javed MA, Khan IA (2002) Yield potential of aromatic rice mutants/varieties in different ecological zones of Sindh. Asian J Plant Sci 1:439–440

Burton KW, Morgan E, Roig A (1984) The influence of heavy metals on the growth of sitka-spruce in South Wales forests. II green house experiments. Plant Soil 78:271–282

Chao Y, Fu D (2004) Kinetic study of the antiport mechanism of an Escherichia coli zinc transporter, ZitB. J Biol Chem 279(13):12043–12050

Clemens S, Palmgren MG, Kramer U (2002) A long way ahead: understanding and engineering plant metal accumulation. Trends Plant Sci 7(7):309–315

Connolly EL, Fett JP, Guerinot ML (2002) Expression of the IRT1 metal transporter is controlled by metals at the levels of transcript and protein accumulation. Plant Cell 14(6):1347–1357

Cousins RJ, McMahon RJ (2000) Integrative aspects of zinc transporters. J Nutr 130:1384S–1387S

Delhaize E, Kataoka T, Hebb DM, White RG, Ryan PR (2003) Genes encoding proteins of the cation diffusion facilitator family that confer manganese tolerance. Plant Cell 15(5):1131–1142

Desbrosses-Fonrouge AG, Voigt K, Schroder A, Arrivault S, Thomine S, Kramer U (2005) Arabidopsis thaliana MTP1 is a Zn transporter in the vacuolar membrane which mediates Zn detoxification and drives leaf Zn accumulation. FEBS Lett 579(19):4165–4174

Dräger DB, Voigt K, Kramer U (2005) Short transcript-derived fragments from the metal hyperaccumulator model species Arabidopsis halleri. Z Naturforsch C 60(3–4):172–178

Eckhardt U, Mas Marques A, Buckhout TJ (2001) Two iron-regulated cation transporters from tomato complement metal uptake-deficient yeast mutants. Plant Mol Biol 45(4):437–448

Eide DJ (1998) The molecular biology of metal ion transport in Saccharomyces cerevisiae. Annu Rev Nutr 18:441–469

Eide D, Broderius M, Fett J, Guerinot ML (1996) A novel iron-regulated metal transporter from plants identified by functional expression in yeast. Proc Natl Acad Sci USA 93(11):5624–5628

Gaxiola RA, Fink GR, Hirschi KD (2002) Genetic manipulation of vacuolar proton pumps and transporters. Plant Physiol 129(3):967–973

Grotz N, Guerinot ML (2002) Limiting nutrients: an old problem with new solutions? Curr Opin Plant Biol 5(2):158–163

Grotz N, Fox T, Connolly E, Park W, Guerinot ML, Eide D (1998) Identification of a family of zinc transporter genes from Arabidopsis that respond to zinc deficiency. Proc Natl Acad Sci USA 95(12):7220–7224

Guerinot ML (2000) To improve nutrition for the world's population. Science 288(5473):1966–1967

Guffanti AA, Wei Y, Rood SV, Krulwich TA (2002) An antiport mechanism for a member of the cation diffusion facilitator family: divalent cations efflux in exchange for K+ and H+. Mol Microbiol 45:145–153

Hampp R, Ziegler H, Ziegler I (1973) Influence of lead ions on the activity of enzymes of reductive pentose phosphate pathway. Biochem Physiol Pflanzen 164:588–595

Heath S, Southworth D, D'Allura JA (1997) Localization of nickel in epidermal subsidiary cells of leaves of Thlaspi montanum var. siskiouense (Brassicaceae) using energy-dispersive X-ray micro- analysis. Int J Plant Sci 158:184–188

Henriques R, Jasik J, Klein M, Martinoia E, Feller U, Schell J, Pais MS, Koncz C (2002) Knockout of Arabidopsis metal transporter gene IRT1 results in iron deficiency accompanied by cell differentiation defects. Plant Mol Biol 50(4–5):587–597

Jing Y, He Z, Yang X (2007) Role of soil rhizobacteria in phytoremediation of heavy metal contaminated soils. J Zhejiang Univ Sci B 8:192–207

Johnson MS, Eaton JW (1980) Environmental contamination through residual trace metal dispersal from a derelict lead-zinc mine. J Environ Qual 9:175–179

Kobae Y, Uemura T, Sato MH, Ohnishi M, Mimura T, Nakagawa T, Maeshima M (2004) Zinc transporter of Arabidopsis thaliana AtMTP1 is localized to vacuolar membranes and implicated in zinc homeostasis. Plant Cell Physiol 45(12):1749–1758

Kohler A, Delaruelle C, Martin D, Encelot N, Martin F (2003) The poplar root transcriptome: analysis of 7000 expressed sequence tags. FEBS Lett 542(1–3):37–41

Korshunova YO, Eide D, Clark WG, Guerinot ML, Pakrasi HB (1999) The IRT1 protein from Arabidopsis thaliana is a metal transporter with a broad substrate range. Plant Mol Biol 40(1):37–44

Krämer U (2005) Phytoremediation: novel approaches to cleaning up polluted soils. Curr Opin Biotechnol 16(2):133–141

Krämer U, Smith RD, Wenzel WW, Raskin I, Salt DE (1997) The role of metal transport and tolerance in nickel hyperaccumulation by Thlaspi goesingense Halacsy. Plant Physiol 115(4):1641–1650

Küpper H, Zhao FJ, McGrath SP (1999) Cellular compartmentation of zinc in leaves of the hyperaccumulator Thlaspi caerulescens. Plant Physiol 119:305–311

Lasat MM, Baker A, Kochian LV (1996) Physiological characterization of root Zn2+ absorption and translocation to shoots in Zn hyperaccumulator and nonaccumulator species of Thlaspi. Plant Physiol 112(4):1715–1722

Lasat MM, Baker AJ, Kochian LV (1998) Altered Zn compartmentation in the root symplasm and stimulated Zn absorption into the leaf as mechanisms involved in Zn hyperaccumulation in Thlaspi caerulescens. Plant Physiol 118:875–883

Lasat MM, Pence NS, Garvin DF, Ebbs SD, Kochian LV (2000) Molecular physiology of zinc transport in the Zn hyperaccumulator Thlaspi caerulescens. J Exp Bot 51:71–79

Li L, Kaplan J (1998) Defects in the yeast high affinity iron transport system result in increased metal sensitivity because of the increased expression of transporters with a broad transition metal specificity. J Biol Chem 273:22181–22187

Li L, Kaplan J (2001) The yeast gene MSC2, a member of the cation diffusion facilitator family, affects the cellular distribution of zinc. J Biol Chem 276:5036–5043

Ling HQ, Bauer P, Bereczky Z, Keller B, Ganal M (2002) The tomato fer gene encoding a bHLH protein controls iron-uptake responses in roots. Proc Natl Acad Sci USA 99:13938–13943

Lombi E, Zhao FJ, Dunhan SJ, McGrath SP (2000) Cadmium accumulation in populations of Thlaspi caerulescens and Thlaspi goesingense. New Phytol 145:11–20

Lombi E, Zhao FJ, Dunham SJ, McGrath SP (2001) Phytoremediation of heavy metal-contaminated soils: natural hyperaccumulation versus chemically enhanced phytoextraction. J Environ Qual 30:1919–1926

MacDiarmid CW, Milanick MA, Eide DJ (2002) Biochemical properties of vacuolar zinc transport systems of Saccharomyces cerevisiae. J Biol Chem 277:39187–39194

Marschner H (1995) Function of mineral nutrients: micronutrients. In: Mineral nutrition of higher plants, 2nd edn. Academic, London

Mäser P, Thomine S, Schroeder JI, Ward JM, Hirschi K, Sze H, Talke IN, Amtmann A, Maathuis FJ, Sanders D, Harper JF, Tchieu J, Gribskov M, Persans MW, Salt DE, Kim SA, Guerinot ML (2001) Phylogenetic relationships within cation transporter families of Arabidopsis. Plant Physiol 126:1646–1667

McGrath SP, Sidoli CMD, Baker AJM, Reeves RD (1993) The potential for the use of metal-accumulating plants for the in situ decontamination of metal-polluted soils. In: Eijsackers HJP, Hamers T (eds) Integrated soil and sediment research: a basis for proper protection. Kluwer, Dordrecht, pp 673–677

McGrath SA, Esquela AF, Lee SJ (1995) Oocyte-specific expression of growth/differentiation factor-9. Mol Endocrinol 9:131–136

Miyabe S, Izawa S, Inoue Y (2001) The Zrc1 is involved in zinc transport system between vacuole and cytosol in Saccharomyces cerevisiae. Biochem Biophys Res Commun 282(1):79–83

Mizuno T, Usui K, Horie K, Nosaka S, Mizuno N, Obata H (2005) Cloning of three ZIP/Nramp transporter genes from a Ni hyperaccumulator plant Thlaspi japonicum and their Ni2+-transport abilities. Plant Physiol Biochem 43:793–801

Moreau S, Thomson RM, Kaiser BN, Trevaskis B, Guerinot ML, Udvardi MK, Puppo A, Day DA (2002) GmZIP1 encodes a symbiosis-specific zinc transporter in soybean. J Biol Chem 277:4738–4746

Nriagu JO, Pacyna JM (1988) Quantitative assessment of worldwide contamination of air, water and soils by trace metals. Nature 333(6169):134–139

Paulsen IT, Saier MH Jr (1997) A novel family of ubiquitous heavy metal ion transport proteins. J Membr Biol 156:99–103

Pence NS, Larsen PB, Ebbs SD, Letham DL, Lasat MM, Garvin DF, Eide D, Kochian LV (2000) The molecular physiology of heavy metal transport in the Zn/Cd hyperaccumulator Thlaspi caerulescens. Proc Natl Acad Sci USA 97:4956–4960

Persans MW, Nieman K, Salt DE (2001) Functional activity and role of cation-efflux family members in Ni hyperaccumulation in Thlaspi goesingense. Proc Natl Acad Sci USA 98:9995–10000

Prasad DDK, Prasad ARK (1987) Altered δ-amino laevulinic acid metabolism by lead and mercury in germinating seedlings of Bajra (Pennisetum typhoideum). J Plant Physiol 127:241–249

Radisky D, Kaplan J (1999) Regulation of transition metal transport across the yeast plasma membrane. J Biol Chem 274:4481–4484

Reeves RD, Brooks RR (1983) European species of Thlaspi L. as indicators of nickel and zinc. J Geochem Explor 18:275–283

Rogers EE, Eide DJ, Guerinot ML (2000) Altered selectivity in an Arabidopsis metal transporter. Proc Natl Acad Sci USA 97:12356–12360

Salt DE, Blaylock M, Kumar NP, Dushenkov V, Ensley BD, Chet I, Raskin I (1995) Phytoremediation: a novel strategy for the removal of toxic metals from the environment using plants. Biotechnology (N Y) 13:468–474

Schat H, Llugany M, Bernhard R (2000) Metal-specific patterns of tolerance, uptake, and transport of heavy metals in hyperaccumulating and non-hyperaccumulating metallophytes. In: Terry N, Banuelos G (eds) Phytoremediation of contaminated soils & water. CRC, Boca Raton, FL, pp 171–188

Shen ZG, Zhao FJ, McGrath SP (1997) Uptake and transport of zinc in the hyperaccumulator *Thlaspi caerulescens* and the non-hyperaccumulator *Thlaspi ochroleucum*. Plant Cell Environ 20:898–906

Simmons RW, Pongasakul P, Saiyasitpanich D, Klinphoklap S (2005) Elevated levels of cadmium and zinc in paddy soils and elevated levels of cadmium in rice grain downstream of a zinc mineralized area in Thailand: implication for public health. Environ Geochem Health 27:501–511

Song WY, Sohn EJ, Martinoia E, Lee YJ, Yang YY, Jasinski M, Forestier C, Hwang I, Lee Y (2003) Engineering tolerance and accumulation of lead and cadmium in transgenic plants. Nat Biotechnol 21:914–919

Tu Shu I, Brouillette JN (1987) Metal ion inhibition of corn root plasma membrane ATPase. Phytochemistry 26:65–69

Vallee BL, Falchuk KH (1993) The biochemical basis of zinc physiology. Physiol Rev 73:79–118

Van de Mortel JE, Almar Villanueva L, Schat H, Kwekkeboom J, Coughlan S, Moerland PD, Loren V, van Themaat E, Koornneef M, Aarts MGM (2006) Large expression differences in genes for iron and zinc homeostasis, stress response, and lignin biosynthesis distinguish roots of Arabidopsis thaliana and the related metal hyperaccumulator Thlaspi caerulescens. Plant Phys 142:1127–1147

van der Zaal BJ, Neuteboom LW, Pinas JE, Chardonnens AN, Schat H, Verkleij JA, Hooykaas PJ (1999) Overexpression of a novel Arabidopsis gene related to putative zinc-transporter genes from animals can lead to enhanced zinc resistance and accumulation. Plant Physiol 119:1047–1055

Varotto C, Maiwald D, Pesaresi P, Jahns P, Salamini F, Leister D (2002) The metal ion transporter IRT1 is necessary for iron homeostasis and efficient photosynthesis in Arabidopsis thaliana. Plant J 31:589–599

Vázquez MD, Lopez J, Carballeira A (1999) Uptake of heavy metals to the extracellular and intracellular compartments in three species of aquatic bryophyte. Ecotoxicol Environ Saf 44:12–24

Vert G, Briat JF, Curie C (2001) Arabidopsis IRT2 gene encodes a root-periphery iron transporter. Plant J 26:181–189

Vert G, Grotz N, Dedaldechamp F, Gaymard F, Guerinot ML, Briat JF, Curie C (2002) IRT1, an Arabidopsis transporter essential for iron uptake from the soil and for plant growth. Plant Cell 14:1223–1233

Wang X, Minasov G, Shoichet BK (2002) The structural bases of antibiotic resistance in the clinically derived mutant beta-lactamases TEM-30, TEM-32, and TEM-34. J Biol Chem 277:32149–32156

Williams LE, Pittman JK, Hall JL (2000) Emerging mechanisms for heavy metal transport in plants. Biochim Biophys Acta 1465:104–126

Wintz H, Fox T, Wu YY, Feng V, Chen W, Chang HS, Zhu T, Vulpe C (2003) Expression profiles of Arabidopsis thaliana in mineral deficiencies reveal novel transporters involved in metal homeostasis. J Biol Chem 278:47644–47653

Zhao H, Eide D (1996) The yeast ZRT1 gene encodes the zinc transporter protein of a high-affinity uptake system induced by zinc limitation. Proc Natl Acad Sci USA 93:2454–2458

Zhao FJ, Hamon RE, Lombi E, McLaughlin MJ, McGrath SP (2002) Characteristics of cadmium uptake in two contrasting ecotypes of the hyperaccumulator Thlaspi caerulescens. J Exp Bot 53 (368):535–543

Index